Spon's Building Costs Guide for Educational Premises

Other titles from E & FN Spon

Green Building Handbook
Woolley, Kimmins, Harrison & Harrison

Green Buildings Pay
B Edwards

Kindergarten Architecture
M Dudek

Programme Management Demystified
G Reiss

Project Management Demystified
2nd Edition
G Reiss

Risk and Safety in Play
Playlink

Spon's Architects' and Builders' Price Book
Davis Langdon & Everest

Spon's Construction Resource Handbook
B Spain

Spon's Landscape and External Works Price Book
Derek Lovejoy Partnership & Davis Langdon & Everest

Spon's Mechanical and Electrical Price Book
Davis Langdon & Everest

Sports Turf
V I Stewart

Understanding JCT Standard Building Contracts
5th Edition
D M Chappell

For more information about these and other titles published by E & FN Spon, please contact:
The Marketing Department, E & FN Spon, 11 New Fetter Lane, London, EC4P 4EE. Tel: 0171 583 9855; Fax: 0171 842 2303; or visit our web site at **www.efnspon.com**

Spon's Building Costs Guide for Educational Premises

Second edition

Edited by
Barnsley and Partners
Quantity Surveyors and Project Managers

E & FN SPON
ALERE FLAMMAM
Taylor & Francis Group

First published 1999 by E & FN Spon
11 New Fetter Lane, London EC4P 4EE

E & FN Spon is an imprint of the Taylor and Francis Group

Printed and bound in Great Britain by T.J. International Ltd, Padstow, Cornwall

Publisher's note:
This book has been reproduced from camera-ready copy supplied by the author.

British Library Cataloguing in Publication Date
A catalogue record for this book is available from the British Library

ISBN 0-419-23310-5

FURNITURE AND FITTINGS

Lists and schedules of prices of all furniture, fittings and other associated equipment are shown on the page facing the appropriate activity area model and the pages following.

The Economic Background.
The demands being made at all levels on the Department for Education and Employment and Local Education Authorities, to provide higher standards of education, coupled with the economic restraints imposed by successive Governments inevitably focus attention upon the use we are making of our school buildings, their value and the standards and costs involved in their operation, management and maintenance.

But how do School Governors, Head Teachers and their Consultants, faced with these responsibilities, cope with the problems of managing their properties and the costs? With time at a premium and the interests of the pupils paramount, they already perform miracles.

How then can they plan to deal with continual fluctuations in the number of pupils on the school roll - extra pupils from the new local housing estates - and/or changes in the curriculum and the effect all of this has on the accommodation?

How can they plan to meet the extra cost of complying with new Health and Safety Regulations and improve revenue to deal with the perpetual increase in the cost of wear and tear?

And do Private Investors understand these problems and then, how can they assess the risk and potential revenue from educational premises?

I have linked up the whole process, from planning the curriculum, calculating the accommodation requirements and its cost, to maintaining the estate after completion and it is hoped that this will provide useful references and a basis for constructive debate. Also, I have given a quick and easy guide to building costs, the use of alternative materials, and help with the administration of building contracts.

The Index
After many years searching laboriously through drawings, schedules and many long lists and files, I have been determined to provide a reference book which provides a comprehensive user friendly guide.

Detailed contents lists are provided at the front of the book and at the beginning of each chapter and every page is clearly referenced which makes the information required easy to find.

The School Village
In addition I have analysed educational premises into activity areas and classified these areas into functional categories, ie. management, hospitality, teaching and sport. The activity areas within each category all have similar management, usage and potential revenue earning characteristics. The area models, cost analyses and the cost of furniture and equipment for each activity area are grouped together within the appropriate category, all of which I have called the 'School Village'. This is intended to offer both a fuller perspective of the use and financial viability of school premises as a whole and help with the development of just individual activity areas.

Schedules and Graphs.
These are a powerful medium for providing data and allowing quick choices to be made; they have been used extensively throughout the book. Some of the graphs provide a unique method of adjusting prices when dealing with continuing changes in the cost of labour and materials.

The Pictorial Glossary.
For clarity the pictorial glossary has been drawn up to combine an easy reference to pictorial and graphic presentations together with the appropriate composite unit costs. This will help with understanding both the technical descriptions for building work and the scope of work covered by the composite rates.

Area Guide Lines
The tables set out in the Appendices C to K showing the overall and individual activity area guide lines are derived directly from the formulae in the D*f*EE Bulletin 82 and these are linked to the key teaching stages described on page 73 of the Bulletin. The Key Teaching Stages are summarised here in Chapter 1 on page 3.

The Cost Model
Inevitably in the technological world in which we live, the cost yardsticks, prices of furniture and equipment, the unit rates for measured building work and the composite unit rates are all derived from a central inter-active cost model. The key unit rates can be used when adjusting cost analyses for the activity areas and all these costs are linked to the regional and inflationary costs indices.

Professional Fees.
The charts shown on pages 234 to 236 show the normal charges raised by the Architect, Quantity Surveyor, Structural Engineer and Services Engineer for their professional services, according to the size and nature of the work involved. These make allowance for all expenses.

Historic Values
A note about our educational premises in which we should take great pride: some may indeed be in need of repair or out of date, but they also include many fine examples of historic buildings.

They reflect the very finest professional and trade skills used at the time in their design and construction and which other nation has such wealth?

But was it easier then to build and maintain school property, and where is the spirit of 19th century philanthropy now?

ACKNOWLEDGEMENTS

I would like to record my appreciation for the assistance provided by many individuals and organisations in the research and preparation necessary to complete this edition.

In particular I would like to thank

Rob Jessop, BA (Hons), Dip.Arch (Birm) who has provided the graphic illustrations forming the basis of the Pictorial Glossary,

Tim Fitzpatrick whose un-stinting efforts in researching the vast market for materials, school furniture and equipment have made such an enormous contribution to the references and prices now included in this edition

and

Margaret Barnsley who helped with the difficult task of editing and correcting the proofs.

Derek Barnsley
Barnsley and Partners
Quantity Surveyors - Project Managers
850, Brighton Road
Purley
Surrey
CR8 2BH

May 1999

Also, I would like to thank the following organisations who have made a positive contribution and a special effort to help us digest all of the copious information they have provided;

Consultants

TPS Consult (Structural Engineers)

Smith and Baxter (Planning Supervisors)

Suppliers

Altro Floors

Amorim (UK) Ltd

Black Cat Music

British Thornton ESF

Burmatex Carpets Ltd

Ecophon Ceiling Systems

Electrolux Food Service

Eternit Ltd

International Food Service Equipment Ltd

Leisure Engineering Projects Ltd

Neighbourhood Electronic Watch Systems Ltd

Marlin Lighting Ltd

Petrarch Claddings Ltd

Pilkington Glass Ltd

Purley Pools (UK) Ltd

Resdev

Ring-Gard (UK) Ltd

Stage Systems Ltd

Wicksteed Leisure Ltd

The background graphic illustrations to the Indexes are reproduced by kind permission of Wicksteed Ltd

QUICK REFERENCES

To plan the costs and compile budgets refer to the following pages:

Method (i) Net Internal Floor Areas

Area = (a x b) + (c x d) = square metres

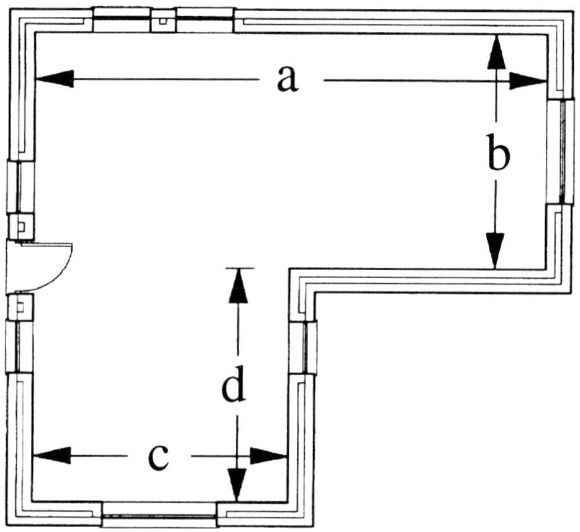

Method (ii) Gross Floor Areas

Area = (e x f) + (g x h) = square metres

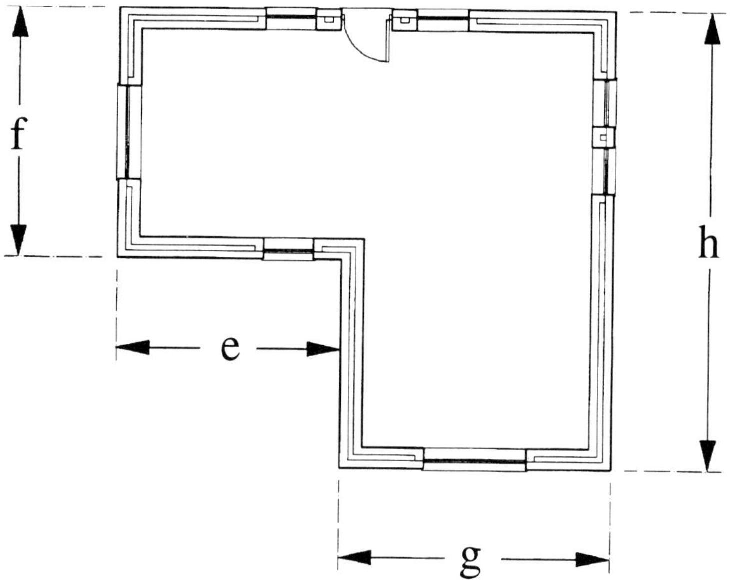

Method (i) to be used for:

- planning building costs, pages 13 to 96;
- unit rates for measured work, pages 104 to 219;
- rating valuations, see Note below.

This method should be used when planning costs and using square metre building prices for preparing estimates.

The measurements are taken inside the building between the internal faces of the external walls and without any deduction for the areas covered by internal partitions, columns and staircases.

Method (ii) to be used for:

- insurance valuations;
- rebuilding costs due to fire etc.,

This method is normally used when making valuations for insurance purposes based upon floor areas.

Measurements are taken outside the building between the external face of external walls and to the centre line of party walls and without any deduction for the areas covered by internal partitions, columns and staircases.

However, some insurance companies may prefer Method (i) and the user should check with the company concerned.

Note: The following Codes of Procedures for measuring can be obtained from The Royal Institution of Chartered Surveyors, 12 Great George Street, London, SW1P 3AD, telephone; 0171 222 7000:

i) Code of Procedure for Measurement of Building Work, 1992 Edition, ISBN 0 85406 361 7;

ii) Code of Measuring Practice, 1994 Edition, ISBN 0 84506 610 1 for valuations.

All measurements of floor areas and units of measurement for building work are based upon the metric method of measurement.

THE COST GUIDES

If a property or estate is to be economically viable and represent ***'value for money'*** over its life cycle, its management must, in addition to controlling building and maintenance costs, take account of its use and market value.

But whereas the economic criteria are clear for commercial properties, this has not always been the case for schools whose use is dedicated to educational activities and whose ownership is generally restricted. This has not made it easy for market forces to influence designs, extend their use and financial viability and improve market values.

Planning and Costing Criteria

- overall area guide lines,
- activity area guide lines,
- curriculum requirements,
- property condition reports,
- building costs,
- maintenance costs,
- running costs,
- revenue.

The Cost Studies endeavour to show how the area guide lines recommended by the D*f*EE and curriculum requirements can be directly related to building costs and provide a 'Benchmark' for a comprehensive development programme.

They also try to show building and maintenance costs in perspective - not as unwelcome expenditure, but as an investment which in due course can provide a rich dividend - and activity areas as a source of additional revenue.

CAPITAL AND MAINTENANCE COSTS

i) **New School Premises**

The Cost Studies explain how the 'Capital Costs' for new schools and colleges can be calculated based upon:

- Overall area guide lines;
- Square metre building costs;
- Professional fee scales.

ii) **Re-modelling and/or extending Existing Premises.**

The Cost Studies also explain how the overall capital, maintenance and operational costs can be calculated based upon:

- condition survey;
- curriculum analysis;
- benchmark target cost analysis;
- activity area guide lines;
- derivative activity cost models;
- price schedules for furniture;
- choice of alternative materials;
- cyclical maintenance programmes;
- professional fees.

iii) **The Property Portfolio**

In Chapter 1, the value of a Portfolio for a commercial property has to be compared with the value for an educational property.

THE COST STUDIES

Chapter 1 Planning the Cost of Educational Premises
Explains how to use the area cost models and cost analyses to complete an accommodation and benchmark analysis. Shows how to estimate the capital cost of building new school premises or re-modelling existing premises. An introduction to the use of private finance and the variations in prices in different LEA regional areas.

Chapter 2 The School Village
Provides a choice of area cost models which fall within the D*f*EE area guidelines together with the derivative cost analyses, and which indicate the types of facilities provided and their revenue potential.

Chapter 3 Cyclical Maintenance Costs
Provide the basis for a formal cyclical maintenance programme to be set up based upon an elemental analysis falling within the D*f*EE activity area guide lines.

Chapter 4 Unit Rates for Measured Building Work
Provide schedules of key rates for the main items of building work on new projects including extensions and refurbishment or re-modelling work together with charts and graphs to help with their adjustment for setting up budgets, checking tenders and choosing alternative materials.

Chapter 5 The Administration of Building Projects
Explains the basis upon which the contractor and professional consultants should be appointed and the types and use of standard building contracts available.

CHAPTER 1

Planning the Cost of Educational Premises

THE COST BASE

The Prices for Planning Costs, the Analyses of Costs and Unit Rates for Measured Building Work are all based upon the cost of wages and prices of materials ruling during the first quarter of 1999 and within the Home Counties.

Planning Requirements.
The economic design and intensive use of new and existing school premises require a very detailed examination of:

a) the size, location and development potential of the site;

b) the requirements of the curriculum;

c) the overall building and external play area requirements necessary to meet the needs of the teaching and non teaching activities;

d) the ongoing maintenance and running costs;

e) the potential for increasing revenue by extending the use of the premises by extra curiculum activities, social activities and by attracting partnering arrangements and private investment.

It is also necessary, when considering the use of an existing site and the facilities being provided by existing buildings, to review their condition and the costs involved in their repair and upkeep.

This chapter provides a method, based upon the D*f*EE recommendations, which can be used in planning, to link together these requirements and to take account of and phase their relative capital, maintenance and running costs. These broad cross links should help those involved in planning to be more aware of the users' requirements, and perhaps for the user, the opportunity to influence and contribute towards the design and choice of finishings, furniture and special technical and security installations.

Readers should also be aware of the high cost of alterations and repairs in existing buildings, particularly historic buildings, and the wide variations in the price of school furniture.

Site Areas.
Bulletin 82 issued by the D*f*EE provides area guidelines for optimum superficial site areas for primary and secondary schools which are shown in table form on Appendix A, page 252. The guide to site values are approximate only and these will vary dramatically according to the location, the present owner and use. A change of use will require planning consent and this subsequently will change the value of the land.

The area guidelines for hard surfaced games and social areas are shown in table form on Appendix B on page 253 and the cost of surfacing, fencing and maintaining these areas are shown on pages 205 to 215 inclusive.

Gross Floor Areas of School Buildings.
Bulletin 82 provides guidelines for the optimum gross superficial floor areas for Primary and Secondary school premises, the proportion of these areas which should be used for teaching and the floor areas required for each pupil. These guidelines are shown in table form on Appendices C to H on pages 254 to 257.

The guidelines indicate the maximum and minimum areas required based upon a series of formulae related to the overall head count and key stages in the pupil's education to be provided for each school to which the National Curriculum will apply.

These areas are considered by the Department to provide ***value for money*** when designing new school premises or re-modelling or extending existing premises in relation to the number of pupils on the school roll. They can also clearly be adapted for use when planning other types of schools and developed to include other activities based upon the community needs when these can be properly associated with the use of educational premises.

Figure 1.1 **AREA GUIDE LINES**
Secondary School Key Stage 3:3, 4:2

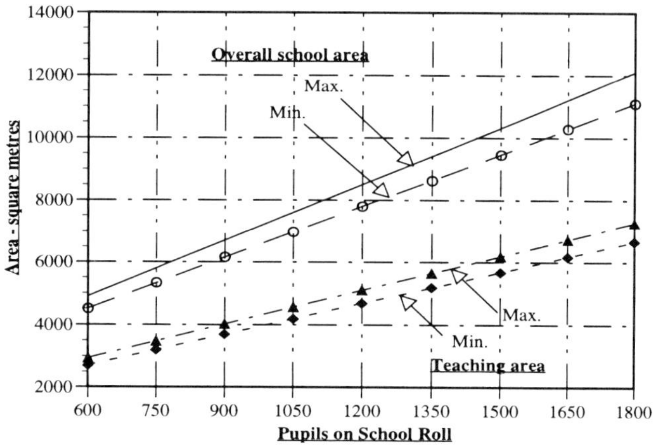

Measuring Net Internal Floor Areas.
The total net internal floor area of buildings is defined as the areas measured inside the buildings between the internal faces of the external walls and without any deduction for the areas covered by internal partitions, columns and staircases as shown on page (xiv).

Teaching and Non Teaching Spaces.
The optimum teaching and non teaching areas are shown on Appendices C to H on pages 254 to 257 and the area guidelines for Classrooms and Practical Rooms on Appendices I to K on pages 260 to 262. The table on Appendix L on page 263 shows teaching periods according to the frequency in their use and related to the size of the teaching group.

The use of these guidelines when setting up the accommodation schedule is explained later.

A guide to the capital costs of new school premises and extending or remodelling existing buildings based upon the internal superficial floor area is shown on page 7.

Curriculum Requirements.
These are, of course, matters for the Head Teacher and Education Authorities to plan, but which clearly need to be more widely understood.

They are governed by the scope of the curriculum, the academic and practical subjects to be taught, the balance to be drawn between group sizes and numbers of teaching staff, the amount of teaching time and the frequency with which the spaces can be put to use. These stem directly from the following:

a) number of pupils on the school roll;
b) analysis of the curriculum;
c) group sizes (general & practical);
d) frequency of use;
e) pupil periods;
f) teacher periods.

The cells in the Accommodation Schedule on page 8 and shown in bold have to be completed by the Head Teacher, the remainder by the Consultant.

There are clearly optimum levels for staffing and the utilization of the teaching spaces, but a broader perspective for the use of teaching, and non teaching areas may well be obtained by listening to the views of the Architect, and perhaps by considering future partnering arrangements.

Private partners will tend to be very objective and commercially minded in their judgements and they will look to extend the whole viability of the accommodation outside normal school hours. They will certainly understand the implications of 'cost and planning' and the implications involved in the use, re-modelling and/or replacement of existing buildings.

Key Teaching Stages.
The four main key stages for the majority of pupils are as follows:

KS 1	up to 7 years old;
KS 2	7 to 11 years;
KS 3	11 to 14 years;
KS 4	14 to school leaving age.

Figure 1.1 sets out the gross floor area requirements for a typical Secondary School (KS 3 & 4) and shows the percentage required for teaching areas.

Frequency of Use
Levels in the region of 90 - 95% in frequency of use will be expected for general subjects. Some practical areas will fall below these high ratings and the guidelines must be followed to meet the Ministry targets - but private partners will look for more extensive use of the premises and facilities to be provided.

Activity Area Requirements
Bulletin 82 provides area guidelines for both general and practical teaching spaces and which fall into eight categories lettered from A to H. (See Figure 4.8 on page 37 of the Bulletin). These are supplemented with guidelines for non teaching areas such as libraries, wash rooms. Both primary and secondary schools are covered.

The area guidelines are again based upon set formulae and as an aid to their use they are represented in table form in the Appendices C to K. The 'area per place' figures to be entered in the Accommodation Analysis on Page 264, column [AI], are a derivative of:

a) the gross areas required for each activity divided by

b) (the number of spaces [AF];
x the average group size [AD];
x adjustment for frequency in use [AG]).

Management and Hospitality Requirements.
The other areas, related to management and hospitality (areas with an additional revenue earning value) and circulation, can be assessed on an overall percentage basis of the gross planned area, but it is better to make a more accurate assessment. Often this is where the best economies can be made, using, for example, general teaching areas with flexible layouts and less circulation space.

Condition Survey.
The Condition Survey should indicate the potential use and development of areas of the existing premises so that it is possible to choose the use to which they are to be put. The survey should also show the cost of making good the dilapidations and these can be entered in column [AV/AW] on page 265 together with the projected maintenance and running costs.

TABLE 'B' ANALYSIS OF COSTS
£ per Square Metre - Based upon Model 3

Elements of the Building (m)	New £ per m2 (n)	Extension £ per m2 (o)	Remodelled £ per m2 (p)
Main Building Work			
* Foundations	109.07	111.00	4.21
* Superstructure	315.07	252.66	134.19
Finishings			
* Floors	47.23	49.68	63.19
* Walls	28.66	29.86	24.58
* Ceilings	27.70	31.94	46.05
Fittings			
* Furniture	53.13	49.37	49.79
* Fittings	94.45	87.76	88.51
* Sanitary Fittings	0.00	0.00	0.00
* Light fittings	23.51	23.25	25.74
Security			
* Protection	2.70	8.34	8.58
* Surveillance	6.44	8.62	6.25
* Communication	3.67	6.23	5.70
FITTING OUT	287.50 £ per m2	295.04 £ per m2	318.39 £ per m2
TOTAL COST	711.65 £ per m2	658.70 £ per m2	456.78 £ per m2

Figure 1.2

The Accommodation Schedule - see page 264
Bulletin 82 shows the method suggested by the D*f*EE for deriving and setting out a school's accommodation requirements stemming directly from these economic planning requirements. This has been sub-divided into the following categories which have been called 'The School Village', in order to focus attention on planning, phasing, funding and income potential:

- M - Management; pages 22 to 39
- H - Hospitality; pages 40 to 63
- A - Academic Teaching Areas; pages 64 to 77
- T - Technology and Science; pages 78 to 91
- S - Sports & Fitness Facilities. pages 92 to 98

The Benchmark Cost Analysis - see page 265
This has been added to the accommodation schedule and is linked, by means of the software package, to the accommodation schedule. The rows are registered horizontally so that each activity area can be followed through and costed, one at a time, allowing a choice to be made and the costs to be benchmarked based upon the following decisions:

- choice of building new, extending or remodelling the existing premises;
- allocation of repair costs from the Condition Survey;
- choice of materials and cyclical maintenance costs;
- operating costs;
- revenue supplements.

The benchmark analysis can now be used to show the combined capital, maintenance and operational costs for each activity area and the accumulated total cost of a proposed development.

Setting up a blank template.
An example of an accommodation schedule and benchmark cost analysis is set out on pages 8 and 9 reflecting the criteria deemed necessary in planning and designing a school development project and against which all future costs would be measured. This model is based upon the gross area for a secondary school with a roll of 1350 pupils.

The schedules shown have been created on a spreadsheet and are intended to follow exactly the method of compilation set out in the Bulletin 82 including 'annual age group analysis' (but with the exception of steps 3 and 4 set out on page 10 of the Bulletin) for general teaching areas, but with the addition of the Benchmark Analysis.

The accommodation schedule and benchmark analysis are two linked and protected inter-active spreadsheets which only allow entry of the critical data in defined cells. The spreadsheets can also be developed to provide alternative more detailed analysis or to simplify the process and alarms which can be built into the software to stop mistakes occurring in the allocation of the areas.

The spreadsheet is too big to be reproduced in its entirety and the legend with the critical data and external works are shown separately. For clarity the full schedules on Appendices M and N may be enlarged on a photocopying machine.

Formatting the Template.
The form taken by the schedule on Appendix M follows the recommendations of Bulletin 82 and shows:

- teaching requirements - set out as a heading;
- the accommodation schedule within the functional categories - set out down the left hand column.

The Formulae.
These are required to be entered into the cells, shown on Appendix M with normal print, before the spreadsheet is protected and these formulae then use the data entered onto the entry platform shown in Figure 1.3 to make the appropriate calculations.

The formulae to be entered into each cell are shown in capital italics at the head of each column. These letters refer to the column references on the accommodation schedule.

The references to be entered into the formulae on the template are "spreadsheet cell references". i.e;

= $D30 x $F30

In order to round these figures up the following formula may be used and similarly in the other calculations:

= CEILING (($D30 x $F30), 5)

Protecting the Template.
When the layout of the analysis is completed on the new spreadsheet and the appropriate formulae have been entered, the spreadsheet should be 'protected' to avoid corruption due to irregular entries or the incorrect alteration or deletion of the formulae.

The cells requiring the entry of new data should be 'unlocked' prior to the protection of the spreadsheet. This leaves them to be used as new entry fields and it will prove helpful to build in labels in each cell to explain their use and the type of data required.

The Password
Make a careful note of the password used when protecting your document. Readers should keep a note book containing directory titles, passwords and other user references.

Data Entry

Data is required to be entered directly into the cells shown with **bold print** after the spreadsheet has been protected or onto the entry platform which can be set up as shown in Figure 1.3 on a linked protected spreadsheet. The cells (indicated on Appendix M with bold figures) require the direct entry of the data on the spreadsheet.

The data to be entered is summarised as follows:

a) Accommodation Schedule [columns - K to AF];
The school roll, teaching, curriculum data and group sizes should be completed by the Head Teacher as shown on the entry platform in the next column and these entries generate in the spreadsheet the area necessary for column [AJ].

The frequency teaching spaces required to be in use can be entered on the spreadsheet or calculated by using the standard built in formula.

b) Area Analysis [column - AI]
The floor areas for teaching space may be taken from the activity area guidelines shown in Appendices I to K on pages 260 to 262 after discussion with the Architect.

The floor areas for non teaching spaces [column - AJ] can be obtained from any of the following sources:

- Bulletin 82 Area Guidelines;
- Table 'A' for each Activity Area;
  (See Summary on page 21)
- The Design Consultant.

The Capital Cost Models

The square metre building costs shown on page 7 are based upon a reasonable assessment of known competitive tender prices over the last ten years and which have been optimised to reflect a balanced view of probable costs of future projects.

The overall building costs are constructed on three elemental cost models reflecting the different costs of new buildings, extensions and re-modelling projects. The superficial building costs of the various activity areas are derivatives of these models

The overall capital cost of a new school can be calculated by multiplying the gross superficial floor area, obtained either from the area guidelines or from the accommodation schedule, by the overall superficial building cost shown in row 1 in the table set out on page 7.

The capital cost of each activity area is calculated by multiplying the floor area set out in column [AJ] of the Accommodation Schedule by the derivative superficial building costs set out in rows 2 to 30 in the table set out on page 7.

The total area requirement summarised in column [AJ] on the Accommodation Analysis can now be transferred to the Benchmark Analysis shown on page 265 and disbursed for costing purposes between the four available building categories as follows:

- existing premises to be retained [AK];
- construction of new buildings/ extensions [AL:AM];
- re-modelling existing buildings [AN].

THE DATA ENTRY PLATFORM

	Curriculum Requirements		
A	Pupils on Roll	Selection	900
B	Periods per Week	Selection	40
C	Pupil Periods	(A x B)	36000
D	Group Size	Selection	27
E	Groups in Year	Selection	7
F	No of Teachers	FTE	47.6
G	Contact Ratio	Selection	0.78
H	Teacher Periods (Total)	(B x F x G)	1485
I	Periods Taught	Selection	5
Management			
	Activity Area	Cost [BC]	Income [BE]
M 1	Reception	£ 3.50	£ 5.00
H 1	Assembly	£ 3.50	£ 5.00
H 2	Library	£ 1.50	£ 2.00
H 3	Conference	£ 15.00	£ 35.00
H 4	Dining	£ 2.50	£ 3.00
H 5	Kitchen	£ 3.50	£ 5.00
	Teaching Spaces		
A 1/2	General	£ 3.50	£ 5.00
T 5	Technology	£ 3.50	£ 5.00
T 6	Food Technology	£ 3.50	£ 5.00
T 7	Multi materials	£ 3.50	£ 5.00
T 7	Graphics	£ 3.50	£ 5.00
T 8	Drama	£ 3.50	£ 5.00
T 9	Art	£ 3.50	£ 5.00
T 11	Music	£ 3.50	£ 5.00
S 1/4	Sport	£ 2.00	£ 3.00

Notes;- This worksheet should be linked to the Accommodation Analysis spreadsheet. The amounts for cost and income for these areas are suggested targets. Exact yardsticks should be calculated for entry into the Benchmarck analysis in Cols BA & BE.

Figure 1.3

Obtaining the Target Capital Cost.
The *'target capital cost'* is obtained by entering the appropriate superficial building rates into the unlocked cells in columns [AO], [AQ] and [AS] in the '*benchmark cost analysis*' shown on page 265. The formulae in columns [AP; AR and AT] then automatically calculate the total building cost for each activity area within each category.

The target cost can be seen to be based upon the following:

a) the total net internal floor area as calculated in column AJ in the accommodation schedule;
b) the apportionments of the building costs between new buildings and the areas within the existing premises, if suitable, which can be re-modelled to meet the requirements of the Condition Survey;

Selecting the Category of Building Work.
The building category must be selected with great care. The costs, shown on page 7, vary as follows:

i) New School Buildings:
The costs will depend upon the extent and design of the buildings and site works entailed. New isolated buildings within a school complex are often an expensive option.
ii) Extending Existing Buildings:
This may be a viable option and should be considered carefully. The building costs will depend on the location, means of access and the availability of specialist services. Economies can be achieved by siting the new building to take advantage of these facilities.
iii) Re-modelling Existing Buildings:
This is generally the cheaper option, but care should be taken to examine each factor. The Condition Survey may show that the buildings are unsuitable, or that the cost of the repairs or extending the services are prohibitive.

Adjusting the Superficial Building Costs.
Whichever category of building work is finallly selected, the costs per square metre for building work must be adjusted to allow for the following:

i) site conditions;
ii) programme for the work;
iii) regional cost variations;
iv) professional fees;
v) inflationary costs;
vi) interest on borrowed capital.

Costing the Activity Areas.
The capital cost of each activity area is calculated by multiplying the following:

a) the areas defined by the application of the appropriate formulae in Bulletin 82 and set out in column [AJ] on the accommodation schedule;
b) derivative unit costs per square metre obtained from the tables shown on page 7.

The superficial rates can be adjusted as described in the examples 3 to 7 set out on pages 18 and 19 to allow for the selection of special materials, finishings, furniture and equipment for each area.

Cyclical Maintenance Programme
Entry into columns [AV and AX] of these costs, which can be obtained from either the yardsticks proposed in Chaper 2 or from the schools' cost records, allows a balance to be drawn with the priorities within the development programme and also to take into account the condition survey.

Phasing the Development Work.
The planning process may include work which, although desirable, may not be possible, either because funds are not available at the time or because it would be impossible to carry out the full development programme in one operation.

The Benchmark Analysis can be extended to make provision for phasing the work, either by adding columns or by linking in additional spreadsheets, say for phases 1, 2 and 3. The danger here is that the proposed area requirements are duplicated and 'alert' buttons can be built into the cell showing the overall area included in the costs should this differ from the total of column AJ on the Accommodation Analysis.

Updating the Costs.
The costs can be updated to accord with any changes that are made with the overall planning costs either during the planning stages, during the building process or, by choice, when re-arranging areas or selecting alternative materials.
A record of each update should be made for reference purposes.

The overall assessment and any updates should be checked by the Quantity Surveyor to ensure costs remain strictly within the budget.

Revenue
Account can be taken by extending the analysis of the potential revenue which it may be possible to earn from each activity area. The operational usages are shown in Table 'A' for each activity area within the school village and the cost and income derived can be entered in columns [BC and BE].

The Benchmark Analysis can be developed in several ways to show:

- the overall economic viability of a development;
- inducements available to attract investment;
- the benefits of staging the works;
- comparison in the cost of re-building or re-modelling;
- the effect of changes in the school roll and curriculum;
- effects of changing the daily utilization of premises.

Ref	ACTIVITY AREA	NEW BUILDING WORK		EXTENSIONS		REMODELLING WORK	
	Note;- All rates INCLUDE an allowance for the cost of roads, paths, site layout, fencing and boundary walls.	**Building Work** Rate per m2	**Fitting Out** Rate per m2	**Building Work** Rate per m2	**Fitting Out** Rate per m2	**Building Work** Rate per m2	**Fitting Out** Rate per m2
1	**Composite Building** * Overall building cost	£ 705	£ 265	£ 655	£ 325	£ 455	£ 300
	Management						
2	* Offices	£ 750	£ 330	£ 705	£ 335	£ 515	£ 365
3	* Reception	£ 630	£ 225	£ 585	£ 240	£ 400	£ 265
4	* Staff Room	£ 705	£ 270	£ 665	£ 280	£ 480	£ 310
5	* Medical Centre	£ 740	£ 295	£ 700	£ 305	£ 520	£ 335
6	* Boiler Room	£ 540	£ 130	£ 500	£ 145	£ 300	£ 160
7	* Laundry Room	£ 705	£ 275	£ 665	£ 285	£ 475	£ 315
	Facilities						
8	* Assembly Hall	£ 670	£ 200	£ 620	£ 215	£ 385	£ 245
9	* Library	£ 705	£ 285	£ 660	£ 290	£ 450	£ 315
10	* Conferencing	£ 790	£ 415	£ 740	£ 420	£ 515	£ 345
11	* Dining Room	£ 685	£ 235	£ 635	£ 250	£ 420	£ 275
12	* Kitchen	£ 1,095	£ 660	£ 1,030	£ 645	£ 835	£ 675
13	* Toilets	£ 1,020	£ 360	£ 1,045	£ 405	£ 930	£ 485
14	* Cloakroom	£ 645	£ 200	£ 605	£ 215	£ 375	£ 235
15	* Changing Room	£ 955	£ 350	£ 960	£ 385	£ 820	£ 450
	Teaching Areas - General						
16	* Single Storey	£ 715	£ 290	£ 660	£ 300	£ 460	£ 320
17	* Two Storey	£ 680	£ 290	£ 630	£ 300	£ 455	£ 320
18	* Nursery School	£ 655	£ 345	£ 535	£ 325	£ 390	£ 365
19	* Studios	£ 750	£ 310	£ 700	£ 315	£ 480	£ 340
	Teaching Areas - Practical						
20	* Single Storey	£ 730	£ 300	£ 680	£ 305	£ 485	£ 330
21	* Two Storey	£ 705	£ 300	£ 655	£ 305	£ 480	£ 330
22	* Pottery	£ 745	£ 305	£ 700	£ 315	£ 490	£ 340
23	* Art	£ 720	£ 280	£ 680	£ 285	£ 465	£ 315
24	* Music Centre	£ 750	£ 325	£ 700	£ 330	£ 490	£ 355
	Teaching Areas - Laboratory						
25	* Single Storey	£ 770	£ 335	£ 715	£ 345	£ 520	£ 370
26	* Two Storey	£ 735	£ 335	£ 685	£ 345	£ 515	£ 370
	Sporting Activities						
27	* Gymnasium	£ 685	£ 200	£ 635	£ 215	£ 375	£ 240
28	* Squash Court	£ 615	£ 160	£ 575	£ 175	£ 330	£ 200
29	* Swimming Pool	£ 730	£ 205	£ 690	£ 225	£ 430	£ 260
	Circulation						
30	* Corridors	£ 605	£ 140	£ 555	£ 170	£ 350	£ 180

See Appendix 'M' for complete schedule layout

Ref	Activity Area	Pupils in Year K	Period Taught L	Maximum Group Size M	Number of Groups N	Pupil Periods O	Teacher Periods P	Average Group Size Q	AGE GROUP ANALYSIS Year 7 Pupil Periods R	Year 7 Teacher Periods S	Year 8 Pupil Periods T	Year 8 Teacher Periods U	Year 9 Pupil Periods V	Year 9 Teacher Periods W	Year 10 Pupil Periods X	Year 10 Teacher Periods Y	Year 11 Pupil Periods Z	Year 11 Teacher Periods AA	TOTALS Pupil Periods AB	TOTALS Teacher Periods AC	Average Group Size AD	Number of Spaces AE	Number of Spaces Adjusted AF	Frequency in use AG	Maximum Group Size AH	Area per Place AI	TOTAL AREAS AJ
	Formulae	A/I or select	Planned	Planned	K/M rounded up	K	LxN rounded up	O/P	← (As Columns K and P for display purposes only on this spread sheet only) (In practice the data in these columns should relate to the curriculum breakdown for year group) →										sum of age groups	sum of age groups	AB/AC	AC/B	AE Rounded up	AE/AF	Curriculum Breakdown	Section 4 Guidelines	AF x AI
M	MANAGEMENT																										
1	Offices																						See Page 22 for area guide lines				45
2	Reception																						See Page 26 for area guide lines				128
3	Staff Room																						See Page 30 for area guide lines				80
4	Medical Centre																						See Page 34 for area guide lines				60
5	Boiler House																						See Page 36 for area guide lines				21
6	Laundry Room																						See Page 36 for area guide lines				25
H	HOSPITALITY																										
7	Assembly Hall																						See Page 40 for area guide lines				260
8	Library																						See Page 44 for area guide lines				142
9	Conferencing																						See Page 48 for area guide lines				
10	Dining Room																						See Page 52 for area guide lines				135
11	Kitchen																						See Page 56 for area guide lines				143
A	TEACHING AREAS - GENERAL																										
12	English	180	5	27	7	900	35	25.71	900	35	900	35	900	35	900	40	900	40	4,500	185	24.32	4.625					
13	Mathematics	180	5	27	7	900	35	25.71	900	35	900	35	900	35	900	40	900	40	4,500	185	24.32	4.625					
14	Modern Languages	180														2	720	32	4,140	169	24.50	4.225					
15	Humanities	180														2	720	32	3,600	148	24.32	3.700					
16	Religion	180														6	360	16	1,800	74	24.32	1.850					
17	Personal and Social	180																	540	21	25.71	0.525					
18	General Studies	180														6	360	16	720	32	22.50	0.800					
A	Total - General	126															3960		19,800	814	24.3	20.350					
	Analysis of Classroom size																										
	Standard size	77.50																	15,345	631	24.3	15.771	18	87.62%	30	50	900
	Large size	22.50																	4,455	183	24.3	4.579	5	91.58%	30	62	310
T	TEACHING AREAS - PRA																										
19	Information Technology	180															180	9	900	45	20.00	1.125	2	56.25%	25	72	144
20	Science	180														4	1440	64	5,580	233	23.95	5.825	7	83.21%	30	85	595
21	Design & Technology																										
	Food	180															180	9	900	45	20.00	1.125	1	112.50%	20	103	103
	m-m/ graphics	180															180	9	900	45	20.00	1.125	2	56.25%	20	103	206
	PECT	90															90	5	450	25	18.00	0.625	1	62.50%	25	88	88
	Textiles	90															90	4	450	20	22.50	0.500	1	50.00%	25	84	84
22	Art																										
	2D Art	180															112	5	764	28	27.29	0.700	1	70.00%	20	91	91
	3S Art/ Textiles	180															68	3	676	24	28.17	0.600	1	60.00%	30	109	109
23	Music	180															180	8	1,440	52	27.69				15	70	140
24	Drama	180															180	8	900	34	26.47				20	91	91
25	PE Indoor	180														2	270	12	1,620	60	27.00				30	260	260
26	PE Outdoor	180														2	270	12	1,620	60	27.00				35		
	Total - Practical	198															3240		16200								3,121
	Total - All Teaching Areas																7200		36000	1,485							
H	CLOAKROOMS																										
27	Toilets																						See Page 58 for area guide lines				105
28	Cloakroom																						See Page 60 for area guide lines				120
29	Changing Rooms																						See Page 61 for area guide lines				83
S	SPORTING FACILITIES																										
30	Gymnasium																						2 for area guide lines				600
31	Squash Court																						3 for area guide lines				
32	Swimming Pool																						4 for area guide lines				
33	Sports Centre																										
M	ANCILLARY SPACES																										
34	Careers																						by School				13
35	Local Resources																						by School				50
36	Special Activities																						6.5% of Teaching Area				79
37	Storage																						6 % of Nett Area				500
38	Circulation																						See Page 37 for area guide lines				1,766
	TOTALS								To enlarge this spreadsheet ; zoom to 150% in landscape format on your photocopying machine.																		6,437

These reference letters follow on from the references to data shown on the Data Entry Platform on Page 5.

Area guide lines for non teaching spaces may be obtained by referring to the cost models set out in the School Village on pages 22 to 96.

The Curriculum Breakdown

A	TEACHING AREAS - GENERAL							
12	English	180	5	27	7	900	35	25.71
13	Mathematics	180	5	27	7	900	35	25.71
14	Modern Languages	180	5	27	7	900	35	25.71
15	Humanities	180	4	27	7	720	28	25.71
16	Religion	180	2	27	7	360	14	25.71
17	Personal and Social	180	1	27	7	180	7	25.71
18	General Studies	180	2	27	7	360	14	25.71
A	Total - General	1260				4320		
	Analysis of Classroom sizes							
	Standard size	77.50%						
	Large size	22.50%						
T	TEACHING AREAS - PRACTICAL							
19	Information Technology	180					9	20.00
20	Science	180					35	25.71
21	Design & Technology							
	Food	180	1	20	9	180	9	20.00
	m-m/ graphics	180					9	20.00
	PECT	90					5	20.00
	Textiles	90					4	30.00
22	Art							
	2D Art	180					6	30.00
	3S Art/ Textiles	180					6	30.00
23	Music	180	2	30	6	360	12	30.00
24	Drama	180	1	30	5	180	6	30.00
25	PE Indoor	180	2	30	6	360	12	30.00
26	PE Outdoor	180	2	30	6	360	12	30.00

This data should be compiled by the Head Teacher based upon the Curriculum Breakdown. See Appendix 2 of D*f*EE Bulletin 82

The data to be entered in columns K to Q may differ for each year group and can be entered separately on an extended linked spread sheet if required.

Check here the total Pupil and Teacher periods agree with Data Entry Platform.

The overall area of the school building should be calculated by reference to the D*f*EE Area Guidelines shown in table form in Appendices C to H on Pages 254 to 259.

See Appendix 'N' for complete schedule layout

Ref	Activity Area	TOTAL AREAS	AREA ANALYSIS (M2)				ANALYSIS OF CAPITAL COSTS							MAINTENANCE WORKS - *per Annum*					ADDITIONAL REVENUE FROM ACCOMMODATION - PER ANNUM					
			Existing Premises	New School	Extension	Re-Modelled Premises	New Buildings		Extension		Re-Modelled		TOTAL BUILDING COST	Existing Premises		New Build		TOTAL MAINT'CE COST	Total Operational Cost				Total Revenue	
							Rate perm2	Capital Cost	Rate perm2	Capital Cost	Rate perm2	Capital Cost		Rate per m2	Total Cost	Rate per m2	Total Cost		Revenue Unit	Uses per annum	Cost per unit	Total Cost per annum	Revenue Unit per unit	Total Revenue per annum
		as AJ	*AK*	*AL*	*AM*	*AN*	*AO*	*AP*	*AQ*	*AR*	*AS*	*AT*	*AU*	*AV*	*AW*	*AX*	*AY*	*AZ*	*BA*	*BB*	*BC*	*BD*	*BE*	*BF*
	Formulae	*AF x AI*	*Allocation of area requirements*				See Page 6	*ALxAO*	*See Page 6*	*AMxAQ*	*See Page 6*	*ANxAS*	*Total*	See *Page 11*	*AKxAV*	AV less *15%*	*ALXAT*	*Total*	*Area Models Table A*		*Area Models Table A*	*(BA X BB x BC)*	*See Page 5*	*(BA X BB x BE)*
M	**MANAGEMENT**																							
1	Offices	45	30.00	15			670	10,050	605		490		10,050	16.67	500	14.17	213	713						
2	Reception	128	85.00			43.00	560		495		375	16,125	16,125	16.67	1,417	14.17		1,417	43	80	3.50	12,040	5.00	17,200
3	Staff Room	80				80.00	630		570		445	35,600	35,600	16.67		14.17								
4	Medical Centre	60		[illegible]			[illegible]	[illegible]	600		485		39,600	16.67		14.17	850	850						
5	Boiler House	21	21.00						420		275			16.67	350	14.17		350						
6	Laundry Room	25							565		[illegible]		[illegible]	[illegible]		14.17	354	354						
H	**HOSPITALITY**																							
7	Assembly Hall	260							520							14.17	3,684	3,684	72	[illegible]			5.00	36,225
8	Library	142							570							14.17			48	[illegible]			2.00	11,520
9	Conferencing								740															
10	Dining Room	135							540														3.00	12,960
11	Kitchen	143							920											80	3.50	20,160	5.00	28,800
A	**TEACHING AREAS - GENERAL**																							
12	English																							
13	Mathematics																							
14	Modern Languages																							
15	Humanities																							
16	Religion																							
17	Personal and Social																							
18	General Studies																							
	Total - General																							
	Analysis of Classroom sizes																							
	Standard size	900	380.00								417	91,740	279,540	16.67	500	14.17	4,251	4,751	180	52	3.50	33,062	5.00	47,232
	Large size	310									417	43,785	172,115	16.67	500	14.17	2,905	3,405						
T	**TEACHING AREAS - PRACTICAL**																							
19	Information Technology	144		144			650	93,600					93,600	16.67	500	14.17	2,040	2,541	48	52	3.50	8,817	5.00	12,595
20	Science	595		395		200.00	650	256,750					256,750	16.67	500	14.17	5,597	6,097	165	52	3.50	30,307	5.00	43,296
21	Design & Technology																							
	Food	103				206.00	655		585		465	95,790	95,790	16.67	500	14.17		500	68	52	3.50	12,490	5.00	17,843
	m-m/ graphics	206				206.00	630		565		455	93,730	93,730	16.67	500	14.17		500	56	52	3.50	10,286	5.00	14,694
	PECT	88	88.00				630		600		455			16.67	500	14.17		500	30	52	3.50	5,510	5.00	7,872
	Textiles	84	84.00				630				455			16.67	500	14.17		500	28	52	3.50	5,143	5.00	7,347
22	Art																							
	2D Art	91	91.00				645		580		440			16.67	500	14.17		500	30	52	3.50	5,510	5.00	7,872
	3S Art/ Textiles	109	109.00				645		580		440			16.67	500	14.17		500	36	52	3.50	6,612	5.00	9,446
23	Music	140		140			675	94,500	605		470		94,500	16.67	500	14.17	1,984	2,484	62	52	3.50	11,388	5.00	16,269
24	Drama	91				260.00	626		553		417	108,420	108,420	16.67	500	14.17		500	72	52	3.50	13,225	5.00	18,893
25	PE Indoor	260												16.67	500	14.17		500						
26	PE Outdoor																							
	Total - Practical																							
	Total - All Teaching Areas																							
H	**CLOAKROOMS**																							
27	Toilets	105				105	900		870		830	87,150	87,150	16.67	500	14.17		500						
28	Cloakroom	120				120	575		510		350	42,000	42,000	16.67	500	14.17		500						
29	Changing Rooms	83				83	850		805		740	61,420	61,420	16.67	500	14.17		500	30	52	1.50	2,362	2.00	3,149
S	**SPORTING FACILITIES**																							
30	Gymnasium	600		600			610	366,000	530		345		366,000	16.67	500	14.17	8,502	9,002	264	300	2.00	158,400	3.00	237,600
31	Squash Court																							
32	Swimming Pool																							
33	Sports Centre																							
M	**ANCILLARY SPACES**																							
34	Careers	13				13	640		570		435	5,655	5,655	16.67	500	14.17		500						
35	Local Resources	50				50	640																	
36	Special Activities	79		79			640	50,560					50,560	16.67	500	14.17	1,119	1,619	35	300	2.00	21,000	3.00	31,500
37	Storage	500	216	35		62					335	20,770	20,770	16.67	500	14.17	496	996						
38	Circulation	1,766	875	116		116					335	38,860	38,860	16.67	500	14.17	1,644	2,144						
	TOTALS	6,437	1,979	2,374		2,289		1,397,640				971,300	2,368,940		12,769		33,638	46,408				401,111		582,314

The areas calculated on the Accommodation Schedule in column AJ are now transferred to the Benchmark Schedule on the linked spread sheet. These areas can now be allocated between columns AK to AN as the Example 1 set out below.

The square metre unit rates to be entered into columns AO, AQ or AS can be obtained by reference to the schedule shown on page 7 or from the cost models set out in the School Village on pages 22 to 96.

Columns AV to AX allow entry of the maintenance costs per m2 which can be obtained from Page 11 or from the School's records. Also add the additional costs set out in the Condition Survey here.

Columns BA to BF allow a comparison of the operational cost and revenue potential of each activity area.

<u>Example</u>

Total area of science accommodation required	595	m2
Apportion to:		
New building work	395	m2
Remodelling existing premises	200	m2

PUBLIC AND PRIVATE PARTNERSHIP

The previous Government's 'Private Finance Initiative' was enacted in 1996, but has had a difficult start because of the restrictions imposed by the Schools' Premises Act and the awkward nature of the deals. Activity areas within school premises which can be regarded as appropriate for funding by private finance and asset transfers are not easy to define.

Basically private initiatives cannot compete with the favourable interest rates available to the Government. The value of such an 'agency' can only be, at any particular point in time, in assisting the Financial Institutions and Development Companies to respond to 'need' and supplement the available state capital funding.

Such ventures therefore call for flexibility by the Government and Institutions in responding to the requirements of commercial investment to meet individual, local and regional needs. They must also accommodate the variety of new uses for school premises which require a new approach to management, operation and maintenance and re-consideration of methods used for valuations on a 'profit' basis to provide the base for investment.

Commercial investors also need to understand in greater depth the basis on which schools operate and are planned and maintained. The stages in this process show how schools have to meet the discipline of economic development, the needs of the curiculum and of the timetable, the required standards in Health and Safety and the proper use of appropriate standards and materials in the construction, re-modelling and maintenance of the premises.

Life cycle costings are needed to make comparisons with PFI tender offers.

THE PROPERTY PORTFOLIO

- *for Business Premises* -

- Plays a prominent part in a healthy company's operational and financial stability.
- The site or premises are chosen because of their:
  - i) affordable price;
  - ii) satisfactory accommodation;
  - iii) suitable location;
  - iv) advantages over other available alternatives;
  - v) accessibility for staff and transport;
  - vi) affordable finance and local taxes.
- Maintenance and running costs are:
  - i) considered at the time of purchase as part of the life cycle cost of the premises,
  - ii) well maintained because they affect working conditions, the operation of the business depends upon them, statutory regulations require them;
  - iii) related to the market value of both the property and the business.
- The 'valuation' of the property:
  - i) on a profit basis, provides benefit from profitable business activities;
  - ii) indicates the condition of the premises and its market value;
  - iii) can easily be compared with the valuation of comparable properties;
  - iv) when set out on the company's balance sheet, indicates the stability of the company and provides collateral.
- For most healthy companies the property portfolios occupy a prominent position on their balance sheets and are closely linked to their profitability.

- *for Educational Premises* -

- Only plays a prominent part in the operational activity and management of a school or university.
- The premises were generally sited within the community decades before and now compete with:
  - i) changes in regional and local conditions;
  - ii) population movements and ethnic variations;
  - iii) other alternative uses not often being available;
  - iv) effects of national curriculum and technology;
  - v) new Health and Safety Regulations.
- Maintenance and running costs are:
  - i) an onerous expenditure; the fine heritage of school and university premises are, for all that, often out of date and unsuitable for the intensive use or demand;
  - ii) expensive to maintain in order to meet new Health and Safety Regulations for public buildings.
- The 'valuation' of the property:
  - i) on a cost basis deprives the governors of benefit from any additional income;
  - ii) may indicate the condition of the premises;
  - iii) set out on the school accounts, a valuation of its premises does not provide a basis for obtaining additional financial resources;
- State schools are funded out of taxation and controlled by The Education (School Premises) Regulations 1996. The restrictions imposed on the use and financing of school buildings obviously contribute to their difficulties and create a very harsh environment in which to improve these premises.

CAPITAL COSTS
(Excluding External Works)

REF	ELEMENT	New School	Extension	Re-Modelling
	STRUCTURE			
1	* Demolition	1.61	8.48	9.78
2	* Foundations	41.28	36.13	3.62
3	* Ground Floor	37.86	30.55	0.00
4	* External Walls	80.16	54.69	3.07
5	* Internal Partitions	37.50	39.72	14.27
6	* Moveable Screens	5.77	6.63	4.67
7	* Frame	5.16	6.47	0.00
8	* Upper Floors	17.31	19.41	0.00
9	* Roof Construction	20.20	24.98	0.00
10	* Staircases	5.44	2.96	0.00
	ROOF COVERINGS			
11	* Roof Coverings	11.54	14.22	1.13
12	* Roof Lights	0.57	0.65	0.00
13	* Rainwater Goods	1.44	1.70	1.41
14	* Canopies and Covered Ways	1.16	0.00	1.51
	DOORS & WINDOWS			
15	* Doors	20.19	20.82	30.83
16	* Windows	29.12	25.63	15.52
	FINISHINGS			
17	* Floors	40.01	45.28	57.11
18	* Walls	25.96	29.11	23.76
19	* Ceilings	23.46	29.11	41.62
20	* Staircases & Balustrading	8.65	5.03	6.23
	FITTINGS & FURNITURE			
21	* Facilities	23.08	19.88	25.98
22	* Teaching Areas	28.76	25.35	31.25
23	* P.E and Sports	11.70	13.11	14.65
24	* Administration	1.74	1.94	1.99
	ENGINEERING INSTALLATION			
25	* Plumbing	20.85	15.75	32.88
26	* Electrical	39.83	42.38	46.53
27	* Heating and Ventilation	44.58	33.01	43.77
28	* Gas	4.97	2.07	6.84
	SPECIALIST INSTALLATIONS			
29	* Lifts & Conveyors	0.00	0.00	0.00
30	* Fire Protection	2.69	8.94	9.12
31	* Security	3.64	5.24	3.77
32	* Communication	3.11	5.68	5.15
	TOTAL (Cost per square metre)	**599.32**	**574.90**	**436.48**

MAINTENANCE AND RUNNING COSTS
(Excluding External Works)

REF	ELEMENT	Annual	Third Year	Seventh Year
	ADMINISTRATION			
1	Estate Management	0.23	0.24	0.27
2	Accountant	0.46	0.80	0.90
3	Caretaker (Accommodation provided)	1.83	1.92	2.15
4	Security (Service Agreement)	0.49	0.51	0.57
	FACILITIES MANAGEMENT			
5	Hospitality	2.29	2.40	2.69
6	Technological (Service Agreements)	0.41	0.43	0.48
7	Library (Service Agreement)	1.22	1.28	1.43
8	Extra Curiculum Activities	1.07	1.12	1.26
	CLEANING & GROUNDSMAN			
9	Internal	3.08	3.57	4.00
10	Windows	0.25	0.26	0.29
11	External	0.34	0.36	0.40
12	Cultivation	0.21	0.23	0.26
	RENOVATION - EXTERNAL			
13	Roof Coverings	0.33	0.85	6.04
14	Roof Lights	0.05	0.10	0.51
15	Rainwater Installation	0.05	0.15	0.63
16	Redecoration - External	0.05	2.56	2.86
	RENOVATION - INTERNAL			
17	Doors	0.15	0.62	1.72
18	Ironmongery	0.06	0.45	0.68
19	Windows	0.11	1.05	7.65
20	Ironmongery	0.03	0.35	0.48
21	Floor coverings	1.01	7.48	11.17
22	Fittings	0.04	1.13	3.58
23	Furniture	0.11	0.43	2.86
24	Redecoration - Internal	0.19	0.20	4.77
	RENOVATION - SERVICES			
25	Plumbing and Drainage	0.18	0.76	1.26
26	Electrical (including charges)	1.16	1.45	2.29
27	Heating and Ventilation	0.91	1.04	2.58
28	Gas	0.03	0.03	0.04
	RENOVATION - SPECIALISTS			
29	Security Alarms	0.10	0.19	0.31
30	Fire Protection	0.08	0.17	0.29
31	Communications (Service Agreement)	0.04	0.34	0.95
32	IT Network (Service charge)	0.15	0.15	0.65
	TOTAL (Cost per square metre)	**16.67**	**32.65**	**66.01**

REGIONAL PRICE VARIATIONS

The price for building work varies according to location.

The square metre building costs and unit rates for measured work included in this cost guide reflect building costs in the Outer London Region.

These should be adjusted to obtain the appropriate costs and rates for the relevent LEA Area by multiplying the constants shown in the two tables set out below.

LEA Code	Area	Constant Multiply by
901	Avon	0.88
902	Bedfordshire	1.01
903	Berkshire	1.01
904	Buckinghamshire	1.01
905	Cambridgeshire	0.91
906	Cheshire	0.92
907	Cleveland	0.92
908	Cornwall	0.88
909	Cumbria	0.92
910	Derbyshire	0.92
911	Devon	0.88
912	Dorset	0.88
913	Durham	0.85
914	East Sussex	0.98
915	Essex	0.91
916	Gloucestershire	0.88
350-359	Greater Manchester	0.96
917	Hampshire	0.92
918	Herefordshire	0.88
918	Worcestershire	0.9
919	Hertfordshire	1.01
920	Humberside	0.85
921	Isle of Wight	0.92
420	Isles of Scilly	0.82
201-213	Inner London	1.1

LEA Code	Area	Constant Multiply by
922	Kent	0.98
923	Lancashire	0.92
924	Leicestershire	0.85
925	Lincolnshire	0.85
340-344	Mersyside	0.92
926	Norfolk	0.85
927	North Yorkshire	0.85
928	Northamptonshire	0.86
929	Northumberland	0.92
930	Nottinghamshire	0.92
301-320	Outer London	1.00
931	Oxfordshire	1.01
932	Shropshire	0.86
933	Somerset	0.88
370-373	South Yorkshire	0.85
934	Staffordshire	0.86
935	Suffolk	0.98
936	Surrey	0.98
390-394	Tyne and Wear	0.85
937	Warwickshire	0.91
330-336	West Midlands	0.86
938	West Sussex	0.98
380-384	West Yorkshire	0.85
939	Wiltshire	0.88

CHAPTER 2

The School Village

The Cost and Use of Activity Areas

FURNITURE AND FITTINGS

Lists and schedules of prices of all furniture, fittings and other associated equipment are shown on the page facing the appropriate activity area model and the pages following.

AN OVERVIEW OF THIS CHAPTER

Ease of Use

This chapter provides the user with a quick and easily understood method of assembling budget estimates for each selected activity area.

The School Village
(Shown on Page 21)

In order to focus attention on management, planning, phasing, funding and income potential, the typical activity areas in schools or universities have been classified into five categories as follows:

- M - Management and Services; pages 22 to 39
- H - Hospitality Facilities; pages 40 to 63
- A - Academic Teaching Areas; pages 64 to 77
- T - Art & Scientific Teaching Areas; pages 78 to 91
- S - Sport and Fitness Facilities. pages 92 to 98

The Activity Areas

Each category contains examples of teaching and non teaching activity areas providing quick references to:

- area planning guides; Models 1 to 4
- total capital costs; Table 'A'
- method for adjusting capital costs; Table 'B'
- cost of furniture and fittings. The schedules

The area models and associated capital building costs set out in Table 'A' also provide an indication of the numbers of spaces available for:

teaching;	Col. g
school meals;	Col. h
assemblies;	Col. i
revenue opportunities;	Col. j
sporting activities;	Col. k
seminars.	Col. l

The examples on pages 16 to 20 show how to use Table 'A' and how to adjust estimates using Table 'B'.

QUICK GUIDE TO COST AND USAGE

Step 1 — Activity Area
Select the Activity Area from page 21, turn to the appropriate page and *select* the ***Model*** (from 1 to 4) showing your preferred ***area planning arrangement***.

Step 2 — Type of School
Under column 'a' in Table 'A', *select* the rows related to the user's type of school (***Primary, Secondary or Higher***).

Step 3 — Area of Planning Arrangement
In the rows related to the user's type of school and under columns 'b', *select* the *Model reference* and under columns 'c', *select* the ***preferred area required***.

Step 4 — Total Building Cost
In the same row selected under step 3 above for the preferred area and under columns 'd', 'e' and 'f', *select* the ***Total Building Cost*** according to the type of school building work to be carried out. (***New, extension or remodelled***).

Step 5 — Regional Variations in Building Cost
Select the LEA Code or Area on page 12 and the adjoining constant and *multiply* the Total Building Cost arrived at in step 4 above by this constant to obtain the adjusted ***Total Regional Building Cost***.

Step 6 — Operational Usage
In the same row selected under step 3 above, *select* under columns 'g', 'h', 'i', 'j', 'k' and 'l', the operational usage which could be available for the selected planning arrangement.

THE MODELS

The models generally show four examples of possible alternative arrangements of each teaching and non teaching activity area.

The user may alter and vary the size and shape of the area models according to the particular teaching and non teaching requirement and then vary the capital costs and usage values as shown in the examples below.

Example

A two storey, double bank classroom block consisting of 12 classrooms with a central corridor on each floor.

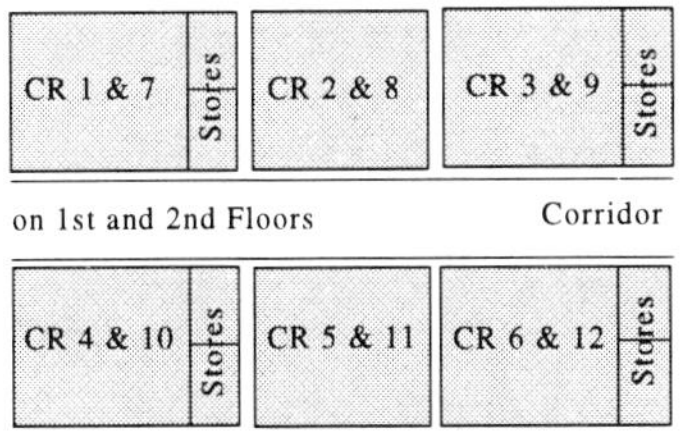

Table 'A'

The tables are set out below the models for each activity area and show examples of alternative sizes, capital costs and the usage facilities which can be obtained from each area appropriate for the type of school and the users' requirements.

The capital costs are set out according to the class of building work to be undertaken including:

d) New work

e) Extending existing buildings or

f) Remodelling existing buildings.

Examples of typical adjustments to:

Example 1 - Capital Building Costs;

Example 2 - Operational Usage Values.

KEY TO THE USE OF TABLE 'A'.

These schedules indicate typical guides to floor areas, total building costs and the resources which would be available based upon the use of the area models and according to the type of school and its special requirements.

School Types
Three types of schools are shown, but the Higher Schools category also makes provision for University Buildings and the requirements of Colleges of Further Education.

Total Building Costs
The Building Costs have been rounded up to the nearest £100 for ease of use and include the cost of furniture, fittings and special services, builder's profit and overheads **but exclude professional fees**.

Regional Variations in Cost
The Building Costs must be adjusted to take account of Regional Variations in Prices as shown on Page 12.

TABLE 'A' CLASSROOMS - TWO STOREY

School Type (a)	Model (b)	Area m2 (c)	Total Building Cost (£) New (d)	Extension (e)	Re-Modelled (f)	Operational Usage (Nos) Teaching Spaces (g)	School Meals (h)	Assembly Spaces (i)	Revenue & Entertainment (j)	Sport (k)	Seminar Spaces (l)
Primary	1	229	178,900	164,800	119,600	92	NIL	77	92	NIL	115
	1	254	198,400	182,800	132,600	102	NIL	85	102	NIL	127
	2	381	278,100	256,300	185,900	153	NIL	127	153	NIL	191
	2	424	309,500	285,200	206,900	170	NIL	142	170	NIL	212
Secondary	2	656	478,900	441,200	320,100	263	NIL	219	263	NIL	328
	2	656	478,900	441,200	320,100	263	NIL	219	263	NIL	328
	3	819	556,100	512,400	371,800	328	NIL	273	328	NIL	410
	3	850	577,200	531,800	385,900	340	NIL	284	340	NIL	425
	4	1,364	856,700	789,300	572,700	546	NIL	455	546	NIL	682
	4	1,411	886,200	816,500	592,400	565	NIL	471	565	NIL	706
Higher	3	847	575,200	529,900	384,500	339	NIL	283	339	NIL	424
	3	866	588,100	541,800	393,100	347	NIL	289	347	NIL	433
	4	1,411	886,200	816,500	592,400	565	NIL	471	565	NIL	706
	4	722	453,500	417,800	303,200	289	NIL	241	289	NIL	361

Models
Each area model may relate to any type of school, but they have been selected here to reflect a guide to average requirements.

Areas
The areas shown are generally considered appropriate for the type of school and the area model arrangement selected.

Teaching Spaces
The numbers of teaching spaces stated are based upon average areas per pupil according to the type of school and key stages. See Appendices I, J and K on pages 260 to 262.

Area Allowances
Operational usage is based upon the following area allowances per pupil:

- School meals; 2 m2
- Assembly; 1.50 m2
- Revenue; 3 m2
- Entertainment; 3 m2
- Sport; 5 m2
- Seminars; 1 m2

The areas shown are generally considered appropriate for the type of school and the model arrangement selected.

Area Guidelines
The above areas and numbers of teaching spaces generally fall within DfEE recommendations. The user should refer to Building Bulletin 82 in order to make more precise calculations

Example 1 — HOW TO ADJUST Capital Building Costs:

The capital building costs are derived from:
(under columns 'd', 'e' and 'f')

a) Floor area (From table 'A' select from col. 'c') multiplied by Cost per m2 (From table 'B' select total from col. 'n', 'o' or 'p')

b) *Adjust* for Regional Variations from page 12.

To revise the floor area for:

a) Teaching Areas,	*refer to*	Appendices 'I to K'
b) Non Teaching Areas,	*refer to*	Appendix 'M'
	or select	suitable area of existing premises.

Example: 12 new classrooms to be added to an existing secondary school with average groups of 25 pupils to be arranged as shown in Model 3 on page 65.

from Appendix 'I' (max size for small classrooms):

select:

12 (classrooms) x 46 m2 = 552 m2
12 (store rooms) x 5 m2 = 60 m2
2 (corridors) x 55 m2 = 110 m2
Total Area = 722 m2

Note;-The areas for corridors and stores are not normally included with the above and are shown separately on the Accommodation Analysis. They are included here to show how the total building cost of a classroom extension can be obtained.

To revised the capital building cost:

multiply;

722 m2 (See above) x **Rate per m2** (From Table 'B' select total rate from Cols 'n' 'o' or 'p') = **Total Building Cost**

Example 2 — HOW TO ADJUST Operational Usage Values:

To Revise Operational Usage Indicators:

a) Area allowances,	*refer to*	page 15
b) Exact allowances,	*refer to*	Appendices 'I to K'
c) Guidance notes,	*refer to*	DfEE Bulletin 82
d) Special selection,	*insert*	User's preferences.

Example: To revise operational usage indicators following the adjustments made in Example 1 in the previous column;

take:

Floor area (From Example 1) divide by Area Allowances (From table on page 15)

The revised operational indicators are:

Teaching Spaces
take: 46 m2 / 25 nr = 1.84 m2
then: 552 m2 / 1.84 m2 = 300 nr.

School meals		=	Nil
Assembly	552 m2 / 1.5 m2	=	368 nr.
Revenue	552 m2 / 3.0 m2	=	184 nr.
Entertainment	552 m2 / 3.0 m2	=	184 nr.
Sports		=	Nil
Seminars	552 m2 / 1.0 m2	=	552 nr.

To adjust the accommodation requirements:

insert the resultants into column [AJ] in the accommodation schedule shown on Appendix 'M'.

To adjust the potential revenue:

insert the resultants into column [BA] in the benchmark cost analysis shown on Appendix 'N'.

KEY TO THE USE OF TABLE 'B'.

Table 'B'

The purpose of the analyses shown in Table 'B' is to provide a basis for adjustment of the total building costs shown in Table 'A' to suit the user's preferred specification for fitting out works.

Note: It is not intended that the user should make adjustments to the cost of the first two elements (1 and 2).

The examples on pages 18 & 19 show how to make the following selections and adjustment to the remaining elements (3 to 12).

- adjust the elemental rates;
- adjust the costs according to the size of the project;
- make provision for regional variations;
- allow for inflation;
- make provision for professional fees;
- revise the total costs.

Examples of typical adjustments to:

Example 3 - floor finishings or ceiling finishings;
Example 4 - wall finishings
Example 5 - furniture or fittings;
Example 6 - sanitary fittings or lighting fittings;
Example 7 - protection, surveillance or communication.

Note:- Elemental rates **include** overheads and profit;
Elemental rates **do not include** professional fees:
Multiply rates by 'Regional Variations' to obtain local rates.

Generally figures have been rounded up to two decimal places causing slight anomalies to occur in the totals in Tables 'B'.

The Analysis

The building cost for each activity area will vary according to the type of building work to be undertaken. **Table 'B'** shows the cost per square metre of floor area for three different types of building work and their analysis into elemental costs as follows:

New	Extension	Remodelling
forming part of a new school development, col. (n)	an addition to existing school premises, col. (o)	altering and fitting out existing school col. (p)

TABLE 'B' ANALYSIS OF COSTS
£ per Square Metre - Based upon Model 3

Elements of the Building (m)	New £ per m2 (n)	Extension £ per m2 (o)	Remodelled £ per m2 (p)
Main Building Work			
* Foundations	57.14	57.60	2.41
* Superstructure	334.33	272.89	133.09
Finishings			
* Floors	47.23	49.68	63.19
* Walls	28.66	29.86	24.58
* Ceilings	27.70	31.94	46.05
Fittings			
* Furniture	53.13	49.37	49.79
* Fittings	94.45	87.76	88.51
* Sanitary Fittings	0.00	0.00	0.00
* Light fittings	23.51	23.25	25.74
Security			
* Protection	2.70	8.34	8.58
* Surveillance	6.44	8.62	6.25
* Communication	3.67	6.23	5.70
FITTING OUT	**287.50** £ per m2	**295.04** £ per m2	**318.39** £ per m2
TOTAL COST	**678.99** £ per m2	**625.53** £ per m2	**453.88** £ per m2

Fitting Out Cost = Total of elements 3 to 12 multiplied by floor area

Total Cost = Total of elements 1 to 12 multiplied by floor area

THE BUILDING ELEMENTS

Grouping of the Elements

A full list of building elements is shown on page 11. These have been condensed into four main groups for use when looking at the costsings for each activity area.

- The first group relates to the main building construction works. These are fixed and should not be altered.
- The other three groups can be adjusted as shown in the Examples on page 18 and 19.

Group 1 (Not to be adjusted)

1	Foundations;	concrete footings; foundation walls; damp proof courses; ground floor construction:
2	Superstructure:	external walls; internal partitions; roofs; coverings; rainwater goods; upper floor construction; staircases; doors; windows; roof lights: electrical; mechanical; plumbing; external roads and paths; drainage; service connections:

Group 2 (See Examples 3 and 4)

Finishings;

3	Floors	floor coverings; screeds; skirtings; damp and sound proof membranes;
4	Walls	internal faced brickwork; plastering; ceramic wall tiling; panelling; dado rails; decorations:
5	Ceilings	boarding; plastering; cornices; suspended ceilings; decorations:

Group 3 (See Example 5 and 6)

Fittings;

6	Furniture	chairs, tables, stools; cupboards; pinboards; blackboards, notice boards
7	Fittings	fitted cupboards; kitchen equipment; gymnasium equipment; staging:
8	Sanitary Fittings;	W.C's; wash basins; shower fittings; sink units; drinking fountains; laboratory sinks:
9	Light Fittings;	ceiling and wall light fittings:

Group 4 (See Example 7)

Security;

10	Alarms	fire alarms; intruder alarms;
11	Surveillance	video cameras; alarms; lighting:
12	Communication	telephones; modems; intercoms; music and sound systems:

Example 3 — HOW TO ADJUST; Elemental Rates
(under columns 'n', 'o' and 'p')
Floor and Ceiling Finishings:

To source new unit rates for building work:

Floor finishings,	*refer to*	pages 156 to 167
Ceiling finishings,	*refer to*	pages 168 to 175
Skirtings,	*refer to*	pages 143
Other materials,	*select*	costs per m2 for other suitable materials.

To calculate the total cost of finishing work:

Example; floor finishings; re-modelling:

		(Unit rate per m2)
Remove existing floor screed	(page 156)	£ 5.37
Damp proof course	(page 125)	£ 6.77
50 mm Cement and sand screed	(page 157)	£ 8.82
Fine ribbed bonded carpet	(page 166)	£ 13.24
	Sub Total	£ 34.20

Total cost = £34.20 x 46 m2 = **£1,573.20**

Example; (i) skirtings & (ii) cover strip; re-modelling:

(i)

		(Unit rate per m)
25 x 100 mm Softwood skirting	(page 142)	£ 2.68
Knot prime and stop < 150 mm	(page 184)	£ 0.39
Oil paint; 1st application	(page 184)	£ 0.29
Oil paint; 2nd application	(page 184)	£ 0.27
Total unit rate per linear metre		£ 3.63

Total cost = perimeter of room x unit rate

Total cost = 29 m x £ 3.63 = **£ 105.27**

(ii)

	(Unit rate per m)
Brass cover strip plugged *(Quotation)*	£ 8.71

cost of cover strip = (2 door openings x unit rate)

Total cost = 2 x 0.75 m x 8.71 = **£ 13.06**

To calculate the revised elemental rate per m2:
(for entry under columns 'p', re-modelling work).

Total costs / floor area = revised elemental rate

(£1.573.20 + £105.27 + £13.06) /46 m2 = **£ 36.77 per m2**

Note: Enter the new rate of £36.77 under column 'p' as shown on page 20

Example 4 — HOW TO ADJUST; Elemental Rates
(under columns 'n', 'o' and 'p')
Wall Finishings:

To source new unit rates for building work:

Plaster work,	*refer to*	pages 168 to 175
Decoration work,	*refer to*	pages 182 to 191
Dado rails	refer to	pages 143
Other materials,	*select*	costs per m2 for other suitable materials.

To calculate the total cost of finishing work:

Example: wall finishings; re-modelling:

		(Unit rate per m2)
Dubbing out existing walls	(page 169)	£ 3.79
Plaster on walls	(page 168)	£ 7.87
Emulsion paint; 1st application	(page 182)	£ 0.43
Emulsion paint; 2nd application	(page 182)	£ 1.33
Emulsion paint; 3rd application	(page 182)	£ 1.10
	Sub Total	£14.52

Total cost = perimeter of room x height of wall above skirting height x unit rate
= 29 m x 2.85 m x £14.52 = £1200.08

Note; No adjustment has been made here for door and window openings.

Example; dado rail; re-modelling:

		(Unit rate per m2)
38 x 75 mm Softwood dado rail	(page 142)	£ 2.61
Knot prime and stop < 150 mm	(page 184)	£ 0.39
Oil paint; 1st application	(page 184)	£ 0.29
Oil paint; 2nd application	(page 184)	£ 0.27
Total unit rate per linear metre		£ 3.56

Total cost = perimeter of room x unit rate

Total cost = 29 m x £ 3.56 = **£ 103.24**

To calculate the revised elemental rate per m2:
(for entry under columns 'p', re-modelling work).

Total costs / floor area = revised elemental rate

(£1200.08 + £103.24) / 46 m2 = **£ 28.34 per m2**

Note: Enter the new rate of £24.03 under column 'p' as shown on page 20

Example 5 HOW TO ADJUST; Elemental Rates
(under columns 'n', 'o' and 'p')
Furniture and Fittings:

Furniture & fittings,	*refer to*	pages 23 to 98 next to the activity area
Other equipment,	*select*	costs for other suitable supplies or, quotations from other suppliers
Composite schedules,	*refer to*	Appendices 'U to X'

Example; new furniture for a classroom:

from Appendix 'V':

Element	Qty	Cost	Total
Blackboards	1	80	80
Book Trolley	1	160	160
Bookcases	2	255	510
Chairs - Standard	35	20	700
Coat & Bag Storage	2	250	500
Computer Workstations	4	250	1,000
Cupboards - High Level	1	350	350
Cupboards - Low Level	2	195	390
Cupboards - Paper Storage	1	180	180
Dry Wipe Board	1	120	120
Mobile Swivel Board	1	240	240
Pinboards	8	58	464
Projection Screen - Wall Mounted	1	120	120
Roller Board - Floor standing	1	400	400
Tables - Standard (two person)	20	80	1,600
			£ 6,814

total cost	x	nr classrooms	/	floor area	=	Revised rate
£6,814	x	12	/	552 m2	=	**£ 148.13**
(See above)		(See above)		(See example no. 1 page 16)		(Enter under cols. 'n', 'o', 'p'

Note;- If the furniture is to be a part of the Builder's Contract add an allowance for overheads and profit - See Page 108.
Otherwise add 5% to the above to allow for ordering, unpacking and placing furniture into position and fixing fittings where required.

Example 6 HOW TO ADJUST; Elemental Rates
(under columns 'n', 'o' and 'p')
Sanitary Fittings:

To source new unit rates for fittings:

Sanitary Fittings,	*refer to*	pages 193
Toilet cubicles,	*refer to*	page 199
Sundries,	*refer to*	page 199
Other equipment,	*select*	costs for other suitable supplies or,
Quotations		from other suppliers

To calculate the cost of the selected sanitary fittings:

Example; selected replacement sanitary fittings:

(Unit rates; nr)		
WC Suites (Key stage 2)	(page 193)	
12 nr @ £241.00 each		£2,892.00
Cubicles range of 12	(page 199)	
12 nr @ £445.00 each		£5,340.00
Toilet roll holders	(page 193)	
12 nr @ £15.00 each		£ 180.00
	Sub Total	£8,412.00
Add 8% for builder's work		672.96
	Total Cost Fittings	£9,084.96

Note: Builder's work includes cutting holes through walls and floors and similar items of work.

To calculate the revised elemental rate per m2:
(for entry under columns 'p', re-modelling work).

floor area of toilets; see page 58 - model 3 = 86 m2

total cost	/	floor area	=	Revised rate
£9,084.96	/	86 m2	=	**£ 105.64**
(See above)		(See Model 3 on page 58)		(Enter under col. 'p')

Note: No adjustment has been made here for the storage cupboard unit for cleaner's equipment or the dressing area.

Example 7 HOW TO ADJUST; Elemental Rates
(under columns 'n', 'o' and 'p')
Alarm Systems:

To source new unit rates for building work:

Alarm System,	*refer to*	page 33
Other equipment,	*select*	costs for other suitable supplies or, quotations from other suppliers

To calculate the cost of the selected alarm systems:

Example; selected alarm system budget pack:

		(Unit rates; nr)
Junior starter pack	(page 33)	£1,995.00
Add 5% for builder's work		99.75
Total Cost Alarm Installation		£2,094.75

Note: Including installation costs such as ductwork and holes through walls and floors.

To calculate the revised elemental rate per m2:
(for entry under columns 'p', re-modelling work).

Floor area of administrative wing:

Office;	model 2, page 22	=	20 m2
Reception;	model 2, page 26	=	24 m2
Staff room;	model 2, page 30	=	27 m2
Corridors;	add 10% of 75 m2	=	8 m2
	Total floor area	=	79 m2

The revised elemental rate per m2 for the above administrative areas would be calculated as follows and the rate inserted in table 'B' for each activity area.

total cost	/	floor area	=	Revised rate
£2,094.75	/	79 m2	=	**£ 26.52**
(See above)				(Enter under col. 'p')

This is a completed table showing the elemental rates for a typical activity area.

A TYPICAL COMPLETED TABLE 'B'.

TABLE 'B' ANALYSIS OF COSTS
£ Per Square Metre - Based upon Model 3

Elements of the Building (m)	New £ per m2 (n)	Extension £ per m2 (o)	Remodelled £ per m2 (p)
Main Building Work			
* Foundations	108.63	111.00	4.18
		69	151.00
		10	78.36
		15	22.17
		94	45.68
		85	46.88
		32	109.38
		99	11.07
		25	25.54
		34	8.51
		62	6.20
		23	5.65
		7	**359.43** £ per m2
		5	**510.44** £ per m2

Elements of the Building (m)	New £ per m2 (n)	Extension £ per m2 (o)	Remodelled £ per m2 (p)
Main Building Work			
* Foundations			
* Superstructure			
Finishings			
* Floors			
* Walls			
* Ceilings			
Fittings			
* Furniture			
* Fittings			
* Sanitary Fittings			
* Light fittings			
Security			
* Protection			
* Surveillance			
* Communication			
FITTING OUT	£ per m2	£ per m2	£ per m2
TOTAL COST	£ per m2	£ per m2	£ per m2

The elemental rates to be added together to provide the total cost of 'FITTING OUT COST.

The elemental rates to be added together to provide the TOTAL COST.

This is a blank table for a typical activity area showing only the elemental rates for the foundations and superstructure which normally should not be adjusted except by reference to a professional consultant.

USING A BLANK TABLE 'B'.

The elemental rates to be entered under col. 'p' from working the typical examples.

Elements of the Building (m)			Remodelled £ per m2 (p)	
Main Building Work				
* Foundations			2.41	
* Superstructure			133.09	
Finishings				
* Floors			**36.77**	Example No. 3
* Walls			**28.34**	Example No. 4
* Ceilings			46.05	Not adjusted
Fittings				
* Furniture			**148.13**	Example No. 5
* Fittings			88.51	Not adjusted
* Sanitary Fittings			**105.64**	Example No. 6
* Light fittings			25.74	Not adjusted
Security				
* Protection			**26.52**	Example No. 7
* Surveillance			6.25	Not adjusted
* Communication			5.70	Not adjusted
FITTING OUT			**517.65** £ per m2	
TOTAL COST			**653.15** £ per m2	

These two columns are left blank when dealing with a remodelled project.

M - MANAGEMENT

H - HOSPITALITY FACILITIES

Academic

A - TEACHING AREAS

Technology & Science

T - TEACHING AREAS

S - SPORTS

FURNITURE AND FITTINGS

Lists and schedules of prices of all furniture, fittings and other associated equipment are shown on the page facing the appropriate activity area model and the following pages.

M 1 - OFFICE AREA ARRANGEMENTS

Model 1

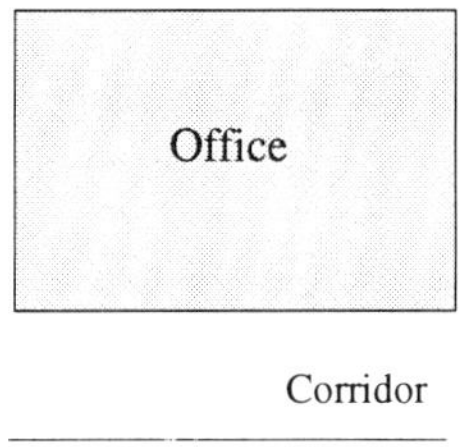

Model 2

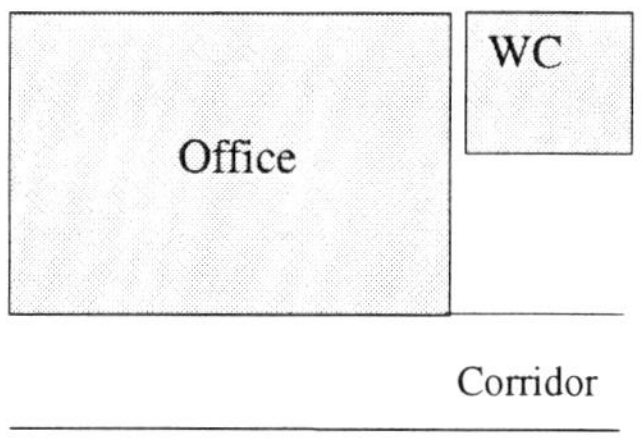

Model 3

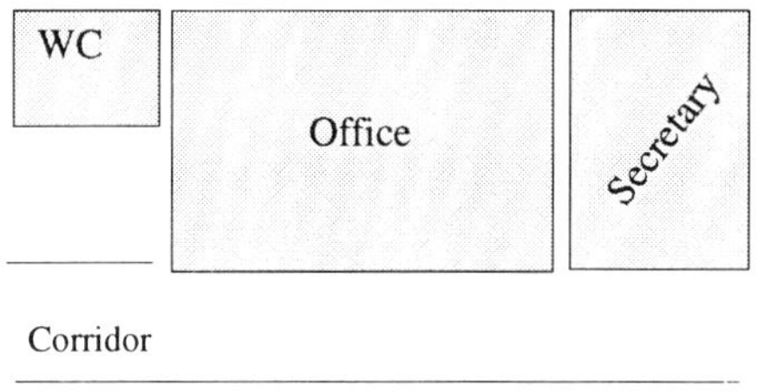

Model 4

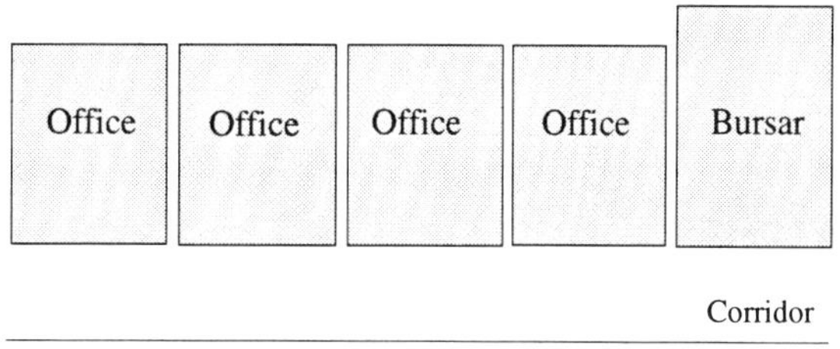

TABLE 'A' OFFICE

School Type (a)	Model (b)	Area m2 (c)	Total Building Cost (£) New (d)	Extension (e)	Re-Modelled (f)	Operational Usage (Nos) Teaching Spaces (g)	School Meals (h)	Assembly Spaces (i)	Revenue & Entertainment (j)	Sport (k)	Seminar Spaces (l)
Primary	1	10	£6,975	£6,525	£4,800	NIL	5	7	NIL	NIL	5
	1	12	£8,375	£7,850	£5,750	NIL	6	8	NIL	NIL	6
	2	14	£10,075	£9,450	£6,925	NIL	7	10	NIL	NIL	7
	2	20	£14,375	£13,475	£9,900	NIL	10	14	NIL	NIL	10
Secondary	2	20	£14,375	£13,475	£9,900	NIL	10	14	NIL	NIL	10
	2	25	£17,975	£16,850	£12,350	NIL	13	17	NIL	NIL	13
	3	26	£19,475	£18,250	£13,400	NIL	13	18	NIL	NIL	13
	3	45	£33,700	£31,575	£23,175	NIL	23	30	NIL	NIL	23
	4	31	£24,600	£23,075	£16,925	NIL	16	21	NIL	NIL	16
	4	35	£27,775	£26,025	£19,100	NIL	18	24	NIL	NIL	18
Higher	3	31	£23,225	£21,750	£15,975	NIL	16	21	NIL	NIL	16
	3	35	£26,200	£24,575	£18,025	NIL	18	24	NIL	NIL	18
	4	48	£38,100	£35,700	£26,200	NIL	24	32	NIL	NIL	24
	4	57	£45,225	£42,400	£31,100	NIL	29	38	NIL	NIL	29

TABLE 'B' ANALYSIS OF COSTS

£ per Square Metre - Based upon Model 3

Elements of the Building (m)	New £ per m2 (n)	Extension £ per m2 (o)	Remodelled £ per m2 (p)
Main Building Work			
* Foundations	109.07	111.00	4.21
* Superstructure	314.33	257.69	148.00
Finishings			
* Floors	59.04	62.10	78.98
* Walls	26.05	27.15	22.35
* Ceilings	27.70	31.94	46.05
Fittings			
* Furniture	50.42	46.85	47.25
* Fittings	117.65	109.32	110.25
* Sanitary Fittings	7.88	8.99	11.16
* Light fittings	23.51	23.25	25.74
Security			
* Protection	2.70	8.34	8.58
* Surveillance	6.44	8.62	6.25
* Communication	3.67	6.23	5.70
FITTING OUT	**325.07** £ per m2	**332.77** £ per m2	**362.31** £ per m2
TOTAL COST	**748.48** £ per m2	**701.46** £ per m2	**514.52** £ per m2

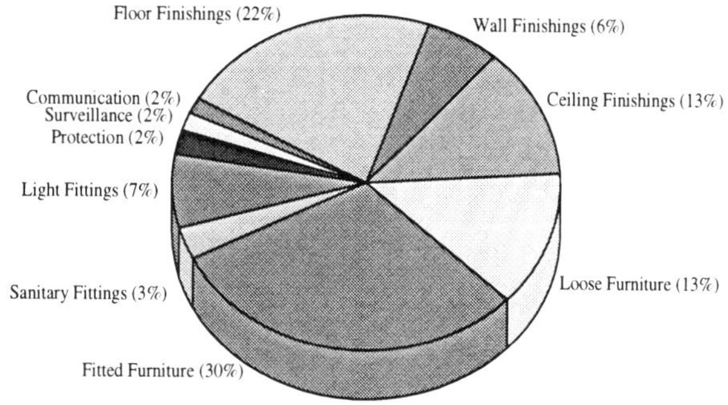

ANALYSIS OF FITTING OUT COSTS

(Costs based upon Table 'B' Column 'p').

CHAIRS

	Price £
MANAGERIAL	
Leather; upholstered; adjustable	
Standard adjustable	255 to 345
Ditto; with arm rest	320 to 430
Ditto; executive	425 to 570
Fabric; upholstered; adjustable	
Standard adjustable	170 to 230
Ditto; with arm rest	310 to 420
Ditto; executive	355 to 480
Fabric; Contract seating; fixed	
Standard	45 to 60
Ditto; with arm rest	60 to 80
Ditto; executive	80 to 110
OPERATORS	
Fabric; upholstered; adjustable	
Standard; manual operated	60 to 80
Ditto; with arm rest	85 to 115
Standard; Synchronous operated	125 to 165
Ditto; with arm rest	145 to 200
RECEPTION	
Leather; upholstered;	
Arm chair;	610 to 825
Sofa; two seater	895 to 1210
Sofa; three seater	1155 to 1565
Fabric; upholstered;	
Standard 1 seater unit	150 to 205
Ditto; with arm rest	185 to 250
Arm chair	215 to 290
VISITORS	
Fabric; chromium plated frame	
Standard	75 to 105
Ditto; with arm rest	100 to 130
Fabric; polished wood frame	
Standard	140 to 190
Ditto; with arm rest	155 to 210

TABLES AND DESKS

	Length x width x height	Price £
TABLES		
Melamine top; bullnosed edging; steel framed legs; epoxy paint finish.		
Oblong; general use	760 x 760 x 736 mm	45 to 60
Oblong; general use	1220 x 760 x 736 mm	57 to 77
Oblong; general use	1524 x 760 x 736 mm	64 to 86
Oblong; general use	1824 x 760 x 736 mm	68 to 92
DESKS		
Laminate or wood veneers; tubular frame; epoxy paint finish.		
Desk unit; oblong	1200 x 730 mm	60 to 80
Desk unit; oblong	1500 x 730 mm	70 to 95
Desk unit; oblong	1800 x 730 mm	75 to 105
Pedestals		
Single; 2 drawer	1200 x 730 mm	100 to 135
Single; 2 drawer	1500 x 730 mm	110 to 145
Single; 2 drawer	1800 x 730 mm	115 to 155
Single; 3 drawer	1200 x 730 mm	100 to 135
Single; 3 drawer	1500 x 730 mm	110 to 145
Single; 3 drawer	1800 x 730 mm	115 to 155
Double; 5 drawer	1500 x 730 mm	145 to 195
Double; 5 drawer	1800 x 730 mm	155 to 205
Traditional; wood veneer top; wood frame; pen tray; top drawer; locking; polished.		
Pedestals		
Single; 3 drawer	1050 x 730 mm	135 to 185
Single; 3 drawer	1200 x 750 mm	140 to 190
Single; 3 drawer & filing drawer	1400 x 750 mm	215 to 290
Double; 4 drawer & filing drawer	1600 x 750 mm	220 to 300

COMPUTERS

	Length x width x height	Price £
WORK STATIONS		
Melamine top; cable management system; cantilever steel frame; epoxy paint finish;		
Fixed		
Staggered front edge	1600 x 800 x 730 mm	390 to 530
Staggered front edge	1800 x 800 x 730 mm	415 to 560
Mobile		
Pedestal; 3 drawer	425 x 575 x 670 mm	185 to 245
Pedestal; 4 drawer	425 x 575 x 670 mm	185 to 245
Pedestal; 3 drawer	425 x 800 x 730 mm	245 to 330
Pedestal; 4 drawer	425 x 800 x 730 mm	245 to 330
Melamine top; sliding keyboard and shelf; adjustable; steel frame; epoxy paint finish;		
Work station	625 x 490 x 755 mm	105 to 140
Work station	625 x 490 x 660 mm	145 to 195

	Width x depth x height	Price £
PRINTER STANDS		
Dot matrix stand; laminated shelves; paper baskets; steel frame; epoxy paint finish		
Stand	632 x 666 x 670 mm	80 to 110
Heavy duty stand; laminated shelves; paper baskets; steel frame; epoxy paint finish		
Stand	700 x 500 x 650 mm	185 to 245
Heavy duty; adjustable; laminated shelves; paper baskets; steel frame; epoxy paint finish		
Stand	600 x 550 x 650 mm	235 to 320

STORAGE UNITS - OPEN

	Width x depth x height	Price £
LOW LEVEL		
Solid beech carcass; polished; static or mobile.		
Static 1 shelf	1040 x 496 x 590 mm	90 to 120
Mobile 1 shelf	1040 x 496 x 650 mm	105 to 140
Static 1 shelf	1040 x 496 x 750 mm	100 to 130
Mobile 1 shelf	1040 x 496 x 810 mm	110 to 145
Static 2 shelves	1040 x 496 x 590 mm	105 to 140
Mobile 2 shelves	1040 x 496 x 650 mm	115 to 155
Static 2 shelves	1040 x 496 x 750 mm	105 to 145
Mobile 2 shelves	1040 x 496 x 810 mm	120 to 160
Static 6 Drawers	1040 x 496 x 590 mm	215 to 290
Mobile 6 Drawers	1040 x 496 x 650 mm	225 to 305
Static 6 Drawers	1040 x 496 x 750 mm	250 to 335
Mobile 6 Drawers	1040 x 496 x 810 mm	260 to 355
Static 1 shelf	360 x 496 x 590 mm	60 to 85
Mobile 1 shelf	360 x 496 x 650 mm	75 to 100
Static 1 shelf	360 x 496 x 750 mm	70 to 95
Mobile 1 shelf	360 x 496 x 810 mm	85 to 110
HIGH LEVEL		
Solid beech carcass; polished; static or mobile.		
3 shelves	1040 x 496 x 1455mm	165 to 225
4 shelves	1040 x 496 x 1455mm	175 to 235
3 shelves, 24 trays	1040 x 496 x 1455mm	215 to 290
48 trays	1040 x 496 x 1455mm	255 to 340
12 drawers, 8 trays	1040 x 496 x 1455mm	405 to 545
3 shelves, 8 drawers	1040 x 496 x 1455mm	320 to 435
Bookcase; oak laminate; MDF carcass.		
3 shelves	950 x 400 x 1800 mm	150 to 200
2 shelves	950 x 400 x 1200 mm	120 to 160
3 shelves, 24 trays	950 x 400 x 750 mm	100 to 135

STORAGE UNITS - WITH DOORS

	Width x depth x height	Price £
LOW LEVEL		
Solid beech carcass; polished; static or mobile; locking.		
Static 1 shelf	1040 x 496 x 590 mm	145 to 200
Mobile 1 shelf	1040 x 496 x 650 mm	160 to 215
Static 1 shelf	1040 x 496 x 750 mm	155 to 205
Mobile 1 shelf	1040 x 496 x 810 mm	165 to 225
Static 2 shelves	1040 x 496 x 590 mm	160 to 215
Mobile 2 shelves	1040 x 496 x 650 mm	170 to 230
Static 2 shelves	1040 x 496 x 750 mm	165 to 220
Mobile 2 shelves	1040 x 496 x 810 mm	175 to 235
Static 6 Drawers	1040 x 496 x 590 mm	270 to 365
Mobile 6 Drawers	1040 x 496 x 650 mm	280 to 380
Static 6 Drawers	1040 x 496 x 750 mm	305 to 410
Mobile 6 Drawers	1040 x 496 x 810 mm	320 to 430
Static 1 shelf	360 x 496 x 590 mm	90 to 120
Mobile 1 shelf	360 x 496 x 650 mm	105 to 140
Static 1 shelf	360 x 496 x 750 mm	100 to 135
Mobile 1 shelf	360 x 496 x 810 mm	110 to 150
HIGH LEVEL		
Solid beech carcass; polished; static or mobile; locking.		
3 shelves	1040 x 496 x 1455 mm	255 to 345
4 shelves	1040 x 496 x 1455 mm	265 to 355
3 shelves, 24 trays	1040 x 496 x 1455 mm	305 to 415
48 trays	1040 x 496 x 1455 mm	345 to 465
12 drawers, 8 trays	1040 x 496 x 1455 mm	505 to 680
3 shelves, 8 drawers	1040 x 496 x 1455 mm	410 to 555
LARGE STEEL CABINET		
Lockable; grey; 3 adjustable shelves.		
	1220 x 457 x 1830 mm	190 to 255
Extra shelf		15 to 20

FILING CABINETS

	Width x depth x height	Price £
STANDARD		
Steel; rollers and sliders; locking; standard files.		
2 drawer	520 x 636 x 739 mm	120 to 160
3 drawer	520 x 636 x 1428 mm	135 to 180
4 drawer	520 x 636 x 1428 mm	145 to 195
FIRE RESISTANT		
Steel; rollers and sliders; fireproof; locking; 25" files.		
2 drawer	520 x 636 x 739 mm	680 to 915
4 drawer	520 x 636 x 1428 mm	1040 to 1405
Steel; rollers and sliders; fireproof; locking; 31" files.		
2 drawer	540 x 800 x 759 mm	735 to 995
3 drawer	540 x 800 x 1103 mm	1000 to 1355
4 drawer	540 x 800 x 1449 mm	1120 to 1515
Budget; steel; anti tilt device; rollers and sliders; 31" files.		
2 drawer		130 to 180
3 drawer		150 to 200
4 drawer		160 to 220

ARTIFICIAL PLANTS

	Price £
Mini Ming Aralia; 1200 mm high	52 to 70
Mini Camelia; 1200 mm high	38 to 52
Mini Aralia; 1800 mm high	70 to 94
Autumn Ficus; 1800 mm high	55 to 75
Golden Ficus; 1800 mm high	55 to 75
Ficus; 2100 mm high	96 to 130
Mango Tree; 2400 mm high	87 to 118

PINBOARDS

	Width x height	Price £
HESSIAN COVERED		
Unframed	600 x 900 mm	17 to 22
Unframed	600 x 1200 mm	19 to 26
Unframed	900 x 1200 mm	26 to 34
Unframed	1200 x 1200 mm	31 to 41
Unframed	1200 x 1500 mm	37 to 49
Unframed	1200 x 1800 mm	42 to 57
Unframed	1200 x 2400 mm	54 to 73
Aluminium Frame	600 x 900 mm	29 to 39
Aluminium Frame	600 x 1200 mm	34 to 46
Aluminium Frame	900 x 1200 mm	44 to 59
Aluminium Frame	1200 x 1200 mm	52 to 70
Aluminium Frame	1200 x 1500 mm	61 to 83
Aluminium Frame	1200 x 1800 mm	70 to 94
Aluminium Frame	1200 x 2400 mm	89 to 120
FELT COVERED		
Unframed	600 x 900 mm	11 to 15
Unframed	600 x 1200 mm	12 to 16
Unframed	900 x 1200 mm	16 to 21
Unframed	1200 x 1200 mm	19 to 26
Unframed	1200 x 1500 mm	22 to 29
Unframed	1200 x 1800 mm	30 to 40
Unframed	1200 x 2400 mm	38 to 51
Aluminium Frame	600 x 900 mm	16 to 21
Aluminium Frame	600 x 1200 mm	18 to 24
Aluminium Frame	900 x 1200 mm	22 to 30
Aluminium Frame	1200 x 1200 mm	26 to 35
Aluminium Frame	1200 x 1500 mm	29 to 39
Aluminium Frame	1200 x 1800 mm	38 to 52
Aluminium Frame	1200 x 2400 mm	48 to 64
Wood Frame	600 x 900 mm	18 to 24
Wood Frame	600 x 1200 mm	21 to 28
Wood Frame	900 x 1200 mm	26 to 35
Wood Frame	1200 x 1200 mm	30 to 40
Wood Frame	1200 x 1500 mm	34 to 46
Wood Frame	1200 x 1800 mm	44 to 60
Wood Frame	1200 x 2400 mm	55 to 75

NOTE;-

See Section relating to Classrooms on pages 66 to 71 for other pinboards, notice boards and the like.

SAFES

	Width x depth x height	Price £
SMALL		
Fire protection; grey		
No alarm	486 x 392 x 359 mm	175 to 235
No alarm	346 x 392 x 499 mm	185 to 245
No alarm	493 x 392 x 629 mm	370 to 500
Alarm fitted	486 x 392 x 359 mm	215 to 290
Alarm fitted	346 x 392 x 499 mm	215 to 295
Alarm fitted	493 x 425 x 629 mm	405 to 545
Electronic lock	486 x 392 x 359 mm	245 to 330
Electronic lock	346 x 392 x 499 mm	245 to 335
Electronic lock	493 x 425 x 629 mm	435 to 590
LARGE		
Fire protection; grey		
No alarm	493 x 515 x 719 mm	445 to 600
No alarm	580 x 515 x 848 mm	535 to 725
No alarm	696 x 608 x 1214 mm	860 to 1165
Alarm fitted	493 x 515 x 719 mm	475 to 645
Alarm fitted	580 x 515 x 848 mm	570 to 770
Alarm fitted	696 x 608 x 1214 mm	895 to 1210
Electronic lock	493 x 515 x 719 mm	510 to 685
Electronic lock	580 x 515 x 848 mm	585 to 785
Electronic lock	696 x 608 x 1214 mm	925 to 1250
Electronic lock	792 x 691 x 1575 mm	1605 to 2170
Ditto; two doors	1442 x 748 x 1716 mm	2325 to 3145
DATA PROTECTION		
Fire protection; grey		
Combination with key	486 x 392 x 359 mm	270 to 360
Combination with key	346 x 392 x 499 mm	270 to 365
Combination with key	493 x 425 x 629 mm	490 to 665
DOCUMENTS		
Fire protection; white		
Key Lock	445 x 395 x 355 mm	200 to 270
Combination Lock	445 x 395 x 355 mm	285 to 385

M 2 - RECEPTION AREA ARRANGEMENTS

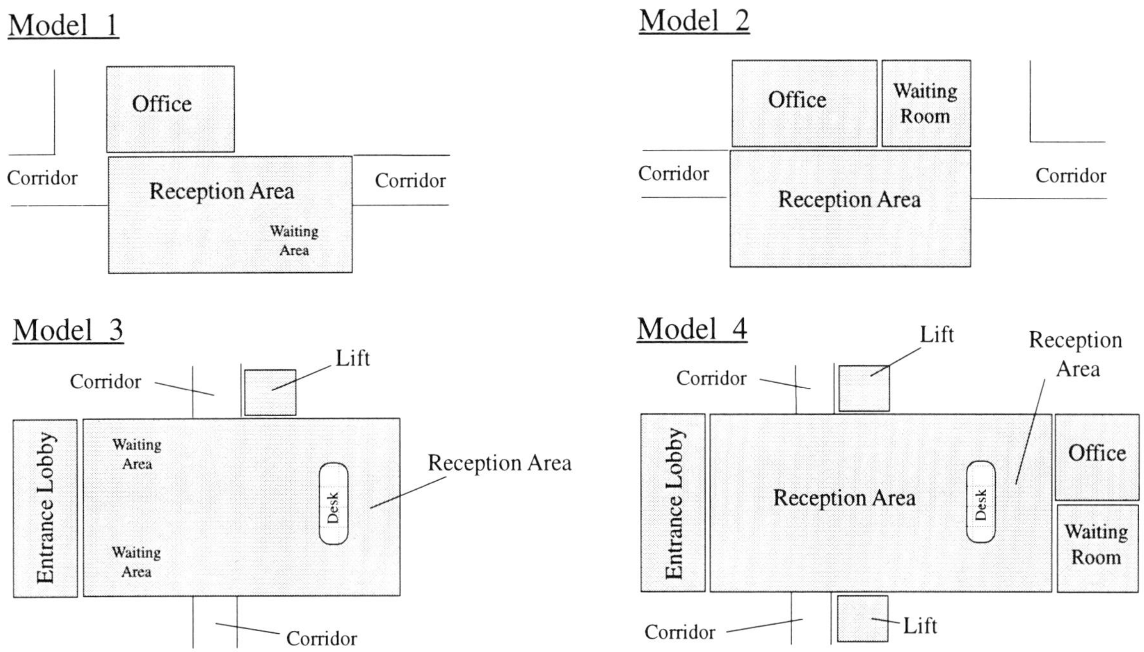

TABLE 'A' **RECEPTION AREA**

School Type (a)	Model (b)	Area m2 (c)	Total Building Cost (£) New (d)	Extension (e)	Re-Modelled (f)	Operational Usage (Nos) Teaching Spaces (g)	School Meals (h)	Assembly Spaces (i)	Revenue & Entertainment (j)	Sport (k)	Seminar Spaces (l)
Primary	1	15	£8,675	£8,075	£5,500	6	10	10	5	NIL	NIL
	1	22	£12,700	£11,850	£8,075	9	15	15	8	NIL	NIL
	2	20	£12,050	£11,250	£7,650	8	14	14	7	NIL	NIL
	2	28	£16,875	£15,725	£10,725	12	19	19	10	NIL	NIL
Secondary	2	24	£14,475	£13,475	£9,200	10	16	16	8	NIL	NIL
	2	35	£21,100	£19,650	£13,400	14	24	24	12	NIL	NIL
	3	69	£43,300	£40,350	£27,500	28	46	46	23	NIL	NIL
	3	<u>128</u>	£80,300	£74,850	£51,000	52	86	86	43	NIL	NIL
	4	50	£41,425	£38,600	£26,300	20	34	34	17	NIL	NIL
	4	73	£60,450	£56,350	£38,400	30	49	49	25	NIL	NIL
Higher	3	38	£23,850	£22,225	£15,150	NIL	NIL	26	13	NIL	NIL
	3	56	£35,150	£32,750	£22,325	NIL	NIL	38	19	NIL	NIL
	4	60	£49,700	£46,325	£31,550	NIL	NIL	40	20	NIL	NIL
	4	89	£73,700	£68,700	£46,800	NIL	NIL	60	30	NIL	NIL

TABLE 'B' ANALYSIS OF COSTS
£ per Square Metre - Based upon Model 3

Elements of the Building (m)	New £ per m2 (n)	Extension £ per m2 (o)	Remodelled £ per m2 (p)
Main Building Work			
* Foundations	109.07	111.00	4.21
* Superstructure	296.02	238.26	132.03
Finishings			
* Floors	59.04	62.10	78.98
* Walls	26.05	27.15	22.35
* Ceilings	27.70	31.94	46.05
Fittings			
* Furniture	50.36	46.79	47.20
* Fittings	22.69	21.08	21.26
* Sanitary Fittings	0.00	0.00	0.00
* Light fittings	23.51	23.25	25.74
Security			
* Protection	2.70	8.34	8.58
* Surveillance	6.44	8.62	6.25
* Communication	3.67	6.23	5.70
FITTING OUT	<u>**222.17**</u> £ per m2	<u>**235.49**</u> £ per m2	<u>**262.11**</u> £ per m2
TOTAL COST	<u>**627.28**</u> £ per m2	<u>**584.76**</u> £ per m2	<u>**398.35**</u> £ per m2

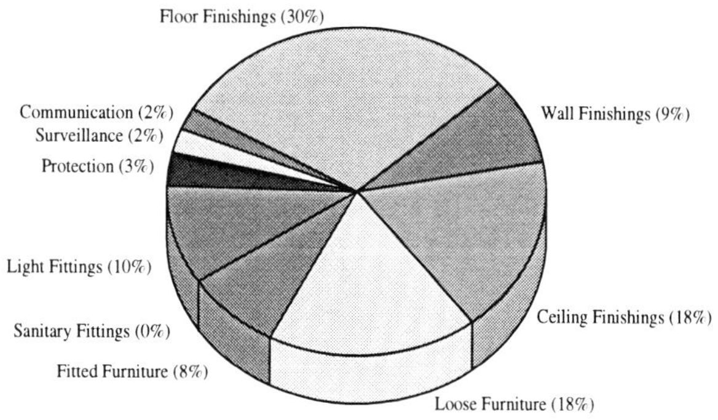

<u>ANALYSIS OF FITTING OUT COSTS</u>
(Costs based upon Table 'B' Column 'p').

SEATING

	Width x depth x height	Price £
MODULAR		
Fully upholstered; tweed covering.		
Chair; standard	660 x 800 x 420 mm	240 to 325
Chair; inside; curved	320 x 800 x 420 mm	220 to 300
Chair; outside; curved	660 x 800 x 420 mm	220 to 300
Table - Wood	710 x 710 x 340 mm	180 to 245
Table - Glass	710 x 710 x 340 mm	200 to 270
Beech frame; polished; designer look; upholstered seat and back; fabric covering.		
Chair; standard		200 to 270
Arm unit (left or right)		220 to 300
Armchair		245 to 330
Metal frame; epoxy paint finish; designer look; upholstered seat and back; fabric covering.		
Chair; standard		185 to 250
Arm unit (left or right)		210 to 280
Armchair		230 to 310
SEPARATES		
Leather suite; wood frame; fully upholstered		
Single armchair	1060 x 940 x 440 mm	850 to 1145
Sofa - 2 seater	1600 x 940 x 440 mm	1170 to 1580
Sofa - 3 seater	2130 x 940 x 440 mm	1680 to 2275
Reception seating; beech frame; polished; upholstered seat and back; tweed covering.		
Lounge chair		150 to 200
Armchair		170 to 230
Sofa - 2 seater		310 to 415
Fabric covered suite; wood frame; fully upholstered; deep cushions.		
Armchair		585 to 790
Sofa - 2 seater		620 to 840
Sofa - 3 seater		775 to 1050
Table; glass top		200 to 270

RECEPTION DESKS

	Length x width x heigh	Price £
MELAMINE FINISH		
Modular; work station; MDF carcass; worktop; modesty panel; melamine finish.		
Standard unit	800 x 727 x 800 mm	225 to 305
Standard unit	1200 x 720 x 800 mm	240 to 320
Standard unit	1600 x 720 x 800 mm	260 to 355
Standard unit	1800 x 720 x 800 mm	275 to 370
Corner unit; 90° curved		375 to 505
Corner unit; triangular		130 to 180
Modular; reception counter; MDF carcass; countertop; melamine finish.		
Standard unit	800 x 727 x 800 mm	155 to 210
Standard unit	1200 x 320 x 800 mm	175 to 235
Standard unit	1600 x 320 x 800 mm	205 to 275
Standard unit	1800 x 320 x 800 mm	225 to 305
Corner unit; 90° curved		235 to 320
Corner unit; triangular		160 to 215
PEDESTAL UNITS		
Pedestal units; single; MDF carcass; melamine finish.		
3 Drawer	1050 x 730 x 800 mm	110 to 145
3 Drawer	1200 x 750 x 800 mm	140 to 190
4 Drawer; filing drawer	1400 x 750 x 800 mm	115 to 150
Pedestal units; double; MDF carcass; melamine finish.		
5 Drawer	1500 x 730 x 800 mm	150 to 200
5 Drawer	1800 x 730 x 800 mm	155 to 210

	Length x width x heigh	Price £
POLISHED HARDWOOD		
Modular; work station; MDF carcass; worktop; modesty panel; wood veneer finish.		
Standard unit	800 x 727 x 800 mm	335 to 450
Standard unit	1200 x 720 x 800 mm	345 to 470
Standard unit	1600 x 720 x 800 mm	380 to 510
Standard unit	1800 x 720 x 800 mm	430 to 580
Corner unit; 90° curved		460 to 625
Corner unit; triangular		170 to 230
Modular; reception counter; MDF carcass; countertop; melamine finish.		
Standard unit	800 x 727 x 800 mm	205 to 280
Standard unit	1200 x 320 x 800 mm	225 to 305
Standard unit	1600 x 320 x 800 mm	265 to 360
Standard unit	1800 x 320 x 800 mm	290 to 390
Corner unit; 90° curved		215 to 290
Corner unit; triangular		320 to 435
PEDESTAL UNITS		
Pedestal units; single; MDF carcass; melamine finish.		
3 Drawer	1050 x 730 x 800 mm	125 to 170
3 Drawer	1200 x 750 x 800 mm	145 to 190
4 Drawer; filing drawer	1400 x 750 x 800 mm	220 to 295
Pedestal units; double; MDF carcass; melamine finish.		
5 Drawer	1500 x 730 x 800 mm	170 to 225
5 Drawer	1800 x 730 x 800 mm	175 to 235

PINBOARDS

	Width x height	Price £
HESSIAN COVERED		
ABS plastic corners, flame resistant and light fast.		
Aluminium frame	600 x 900 mm	32 to 43
Aluminium frame	600 x 1200 mm	38 to 51
Aluminium frame	900 x 1200 mm	44 to 59
Aluminium frame	1200 x 1500 mm	57 to 77
Aluminium frame	1200 x 1800 mm	66 to 89
Wood frame, flame resistant and light fast.		
Wood frame	600 x 900 mm	32 to 43
Wood frame	600 x 1200 mm	38 to 51
Wood frame	900 x 1200 mm	44 to 59
Wood frame	1200 x 1500 mm	57 to 77
Wood frame	1200 x 1800 mm	66 to 89
FELT COVERED		
ABS plastic corners, flame resistant and light fast.		
Aluminium frame	600 x 900 mm	36 to 49
Aluminium frame	600 x 1200 mm	43 to 58
Aluminium frame	900 x 1200 mm	50 to 67
Aluminium frame	1200 x 1500 mm	64 to 86
Aluminium frame	1200 x 1800 mm	74 to 100
Wood frame, flame resistant and light fast.		
Wood frame	600 x 900 mm	36 to 49
Wood frame	600 x 1200 mm	43 to 58
Wood frame	900 x 1200 mm	50 to 67
Wood frame	1200 x 1500 mm	64 to 86
Wood frame	1200 x 1800 mm	74 to 100

LETTER BOARDS

	Width x height	Price £
LETTER BOARDS		
Glass; hinged toughened glass doors; single or double; grooved felt backboard		
Hardwood cabinet	600 x 450 mm	235 to 317
Hardwood cabinet	900 x 600 mm	332 to 449
Hardwood cabinet	1200 x 900 mm	527 to 712
Open board; grooved felt.		
Open	450 x 600 mm	64 to 87
Open	600 x 900 mm	73 to 98
Open	1200 x 900 mm	85 to 114
Poster frame; grooved felt.		
Open	A4	36 to 48
Open	A3	45 to 60
Open	A2	62 to 84
Open	A1	89 to 120
Free standing; grooved felt; tubular steel frame; epoxy finish.		
Gold frame	450 x 600 mm	135 to 183
Gold frame	600 x 900 mm	165 to 223
Gold frame; cover	600 x 900 mm	293 to 397
STANDS		
Single; brass		365 to 494
Single; black		199 to 269
Double; brass		663 to 897
LETTERING		
Alpha numeric	19 mm	14 to 19
Alpha numeric	28 mm	18 to 24
Alpha numeric	50 mm	24 to 32

DISPLAY BOARDS

	Width x height	Price £
NOTICE BOARDS		
Glazed; lockable; single or double; toughened hinged glass doors; choice of backboard.		
Wood cabinet	900 x 600 mm	130 to 176
Wood cabinet	900 x 1200 mm	201 to 272
Wood cabinet	1200 x 1200 mm	234 to 316
Wood cabinet	1000 x 1500 mm	240 to 324
Wood cabinet	1200 x 1800 mm	324 to 438
Aluminium frame	900 x 600 mm	207 to 280
Aluminium frame	900 x 1200 mm	286 to 387
Aluminium frame	1200 x 1200 mm	292 to 395
Aluminium frame	1000 x 1500 mm	293 to 396
Aluminium frame	1200 x 1800 mm	379 to 512
Glazed; lockable; single or double; toughened sliding glass doors; choice of backboard.		
Aluminium frame	900 x 600 mm	114 to 154
Aluminium frame	1200 x 600 mm	130 to 176
Aluminium frame	1200 x 900 mm	150 to 203
Aluminium frame	1200 x 1200 mm	177 to 240
Aluminium frame	1500 x 1200 mm	212 to 286
Aluminium frame	1800 x 1200 mm	249 to 337
Aluminium frame	2400 x 1200 mm	307 to 415
PRESENTATION BOARDS		
Display board; double		
Hessian covered	1200 x 1200 mm	325 to 439
Felt covered	1200 x 1200 mm	350 to 473
Cork covered	1200 x 1200 mm	350 to 473

DISPLAYS

	Width x depth x height	Price £
DISPLAY CABINETS		
Tower case; aluminium; heavy duty; epoxy coated; 4 mm float glass panels; adjustable glass shelves.		
With lighting	600 x 600 x 2100 mm	737 to 997
Without lighting	600 x 600 x 2100 mm	690 to 933
SHOW CASE		
Aluminium; heavy duty; epoxy coated; solid back; 3 glass shelves; bottom storage section; sliding doors; lockable.		
With lighting	1200x450x2100 mm	850 to 1150
Without lighting	1200x450x2100 mm	775 to 1050
COUNTER SHOWCASE		
Aluminium; epoxy coated; 2 glass shelves; sliding lockable rear doors.		
Without lighting	1200 x 600 x 900 mm	550 to 750
WALL DISPLAY CABINET		
Hardwood frame; toughened sliding glass doors; lockable; adjustable shelf; felt covered backboard.		
Trophy display	1200 x 200 x 900 mm	400 to 525

SCREENS

	Width x depth x height	Price £
ROOM DIVIDERS		
Display board; double sided; aluminium frame; free standing.		
Hessian covered		
Board only	1200 x 1200 mm	115 to 155
Board only	1500 x 1200 mm	125 to 170
Board only	1800 x 1200 mm	170 to 225
Felt Covered		
Board only	1200 x 1200 mm	140 to 190
Board only	1500 x 1200 mm	165 to 220
Board only	1800 x 1200 mm	190 to 260
Leg 185 cm		45 to 60
FOLDING; MOBILE		
Display board; hinged panels; aluminium frame; lockable castors		
Hessian covered		
3 panels	900 x 1200 mm	420 to 565
3 panels	1200 x 1500 mm	485 to 660
3 panels	1200 x 1800 mm	520 to 700
Felt Covered		
3 panels	900 x 1200 mm	450 to 610
3 panels	1200 x 1500 mm	520 to 705
3 panels	1200 x 1800 mm	560 to 755
DISPLAY SYSTEM		
Display board; double sided; panels size 600 x 900 mm; aluminium frame; poles 1800 mm.		
Starter kit; 1 panel		130 to 175
Mini kit; 3 panels		290 to 390
Mini kit; 3 panels; ext		350 to 475
Mini kit; 6 panels		605 to 820

PLANTERS

	Width x height	Price £
TRAYS AND POTS		
Trays; plastic		
Square	600 x 600 mm	16 to 21
Square	1200 x 1200 mm	20 to 27
Square	1500 x 1500 mm	24 to 33
Oblong	600 x 1200 mm	19 to 26
Oblong	600 x 1500 mm	24 to 33
Oblong	1200 x 2000 mm	30 to 41
Pots; terra cotta		
300 mm diameter		19 to 26
450 mm diameter		24 to 33
600 mm diameter		30 to 41
750 mm diameter		39 to 52
Pots; ceramic		
300 mm diameter		43 to 57
450 mm diameter		54 to 73
600 mm diameter		67 to 91
750 mm diameter		87 to 117
ARTIFICIAL PLANTS		
Mini Ming Aralia; 1200 mm high		52 to 71
Mini Camelia; 1200 mm high		39 to 52
Mini Aralia; 1800 mm high		70 to 95
Autumn Ficus; 1800 mm high		56 to 75
Golden Ficus; 1800 mm high		56 to 75
Ficus; 2100 mm high		97 to 132
Mango Tree; 2400 mm high		88 to 119

M 3 - STAFF ROOM AREA ARRANGEMENTS

Model 1

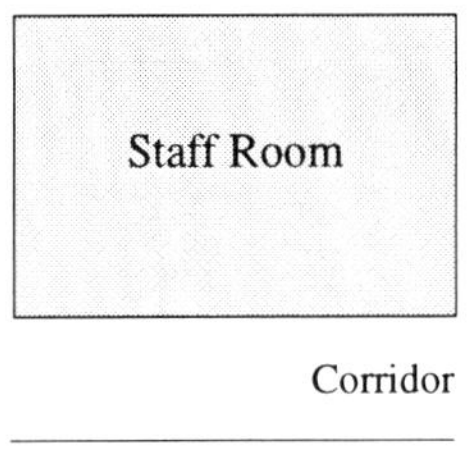

Model 2

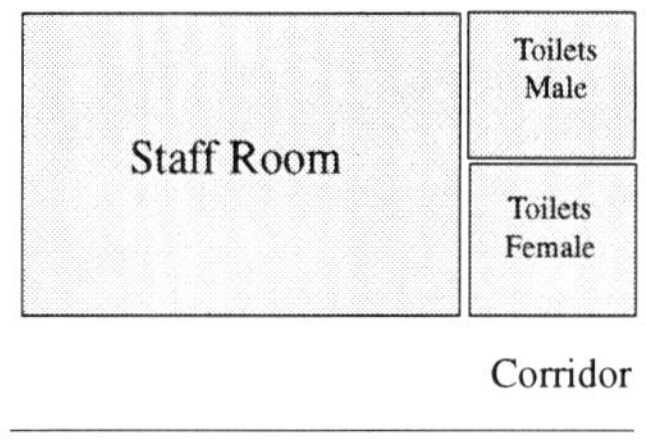

Model 3

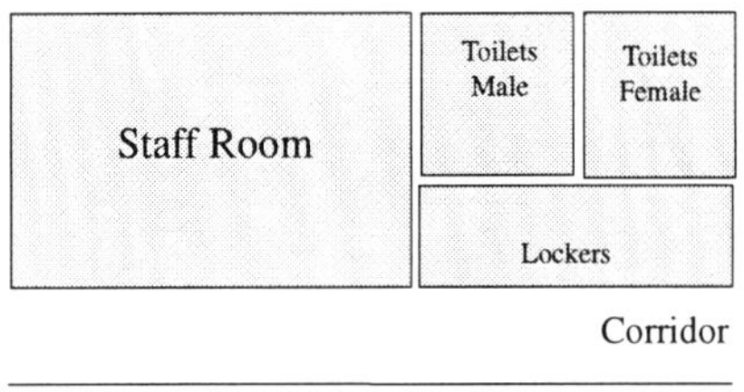

Model 4

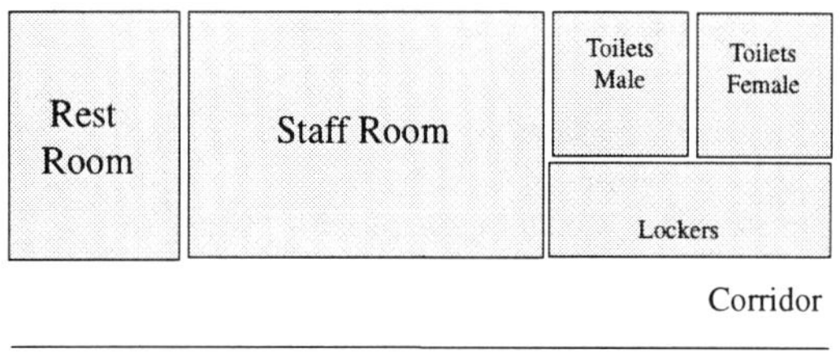

TABLE 'A' STAFF ROOM

School Type (a)	Model (b)	Area m2 (c)	Total Building Cost (£) New (d)	Extension (e)	Re-Modelled (f)	Operational Usage (Nos) Teaching Spaces (g)	School Meals (h)	Assembly Spaces (i)	Revenue & Entertainment (j)	Sport (k)	Seminar Spaces (l)
Primary	1	8	£5,150	£4,850	£3,475	3	4	6	NIL	NIL	6
	1	11	£7,050	£6,650	£4,775	4	6	8	NIL	NIL	8
	2	18	£11,925	£11,225	£8,050	6	9	12	NIL	NIL	12
	2	27	£17,875	£16,850	£12,075	9	14	18	NIL	NIL	18
Secondary	2	18	£11,925	£11,225	£8,050	6	9	12	NIL	NIL	12
	2	27	£17,875	£16,850	£12,075	9	14	18	NIL	NIL	18
	3	26	£18,325	£17,250	£12,350	9	13	18	NIL	NIL	18
	3	80	£56,325	£53,050	£38,000	27	40	54	NIL	NIL	54
	4	32	£23,675	£22,300	£15,975	11	16	22	NIL	NIL	22
	4	46	£34,025	£32,025	£22,950	16	23	31	NIL	NIL	31
Higher	3	29	£20,425	£19,250	£13,775	10	15	20	NIL	NIL	20
	3	43	£30,275	£28,525	£20,425	15	22	29	NIL	NIL	29
	4	36	£26,625	£25,075	£17,975	12	18	24	NIL	NIL	24
	4	52	£38,450	£36,200	£25,950	18	26	35	NIL	NIL	35

TABLE 'B' ANALYSIS OF COSTS

£ per Square Metre - Based upon Model 3

Elements of the Building (m)	New £ per m2 (n)	Extension £ per m2 (o)	Remodelled £ per m2 (p)
Main Building Work			
* Foundations	109.08	111.01	4.21
* Superstructure	328.89	273.80	165.34
Finishings			
* Floors	42.51	44.71	56.87
* Walls	26.05	27.15	22.35
* Ceilings	27.70	31.94	46.05
Fittings			
* Furniture	34.54	32.09	32.36
* Fittings	80.58	74.87	75.52
* Sanitary Fittings	18.38	20.99	26.04
* Light fittings	23.51	23.25	25.74
Security			
* Protection	2.70	8.34	8.58
* Surveillance	6.44	8.62	6.25
* Communication	3.67	6.23	5.70
FITTING OUT	**266.09** £ per m2	**278.18** £ per m2	**305.45** £ per m2
TOTAL COST	**704.06** £ per m2	**662.99** £ per m2	**475.00** £ per m2

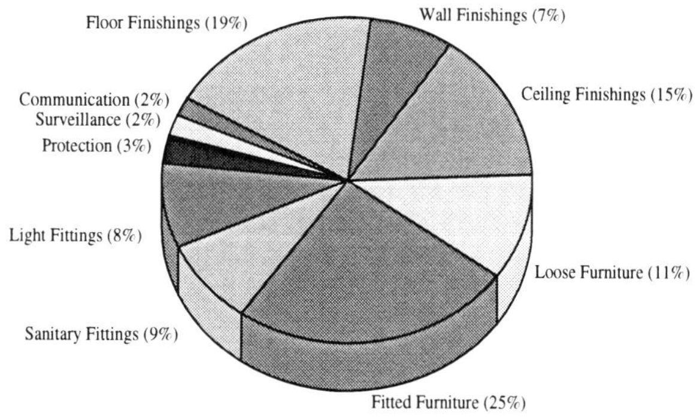

ANALYSIS OF FITTING OUT COSTS

(Costs based upon Table 'B' Column 'p').

SEATING

EASY CHAIRS

Standard polypropylene; back and wide seat; steel square tube frame; epoxy paint finish; padded; vinyl or tweed covering.

	Width x depth x height	Price £
Chair	605 x 695 x 370 mm	55 to 75
Armchair	605 x 695 x 370 mm	70 to 95
Ski framed chair	605 x 695 x 370 mm	60 to 80

Puffin; ergonomical seating; steel oval tube frame; epoxy paint finish; padded; vinyl or tweed covers

	Width x depth x height	Price £
Chair	550 x 740 x 400 mm	85 to 115
Armchair	550 x 740 x 400 mm	105 to 140

MODULAR CHAIRS

Modular; metal frame; epoxy paint finish; fully upholstered seat and back; fabric covering.

	Width x depth x height	Price £
Chair; standard	660 x 800 x 420 mm	240 to 325
Chair; inside; curve	320 x 800 x 420 mm	220 to 300
Chair; outside; curve	660 x 800 x 420 mm	220 to 300
Table - Wood	710 x 710 x 340 mm	180 to 245
Table - Glass	710 x 710 x 340 mm	200 to 270

Beech frame; polished; designer look; upholstered seat and back; fabric covering.

	Width x depth x height	Price £
Chair		200 to 270
Armchair		245 to 330
Arm unit (left or right)		220 to 300

Metal frame; epoxy paint finish; designer look; upholstered seat and back; fabric covering.

	Width x depth x height	Price £
Chair		185 to 250
Armchair		230 to 310
Arm unit (left or right)		210 to 280

SEATING

SUITES OF FURNITURE

Leather suite; wood frame; fully upholstered

	Width x depth x height	Price £
Single armchair	1060 x 940 x 440 mm	850 to 1145
Sofa - 2 seater	1600 x 940 x 440 mm	1170 to 1580
Sofa - 3 seater	2130 x 940 x 440 mm	1680 to 2275

Reception seating; beech frame; polished; upholstered seat and back; tweed covering.

	Width x depth x height	Price £
Lounge chair		150 to 200
Armchair		170 to 230
Sofa - 2 seater		310 to 415

Fabric covered suite; wood frame; fully upholstered; deep cushions.

	Width x depth x height	Price £
Armchair		585 to 790
Sofa - 2 seater		415 to 560
Sofa - 3 seater		775 to 1050
Table; glass top		200 to 270

STACKING CHAIRS

Standard polypropylene easy chair; waterfall front; ergonomical seating; steel oval tubular frame; epoxy paint finish; padded; vinyl or tweed covering.

	Width x depth x height	Price £
Chair	560 x 600 x 405 mm	65 to 85
Armchair	635 x 600 x 405 mm	75 to 105

Conventional polypropylene back and seat; heavy duty; steel square tube frame; epoxy paint finish; padded; vinyl covering.

	Width x depth x height	Price £
Chair	475 x 395 x 465 mm	25 to 35

TABLES

TABLES; WRITING

General purpose; melamine tops; steel frame and legs; epoxy paint finish.

	Length x width x height	Price £
Stacking	760 x 395 x 760 mm	45 to 60
Stacking	1220 x 395 x 760 mm	60 to 80
Stacking	1524 x 395 x 760 mm	65 to 90
Stacking	1824 x 395 x 760 mm	70 to 95

General purpose; hardwood tops; steel frame and legs; epoxy paint finish.

	Length x width x height	Price £
Stacking	760 x 395 x 760 mm	60 to 80
Stacking	1220 x 395 x 760 mm	75 to 100
Stacking	1524 x 395 x 760 mm	85 to 115
Stacking	1824 x 395 x 760 mm	90 to 120

TABLES; COFFEE

Coffee; melamine top; steel frame and legs; epoxy paint finish.

	Length x width x height	Price £
Rectangle	900 x 600 x 380 mm	105 to 145
Square	600 x 600 x 380 mm	100 to 135

Coffee; melamine top; softwood frame and legs; paint finish.

	Length x width x height	Price £
Rectangle	600 x 600 x 380 mm	105 to 145

Coffee; melamine top; hardwood frame and legs; polished.

	Length x width x height	Price £
Rectangle	600 x 600 x 380 mm	125 to 170

Coffee; hardwood top; hardwood frame and legs; polished.

	Length x width x height	Price £
Rectangle	900 x 600 x 380 mm	145 to 195

STORAGE

	Width x depth x height	Price £
LOW UNITS; BEECH FINISH		
Static; open		
1 Shelf	360 x 496 x 590 mm	55 to 75
1 Shelf	1040 x 496 x 590 mm	80 to 110
2 Shelves	1040 x 496 x 590 mm	95 to 130
2 Shelves	1040 x 496 x 750 mm	95 to 130
6 Drawers	1040 x 496 x 590 mm	195 to 260
Static with doors		
1 Shelf	360 x 496 x 590 mm	80 to 110
1 Shelf	1040 x 496 x 590 mm	135 to 180
2 Shelves	1040 x 496 x 590 mm	145 to 195
2 Shelves	1040 x 496 x 750 mm	150 to 200
6 Drawers	1040 x 496 x 590 mm	245 to 330
Mobile; open		
1 Shelf	360 x 496 x 590 mm	70 to 90
1 Shelf	1040 x 496 x 590 mm	95 to 125
2 Shelves	1040 x 496 x 590 mm	105 to 140
2 Shelves	1040 x 496 x 750 mm	110 to 145
6 Drawers	1040 x 496 x 590 mm	205 to 275
Mobile with doors		
1 Shelf	360 x 496 x 590 mm	95 to 125
1 Shelf	1040 x 496 x 590 mm	145 to 195
2 Shelves	1040 x 496 x 590 mm	155 to 210
2 Shelves	1040 x 496 x 750 mm	160 to 215
6 Drawers	1040 x 496 x 590 mm	255 to 345
TALL UNITS; BEECH FINISH		
Static; open		
3 Shelf	360 x 496 x 1455 mm	150 to 200
4 Shelf	1040 x 496 x 1455 mm	160 to 215
3 Shelf; 24 trays	1040 x 496 x 1455 mm	195 to 265
3 Shelf, 8 drawers	1040 x 496 x 1455 mm	290 to 395
Static with doors		
3 Shelf	360 x 496 x 1455 mm	230 to 315
4 Shelf	1040 x 496 x 1455 mm	240 to 325
3 Shelf; 24 trays	1040 x 496 x 1455 mm	280 to 375
3 Shelf, 8 drawers	1040 x 496 x 1455 mm	375 to 505

PIN BOARDS

	Width x height	Price £
ABS FRAMES		
Plastic corners, flame resistant and light fast; hessian covered.		
Aluminium frame	600 x 600 mm	30 to 35
Aluminium frame	600 x 900 mm	35 to 45
Aluminium frame	600 x 1200 mm	40 to 55
Aluminium frame	900 x 1200 mm	45 to 60
Aluminium frame	1200 x 1500 mm	60 to 80
Aluminium frame	1200 x 1800 mm	70 to 90
Plastic corners, flame resistant and light fast; felt covered.		
Aluminium frame	600 x 600 mm	30 to 40
Aluminium frame	600 x 900 mm	40 to 50
Aluminium frame	600 x 1200 mm	45 to 60
Aluminium frame	900 x 1200 mm	50 to 70
Aluminium frame	1200 x 1500 mm	65 to 90
Aluminium frame	1200 x 1800 mm	75 to 100
WOOD FRAMES		
Flame resistant and light fast; hessian covered.		
Wood frame	600 x 600 mm	30 to 35
Wood frame	600 x 900 mm	35 to 45
Wood frame	600 x 1200 mm	40 to 55
Wood frame	900 x 1200 mm	45 to 60
Wood frame	1200 x 1500 mm	60 to 80
Wood frame	1200 x 1800 mm	70 to 90
Flame resistant and light fast; felt covered.		
Wood frame	600 x 600 mm	30 to 40
Wood frame	600 x 900 mm	40 to 50
Wood frame	600 x 1200 mm	45 to 60
Wood frame	900 x 1200 mm	50 to 70
Wood frame	1200 x 1500 mm	65 to 90
Wood frame	1200 x 1800 mm	75 to 100

NOTICE BOARDS / ROOM DIVIDORS

	Width x height	Price £
NOTICE BOARDS		
Glazed Lockable; toughened hinged glass doors, felt or hessian.		
Aluminium frame	900 x 600 mm	210 to 280
Aluminium frame	900 x 1200 mm	290 to 390
Aluminium frame	1000 x 1500 mm	295 to 400
Aluminium frame	1200 x 1800 mm	380 to 515
Wood frame	900 x 600 mm	130 to 180
Wood frame	900 x 1200 mm	205 to 275
Wood frame	1000 x 1500 mm	240 to 325
Wood frame	1200 x 1800 mm	325 to 440
ROOM DIVIDERS		
Mobile folding screen; lockable castors; aluminium frame.		
Hessian covered		
3 panels	1200 x 1500 mm	485 to 660
3 panels	1200 x 1800 mm	520 to 700
Nylon covered		
3 panels	1200 x 1500 mm	520 to 705
3 panels	1200 x 1800 mm	560 to 755
Display screen; metal legs;		
Hessian covered		
1 panel	1200 x 1200 mm	290 to 390
1 panel	1500 x 1200 mm	300 to 405
1 panel	1800 x 1200 mm	340 to 460
Nylon covered		
1 panel	1200 x 1200 mm	310 to 420
1 panel	1500 x 1200 mm	335 to 455
1 panel	1800 x 1200 mm	365 to 490
Budget presentation panel; metal legs		
Hessian; 2 panels	1200 x 2000 mm	355 to 475
Cork; 2 panels	1200 x 2000 mm	325 to 440

LUMINAIRES

	Price £
Tubular Fluorescent; single luminaire;	
Batten only	
600 mm long	15 to 20
1200 mm long	18 to 24
1800 mm long	21 to 28
Batten with deflector	
600 mm long	16 to 22
1200 mm long	20 to 27
1800 mm long	24 to 32
Batten with prismatic diffuser	
600 mm long	17 to 23
1200 mm long	21 to 28
1800 mm long	26 to 34
Batten with acrylic prismatic diffuser	
600 mm long	23 to 31
1200 mm long	28 to 38
1800 mm long	34 to 46
58 w; reflector; wire covers.	
1200 mm long	46 to 62
Tubular Fluorescent; twin luminaires;	
Batten only	
600 mm long	24 to 32
1200 mm long	32 to 43
1800 mm long	39 to 53
Batten with deflector	
600 mm long	26 to 35
1200 mm long	36 to 49
1800 mm long	45 to 61
Batten with prismatic diffuser	
600 mm long	28 to 37
1200 mm long	38 to 52
1800 mm long	49 to 66
Batten with acrylic prismatic diffuser	
600 mm long	37 to 50
1200 mm long	52 to 70
1800 mm long	66 to 89

LUMINAIRES

	Price £
Modular ceiling luminaires;	
Recessed fittings;	
3 x 18 watt	31 to 42
4 x 18 watt	34 to 45
3 x 36 watt	38 to 51
4 x 36 watt	41 to 55
Recessed fittings;	
3 x 18 watt	36 to 49
4 x 18 watt	38 to 52
3 x 36 watt	42 to 57
4 x 36 watt	46 to 61
Recessed fittings;	
3 x 18 watt	39 to 52
4 x 18 watt	41 to 55
3 x 36 watt	45 to 61
4 x 36 watt	49 to 66
Extra for 3 hour integral emergency maintained packs;	
3 x 18 watt	77 to 104
3 x 36 watt	93 to 125
Compact fluorescent luminaires.	
Down lighters; recessed; low energy	0 to 0
18 watt x 2	33 to 45
28 watt	39 to 53
Down lighters; recessed; enery masters	
21 watt	33 to 45
28 watt	39 to 53
Ceiling fittings; perspect covers.	
recessed	37 to 50
surface fixed	39 to 53
Bulkhead Luminaires	
28 Watt	34 to 46
28 Watt; maintained	50 to 68
250 w; loading bay; wire guard.	
Security Floodlights; external	
Photocell	48 to 65

PERSONAL PROTECTION ALARM SYSTEM

	Price £
ALARM SYSTEMS	
Uniquely coded transmitters; hand held; display system; up to 255 transmitters; push button triggers; Infra red and radio Keyfob triggers; portable area alarms; PIR detectors; Radio magnetic door switches operators; printer record output	
Transmitters; battery operated	
Push button	135 to 180
Infra red	240 to 325
Receivers	
Base unit	170 to 230
Loud receiver	190 to 260
Receiver with printer	400 to 540
Stand alone alarm	315 to 425
BUDGET COSTS	
Infant starter pack	
6 push button transmitters; 1 receiver unit.	845 to 1140
Junior starter pack	
10 push button transmitters; 2 receiver units.	1475 to 1995
Senior starter pack	
10 infra red transmitters; 10 push button transmitters; 2 receiver units.	2525 to 3420

M 4 - MEDICAL CENTRE AREA ARRANGEMENTS

Model 1

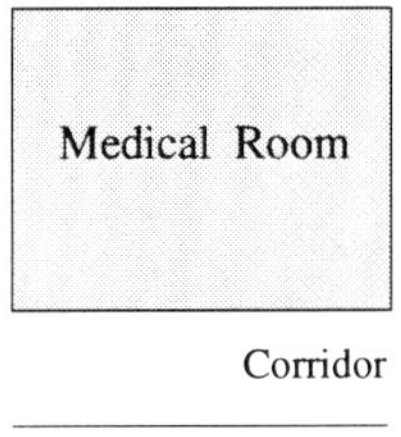

Model 2

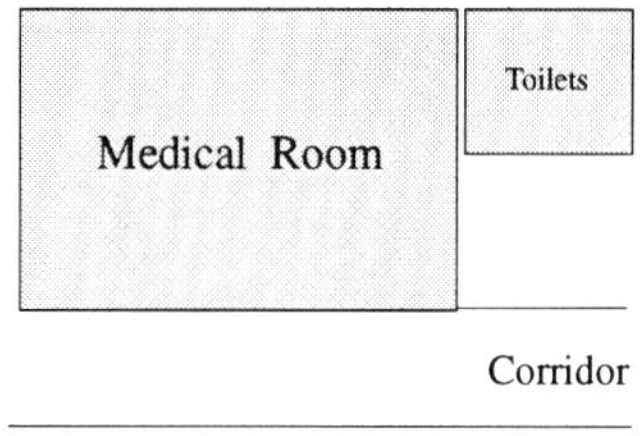

Model 3

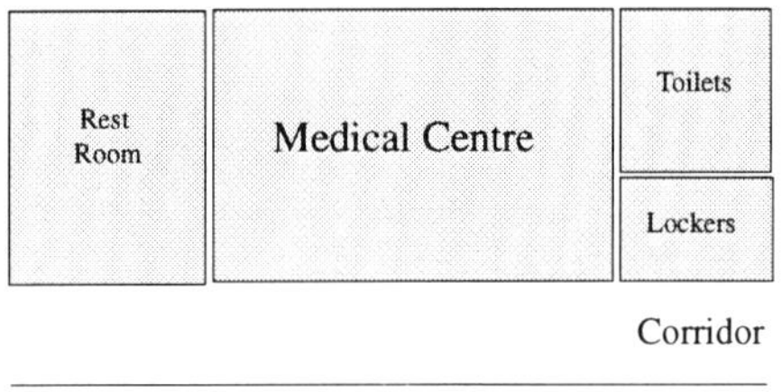

Model 4

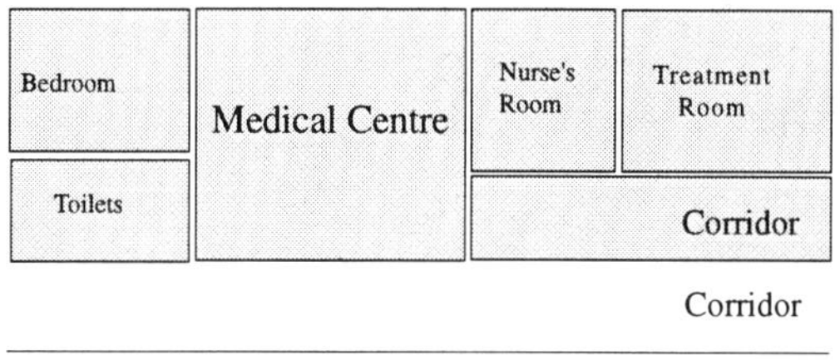

TABLE 'A' **MEDICAL CENTRE**

School Type (a)	Model (b)	Area m2 (c)	Total Building Cost (£) New (d)	Extension (e)	Re-Modelled (f)	Operational Usage (Nos) Teaching Spaces (g)	School Meals (h)	Assembly Spaces (i)	Revenue & Entertainment (j)	Sport (k)	Seminar Spaces (l)
Primary	1	7	£4,675	£4,400	£3,250	3	1	NIL	NIL	NIL	2
	1	11	£7,325	£6,925	£5,125	4	2	NIL	NIL	NIL	4
	2	11	£7,650	£7,225	£5,350	4	2	NIL	NIL	NIL	4
	2	16	£11,125	£10,525	£7,775	6	2	NIL	NIL	NIL	5
Secondary	2	13	£9,025	£8,550	£6,300	5	2	NIL	NIL	NIL	4
	2	18	£12,500	£11,825	£8,725	6	2	NIL	NIL	NIL	6
	3	40	£29,550	£27,950	£20,625	14	4	NIL	NIL	NIL	12
	3	60	£44,300	£41,925	£30,925	20	6	NIL	NIL	NIL	18
	4	37	£29,500	£27,925	£20,600	13	4	NIL	NIL	NIL	11
	4	55	£43,850	£41,500	£30,625	19	6	NIL	NIL	NIL	16
Higher	3	27	£19,950	£18,875	£13,925	9	3	NIL	NIL	NIL	8
	3	37	£27,325	£25,850	£19,075	13	4	NIL	NIL	NIL	11
	4	41	£32,700	£30,950	£22,825	14	5	NIL	NIL	NIL	12
	4	60	£47,850	£45,275	£33,400	20	6	NIL	NIL	NIL	18

TABLE 'B' **ANALYSIS OF COSTS**
£ per Square Metre - Based upon Model 3

Elements of the Building (m)	New £ per m2 (n)	Extension £ per m2 (o)	Remodelled £ per m2 (p)
Main Building Work			
* Foundations	109.07	111.00	4.21
* Superstructure	338.78	285.35	180.22
Finishings			
* Floors	40.15	42.23	53.71
* Walls	26.05	27.15	22.35
* Ceilings	27.70	31.94	46.05
Fittings			
* Furniture	46.05	42.78	43.15
* Fittings	87.76	81.54	82.24
* Sanitary Fittings	26.25	29.98	37.19
* Light fittings	23.51	23.25	25.74
Security			
* Protection	2.70	8.34	8.58
* Surveillance	6.44	8.62	6.25
* Communication	3.67	6.23	5.70
FITTING OUT	**290.29** £ per m2	**302.05** £ per m2	**330.96** £ per m2
TOTAL COST	**738.16** £ per m2	**698.40** £ per m2	**515.39** £ per m2

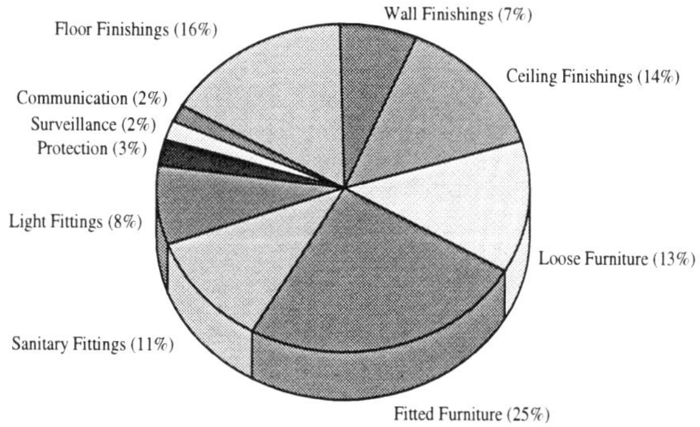

ANALYSIS OF FITTING OUT COSTS
(Costs based upon Table 'B' Column 'p').

COUCHES AND SCREENS

	Length x width x height	Price £
EXAMINATION COUCH		
Vinyl flame retardant padded top; tubular steel frame; epoxy paint finish.		
Adjustable back rest	1840 x 622 x 765 mm	160 to 220
Over bed table; melamine top; tubular steel frame; epoxy paint finish.		
Table	540 x 620 x 940 mm	35 to 45
BEDS		
Single; fire retardent padding; sprung; tubular steel frame; epoxy paint finish; castors.		
Single bed	1840 x 900 x 650 mm	140 to 190
Divan type	1840 x 900 x 650 mm	255 to 345
WARD SCREENS		
Tubular steel frame; 100 mm castors; epoxy paint finish; flame retardant polyester curtains.		
Three folds	1830 x 1745 mm	100 to 130
Four folds	2440 x 1745 mm	125 to 165
Five folds	3050 x 1745 mm	130 to 180
HEIGHT SCALE		
Polished hardwood frame, brass fittings.		
Floor standing apparatus		35 to 45

TROLLEYS

	Width x depth x height	Price £
TROLLEY		
Tubular steel frame; 100 mm castors; epoxy paint finish; laminate shelves; hardwood lipping; tray 915 x 510 mm.		
Two tier	1050 x 510 x 1020 mm	145 to 195
Three tier	1050 x 510 x 1020 mm	205 to 275
Tubular steel frame; 100 mm castors; epoxy paint finish; removable laminate shelves; hardwood lipping; tray 915 x 510 mm.		
Two tier	880 x 485 x 925 mm	110 to 150
Three tier	1050 x 510 x 1020 mm	140 to 190
Tubular steel frame; 100 mm castors; epoxy paint finish; removable aluminium shelves; soiled linen collection bag.		
	1100 x 570 x 1050 mm	150 to 225
NURSES' LOCKERS		
Steel carcass; coloured doors; 6 compartments (two hanging); ventilated; heavy duty; epoxy paint finish.		
1 Door	457 x 457 x 1178 mm	100 to 125

FURNITURE

	Width x depth x height	Price £
CHAIRS		
Easy; square back and base; padded back and seat; tubular steel frame; epoxy paint finish;		
Chair	605 x 695 x 370 mm	55 to 75
Arm chair	1840 x 622 x 765 mm	70 to 95
Ski frame	1840 x 622 x 765 mm	60 to 80
Conventional	475 x 395 x 465 mm	25 to 35
TABLES		
Melamine work surface; polyester edging; rigid tubular steel frame; epoxy paint finish.		
Stacking	760 x 760 x 736 mm	45 to 60
Stacking	1220 x 760 x 736 mm	60 to 80
Stacking	1524 x 760 x 736 mm	65 to 90
Stacking	1824 x 760 x 736 mm	70 to 95
FILING CABINETS		
Budget; steel; anti tilt device; epoxy paint finish.		
2 Drawer		130 to 180
3 Drawer		150 to 200
4 Drawer		160 to 220
Fire resistant; theft proof.		
2 Drawer	540 x 800 x 759 mm	735 to 995
3 Drawer	540 x 800 x 1103 mm	1000 to 1355
4 Drawer	540 x 800 x 1449 mm	1120 to 1515

M 5 & 6 - BOILER HOUSE AND LAUNDRY ROOM ARRANGEMENTS

Model 1

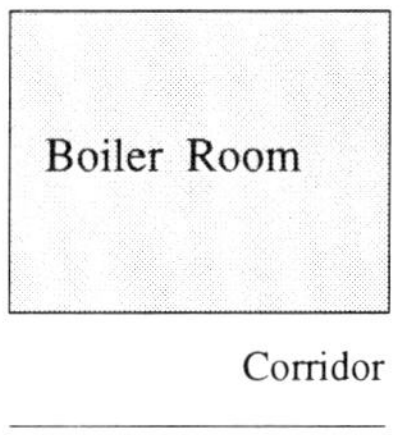

Model 2

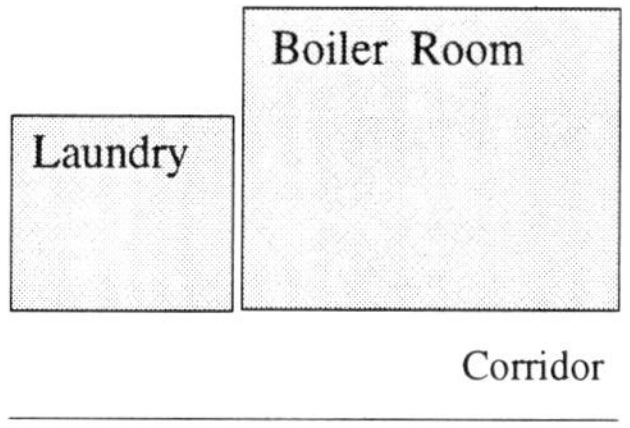

Model 3

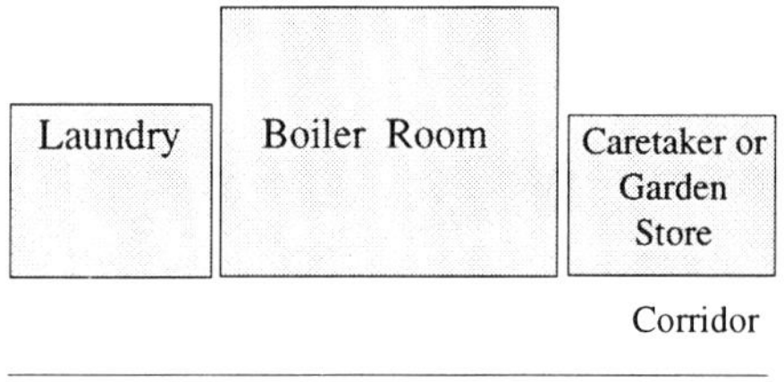

Model 4

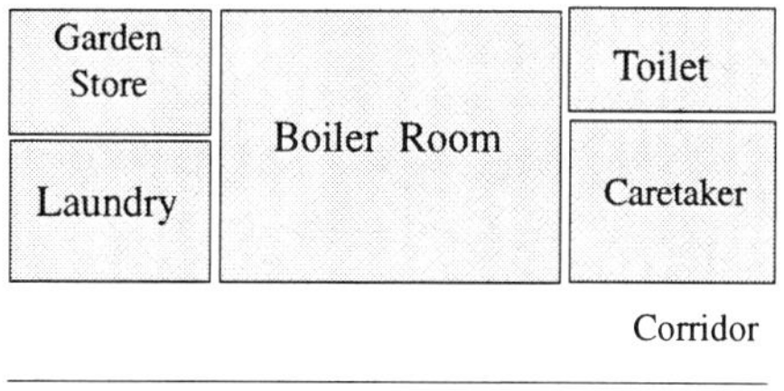

TABLE 'A' **SERVICE ROOMS**

School Type (a)	Model (b)	Area m2 (c)	Total Building Cost (£) New (d)	Extension (e)	Re-Modelled (f)	Operational Usage (Nos) Teaching Spaces (g)	School Meals (h)	Assembly Spaces (i)	Revenue & Entertainment (j)	Sport (k)	Seminar Spaces (l)
Primary	1	8	£3,900	£3,600	£2,150	NIL	NIL	NIL	NIL	NIL	NIL
	1	14	£6,825	£6,300	£3,750	NIL	NIL	NIL	NIL	NIL	NIL
	2	11	£5,600	£5,175	£3,075	NIL	NIL	NIL	NIL	NIL	NIL
	2	18	£9,150	£8,450	£5,025	NIL	NIL	NIL	NIL	NIL	NIL
Secondary	2	16	£8,125	£7,525	£4,475	NIL	NIL	NIL	NIL	NIL	NIL
	2	23	£11,675	£10,800	£6,425	NIL	NIL	NIL	NIL	NIL	NIL
	3	35	£18,900	£17,475	£10,400	NIL	NIL	NIL	NIL	NIL	NIL
	3	46	£24,850	£22,975	£13,650	NIL	NIL	NIL	NIL	NIL	NIL
	4	48	£28,000	£25,875	£15,375	NIL	NIL	NIL	NIL	NIL	NIL
	4	56	£32,675	£30,200	£17,950	NIL	NIL	NIL	NIL	NIL	NIL
Higher	3	40	£21,600	£19,975	£11,875	NIL	NIL	NIL	NIL	NIL	NIL
	3	51	£27,550	£25,450	£15,125	NIL	NIL	NIL	NIL	NIL	NIL
	4	54	£31,500	£29,100	£17,300	NIL	NIL	NIL	NIL	NIL	NIL
	4	61	£35,575	£32,875	£19,550	NIL	NIL	NIL	NIL	NIL	NIL

TABLE 'B' **ANALYSIS OF COSTS**
£ per Square Metre - Based upon Model 3

Elements of the Building (m)	New £ per m2 (n)	Extension £ per m2 (o)	Remodelled £ per m2 (p)
Main Building Work			
* Foundations	109.07	111.00	4.21
* Superstructure	301.97	243.83	132.77
Finishings			
* Floors	21.26	22.35	28.43
* Walls	26.05	27.15	22.35
* Ceilings	27.70	31.94	46.05
Fittings			
* Furniture	13.61	12.65	12.76
* Fittings	3.89	3.61	3.64
* Sanitary Fittings	0.00	0.00	0.00
* Light fittings	23.51	23.25	25.74
Security			
* Protection	2.70	8.34	8.58
* Surveillance	6.44	8.62	6.25
* Communication	3.67	6.23	5.70
FITTING OUT	**128.83** £ per m2	**144.13** £ per m2	**159.50** £ per m2
TOTAL COST	**539.89** £ per m2	**498.97** £ per m2	**296.47** £ per m2

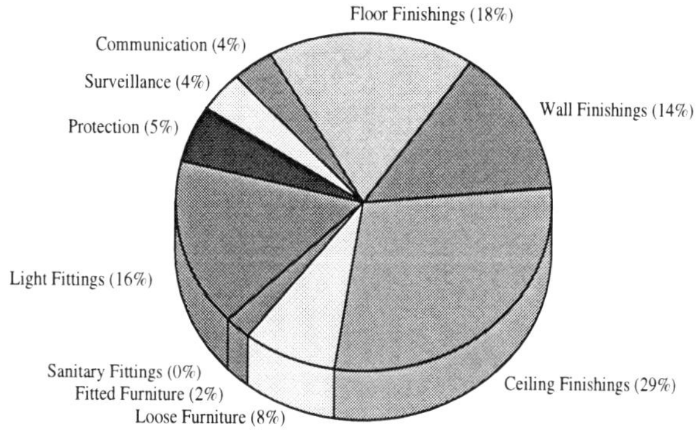

ANALYSIS OF FITTING OUT COSTS
(Costs based upon Table 'B' Column 'p').

M 7 - CIRCULATION AREA ARRANGEMENTS

Model 1

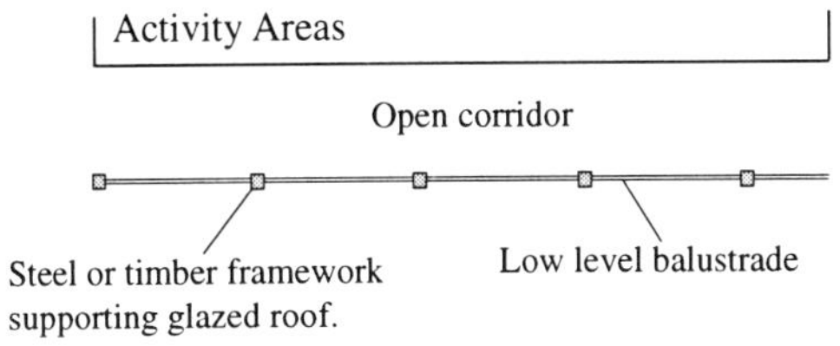

Model 2

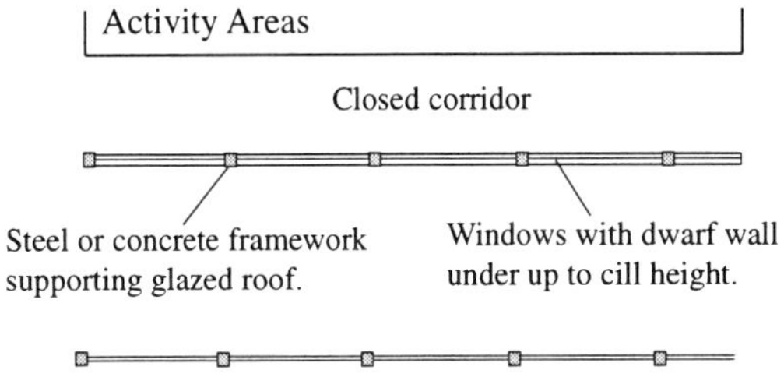

Model 3

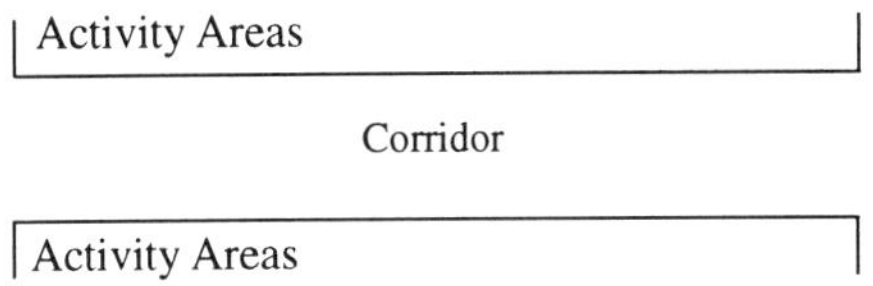

Model 4

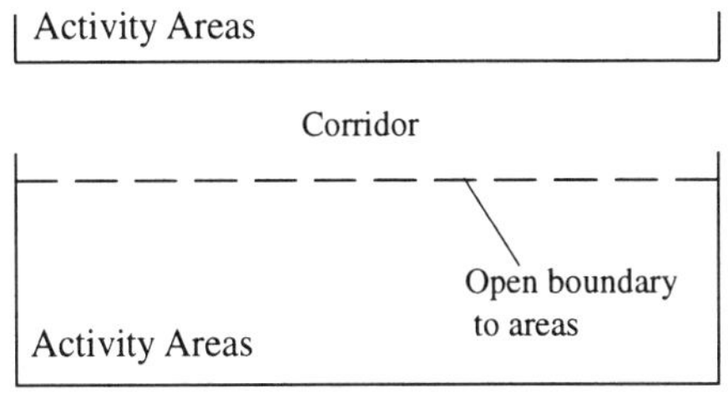

TABLE 'B' ANALYSIS OF COSTS
£ per Square Metre - Based upon Model 3

Elements of the Building (m)	New £ per m2 (n)	Extension £ per m2 (o)	Remodelled £ per m2 (p)
Main Building Work			
* Foundations	109.07	111.00	4.21
* Superstructure	333.89	259.17	153.15
Finishings			
* Floors	44.87	47.19	60.03
* Walls	26.05	27.15	22.35
* Ceilings	27.70	31.94	46.05
Fittings			
* Furniture	1.77	1.65	1.66
* Fittings	0.53	0.49	0.50
* Sanitary Fittings	1.58	11.99	2.23
* Light fittings	19.99	23.25	21.88
Security			
* Protection	2.70	8.34	8.58
* Surveillance	6.44	8.62	6.25
* Communication	3.67	6.23	5.70
FITTING OUT	**135.30** £ per m2	**166.84** £ per m2	**175.22** £ per m2
TOTAL COST	**578.26** £ per m2	**537.01** £ per m2	**332.58** £ per m2

TABLE 'A' CIRCULATION AREAS

School Type (a)	Model (b)	Area m2 (c)	Total Building Cost (£) New (d)	Extension (e)	Re-Modelled (f)	Operational Usage (Nos) Teaching Spaces (g)	School Meals (h)	Assembly Spaces (i)	Revenue & Entertainment (j)	Sport (k)	Seminar Spaces (l)
Primary	1	100	£21,575	£20,025	£12,400	40	50	67	34	20	NIL
	2	100	£49,275	£45,775	£28,350	40	50	67	34	20	NIL
	3	<u>100</u>	£61,600	£57,200	£35,425	40	50	67	34	20	NIL
	4	100	£52,350	£48,625	£30,125	40	50	67	34	20	NIL
Secondary	1	100	£20,250	£18,800	£11,650	40	50	67	34	20	NIL
	2	100	£46,275	£42,975	£26,625	40	50	67	34	20	NIL
	3	<u>100</u>	£57,850	£53,725	£33,275	40	50	67	34	20	NIL
	4	100	£49,175	£45,650	£28,275	40	50	67	34	20	NIL
Higher	1	100	£20,850	£19,375	£12,000	40	50	67	34	20	NIL
	2	100	£47,650	£44,250	£27,425	40	50	67	34	20	NIL
	3	<u>100</u>	£59,575	£55,325	£34,275	40	50	67	34	20	NIL
	4	100	£50,650	£47,025	£29,125	40	50	67	34	20	NIL

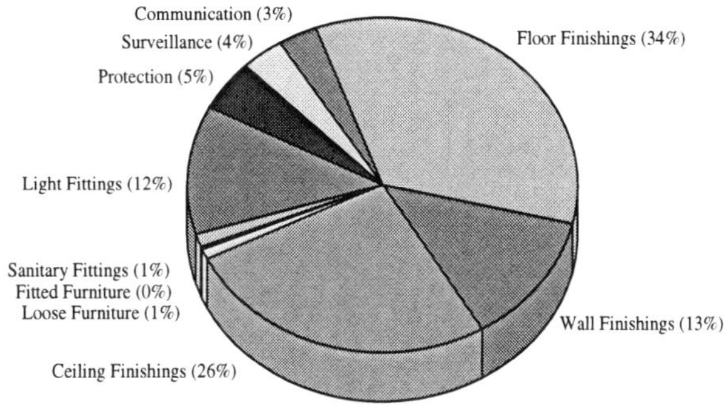

ANALYSIS OF FITTING OUT COSTS

(Costs based upon Table 'B' Column 'p').

CARPET AND FLOOR CLEANERS.

	Price £
Carpet Cleaners; recovery tanks; electric motor.	
Scrubbers	
stain and spot remover	430 to 585
350 mm cleaner; tools	430 to 585
425 mm cleaner; tools	430 to 585
heavy duty; self propelled	430 to 585
Automatic scrubbers	
170 mm cleaner	430 to 585
200 mm cleaner	430 to 585
260 mm cleaner	430 to 585
320 mm cleaner	430 to 585
Rider scrubbers	
340 mm cleaner	430 to 585
Vacuum Cleaners; refuse bags; tools; electric motor.	
Light duty	430 to 585
Standard	430 to 585
Heavy duty	430 to 585
Large area	430 to 585
Wet/ dry vacuums	430 to 585
Wide area squeegee	430 to 585
Heavy duty - wet/dry	430 to 585
Pump out vacuum	430 to 585
Tip out	430 to 585
Back pack cleaner; tools	430 to 585
Compressed air cleaner	430 to 585

BURNISHERS AND POLISHERS.

	Price £
Burnishers; electric motor.	
Light duty	430 to 585
Standard	430 to 585
Heavy duty	430 to 585
Polishers; electric motor.	
Standard; single speed	
425 mm base	430 to 585
500 mm base	430 to 585
Standard; two speed	
425 mm base	430 to 585
500 mm base	430 to 585
Multi use; gear driven	
400 mm base	430 to 585
475 mm base	430 to 585
550 mm base	430 to 585
Parker Sweepers	
Vacuum litter collector	430 to 585
Scavenger vacuum cleaner	430 to 585
Walk behind sweeper	430 to 585
Manual sweeper	430 to 585

NOTE; The prices for maintenance equipment is for industrial quality only.

RACKING AND TRUCKING.

	Price £
Racking; steel frame; chipboard shelving; bolted or rivetted connections.	
1200 x 450 mm	
three tiers	75 to 95
four tiers	90 to 120
five tiers	110 to 145
1830 x 450 mm	
three tiers	85 to 110
four tiers	105 to 140
five tiers	125 to 165
2440 x 450 mm	
three tiers	105 to 145
four tiers	135 to 180
five tiers	165 to 220
1200 x 610 mm	
three tiers	80 to 110
four tiers	100 to 135
five tiers	120 to 160
1830 x 610 mm	
three tiers	95 to 125
four tiers	120 to 160
five tiers	140 to 190
2440 x 610 mm	
three tiers	115 to 155
four tiers	145 to 200
five tiers	180 to 240
Sundries;	
Sack truck	45 to 60
Cylinder truck	85 to 115
Multi cylinder truck	130 to 170
Safety steps; mobile	300 to 405

ROTARY MOWERS; PETROL.

	Size	Price £
Briggs & Stratton; self propelled; front wheels; 2 piece power driven rear roller; fan assisted grass box.		
Recoil start		
4 HP engine;		
475 mm cutting width	13 - 63 mm cut height	430 to 585
Electric key start		
4 HP engine;		
475 mm cutting width	13 - 63 mm cut height	455 to 610
Briggs & Stratton; aluminium deck; front and rear wheels; polymer/ fabric grass box.		
Recoil start		
4 HP engine;		
475 mm cutting width	30 - 80 mm cut height	360 to 485
Tecumseh Centura; self propelled; front wheels; rear roller; fan assisted grass box.		
Recoil start		
3.8 HP engine;		
400 mm cutting width	10 - 60 mm cut height	310 to 420
Tecumseh Centura; steel deck; self propelled; front wheels rear roller; fan assisted grass box.		
Recoil start		
4 HP engine;		
450 mm cutting width	30 - 75 mm cut height	165 to 225

ROTARY MOWERS; PETROL.

	Size	Price £
Briggs & Stratton; front and rear wheels; polymer/ fabric grass box.		
Recoil start		
4 HP engine;		
400 mm cutting width	30 - 75 mm cut height	145 to 195
4.5 HP engine;		
475 mm cutting width	30 - 80 mm cut height	360 to 485
Electric key start		
3.5 HP engine;		
400 mm cutting width	10 - 60 mm cut height	340 to 455
Briggs & Stratton; front wheels; rear roller; fan assisted grass box.		
Recoil start		
3.5 HP engine;		
400 mm cutting width	10 - 60 mm cut height	265 to 360
Electric key start		
3.5 HP engine;		
400 mm cutting width	10 - 60 mm cut height	340 to 455
Briggs & Stratton; self propelled; front wheels; rear roller; fan assisted grass box.		
Recoil start		
3.8 HP engine;		
400 mm cutting width	10 - 60 mm cut height	310 to 420
Electric key start		
3.5 HP engine;		
400 mm cutting width	10 - 60 mm cut height	405 to 545
Briggs & Stratton; self propelled; front wheels; rear roller; fan assisted grass box.		
Recoil start		
3.8 HP engine;		
400 mm cutting width	10 - 60 mm cut height	310 to 420

CYLINDER MOWERS; PETROL.

	Size	Price £
148 cc; 4 stroke; self propelled; 6 - 36 mm cut height; front and rear rollers; quick release blades; polymer grass box.		
Recoil start		
350 mm cutting width	1275 mm max width	460 to 620
425 mm cutting width	1500 mm max width	505 to 680
500 mm cutting width	1700 mm max width	645 to 875
Electric key start		
350 mm cutting width	1275 mm max width	530 to 720
425 mm cutting width	1500 mm max width	575 to 775
500 mm cutting width	1700 mm max width	720 to 970
172cc; overhead valve; self propelled; 5 - 35 mm cut height; front and rear rollers; polymer grass box.		
Recoil start		
600 mm cutting width	1925 mm max width	1325 to 1795
Extra for ride on seat		255 to 340
Electric key start		
500 mm cutting width	1675 mm max width	1255 to 1695
600 mm cutting width	1925 mm max width	1385 to 1870
750 mm cutting width	2275 mm max width	1325 to 1795
Extra for ride on seat		255 to 340
172cc; overhead valve; self propelled; 3 - 18 mm cut height; front and rear rollers; aluminium grass box.		
Recoil start		
500 mm cutting width	1925 mm max width	1400 to 1890

H 1 - ASSEMBLY HALL AREA ARRANGEMENTS

Model 1

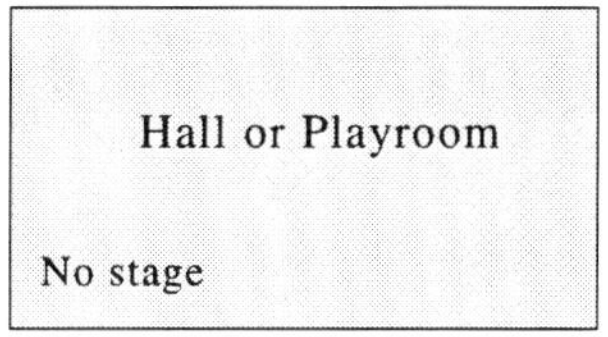

Model 2

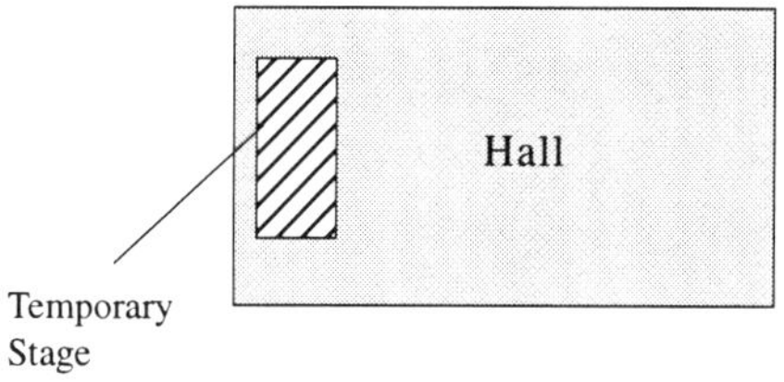

Model 3

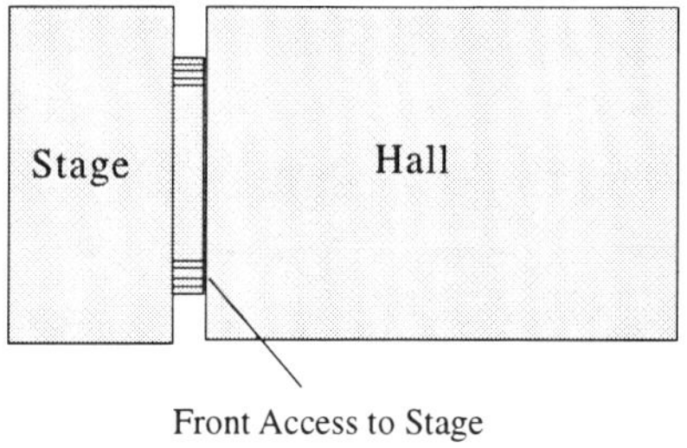

Model 4

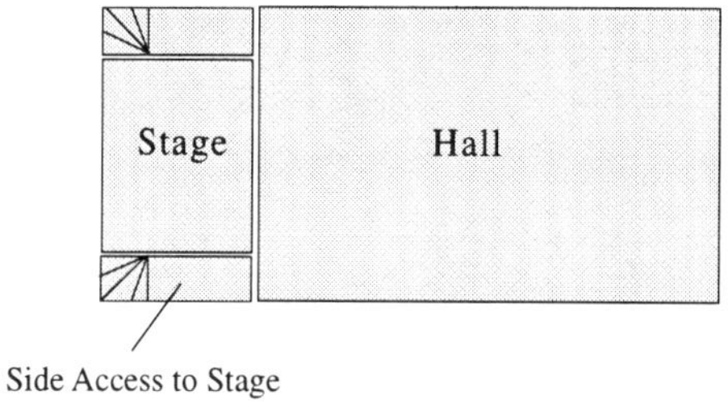

TABLE 'A' ASSEMBLY HALL

School Type (a)	Model (b)	Area m2 (c)	Total Building Cost (£) New (d)	Extension (e)	Re-Modelled (f)	Operational Usage (Nos) Teaching Spaces (g)	School Meals (h)	Assembly Spaces (i)	Revenue & Entertainment (j)	Sport (k)	Seminar Spaces (l)
Primary	1	82	£50,275	£46,725	£28,775	33	41	55	28	17	33
	1	112	£68,675	£63,825	£39,300	45	56	75	38	23	45
	2	93	£58,275	£54,150	£33,350	38	47	62	31	19	38
	2	126	£78,950	£73,350	£45,175	51	63	84	42	26	51
Secondary	2	112	£70,175	£65,200	£40,150	45	56	75	38	23	45
	2	147	£92,100	£85,575	£52,700	59	74	98	49	30	59
	3	153	£101,975	£94,750	£58,350	62	77	102	51	31	62
	3	266	£177,275	£164,725	£101,450	107	133	178	89	54	107
	4	172	£120,350	£111,850	£68,875	69	86	115	58	35	69
	4	229	£160,250	£148,900	£91,700	92	115	153	77	46	92
Higher	3	190	£126,625	£117,675	£72,475	76	95	127	64	38	76
	3	287	£191,275	£177,725	£109,450	115	144	192	96	58	115
	4	235	£164,450	£152,800	£94,100	94	118	157	79	47	94
	4	369	£258,200	£239,950	£147,775	148	185	246	123	74	148

TABLE 'B' ANALYSIS OF COSTS
£ per Square Metre - Based upon Model 3

Elements of the Building (m)	New £ per m2 (n)	Extension £ per m2 (o)	Remodelled £ per m2 (p)
Main Building Work			
* Foundations	109.07	111.00	4.21
* Superstructure	361.62	295.33	136.69
Finishings			
* Floors	68.49	72.03	91.62
* Walls	33.87	35.29	29.05
* Ceilings	27.70	31.94	46.05
Fittings			
* Furniture	10.43	9.69	9.77
* Fittings	18.86	17.53	17.68
* Sanitary Fittings	0.00	0.00	0.00
* Light fittings	23.51	23.25	25.74
Security			
* Protection	2.70	8.34	8.58
* Surveillance	6.44	8.62	6.25
* Communication	3.67	6.23	5.70
FITTING OUT	**195.68** £ per m2	**212.91** £ per m2	**240.44** £ per m2
TOTAL COST	**666.39** £ per m2	**619.24** £ per m2	**381.34** £ per m2

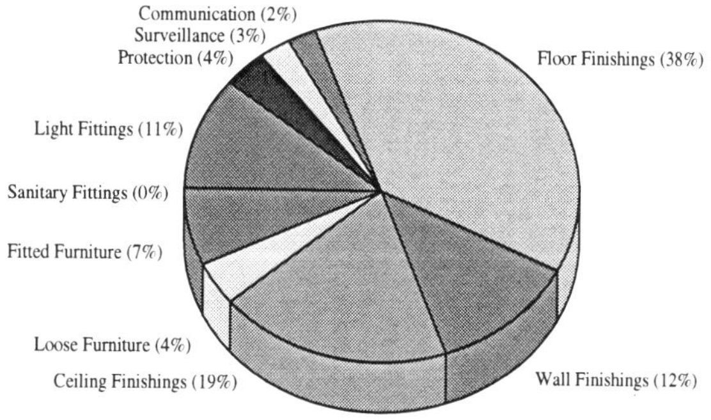

ANALYSIS OF FITTING OUT COSTS
(Costs based upon Table 'B' Column 'p').

CHAIRS

STACKING

Standard Polypropylene; back and seat; coloured; steel frame; epoxy paint finish.

	Width x depth x height	Price £
Group - Nursery	525 x 400 x 360 mm	8 to 11
Group - Pre-school	545 x 400 x 385 mm	8 to 11
Group - Primary	585 x 400 x 410 mm	8 to 11
Group - Junior	680 x 485 x 450 mm	8 to 11
Group - Adult	720 x 485 x 450 mm	9 to 11
Group - Adult	795 x 463 x 480 mm	13 to 17

Designer Polypropylene; back and seat; coloured; steel frame; epoxy paint finish.

	Width x depth x height	Price £
Group - Nursery	515 x 350 x 350 mm	7 to 10
Group - Pre-school	590 x 375 x 360 mm	8 to 10
Group - Primary	660 x 410 x 400 mm	9 to 12
Group - Junior	710 x 470 x 500 mm	9 to 12
Group - Adult	750 x 470 x 500 mm	9 to 12

Two piece Polypropylene; back and seat; coloured; steel frame; epoxy paint finish.

	Width x depth x height	Price £
Group - Pre-school	584 x 336 x 375 mm	10 to 13
Group - Primary	622 x 350 x 375 mm	10 to 13
Group - Junior	720 x 412 x 460 mm	11 to 15
Group - Adult	760 x 412 x 460 mm	12 to 16

Extra cost for the following

	Price £
Lock link	3 to 3
Linking clip	2 to 2
Fire resistant chairs	1 to 2
Drop links	1 to 2

BANQUETING

Upholstered two piece stacking chair; steel frame; epoxy paint finish.

	Width x depth x height	Price £
Aluminium frame		
Tweed finish	870 x 420 x 580 mm	46 to 62
Velvet finish	870 x 420 x 580 mm	47 to 63
Steel frame		
Tweed finish	870 x 420 x 590 mm	39 to 53
Velvet finish	870 x 420 x 590 mm	41 to 55

Economy upholstered two piece stacking chair; steel frame; epoxy paint finish.

	Width x depth x height	Price £
Aluminium frame		
Velvet finish	880 x 410 x 510 mm	24 to 33
Tweed finish	880 x 410 x 510 mm	25 to 34
Vinyl finish	880 x 410 x 510 mm	29 to 39
Steel frame		
Velvet finish	895 x 450 x 535 mm	27 to 36
Tweed finish	880 x 410 x 510 mm	28 to 37
Vinyl finish	880 x 410 x 510 mm	32 to 43

Upholstered two piece conference stacking chair; steel frame; epoxy paint finish.

	Width x depth x height	Price £
Steel frame		
Tweed finish	795 x 457 x 565 mm	35 to 47
Tweed finish	790 x 505 x 510 mm	54 to 73
Steel frame, armchair		
Tweed finish	795 x 560 x 565 mm	53 to 71
Tweed finish	790 x 585 x 510 mm	67 to 91
Steel frame, linking		
Tweed finish	780 x 458 x 540 mm	40 to 54

Extra cost for the following

	Price £
Book boxes	8 to 11
Book shelves	6 to 8
Book rest	11 to 15
Linking device	1 to 2

FOLDING

Polypropelene coloured folding seat; steel frame; epoxy paint finish.

	Width x depth x height	Price £
Group - Adult		
Two piece	820 x 450 x 490 mm	14 to 19
Two piece all steel	750 x 460 x 540 mm	13 to 18
Ditto; permament link	720 x 485 x 450 mm	19 to 26
Upholstered 2 piece	720 x 485 x 450 mm	30 to 41
Ditto; permament link	720 x 485 x 450 mm	36 to 48
Comfort back 2 piece	890 x 460 x 540 mm	35 to 47

BENCHES

Laminate covered aluminium frame and edges.

	Width x depth x height	Price £
Stacking & Folding - economy		
Group - Primary	432 x 254 x 1220 mm	32 to 42
Group - Junior	432 x 254 x 1830 mm	40 to 54
Stacking & Folding		
Group - Primary	381 x 354 x 1830 mm	46 to 62
Group - Junior	432 x 254 x 1520 mm	43 to 57
Group - Adult	483 x 254 x 1220 mm	38 to 51

TABLES

ECONOMY RANGE

Laminate finish; folding table; steel legs; epoxy paint finish..

	Length x width	Price £
Table size	915 x 610 mm	40 to 55
Table size	915 x 685 mm	41 to 56
Table size	1220 x 610 mm	47 to 64
Table size	1220 x 685 mm	48 to 65
Table size	1830 x 610 mm	59 to 80
Table size	1830 x 685 mm	60 to 81

Coloured laminate finish; folding table; chromium plated steel legs.

	Length x width	Price £
Table size	915 x 610 mm	49 to 67
Table size	915 x 685 mm	51 to 68
Table size	915 x 760 mm	52 to 71
Table size	1220 x 610 mm	56 to 76
Table size	1220 x 685 mm	58 to 78
Table size	1220 x 760 mm	60 to 81
Table size	1520 x 610 mm	61 to 82
Table size	1520 x 685 mm	64 to 87
Table size	1520 x 760 mm	67 to 90
Table size	1830 x 480 mm	64 to 86
Table size	1830 x 610 mm	70 to 95
Table size	1830 x 685 mm	72 to 97
Table size	1830 x 760 mm	74 to 100

STANDARD RANGE

Coloured laminate finish; folding table; chromium plated legs.

	Length x width	Price £
Table size	915 x 610 mm	69 to 93
Table size	915 x 685 mm	70 to 95
Table size	915 x 760 mm	73 to 99
Table size	1220 x 610 mm	82 to 111
Table size	1220 x 685 mm	84 to 114
Table size	1220 x 760 mm	87 to 118
Table size	1520 x 610 mm	91 to 122
Table size	1520 x 685 mm	95 to 129
Table size	1520 x 760 mm	99 to 134
Table size	1830 x 610 mm	106 to 143
Table size	1830 x 685 mm	108 to 146
Table size	1830 x 760 mm	111 to 150

Coloured laminate finish; folding table; aluminium legs.

	Length x width	Price £
Table size	915 x 610 mm	53 to 72
Table size	915 x 685 mm	55 to 74
Table size	915 x 760 mm	56 to 76
Table size	1220 x 610 mm	61 to 83
Table size	1220 x 685 mm	63 to 84
Table size	1220 x 760 mm	65 to 87
Table size	1520 x 610 mm	66 to 90
Table size	1520 x 685 mm	68 to 92
Table size	1520 x 760 mm	71 to 95
Table size	1830 x 480 mm	68 to 92
Table size	1830 x 610 mm	74 to 100
Table size	1830 x 685 mm	75 to 101
Table size	1830 x 760 mm	78 to 105

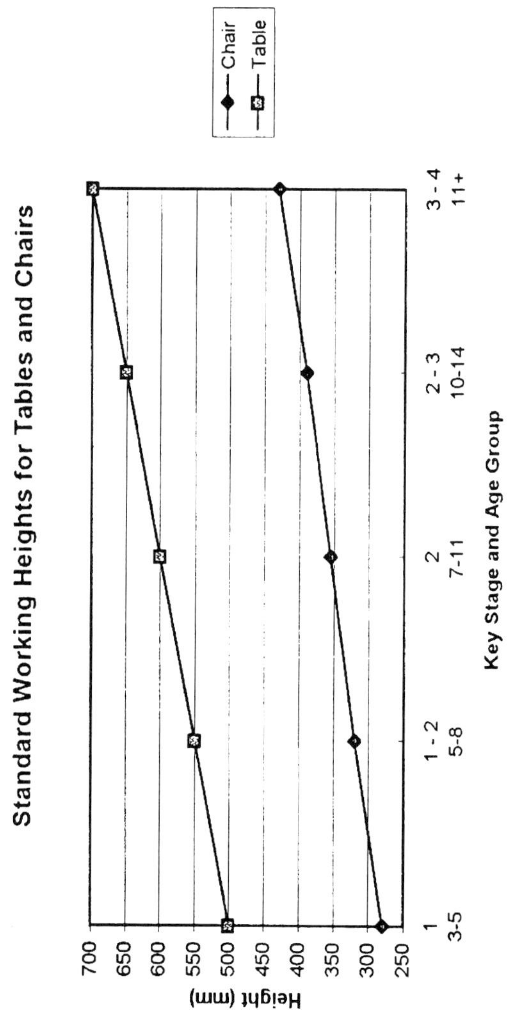

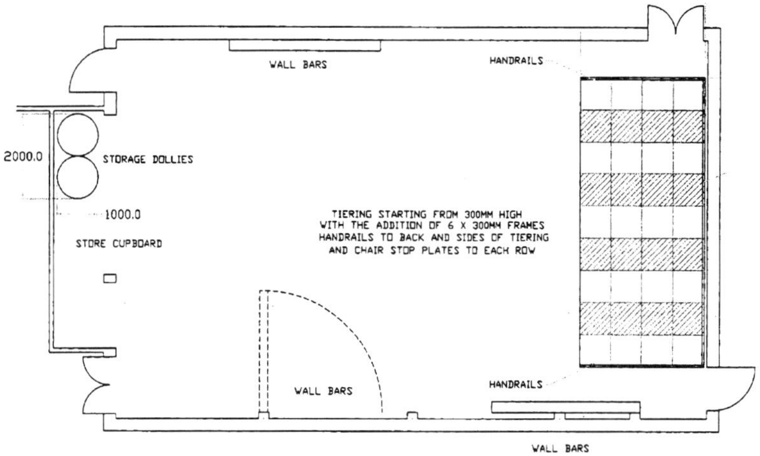

Typical Primary School Assembly Hall
(also for P.E Training)

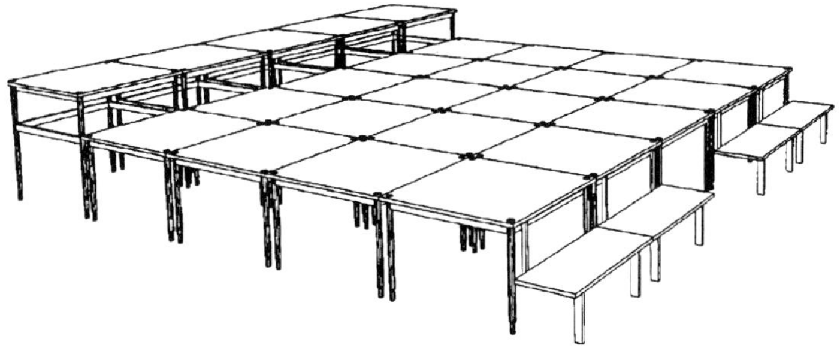

Typical Arrangement of Flat Staging

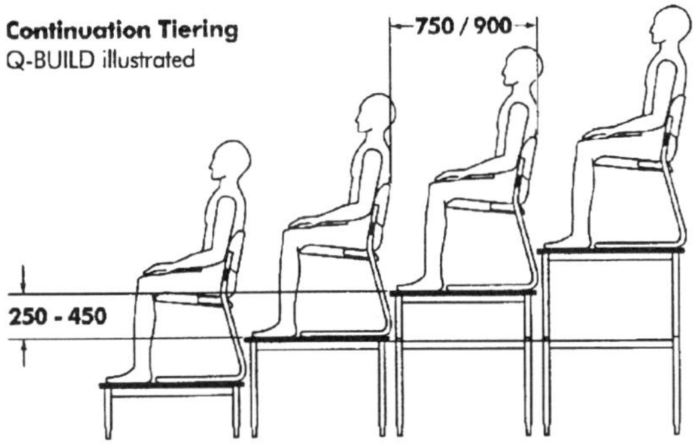

Typical Seating Tiers

Drawings by Stage Systems

COMPARATIVE PRICES FOR FLAT STAGING
(Areas are approximate and depend upon the exact size of the staging components)

Size of Stage			Metal Tube Construction		Wood Modular Construction			
Width	Length	Height	Units	Standard	Units	Standard	Units	Lightweight
2.40 m	2.40 m	300 mm	9	£ 918.00	4	£ 706.00	6	£ 732.00
2.40 m	2.40 m	500 mm	9	£ 981.00	4	£ 802.00	6	£ 876.00
2.40 m	2.40 m	750 mm	9	£ 1,134.00	4	£ 1,014.00		
3.75 m	2.40 m	300 mm	15	£ 1,530.00	6	£ 1,017.00	12	£ 1,356.00
3.75 m	2.40 m	500 mm	15	£ 1,635.00	6	£ 1,149.00	12	£ 1,548.00
3.75 m	2.40 m	750 mm	15	£ 1,890.00	6	£ 1,443.00		
4.50 m	2.40 m	300 mm	18	£ 1,836.00	8	£ 1,328.00	15	£ 1,695.00
4.50 m	2.40 m	500 mm	18	£ 1,962.00	8	£ 1,496.00	15	£ 1,935.00
4.50 m	2.40 m	750 mm	18	£ 2,268.00	8	£ 1,872.00		
3.75 m	3.75 m	300 mm	25	£ 2,550.00	9	£ 1,494.00	16	£ 1,808.00
3.75 m	3.75 m	500 mm	25	£ 2,725.00	9	£ 1,683.00	16	£ 2,064.00
3.75 m	3.75 m	750 mm	25	£ 3,150.00	9	£ 2,106.00	0	
3.75 m	4.80 m	300 mm	30	£ 3,060.00	12	£ 1,950.00	20	£ 2,260.00
3.75 m	4.80 m	500 mm	30	£ 3,270.00	12	£ 2,190.00	20	£ 2,580.00
3.75 m	4.80 m	750 mm	30	£ 3,780.00	12	£ 2,730.00		

H 2 - LIBRARY AREA ARRANGEMENTS

Model 1

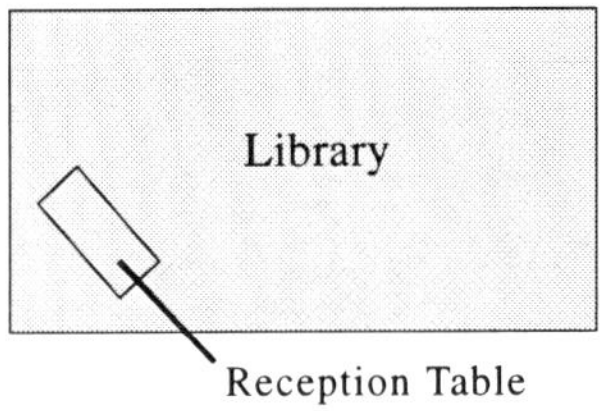

Model 2

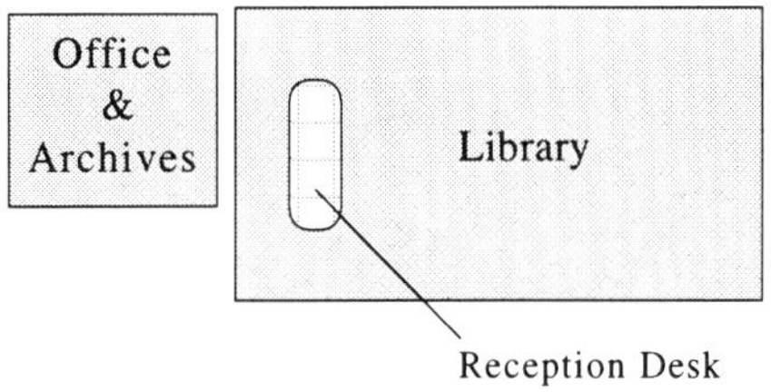

Model 3

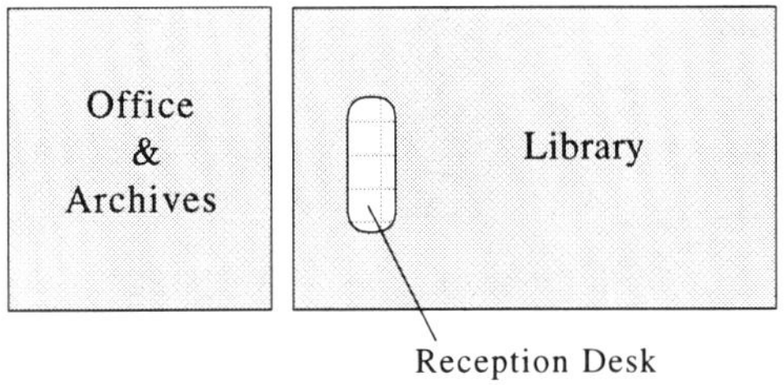

Model 4

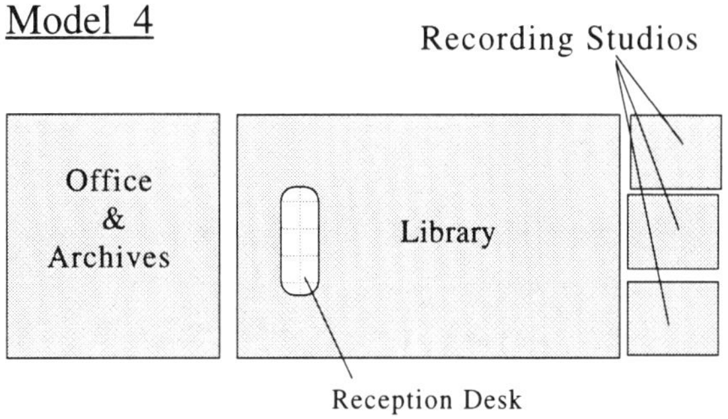

TABLE 'A' LIBRARY

School Type (a)	Model (b)	Area m2 (c)	Total Building Cost (£) New (d)	Extension (e)	Re-Modelled (f)	Operational Usage (Nos) Teaching Spaces (g)	School Meals (h)	Assembly Spaces (i)	Revenue & Entertainment (j)	Sport (k)	Seminar Spaces (l)
Primary	1	16	£10,375	£9,700	£6,625	7	NIL	11	6	NIL	7
	1	30	£19,450	£18,175	£12,425	12	NIL	20	10	NIL	12
	2	24	£16,075	£15,025	£10,275	10	NIL	16	8	NIL	10
	2	49	£32,800	£30,650	£20,950	20	NIL	33	17	NIL	20
Secondary	2	35	£23,425	£21,900	£14,975	14	NIL	24	12	NIL	14
	2	50	£33,450	£31,275	£21,375	20	NIL	34	17	NIL	20
	3	81	£57,050	£53,325	£36,450	33	NIL	54	27	NIL	33
	3	142	£100,000	£93,500	£63,900	57	NIL	95	48	NIL	57
	4	105	£79,125	£73,975	£50,550	42	NIL	70	35	NIL	42
	4	171	£128,850	£120,475	£82,325	69	NIL	114	57	NIL	69
Higher	3	87	£61,275	£57,275	£39,150	35	NIL	58	29	NIL	35
	3	160	£112,675	£105,350	£72,000	64	NIL	107	54	NIL	64
	4	139	£104,725	£97,925	£66,925	56	NIL	93	47	NIL	56
	4	198	£149,175	£139,475	£95,325	80	NIL	132	66	NIL	80

TABLE 'B' ANALYSIS OF COSTS
£ per Square Metre - Based upon Model 3

Elements of the Building (m)	New £ per m2 (n)	Extension £ per m2 (o)	Remodelled £ per m2 (p)
Main Building Work			
* Foundations	109.07	111.00	4.21
* Superstructure	314.68	259.24	133.78
Finishings			
* Floors	47.23	49.68	63.19
* Walls	26.05	27.15	22.35
* Ceilings	27.70	31.94	46.05
Fittings			
* Furniture	99.78	92.70	93.50
* Fittings	43.24	40.17	40.52
* Sanitary Fittings	0.00	0.00	0.00
* Light fittings	23.51	23.25	25.74
Security			
* Protection	2.70	8.34	8.58
* Surveillance	6.44	8.62	6.25
* Communication	3.67	6.23	5.70
FITTING OUT	**280.32** £ per m2	**288.07** £ per m2	**311.87** £ per m2
TOTAL COST	**704.09** £ per m2	**658.32** £ per m2	**449.86** £ per m2

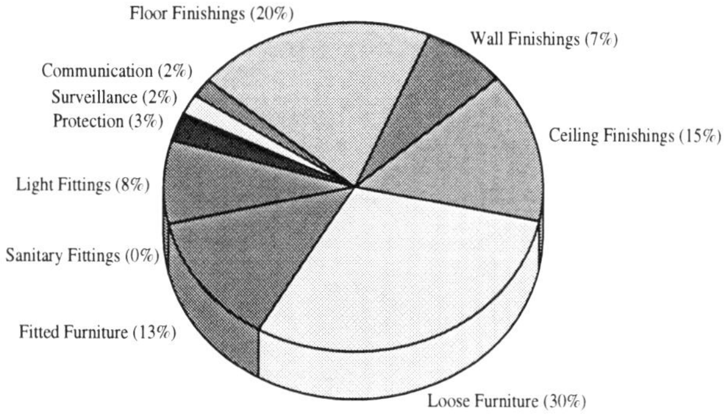

ANALYSIS OF FITTING OUT COSTS
(Costs based upon Table 'B' Column 'p').

HEIGHT	SHELF	WOOD			
		SINGLE SIDED		DOUBLE SIDED	
	Number	Starter Price £	Extension Price £	Starter Price £	Extension Price £
700 mm long units					
High Level Units					
1200 mm high	2	160 to 215	115 to 155	290 to 395	220 to 295
1500 mm high	3	175 to 235	125 to 165	300 to 405	115 to 155
1800 mm high	4	200 to 275	140 to 190	355 to 480	250 to 340
2000 mm high	5	230 to 305	160 to 215	405 to 545	290 to 395
Low Level Units					
900 mm high	2	160 to 215	115 to 150	290 to 390	220 to 295
1200 mm high	3	170 to 230	125 to 165	300 to 405	215 to 290
1500 mm high	4	200 to 270	140 to 185	355 to 480	255 to 345
1700 mm high	5	225 to 305	155 to 210	400 to 540	290 to 390
800 mm long units					
High Level Units					
1200 mm high	2	160 to 220	115 to 155	295 to 395	220 to 300
1500 mm high	3	115 to 155	125 to 170	115 to 155	115 to 155
1800 mm high	4	205 to 275	140 to 190	360 to 485	250 to 340
2000 mm high	5	230 to 310	160 to 215	405 to 545	295 to 395
Low Level Units					
900 mm high	2	160 to 215	115 to 155	290 to 395	220 to 295
1200 mm high	3	175 to 235	125 to 165	305 to 410	235 to 320
1500 mm high	4	200 to 275	140 to 190	355 to 480	255 to 345
1700 mm high	5	230 to 305	160 to 215	405 to 545	290 to 395
900 mm long units					
High Level Units					
1200 mm high	2	170 to 230	125 to 165	305 to 415	235 to 320
1500 mm high	3	190 to 255	130 to 180	320 to 430	250 to 335
1800 mm high	4	215 to 290	150 to 200	375 to 505	265 to 355
2000 mm high	5	240 to 325	170 to 230	420 to 570	305 to 415
Low Level Units					
900 mm high	2	170 to 230	120 to 165	305 to 410	235 to 320
1200 mm high	3	185 to 250	130 to 175	315 to 425	245 to 335
1500 mm high	4	215 to 285	150 to 200	375 to 505	265 to 355
1700 mm high	5	240 to 320	170 to 230	420 to 570	305 to 410

HEIGHT	SHELF	METAL			
		SINGLE SIDED		DOUBLE SIDED	
	Number	Starter Price £	Extension Price £	Starter Price £	Extension Price £
700 mm long units					
High Level Units					
1200 mm high	3	85 to 115	65 to 90	130 to 175	105 to 140
1500 mm high	4	95 to 130	75 to 100	150 to 205	115 to 155
1800 mm high	5	110 to 145	85 to 115	175 to 235	145 to 195
2000 mm high	6	125 to 165	95 to 130	195 to 265	165 to 220
800 mm long units					
High Level Units					
1200 mm high	3	90 to 120	70 to 90	130 to 180	105 to 145
1500 mm high	4	115 to 155	80 to 105	115 to 155	115 to 155
1800 mm high	5	110 to 150	90 to 120	175 to 235	145 to 195
2000 mm high	6	125 to 165	100 to 130	200 to 270	165 to 220
900 mm long units					
High Level Units					
1200 mm high	3	95 to 130	75 to 100	145 to 195	120 to 160
1500 mm high	4	105 to 145	85 to 115	165 to 225	140 to 185
1800 mm high	5	120 to 160	95 to 130	190 to 255	160 to 215
2000 mm high	6	135 to 180	110 to 145	215 to 290	180 to 245

STORAGE

	Width x depth x height	Price £
PERIODICAL DISPLAYS		
MDF carcass; laminated; maple, beech, ash or oak finish; matching hardwood lipping; back storage; adjustable feet.		
Key stage 3 - 4	1015 x 400 x 1800 mm	322 to 436
CD OR CASSETTE STORAGE		
MDF carcass; laminated; maple, beech, ash or oak finish; matching hardwood lipping; back storage; adjustable feet.		
Key stage 3 - 4	1000 x 565 x 1500 mm	409 to 554
PAPER STORAGE		
MDF carcass; five drawers; laminated; maple, beech, ash or oak finish; matching hardwood lipping; polyurethene finish; steel frame; epoxy paint finish.		
A1	970 x 700 x 920 mm	333 to 451
AO	1315 x 940 x 920 mm	409 to 553
BOOK TROLLEY		
MDF carcass; two splayed shelves; laminated; maple, beech, ash or oak finish; matching hardwood lipping; polyurethene finish; castors.		
Single sided	1015 x 400 x 1800 mm	323 to 436
MDF carcass; two splayed shelves; laminated; maple, beech, ash or oak finish; matching hardwood lipping; polyurethene finish; steel frame; epoxy paint finish; castors.		
Single sided	900 x 400 x 900 mm	143 to 193
KINDERBOX		
MDF carcass; laminated; maple, beech, ash or oak finish; matching hardwood lipping; polyurethene finish; steel frame; epoxy paint finish.		
4 compartments	610 x 610 x 300 mm	76 to 103

CARRELS

	Width x depth x height	Price £
IT/AV STUDY CARRELS		
MDF worktop; privacy panels; laminated; maple, beech, ash or oak finish; matching hardwood lipping; polyurethene finish; steel frame; epoxy paint finish.		
Single sided	800 x 750 x 700 mm	235 to 318
extension piece	775 x 750 x 700 mm	209 to 282
optional shelf	775 x 250 mm	22 to 29
Single sided	1000 x 750 x 700 mm	253 to 342
extension piece	975 x 750 x 700 mm	233 to 315
optional shelf	975 x 250 mm	27 to 37
Double sided	800 x 750 x 700 mm	402 to 544
extension piece	775 x 750 x 700 mm	294 to 397
optional shelf	775 x 250 mm	22 to 29
Double sided	1000 x 750 x 700 mm	466 to 630
extension piece	975 x 750 x 700 mm	356 to 482
optional shelf	975 x 250 mm	27 to 37
Modular carrel cluster; hardwood worktop; side panels and shelves; linked; solid hardwood lipping; polyurethene finish; steel frame; epoxy paint finish.		
4 person; 12 dividers		856 to 1158
6 person; 18 dividers		1151 to 1558
STUDY BAYS		
MDF carcass; worktop; back; pivacy panels and shelf; laminated; maple, beech, ash or oak finish; matching hardwood lipping; polyurethene finish.		
Study Bays		
Single sided	850 x 600 x 700 mm	210 to 284
extension piece	825 x 600 x 700 mm	179 to 242
Double sided	850 x 1140 x 700 mm	388 to 525
extension piece	825 x 1140 x 700 mm	278 to 376
Extra for light fitting		17 to 23

DISPLAYS

	Width x depth x height	Price £
HESSIAN COVERED		
ABS plastic corners, flame resistant and light fast.		
Aluminium frame	600 x 900 mm	33 to 44
Aluminium frame	600 x 1200 mm	39 to 52
Aluminium frame	900 x 1200 mm	44 to 60
Aluminium frame	1200 x 1500 mm	57 to 78
Aluminium frame	1200 x 1800 mm	66 to 90
Wood frame, flame resistant and light fast.		
Wood frame	600 x 900 mm	33 to 44
Wood frame	600 x 1200 mm	39 to 52
Wood frame	900 x 1200 mm	44 to 60
Wood frame	1200 x 1500 mm	57 to 78
Wood frame	1200 x 1800 mm	66 to 90
FELT COVERED		
ABS plastic corners, flame resistant and light fast.		
Aluminium frame	600 x 900 mm	37 to 49
Aluminium frame	600 x 1200 mm	43 to 58
Aluminium frame	900 x 1200 mm	50 to 67
Aluminium frame	1200 x 1500 mm	65 to 87
Aluminium frame	1200 x 1800 mm	75 to 101
Wood frame, flame resistant and light fast.		
Wood frame	600 x 900 mm	37 to 49
Wood frame	600 x 1200 mm	43 to 58
Wood frame	900 x 1200 mm	50 to 67
Wood frame	1200 x 1500 mm	65 to 87
Wood frame	1200 x 1800 mm	75 to 101

CHAIRS

	Width x depth x height	Price £
STANDARD; METAL FRAMED		
Polypropylene seat; one piece plastic shell; steel frame; epoxy paint finish; stacking.		
Key Stage 1	350 x 350 x 280 mm	7 to 9
Key Stage 2	375 x 360 x 320 mm	7 to 10
Key Stage 3	470 x 500 x 430 mm	8 to 11
Key Stage 4	470 x 500 x 430 mm	8 to 11
Polypropylene back and seat; tubular skid frame; epoxy paint finish; stacking.		
Key Stage 1	350 x 350 x 280 mm	8 to 11
Key Stage 2	375 x 360 x 320 mm	8 to 11
Key Stage 2 - 3	470 x 500 x 390 mm	10 to 13
Key Stage 3 - 4	470 x 500 x 430 mm	10 to 13
Economy upholstered two piece stacking chair; steel frame; epoxy paint finish.		
Aluminium frame		
Velvet finish	880 x 410 x 510 mm	24 to 33
Tweed finish	880 x 410 x 510 mm	25 to 34
Vinyl finish	880 x 410 x 510 mm	29 to 39
Steel frame		
Velvet finish	895 x 450 x 535 mm	27 to 37
Tweed finish	880 x 410 x 510 mm	28 to 38
Vinyl finish	880 x 410 x 510 mm	32 to 43
TIMBER FRAMED		
Standard; upholstered seat and back; beech frame; polished; linked; stacking.		
Key Stage 2	355 x 290 x 320 mm	30 to 40
Key Stage 2	360 x 320 x 355 mm	32 to 43
Key Stage 2 - 3	410 x 390 x 390 mm	33 to 45
Key Stage 3 - 4	410 x 390 x 445 mm	34 to 46

TABLES

	Length x width x height	Price £
NON STACKING		
Melamine top; bullnosed edging; steel framed legs; epoxy paint finish.		
Square		
Key Stage 1	1100 x 1100 x 500 mm	49 to 66
Key Stage 1 - 2	1100 x 1100 x 550 mm	49 to 66
Key Stage 2	1100 x 1100 x 600 mm	49 to 66
Key Stage 2 - 3	1100 x 1100 x 650 mm	49 to 66
Key Stage 3 - 4	1100 x 1100 x 700 mm	49 to 66
Circular		
Key Stage 1	550 dia x 500 mm	27 to 36
Key Stage 1 - 2	550 dia x 550 mm	27 to 36
Key Stage 1	750 dia x 500 mm	37 to 50
Key Stage 1 - 2	750 dia x 550 mm	37 to 50
Key Stage 2	750 dia x 600 mm	37 to 50
Key Stage 2 - 3	750 dia x 650 mm	37 to 50
Key Stage 3 - 4	750 dia x 700 mm	37 to 50
Key Stage 1	1000 dia x 500 mm	45 to 60
Key Stage 1 - 2	1000 dia x 550 mm	45 to 60
Key Stage 2	1000 dia x 600 mm	45 to 60
Key Stage 2 - 3	1000 dia x 650 mm	45 to 60
Key Stage 3 - 4	1000 dia x 700 mm	45 to 60
Melamine top; colour range; polyurethene edging; steel framed legs; epoxy paint finish.		
Rectangular		
Key Stage 1	1200 x 600 x 500 mm	43 to 58
Key Stage 1 - 2	1200 x 600 x 550 mm	43 to 58
Key Stage 2	1200 x 600 x 600 mm	43 to 58
Key Stage 2 - 3	1200 x 600 x 650 mm	43 to 58
Key Stage 3 - 4	1200 x 600 x 700 mm	43 to 58
Trapezoidal		
Key Stage 1 - 2	1000 x trap x 550 mm	42 to 57
Key Stage 2	1000 x trap x 600 mm	42 to 57
Key Stage 2 - 3	1100 x trap x 650 mm	42 to 57
Key Stage 3 - 4	1100 x trap x 700 mm	42 to 57

RECEPTION DESKS

	Length x width x height	Price £
MELAMINE FINISH		
Modular; reception counter; MDF carcass; countertop; melamine finish.		
Standard unit	800 x 727 x 800 mm	153 to 207
Standard unit	1200 x 320 x 400 mm	175 to 236
Standard unit	1600 x 320 x 400 mm	204 to 276
Standard unit	1800 x 320 x 400 mm	226 to 305
Corner unit; 90° curved		236 to 319
Corner unit; triangular		158 to 213
Pedestal units; single; MDF carcass; melamine finish.		
3 drawer	1050 x 730 mm	108 to 145
3 drawer	1200 x 750 mm	138 to 187
4 drawer; filing drawer	1400 x 750 mm	111 to 150
Pedestal units; double; MDF carcass; melamine finish.		
5 drawer	1500 x 730 mm	147 to 198
5 drawer	1800 x 730 mm	153 to 207
POLISHED HARDWOOD		
Modular; reception counter; MDF carcass; hardwood countertop; polished.		
Standard unit	800 x 727 x 800 mm	192 to 259
Standard unit	1200 x 320 x 400 mm	218 to 295
Standard unit	1600 x 320 x 400 mm	255 to 345
Standard unit	1800 x 320 x 400 mm	282 to 381
Corner unit; 90° curved		295 to 399
Corner unit; triangular		197 to 266
Pedestal units; single; MDF carcass; hardwood top; polished.		
3 drawer	1050 x 730 mm	140 to 189
3 drawer	1200 x 750 mm	180 to 243
4 drawer; filing drawer	1400 x 750 mm	144 to 195
Pedestal units; double; MDF carcass; hardwood top; polished.		
5 drawer	1500 x 730 mm	191 to 258
5 drawer	1800 x 730 mm	199 to 270

H 3 - CONFERENCE HALL AREA ARRANGEMENTSH

Model 1

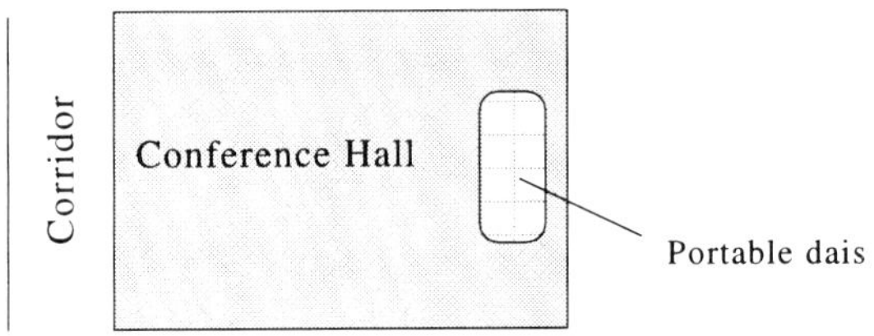

Model 2

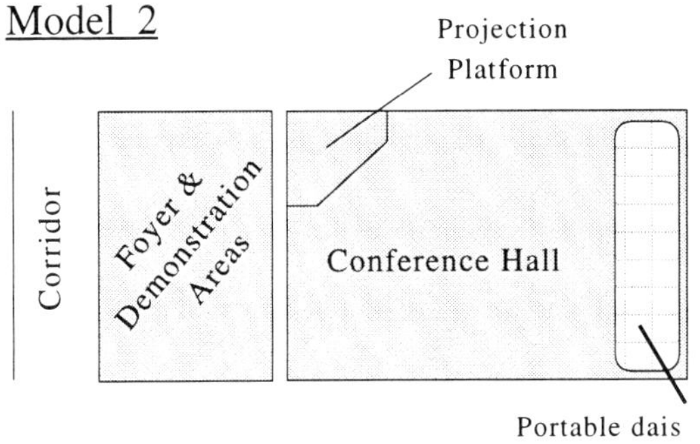

Model 3

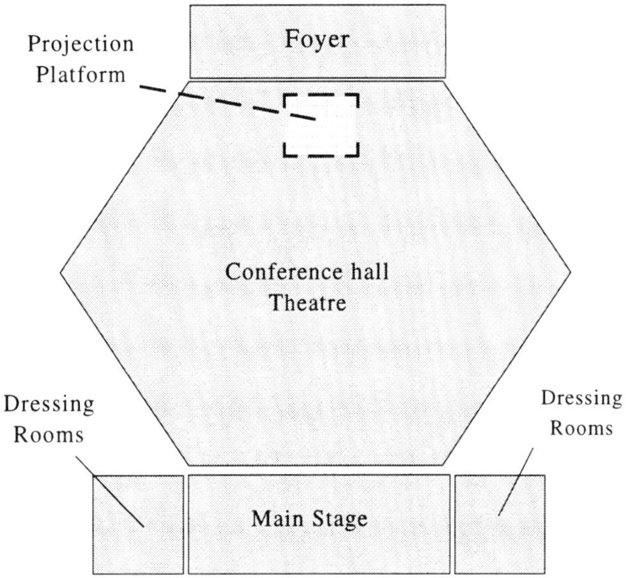

TABLE 'A' **CONFERENCE HALL**

School Type (a)	Model (b)	Area m2 (c)	Total Building Cost (£) New (d)	Extension (e)	Re-Modelled (f)	Operational Usage (Nos) Teaching Spaces (g)	School Meals (h)	Assembly Spaces (i)	Revenue & Entertainment (j)	Sport (k)	Seminar Spaces (l)
Primary		0				NIL	NIL	NIL	NIL	NIL	NIL
		0				NIL	NIL	NIL	NIL	NIL	NIL
		0				NIL	NIL	NIL	NIL	NIL	NIL
		0				NIL	NIL	NIL	NIL	NIL	NIL
Secondary	1	250	282,725	249,650	130,200	100	125	167	125	63	72
	2	450	521,650	460,600	240,225	180	225	300	225	113	129
	3	900	1,170,500	1,033,500	539,000	360	450	600	450	225	258
		0				NIL	NIL	NIL	NIL	NIL	NIL
		0				NIL	NIL	NIL	NIL	NIL	NIL
		0				NIL	NIL	NIL	NIL	NIL	NIL
Higher	1	288	325,150	287,100	149,725	115	144	192	144	72	83
	2	518	599,875	529,675	276,250	207	259	345	259	130	148
	3	1,035	1,346,075	1,188,525	619,850	414	518	690	518	259	296
		0				NIL	NIL	NIL	NIL	NIL	NIL

TABLE 'B' **ANALYSIS OF COSTS**
£ per Square Metre - Based upon Model 3

Elements of the Building (m)	New £ per m2 (n)	Extension £ per m2 (o)	Remodelled £ per m2 (p)
Main Building Work			
* Foundations	196.92	161.89	12.45
* Superstructure	529.35	448.56	155.15
Finishings			
* Floors	73.17	66.94	62.30
* Walls	39.98	36.58	21.15
* Ceilings	47.04	43.03	47.28
Fittings			
* Furniture	122.08	111.68	109.87
* Fittings	52.90	48.40	47.61
* Sanitary Fittings	17.27	19.16	15.55
* Light fittings	27.59	23.95	24.83
Security			
* Protection	10.45	13.22	9.55
* Surveillance	9.22	14.55	9.16
* Communication	4.92	10.59	5.86
FITTING OUT	**404.63** £ per m2	**388.10** £ per m2	**353.17** £ per m2
TOTAL COST	**1,130.90** £ per m2	**998.55** £ per m2	**520.77** £ per m2

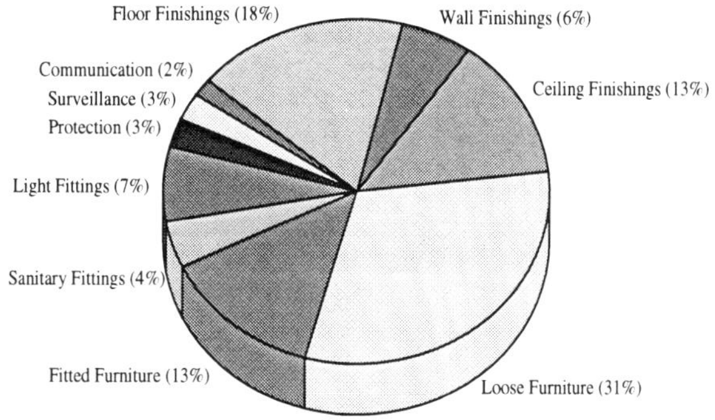

ANALYSIS OF FITTING OUT COSTS
(Costs based upon Table 'B' Column 'p').

CHAIRS

LECTURE

Upholstered back and seat; tubular frame; MDF tablet; wire storage under tray.

	Width x depth x height	Price £
Black frame		
Key stage 3 - 4	525 x 570 x 465 mm	106 to 143
Chromium plated		
Key stage 3 - 4	525 x 570 x 465 mm	118 to 159

Upholstered back and seat; Samsonite neutral steel frame; MDF tablet.

	Width x depth x height	Price £
Key stage 3 - 4	460 x 560 x 465 mm	56 to 76

Upholstered back and seat; tubular steel frame; swivel lecture tablet.

	Width x depth x height	Price £
Key stage 3 - 4	590 x 550 x 455 mm	102 to 138

Upholstered back and seat; curved beech frame; beech tablet.

	Width x depth x height	Price £
Key stage 3 - 4	625 x 570 x 440 mm	192 to 259

Upholstered; stylish; back and seat on moulded shell; steel frame; epoxy paint finish.

	Width x depth x height	Price £
Chair		
Fabric	510 x 570 x 500 mm	36 to 49
Arm chair		
Fabric	640 x 570 x 500 mm	51 to 69

Upholstered; stylish; back and seat on moulded shell; steel frame; chromium plated finish.

	Width x depth x height	Price £
Chair		
Fabric	510 x 570 x 500 mm	39 to 52
Arm chair		
Fabric	640 x 570 x 500 mm	54 to 72

Purpose built conference chair; upholstered; stylish; back and seat on moulded shell; steel frame; chromium plated finish; stacks 12 high.

	Width x depth x height	Price £
Chair		
Fabric	500 x 510 x 480 mm	44 to 59
Vinyl	500 x 510 x 480 mm	50 to 68
Arm chair		
Fabric	605 x 510 x 480 mm	50 to 68
Vinyl	605 x 510 x 480 mm	63 to 85

Purpose built conference chair; upholstered; stylish; back and seat on moulded shell; steel frame; epoxy paint finish; stacks 12 high.

	Width x depth x height	Price £
Chair		
Fabric	480 x 525 x 445 mm	51 to 69
Arm chair		
Fabric	580 x 525 x 445 mm	71 to 96

STACKING

Upholstered back and seat; laminated beech frame; polished; heavy duty.

	Width x depth x height	Price £
Fabric	480 x 520 x 480 mm	40 to 55
Vinyl	625 x 570 x 440 mm	45 to 60

High quality; upholstered back and seat; chromium plated tubular steel frame; stacks 5 high.

	Width x depth x height	Price £
Chair		
Fabric	500 x 510 x 480 mm	44 to 59
Vinyl	500 x 510 x 480 mm	50 to 68
Arm chair		
Fabric	605 x 510 x 480 mm	56 to 76
Vinyl	605 x 510 x 480 mm	63 to 85

Upholstered; round back; large seat; laminated beech frame; polished.

	Width x depth x height	Price £
Chair		
Fabric	460 x 590 x 470 mm	81 to 109
Vinyl	480 x 520 x 480 mm	91 to 124
Arm chair		
Fabric	550 x 590 x 470 mm	90 to 121
Vinyl	480 x 520 x 480 mm	98 to 133

Upholstered; back and seat; high quality; laminated beech frame; hand finished; polished.

	Width x depth x height	Price £
Chair		
Fabric	470 x 610 x 450 mm	112 to 152
Vinyl	470 x 610 x 450 mm	120 to 162
Arm chair		
Fabric	540 x 610 x 450 mm	118 to 159
Vinyl	540 x 610 x 450 mm	125 to 169

Upholstered; back and seat; high quality; steel frame; chromium plated.

	Width x depth x height	Price £
Chair		
Fabric	540 x 560 x 460 mm	56 to 76
Chair with arm		
Fabric	550 x 560 x 460 mm	75 to 102
Arm chair		
Fabric	570 x 560 x 460 mm	75 to 102

FOLDING TABLES

	Length x width x height	Price £
Solid top; melamine surface; polyurethene moulded edges; steel frame with locking rings; epoxy paint finish.		
Rectangular	1220 x 760 x 740 mm	63 to 85
Rectangular	1525 x 760 x 740 mm	63 to 86
Rectangular	1830 x 760 x 740 mm	66 to 89
Circular	1525 dia x 740 mm	122 to 165
Table truck.		
10 rectangular tables		168 to 227
6 round tables		244 to 330
Solid cored top; melamine work surface; polyurethene moulded edges; steel frame with locking rings; adjustable height; epoxy paint finish.		
Rectangular	1525 x 760 x 585 mm	86 to 117
Rectangular	1830 x 760 x 585 mm	89 to 120
Solid cored top; light oak, beech or grey scratch resistant melamine finish; light oak, cherry or beech veneers; polyurethene moulded edges; steel frame; epoxy paint finish		
Rectangular; straight legs		
MFC finish	675 x 675 x 730 mm	132 to 178
Wood veneer	675 x 675 x 730 mm	166 to 225
MFC finish	750 x 750 x 730 mm	139 to 187
Wood veneer	750 x 750 x 730 mm	183 to 248
MFC finish	1350 x 675 x 730 mm	168 to 227
Wood veneer	1350 x 675 x 730 mm	218 to 294
MFC finish	1500 x 750 x 730 mm	180 to 243
Wood veneer	1500 x 750 x 730 mm	239 to 323
Semi circular; straight legs		
MFC finish	1350 dia x 675 mm	187 to 252
Wood veneer	1350 dia x 675 mm	242 to 327
MFC finish	1500 dia x 750 mm	198 to 267
Wood veneer	1500 dia x 750 mm	252 to 341
Circular; straight legs		
MFC finish	1200 dia x 730 mm	187 to 252
Wood veneer	1200 dia x 730 mm	242 to 327

	Length x width x height	Price £
Solid cored top; light oak, beech or grey scratch resistant melamine finish; light oak, cherry or beech veneers; polyurethene moulded edges; steel frame; chromium plated finish.		
Rectangular; curved legs		
MFC finish	675 x 675 x 730 mm	151 to 204
Wood veneer	675 x 675 x 730 mm	188 to 255
MFC finish	750 x 750 x 730 mm	160 to 216
Wood veneer	750 x 750 x 730 mm	197 to 266
MFC finish	1350 x 675 x 730 mm	192 to 260
Wood veneer	1350 x 675 x 730 mm	240 to 324
MFC finish	1500 x 750 x 730 mm	203 to 275
Wood veneer	1500 x 750 x 730 mm	261 to 352
Trapezoidal; curved legs		
MFC finish	1350 x 675 x 730 mm	200 to 270
Wood veneer	1350 x 675 x 730 mm	251 to 339
Semi circular; curved legs		
MFC finish	1350 dia x 675 mm	210 to 284
Wood veneer	1350 dia x 675 mm	261 to 352
MFC finish	1500 dia x 750 mm	221 to 299
Wood veneer	1500 dia x 750 mm	283 to 383
Circular; curved legs		
MFC finish	1200 dia x 730 mm	203 to 275
Wood veneer	1200 dia x 730 mm	269 to 364
High quality; folding; melamine work surface; polyurethane edging; aluminium clad tubular steel legs.		
Key Stage 3 - 4	915 dia x 685 mm	67 to 91
Key Stage 3 - 4	1220 dia x 685 mm	80 to 108
Key Stage 3 - 4	1830 dia x 685 mm	103 to 139
Table truck.		
6 rectangular tables		252 to 341

	Length x width x height	Price £
Modular; Solid cored top; detachable; light oak, beech or grey melamine scratch resistant finish; light oak, cherry or beech veneers; polyurethene moulded edges; steel frame; chromium plated finish .		
Rectangular;		
MFC finish	750 x 750 x 730 mm	163 to 220
Wood veneer	750 x 750 x 730 mm	214 to 290
MFC finish	1500 x 750 x 730 mm	189 to 256
Wood veneer	1500 x 750 x 730 mm	276 to 373
Trapezoidal;		
MFC finish	1500 x 675 x 730 mm	203 to 274
Wood veneer	1500 x 675 x 730 mm	283 to 383
Semi circular;		
MFC finish	1500 dia x 750 mm	221 to 299
Wood veneer	1500 dia x 750 mm	336 to 455
Circular;		
MFC finish	1200 dia x 730 mm	189 to 256
Wood veneer	1200 dia x 730 mm	363 to 490
38 mm hardwood veneered top; solid lipping; protective lacquer; steel frame; chromium plate or epoxy paint finish.		
Rectangular;		
Light oak/ mahogany	1800x 1200 x 730mm	370 to 501
Cherry/ Walnut	1800x 1200 x 730mm	407 to 551
Light oak/ mahogonay	2400x 1200 x 730mm	446 to 603
Cherry/ Walnut	2400x 1200 x 730mm	491 to 665
Light oak/ mahogany	3000x 1200 x 730mm	609 to 824
Cherry/ Walnut	3000x 1200 x 730mm	671 to 907
Rectangular; D'end top		
Light oak/ mahogany	1800x 1200 x 730mm	457 to 618
Cherry/ Walnut	1800x 1200 x 730mm	503 to 681
Light oak/ mahogonay	2400x 1200 x 730mm	528 to 715
Cherry/ Walnut	2400x 1200 x 730mm	581 to 785
Light oak/ mahogany	3000x 1200 x 730mm	690 to 933
Cherry/ Walnut	3000x 1200 x 730mm	757 to 1025

CONFERENCE TABLES

Leg based Conference table; 38 mm top; hardwood veneers; solid wood lipping; protective lacquer; steel legs; epoxy paint finish.

	Length x width x height	Price £
Rectangular;		
Light oak/ mahogany	1800 x 1200 x 730mm	370 to 501
Cherry/ walnut	1800 x 1200 x 730mm	407 to 551
Light oak/ mahogany	2400 x 1200 x 730mm	446 to 603
Cherry/ walnut	2400 x 1200 x 730mm	491 to 665
Light oak/ mahogany	3000 x 1200 x 730mm	609 to 824
Cherry/ walnut	3000 x 1200 x 730mm	671 to 907
D' End		
Light oak/ mahogany	1800 x 1200 x 730mm	457 to 618
Cherry/ walnut	1800 x 1200 x 730mm	503 to 681
Light oak/ mahogany	2400 x 1200 x 730mm	528 to 715
Cherry/ walnut	2400 x 1200 x 730mm	581 to 785
Light oak/ mahogany	3000 x 1200 x 730mm	690 to 933
Cherry/ walnut	3000 x 1200 x 730mm	757 to 1025
Boat top		
Light oak/ mahogany	1800 x 1200 x 730mm	564 to 763
Cherry/ walnut	1800 x 1200 x 730mm	623 to 842
Light oak/ mahogany	2400 x 1200 x 730mm	637 to 862
Cherry/ walnut	2400 x 1200 x 730mm	700 to 947
Light oak/ mahogany	3000 x 1200 x 730mm	799 to 1080
Cherry/ walnut	3000 x 1200 x 730mm	879 to 1189

GENERAL TABLES

Melamine laminate work top; colour co-ordinated; preformed edge; damage resistant and impervious to water; square steel tubular legs; epoxy paint finish.

	Length x width x height	Price £
Square		
Key Stage 1	600 x 600 x 500mm	31 to 42
Key Stage 1 - 2	600 x 600 x 550mm	32 to 43
Key Stage 2	600 x 600 x 600mm	32 to 43
Key Stage 2 - 3	600 x 600 x 650mm	32 to 43
Key Stage 3 - 4	600 x 600 x 700mm	32 to 44
Rectangular		
Key Stage 1	1200 x 600 x 500mm	47 to 63
Key Stage 1 - 2	1200 x 600 x 550mm	47 to 63
Key Stage 2	1200 x 600 x 600mm	47 to 64
Key Stage 2 - 3	1200 x 600 x 650mm	48 to 64
Key Stage 3 - 4	1200 x 600 x 700mm	48 to 64
Trapezoidal		
Key Stage 1	1200 x 520 x 500mm	45 to 61
Key Stage 1 - 2	1200 x 520 x 550mm	46 to 62
Key Stage 2	1200 x 520 x 600mm	46 to 62
Key Stage 2 - 3	1200 x 520 x 650mm	46 to 62
Key Stage 3 - 4	1200 x 520 x 700mm	46 to 62

DISPLAY BOARDS

FLIP CHARTS

	Width x depth	Price £
Brentford mobile flip chart		674 to 912
Leitz easy flip unit/ foil		51 to 69
Leitz easy flip bag		16 to 21
Dry wipe flip chart		157 to 212
HPL flip chart		79 to 106
Clip on flip chart		27 to 36

CONFERENCE BOARDS

White board; hinged

	Width x depth	Price £
Vitreous enamel;		
Magnetic	600 x 1200 mm	139 to 188
Magnetic	900 x 1200 mm	203 to 274
Magnetic	1000 x 1500 mm	270 to 365
High pressure laminate		
Non Magnetic	600 x 1200 mm	119 to 160
Non Magnetic	900 x 1200 mm	172 to 233
Non Magnetic	1000 x 1500 mm	230 to 310

PROJECTION SCREENS

Tripod style; free standing.

	Width x depth	Price £
Pull down lightweight	1200 x 1200 mm	55 to 75
Pull down lightweight	1500 x 1500 mm	64 to 86
Pull down lightweight	1800 x 1800 mm	93 to 126

Wall mounted

	Width x depth	Price £
Pull down	1200 x 1200 mm	51 to 69
Pull down	1524 x 1524 mm	59 to 80
Pull down	1778 x 1778 mm	72 to 97
Pull down	2134 x 2134 mm	114 to 154
Pull down	2400 x 1800 mm	89 to 120
Pull down	2700 x 2400 mm	160 to 217

H 4 - DINING ROOM AREA ARRANGEMENTS

Model 1

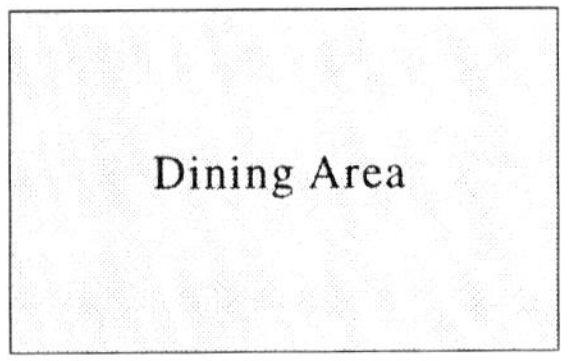

Model 2

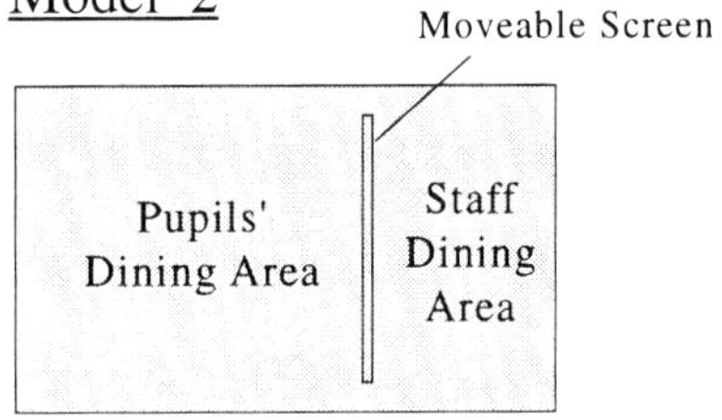

Model 3

Model 4

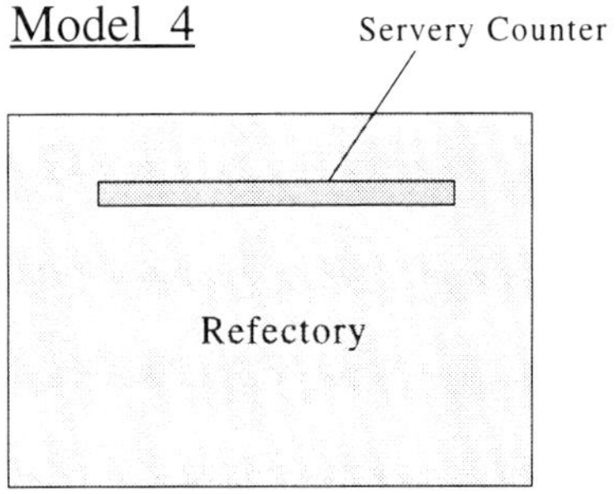

TABLE 'A' **DINING ROOM**

School Type (a)	Model (b)	Area m2 (c)	Total Building Cost (£) New (d)	Extension (e)	Re-Modelled (f)	Operational Usage (Nos) Teaching Spaces (g)	School Meals (h)	Assembly Spaces (i)	Revenue & Entertainment (j)	Sport (k)	Seminar Spaces (l)
Primary	1	17	£11,000	£10,250	£6,725	7	9	12	6	4	7
	1	24	£15,550	£14,475	£9,500	10	12	16	8	5	10
	2	30	£19,825	£18,475	£12,125	12	15	20	10	6	12
	2	33	£21,800	£20,325	£13,350	14	17	22	11	7	14
Secondary	2	40	£26,425	£24,625	£16,175	16	20	27	14	8	16
	2	51	£33,700	£31,375	£20,600	21	26	34	17	11	21
	3	94	£64,025	£59,625	£39,150	38	47	63	32	19	38
	3	135	£91,950	£85,625	£56,225	54	68	90	45	27	54
	4	168	£138,450	£128,925	£84,625	68	84	112	56	34	68
	4	196	£161,500	£150,425	£98,750	79	98	131	66	40	79
Higher	3	103	£70,150	£65,325	£42,900	42	52	69	35	21	42
	3	146	£99,425	£92,600	£60,800	59	73	98	49	30	59
	4	179	£147,500	£137,375	£90,175	72	90	120	60	36	72
	4	207	£170,575	£158,875	£104,275	83	104	138	69	42	83

TABLE 'B' **ANALYSIS OF COSTS**
£ per Square Metre - Based upon Model 3

Elements of the Building (m)	New £ per m2 (n)	Extension £ per m2 (o)	Remodelled £ per m2 (p)
Main Building Work			
* Foundations	109.07	111.00	4.21
* Superstructure	338.48	277.22	141.79
Finishings			
* Floors	49.60	52.16	66.35
* Walls	29.96	31.22	25.70
* Ceilings	27.70	31.94	46.05
Fittings			
* Furniture	25.84	24.01	24.22
* Fittings	60.30	56.03	56.51
* Sanitary Fittings	3.68	4.20	5.21
* Light fittings	23.51	23.25	25.74
Security			
* Protection	2.70	8.34	8.58
* Surveillance	6.44	8.62	6.25
* Communication	3.67	6.23	5.70
FITTING OUT	**233.40** £ per m2	**245.99** £ per m2	**270.30** £ per m2
TOTAL COST	**680.96** £ per m2	**634.22** £ per m2	**416.30** £ per m2

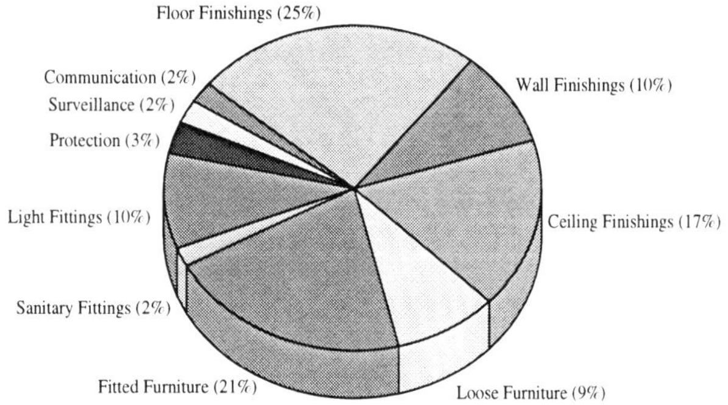

ANALYSIS OF FITTING OUT COSTS
(Costs based upon Table 'B' Column 'p').

CHAIRS

STACKING

<u>Standard; coloured polypropylene seat; steel frame; epoxy paint finish.</u>

	Width x depth x height	Price £
Key Stage 1	350 x 350 x 280 mm	7 to 9
Key Stage 2	375 x 360 x 320 mm	7 to 9
Key Stage 2 - 3	410 x 400 x 355 mm	8 to 10
Key Stage 3	470 x 500 x 390 mm	8 to 10
Key Stage 3 - 4	470 x 500 x 430 mm	8 to 11

<u>Traditional; plywood seat and back; linked; laminated Beech frame, polished.</u>

	Width x depth x height	Price £
Key Stage 1 - 2	355 x 290 x 320 mm	19 to 26
Key Stage 2	360 x 320 x 355 mm	21 to 28
Key Stage 2 - 3	410 x 390 x 390 mm	22 to 29
Key Stage 3 - 4	410 x 390 x 445 mm	22 to 30

<u>Traditional; plywood back and upholstered seat; laminated Beech frame, polished.</u>

	Width x depth x height	Price £
Key Stage 2	355 x 290 x 320 mm	26 to 34
Key Stage 2	360 x 320 x 355 mm	27 to 36
Key Stage 2 - 3	410 x 390 x 390 mm	29 to 38
Key Stage 3 - 4	410 x 390 x 445 mm	29 to 39

<u>Banqueting chair; upholstered; Tweed/Velvet; deep padded seat/back; square or round top; aluminium frame.</u>

	Width x depth x height	Price £
Key Stage 3 - 4	420 x 580 x 460 mm	43 to 57
Key Stage 3 - 4	420 x 580 x 460 mm	44 to 60

CHAIRS

FOLDING

<u>Folding; oversize seat; curved back; steel frame; epoxy paint finish; heavy use.</u>

	Width x depth x height	Price £
Key Stage 3 - 4	460 x 540 x 450 mm	13 to 17
Ditto permament link	460 x 540 x 450 mm	18 to 25

<u>Folding; lightweight; polypropylene back and seat; steel frame; epoxy paint finish; heavy use.</u>

	Width x depth x height	Price £
Key Stage 3 - 4	460 x 540 x 450 mm	13 to 18

<u>Folding; upholstered; tweed or vinyl; oversize seat with curved back; steel frame; epoxy paint finish.</u>

	Width x depth x height	Price £
Key Stage 3 - 4	460 x 540 x 450 mm	29 to 39
Ditto permament link	460 x 540 x 450 mm	34 to 46

<u>Folding; comfort radius back; padded seat; steel frame; epoxy paint finish.</u>

	Width x depth x height	Price £
Key Stage 3 - 4	460 x 540 x 450 mm	33 to 44

<u>Folding; chromium plated frame, grained polyproylene seat/back, articulating back rest.</u>

	Width x depth x height	Price £
Key Stage 3 - 4	490 x 460 x 480 mm	53 to 72

BENCHES

STACKING

<u>Laminate ribbed seat; aluminium edging; polypropylene corners; aluminium frame; folding; heavy use.</u>

	Width x depth x height	Price £
Ket Stage 2	1220 x 254 x 381 mm	36 to 49
Ket Stage 2	1520 x 254 x 381 mm	41 to 55
Ket Stage 2	1830 x 254 x 381 mm	44 to 59
Ket Stage 3 - 4	1220 x 254 x 432 mm	36 to 49
Ket Stage 3 - 4	1520 x 254 x 432 mm	41 to 55
Ket Stage 3 - 4	1830 x 254 x 432 mm	44 to 59

<u>Economy; laminate ribbed seat; aluminium edging; polypropylene corners; aluminium frame; folding.</u>

	Width x depth x height	Price £
Ket Stage 3 - 4	1220 x 254 x 432 mm	30 to 40
Ket Stage 3 - 4	1830 x 254 x 432 mm	38 to 51
Ditto; executive	1830 x 254 x 432 mm	46 to 62

CHAIR TROLLEY

<u>Steel tubular frame, strong rubber wheels, this item can be used to move any type of stacking chair.</u>

	Price £
Trolley	50 to 67

TABLES

FOLDING

Lightweight; laminate top; ribbed extrusion; polypropylene moulded corners; steel tubular legs; epoxy paint finish.

	Length x width x height	Price £
Key Stage 1	915 x 610 x 508 mm	47 to 63
Key Stage 1	1220 x 685 x 508 mm	55 to 74
Key Stage 1	1830 x 685 x 508 mm	68 to 92
Key Stage 2	915 x 610 x 546 mm	47 to 63
Key Stage 2	1220 x 685 x 546 mm	55 to 74
Key Stage 2 - 3	1830 x 685 x 546 mm	68 to 92
Key Stage 2 - 3	915 x 685 x 584 mm	48 to 65
Key Stage 2 - 3	1220 x 760 x 584 mm	57 to 77
Key Stage 2 - 3	1830 x 760 x 584 mm	70 to 95
Key Stage 3	915 x 685 x 635 mm	48 to 65
Key Stage 3	1220 x 760 x 635 mm	57 to 77
Key Stage 3	1830 x 760 x 635 mm	70 to 95
Key Stage 4	915 x 760 x 698 mm	50 to 67
Key Stage 4	1220 x 760 x 698 mm	57 to 77
Key Stage 4	1830 x 760 x 698 mm	70 to 95

Samsonite - Solid core top, melamine finish, polyurethane moulded edges, steel frame, locking rings on leg mechanism; epoxy paint finish.

	Length x width x height	Price £
Key Stage 3 - 4	1220 x 760 x 740 mm	63 to 85
Key Stage 3 - 4	1525 x 760 x 740 mm	63 to 86
Key Stage 3 - 4	1830 x 685 x 508 mm	66 to 89
Key Stage 3 - 4	1525 Dia x 740 mm	122 to 165

TABLES

FOLDING

Heavyweight; laminate top; ribbed extrusion; polypropylene moulded corners; steel tubular legs.

	Length x width x height	Price £
Key Stage 1	915 x 610 x 508 mm	51 to 69
Key Stage 1	1220 x 685 x 508 mm	60 to 80
Key Stage 1	1830 x 685 x 508 mm	71 to 96
Key Stage 2	915 x 610 x 546 mm	51 to 69
Key Stage 2	1220 x 685 x 546 mm	60 to 80
Key Stage 2 - 3	1830 x 685 x 546 mm	71 to 96
Key Stage 2 - 3	915 x 685 x 584 mm	52 to 70
Key Stage 2 - 3	1220 x 760 x 584 mm	62 to 83
Key Stage 2 - 3	1830 x 760 x 584 mm	74 to 100
Key Stage 3	915 x 685 x 635 mm	52 to 70
Key Stage 3	1220 x 760 x 635 mm	62 to 83
Key Stage 3	1830 x 760 x 635 mm	74 to 100
Key Stage 4	915 x 760 x 698 mm	54 to 72
Key Stage 4	1220 x 760 x 698 mm	62 to 83
Key Stage 4	1830 x 760 x 698 mm	74 to 100

Slimfold banquet table; beige nylon-flock top; aluminium edging; folding steel legs.

	Length x width x height	Price £
Key Stage 4	1830 dia x 720 mm	133 to 179
Key Stage 4	1830 dia x 760 mm	133 to 179
Key Stage 4	1520 dia x 720 mm	108 to 146
Key Stage 4	1520 dia x 760 mm	108 to 146

TABLES

FOLDING

Top quality; laminate top; ribbed extrusion; polypropylene moulded corners; steel tubular legs; epoxy paint finish.

	Length x width x height	Price £
Key Stage 1	915 x 610 x 508 mm	65 to 88
Key Stage 1	1220 x 685 x 508 mm	80 to 108
Key Stage 1	1830 x 685 x 508 mm	103 to 139
Key Stage 2	915 x 610 x 546 mm	65 to 88
Key Stage 2	1220 x 685 x 546 mm	80 to 108
Key Stage 2	1830 x 685 x 546 mm	103 to 139
Key Stage 2 - 3	915 x 685 x 584 mm	67 to 91
Key Stage 2 - 3	1220 x 760 x 584 mm	83 to 112
Key Stage 2 - 3	1830 x 760 x 584 mm	106 to 143
Key Stage 3	915 x 685 x 635 mm	67 to 91
Key Stage 3	1220 x 760 x 635 mm	83 to 112
Key Stage 3	1830 x 760 x 635 mm	106 to 143
Key Stage 4	915 x 760 x 698 mm	70 to 94
Key Stage 4	1220 x 760 x 698 mm	83 to 112
Key Stage 4	1830 x 760 x 698 mm	106 to 143

TABLE TROLLEY

Galvanized steel frame, MDF base, specially designed for the transportation of folding tables.

	Price £
Max 5 feet	142 to 192
Max 6 feet	152 to 206

TABLES

	Length x width x height	Price £
STACKING		
Laminate top; grey; colour co-ordinated proform edge; steel frame; epoxy paint finish; damage resistant and impervious to water.		
Key Stage 1	1200 x 600 x 500 mm	47 to 63
Key Stage 1 - 2	1200 x 600 x 550 mm	47 to 63
Key Stage 2	1200 x 600 x 600 mm	47 to 64
Key Stage 2 - 3	1200 x 600 x 650 mm	48 to 64
Key Stage 3 - 4	1200 x 600 x 700 mm	48 to 64
Spiral stacking; melamine top; melamine edging; steel frame; epoxy paint finish.		
Key Stage 4	601 x 610 x 762 mm	23 to 31
Key Stage 4	762 x 762 x 762 mm	24 to 32
Key Stage 4	915 x 610 x 762 mm	26 to 35
Key Stage 4	1220 x 762 x 762 mm	32 to 43
Key Stage 4	1520 x 762 x 762 mm	36 to 48
Key Stage 4	1830 x 762 x 762 mm	37 to 49
Goal post tivoli table; laminate top; laminate edging; chromium plated frame.		
Key Stage 4	762 x 762 x 762 mm	93 to 126
Key Stage 4	1220 x 762 x 762 mm	114 to 154
Key Stage 4	1677 x 762 x 762 mm	148 to 200

	Price Range; including installation		
	Economic	Standard	Good
Kitchen Equipment			
	£	£	£
Air break tank	190	215	240
Bench; mobile wall unit; c/w upstand	370	410	455
Bench; mobile wall unit; c/w upstand; shelf	445	580	710
Bench; wall; stainless steel	195	270	345
Bench; work; stainless steel	240	305	370
Benches; centre; c/w undershelf	385	440	490
Canopy; aluminium; extract ventilation	2935	3990	5045
Canopy; stainless steel; condensation	530	925	1320
Cash registers	180	295	410
Cashiers unit; two person	970	1290	1615
Cold Room	5160	6235	7310
Cooker range; solid top; six burner	1040	1755	2475
Cutlery/condiment units	165	215	270
Dishwash table; cleans	165	260	350
Dishwashing table; dirties	885	1240	1595
Double pan fryer	1290	2635	3980
Food mixer (with stand) 20 qrt	990	2915	4840
Freezer; freezer	1290	2560	3830
Hot cupboard; pass through type	1025	1320	1615
Insect control unit	105	140	175
Menu boards	55	95	130
Mobile bin	50	105	165
Mobile tray stand	165	215	270
Oven; convection; 20	6990	8280	9570
Oven; ten tray combination	4085	7450	10815
Pan; brat	1640	2960	4275
Pass thru' dishwasher	1905	2485	3065
Rack; zinc chromate; four tier rack	130	240	345
Refrigerator	710	2240	3765
Servery counter; mobile Bain Marie	2420	3530	4635
Servery counter; refrigerated	1720	3010	4300
Shutters; 4 metres; roller; servery counter	645	810	970
Sink; preparation; stainless steel	535	765	995
Sink; sterilizing; double bowl	530	600	670
Tray return trolley	270	325	380
Wash hand basin	130	190	250
Water softener	280	560	835
Worktop (700mm) with base (1400mm)	860	1025	1185

H 5 - KITCHEN AREA ARRANGEMENTS

TABLE 'A'			KITCHEN									
			Total Building Cost (£)			Operational Usage (Nos)						
School Type (a)	Model (b)	Area m2 (c)	New (d)	Extension (e)	Re-Modelled (f)	Teaching Spaces (g)	School Meals (h)	Assembly Spaces (i)	Revenue & Entertainment (j)	Sport (k)	Seminar Spaces (l)
Primary	1	17	£15,800	£14,825	£12,025	NIL	2	NIL	9	NIL	NIL
	1	24	£22,300	£20,925	£16,975	NIL	2	NIL	12	NIL	NIL
	2	48	£54,025	£50,700	£41,100	NIL	4	NIL	24	NIL	NIL
	2	66	£74,275	£69,700	£56,525	NIL	5	NIL	33	NIL	NIL
Secondary	2	60	£67,525	£63,375	£51,375	NIL	5	NIL	30	NIL	NIL
	2	89	£100,175	£94,000	£76,200	NIL	7	NIL	45	NIL	NIL
	3	110	£120,200	£112,775	£91,450	NIL	8	NIL	55	NIL	NIL
	3	143	£156,250	£146,625	£118,875	NIL	11	NIL	72	NIL	NIL
	4	131	£164,600	£154,475	£125,225	NIL	10	NIL	66	NIL	NIL
	4	164	£206,075	£193,375	£156,775	NIL	12	NIL	82	NIL	NIL
Higher	3	133	£145,325	£136,375	£110,550	NIL	10	NIL	67	NIL	NIL
	3	173	£189,025	£177,375	£143,800	NIL	13	NIL	87	NIL	NIL
	4	158	£198,525	£186,300	£151,050	NIL	12	NIL	79	NIL	NIL
	4	198	£248,775	£233,450	£189,275	NIL	15	NIL	99	NIL	NIL

TABLE 'B' ANALYSIS OF COSTS
£ per Square Metre - Based upon Model 3

Elements of the Building (m)	New £ per m2 (n)	Extension £ per m2 (o)	Remodelled £ per m2 (p)
Main Building Work			
* Foundations	109.07	111.00	4.21
* Superstructure	326.97	272.63	152.10
Finishings			
* Floors	59.04	62.10	78.98
* Walls	26.05	27.15	22.35
* Ceilings	27.70	31.94	46.05
Fittings			
* Furniture	346.70	322.13	324.90
* Fittings	148.59	138.05	139.24
* Sanitary Fittings	12.08	13.79	17.11
* Light fittings	23.51	23.25	25.74
Security			
* Protection	2.70	8.34	8.58
* Surveillance	6.44	8.62	6.25
* Communication	3.67	6.23	5.70
FITTING OUT	**656.48** £ per m2	**641.59** £ per m2	**674.89** £ per m2
TOTAL COST	**1,092.53** £ per m2	**1,025.23** £ per m2	**831.20** £ per m2

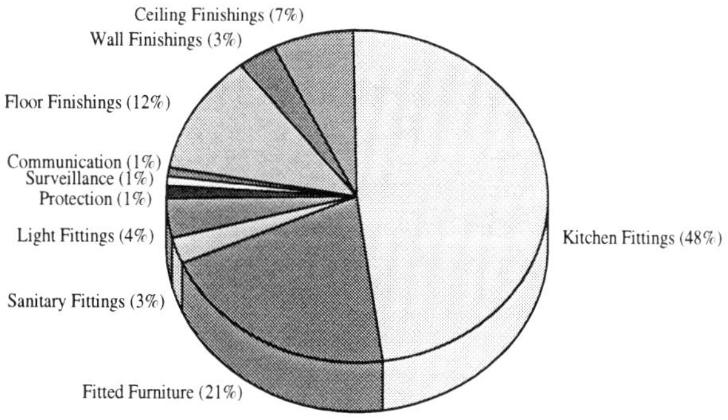

ANALYSIS OF FITTING OUT COSTS
(Costs based upon Table 'B' Column 'p').

KITCHEN PLAN TO FEED UP TO 200 PERSONS

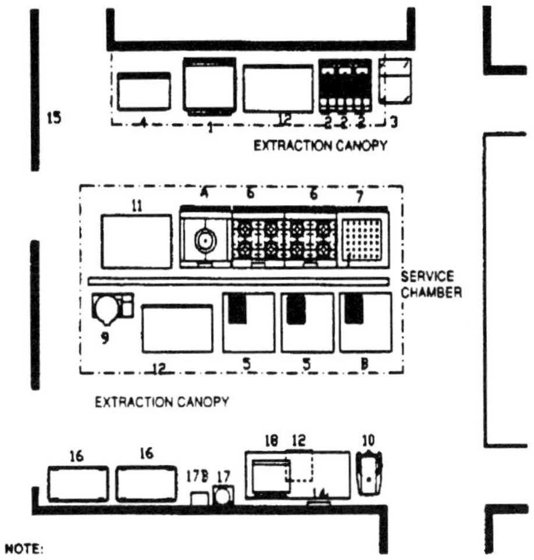

NOTE:
THE BOILING TABLE (A) AND COMBINATION OVEN (B) ARE SCALED IN PASTRY PREP. THEY ARE LOCATED IN THE KITCHEN TO GIVE FLEXIBILITY OF USE

KITCHEN PLAN TO FEED UP TO 300 PERSONS

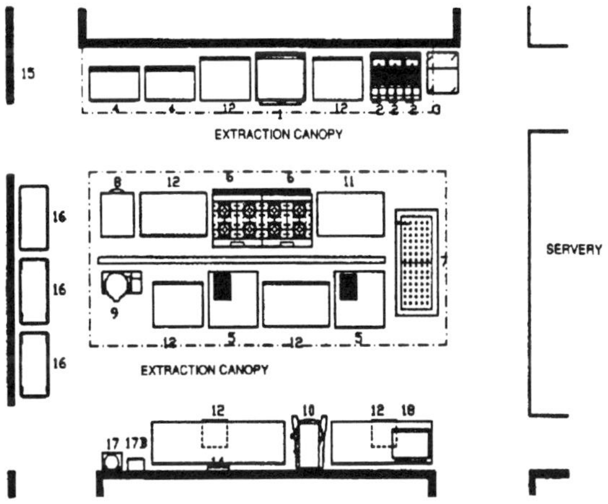

1 BRATT PAN
2 DEEP FAT FRYER (FROM SERVERY SCALE)
3 FILTRATION UNIT
4 GRILL ON FLOOR STAND
5 COMBINATION OVEN
6 OVEN RANGE
7 BAIN MARIE
8 PRESSURE STEAMER
9 TILTING KETTLE
10 FLOOR STANDING MIXER
11 HOT CUPBOARD
12 BENCH
14 KNIFE RACK
15 NOTICE BOARD
16 TROLLEY
17 WASH HAND BASIN
17B TOWEL DISPENSER
18 MICROWAVE OVEN

KITCHEN PLAN TO FEED UP TO 500 PERSONS

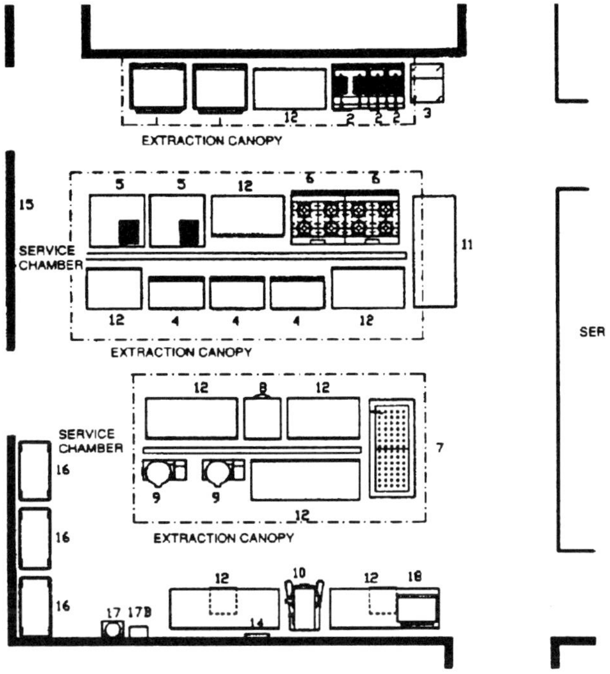

1 BRATT PAN
2 DEEP FAT FRYER (FROM SERVERY SCALE)
3 FILTRATION UNIT
4 GRILL ON FLOOR STAND
5 COMBINATION OVEN
6 OVEN RANGE
7 BAIN MARIE HOT CUPBOARD
8 PRESSURE STEAMER
9 TILTING KETTLE
10 FLOOR STANDING MIXER
11 HOT CUPBOARD
12 BENCH
14 KNIFE RACK
15 NOTICE BOARD
16 TROLLEY
17 WASH HAND BASIN
17B TOWEL DISPENSER
18 MICROWAVE OVEN

H 6 - TOILET AREA ARRANGEMENTS

Model 1

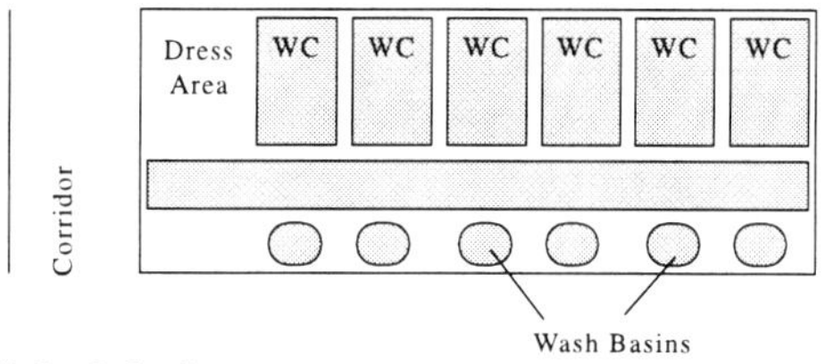

Model 2

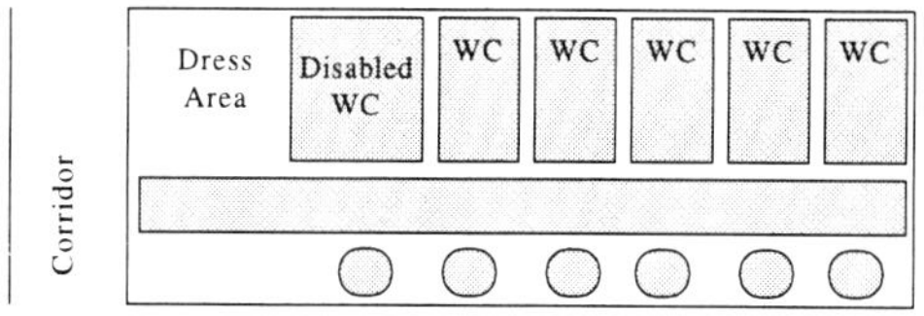

Model 3

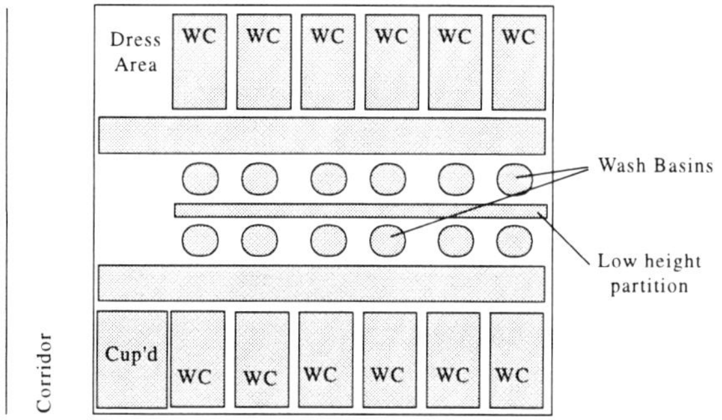

Model 4

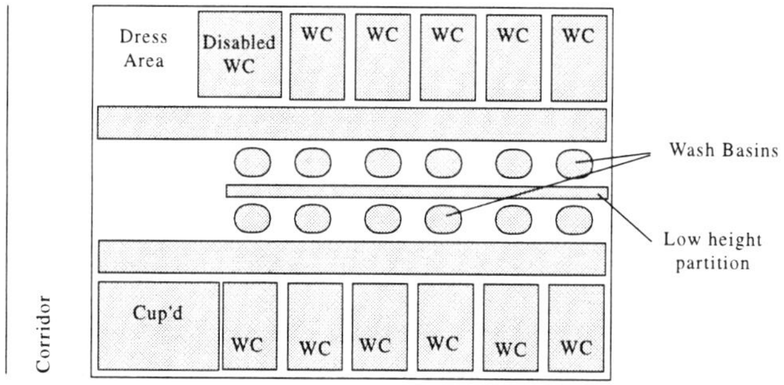

TABLE 'A' TOILETS

School Type (a)	Model (b)	Area m2 (c)	Total Building Cost (£) New (d)	Extension (e)	Re-Modelled (f)	Operational Usage (Nos) Teaching Spaces (g)	School Meals (h)	Assembly Spaces (i)	Revenue & Entertainment (j)	Sport (k)	Seminar Spaces (l)
Primary	1	17	£14,750	£15,075	£13,425	NIL	NIL	NIL	NIL	6	NIL
	1	28	£24,275	£24,800	£22,125	NIL	NIL	NIL	NIL	10	NIL
	2	27	£26,150	£26,725	£23,850	NIL	NIL	NIL	NIL	9	NIL
	2	42	£40,675	£41,575	£37,075	NIL	NIL	NIL	NIL	14	NIL
Secondary	2	53	£51,325	£52,475	£46,775	NIL	NIL	NIL	NIL	18	NIL
	2	76	£73,600	£75,225	£67,075	NIL	NIL	NIL	NIL	26	NIL
	3	86	£87,675	£89,600	£79,900	NIL	NIL	NIL	NIL	29	NIL
	3	<u>105</u>	£107,025	£109,400	£97,550	NIL	NIL	NIL	NIL	35	NIL
	4	115	£134,800	£137,800	£122,875	NIL	NIL	NIL	NIL	39	NIL
	4	139	£162,925	£166,550	£148,500	NIL	NIL	NIL	NIL	47	NIL
Higher	3	94	£95,825	£97,950	£87,325	NIL	NIL	NIL	NIL	32	NIL
	3	114	£116,200	£118,775	£105,925	NIL	NIL	NIL	NIL	38	NIL
	4	126	£147,700	£150,975	£134,625	NIL	NIL	NIL	NIL	42	NIL
	4	150	£175,825	£179,725	£160,250	NIL	NIL	NIL	NIL	50	NIL

TABLE 'B' **ANALYSIS OF COSTS**
£ per Square Metre - Based upon Model 3

Elements of the Building (m)	New £ per m2 (n)	Extension £ per m2 (o)	Remodelled £ per m2 (p)
Main Building Work			
* Foundations	109.07	111.00	4.21
* Superstructure	552.68	528.36	442.94
Finishings			
* Floors	59.04	62.10	78.98
* Walls	26.05	27.15	22.35
* Ceilings	27.70	31.94	46.05
Fittings			
* Furniture	9.56	8.88	8.96
* Fittings	5.06	4.70	4.74
* Sanitary Fittings	193.73	221.25	274.49
* Light fittings	23.51	23.25	25.74
Security			
* Protection	2.70	8.34	8.58
* Surveillance	6.44	8.62	6.25
* Communication	3.67	6.23	5.70
FITTING OUT	**357.47** £ per m2	**402.45** £ per m2	**481.83** £ per m2
TOTAL COST	**1,019.23** £ per m2	**1,041.81** £ per m2	**928.98** £ per m2

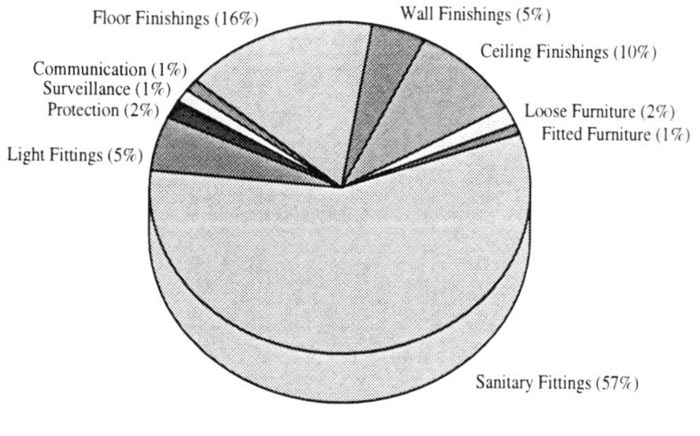

ANALYSIS OF FITTING OUT COSTS
(Costs based upon Table 'B' Column 'p').

W.C. CUBICLES

LAMINATED SOLID CORE PANELS

WC Cubicle; standard; aluminium frames; self finish.

	Width x depth x height	Price £
Key Stage 1 - 2		
Corner cubicle		
(1 side; 1 door)	850 x 1500 x 1150 mm	220 to 300
Corner cubicle; wide		
(1 side; 1 front; 1 door)	1150 x 1500 x 1150 mm	250 to 335
Cubicle between walls		
(1 door)	850 x 1500 x 1150 mm	140 to 190
(1 door; 1 front)	1150 x 1500 x 1150 mm	195 to 260
Island cubicle		
(2 sides; 1 door)	850 x 1500 x 1150 mm	250 to 335
Key Stage 2 - 3		
Corner cubicle		
(1 side; 1 door)	850 x 1500 x 1500 mm	250 to 335
Corner cubicle; wide		
(1 side; 1 front; 1 door)	1150 x 1500 x 1500 mm	275 to 375
Cubicle between walls		
(1 door)	850 x 1500 x 1500 mm	165 to 225
(1 door; 1 front)	1150 x 1500 x 1500 mm	195 to 260
Island cubicle		
(2 sides; 1 door)	850 x 1500 x 1500 mm	275 to 375
Key Stage 4 and Adult		
Corner cubicle		
(1 side; 1 door)	850 x 1500 x 2100 mm	305 to 410
Corner cubicle; wide		
(1 side; 1 front; 1 door)	1150 x 1500 x 2100 mm	305 to 410
Cubicle between walls		
(1 door)	850 x 1500 x 2100 mm	165 to 225
(1 door; 1 front)	1150 x 1500 x 2100 mm	220 to 300
Island cubicle		
(2 sides; 1 door)	850 x 1500 x 2100 mm	440 to 595

RESIN BONDED PANELS

WC Cubicle; self finish; steel frame; epoxy paint finish.

	Width x depth x height	Price £
Key Stage 1 - 2		
Corner cubicle		
(1 side; 1 door)	850 x 1500 x 1150 mm	320 to 435
Corner cubicle; wide		
(1 side; 1 front; 1 door)	1150 x 1500 x 1150 mm	340 to 460
Cubicle between walls		
(1 door)	850 x 1500 x 1150 mm	210 to 280
(1 door; 1 front)	1150 x 1500 x 1150 mm	260 to 350
Island cubicle		
(2 sides; 1 door)	850 x 1500 x 1150 mm	365 to 495
Key Stage 2 - 3		
Corner cubicle		
(1 side; 1 door)	850 x 1500 x 1500 mm	340 to 460
Corner cubicle; wide		
(1 side; 1 front; 1 door)	1150 x 1500 x 1500 mm	365 to 490
Cubicle between walls		
(1 door)	850 x 1500 x 1500 mm	205 to 280
(1 door; 1 front)	1150 x 1500 x 1500 mm	285 to 385
Island cubicle		
(2 sides; 1 door)	850 x 1500 x 1500 mm	390 to 530
Key Stage 4 and Adult		
Corner cubicle		
(1 side; 1 door)	850 x 1500 x 2100 mm	440 to 595
Corner cubicle; wide		
(1 side; 1 front; 1 door)	1150 x 1500 x 2100 mm	440 to 595
Cubicle between walls		
(1 door)	850 x 1500 x 2100 mm	235 to 320
(1 door; 1 front)	1150 x 1500 x 2100 mm	335 to 455
Island cubicle		
(2 sides; 1 door)	850 x 1500 x 2100 mm	625 to 845

SHOWER CUBICLES

RESIN BONDED PANELS

Shower Cubicle; self finish; plastic curtains and rails; steel frame; epoxy paint finish.

	Width x depth x height	Price £
Key Stage 1 - 2		
Corner cubicle		
(1 side; 1 curtain)	850 x 1500 x 1150 mm	360 to 490
Corner cubicle; wide		
(1 side; 1 front; 1 door)	1150 x 1500 x 1150 mm	380 to 515
Cubicle between walls		
(1 curtain)	850 x 1500 x 1150 mm	115 to 155
(1 curtain; 1 front)	1150 x 1500 x 1150 mm	205 to 280
Island cubicle		
(2 sides; 1 curtain)	850 x 1500 x 1150 mm	350 to 475

DISABLED EQUIPMENT

	Price £
Toilet support arm; operational with one hand; when folded remains in upright position; wall mounted tubular steel construction; chromium plated.	
Support arm	80 to 110
Support arm with leg	80 to 110
Toilet support rails; 32 mm diameter; PVC construction.	
150 mm Rail (Pack of 3)	30 to 40
300 mm Rail (Pack of 3)	35 to 45
450 mm Rail (Pack of 3)	40 to 50
600 mm Rail (Pack of 3)	45 to 60
Shower chair; folding steel tubular frame; epoxy paint finish; embossed padded seat; remains in upright position when folded.	
Wall mounted	135 to 180
Shower chair; folding steel tubular frame; epoxy paint finish; embossed padded seat with cut out; remains in upright position when folded; adjustable legs.	
Floor mounted	180 to 240

H 7 - CLOAKROOM AREA ARRANGEMENTS

Model 1

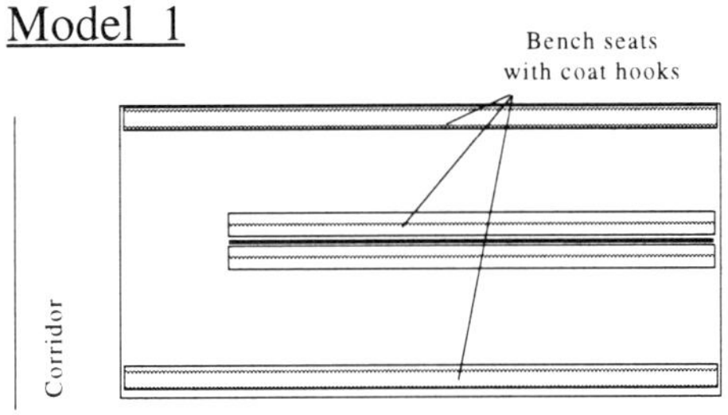

Model 2

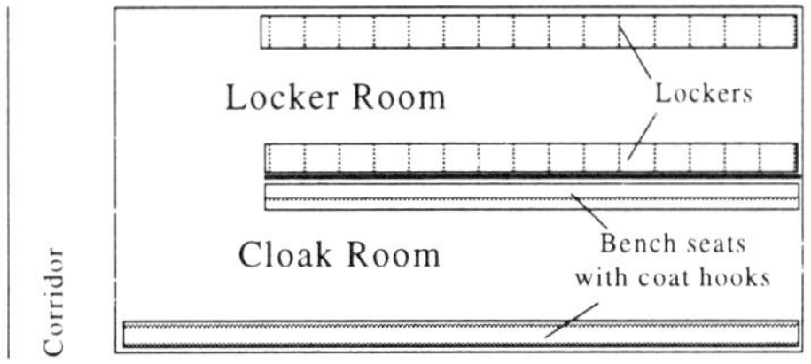

Model 3

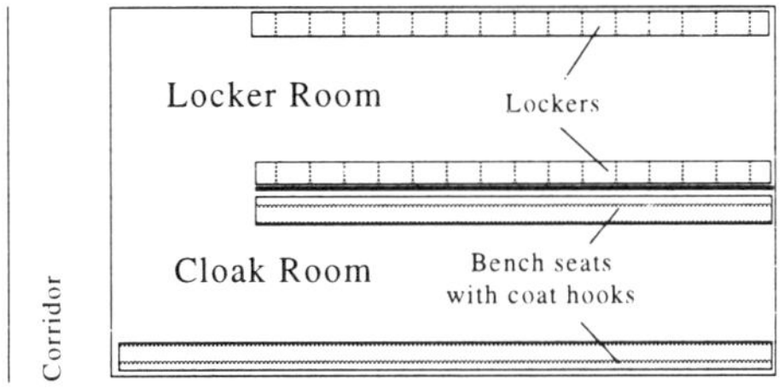

TABLE 'A' **CLOAK ROOM**

School Type (a)	Model (b)	Area m2 (c)	Total Building Cost (£) New (d)	Extension (e)	Re-Modelled (f)	Operational Usage (Nos) Teaching Spaces (g)	School Meals (h)	Assembly Spaces (i)	Revenue & Entertainment (j)	Sport (k)	Seminar Spaces (l)
Primary	1	13	£7,500	£7,050	£4,375	NIL	NIL	NIL	NIL	6	NIL
	1	21	£12,125	£11,375	£7,050	NIL	NIL	NIL	NIL	9	NIL
	2	16	£10,075	£9,425	£5,850	NIL	NIL	NIL	NIL	7	NIL
	2	25	£15,725	£14,750	£9,125	NIL	NIL	NIL	NIL	10	NIL
Secondary	2	21	£13,200	£12,375	£7,675	NIL	NIL	NIL	NIL	9	NIL
	2	33	£20,750	£19,450	£12,050	NIL	NIL	NIL	NIL	14	NIL
	3	81	£51,925	£48,700	£30,125	NIL	NIL	NIL	NIL	33	NIL
	3	120	£76,925	£72,150	£44,625	NIL	NIL	NIL	NIL	48	NIL
						NIL	NIL	NIL	NIL	NIL	NIL
						NIL	NIL	NIL	NIL	NIL	NIL
Higher	3	97	£62,200	£58,325	£36,075	NIL	NIL	NIL	NIL	39	NIL
	3	142	£91,025	£85,375	£52,825	NIL	NIL	NIL	NIL	57	NIL
						NIL	NIL	NIL	NIL	NIL	NIL
						NIL	NIL	NIL	NIL	NIL	NIL

TABLE 'B' **ANALYSIS OF COSTS**

£ per Square Metre - Based upon Model 3

Elements of the Building (m)	New £ per m2 (n)	Extension £ per m2 (o)	Remodelled £ per m2 (p)
Main Building Work			
* Foundations	109.07	111.00	4.21
* Superstructure	333.76	278.34	132.77
Finishings			
* Floors	47.23	49.68	63.19
* Walls	26.05	27.15	22.35
* Ceilings	27.70	31.94	46.05
Fittings			
* Furniture	59.38	55.17	55.64
* Fittings	1.48	1.37	1.38
* Sanitary Fittings	0.00	0.00	0.00
* Light fittings	23.51	23.25	25.74
Security			
* Protection	2.70	8.34	8.58
* Surveillance	6.44	8.62	6.25
* Communication	3.67	6.23	5.70
FITTING OUT	**198.17** £ per m2	**211.74** £ per m2	**234.88** £ per m2
TOTAL COST	**641.01** £ per m2	**601.08** £ per m2	**371.85** £ per m2

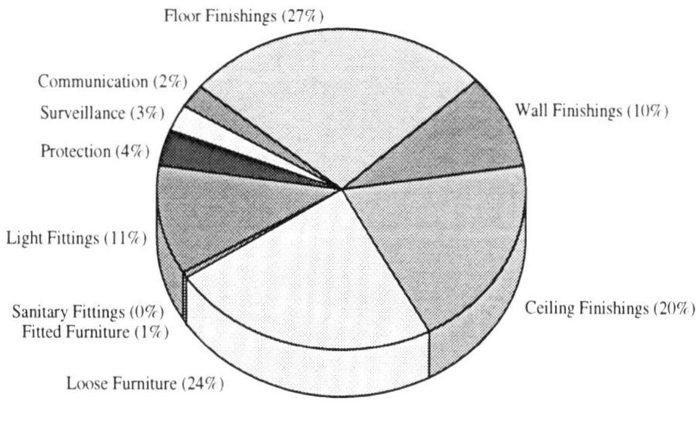

ANALYSIS OF FITTING OUT COSTS

(Costs based upon Table 'B' Column 'p').

H 8 - CHANGING ROOM AREA ARRANGEMENTS

Model 1 (Female)

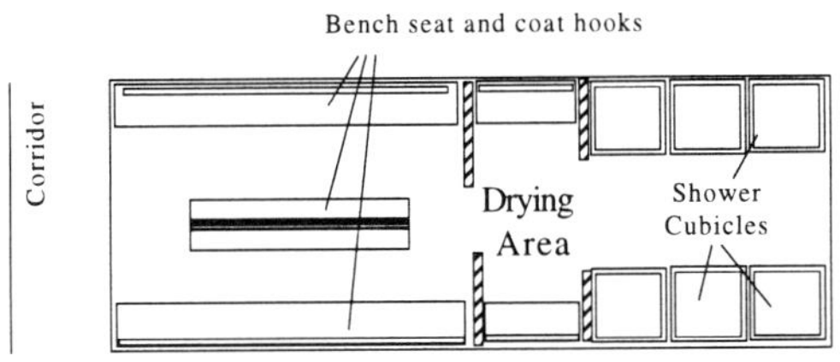

Model 2 (Male)

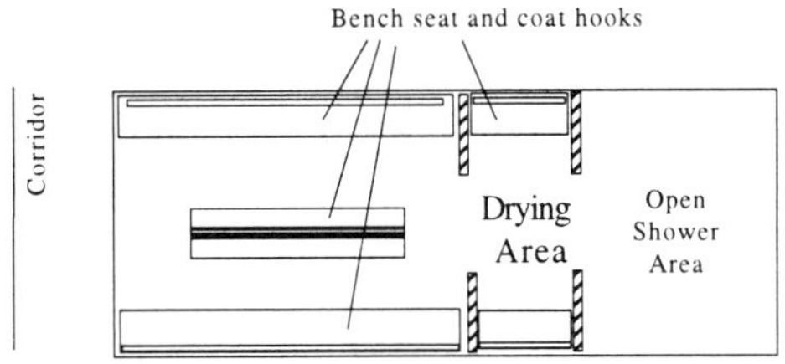

Model 3 (Female)

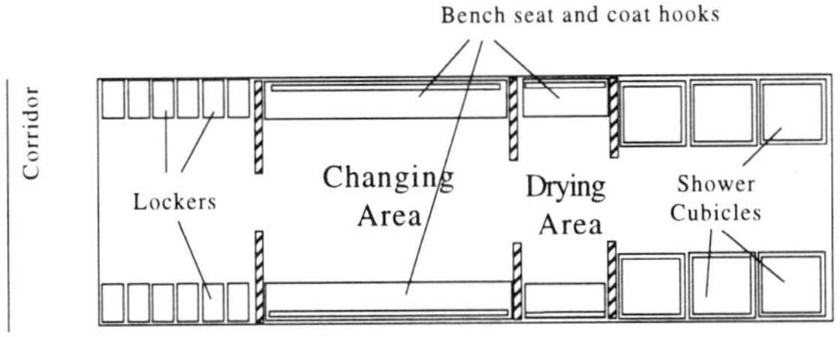

Model 4 (Male)

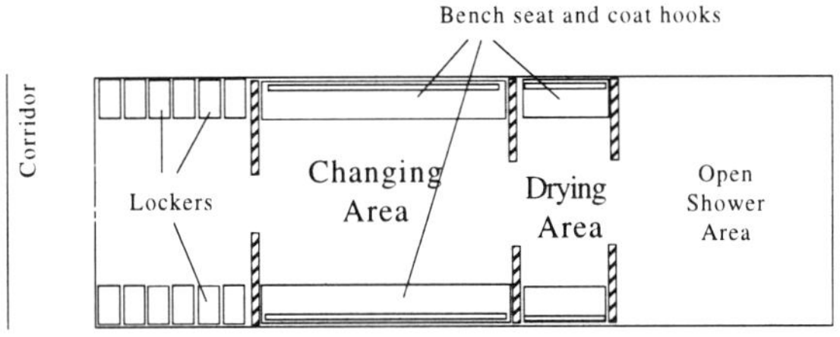

TABLE 'A' **CHANGING ROOM**

School Type (a)	Model (b)	Area m2 (c)	Total Building Cost (£) New (d)	Extension (e)	Re-Modelled (f)	Operational Usage (Nos) Teaching Spaces (g)	School Meals (h)	Assembly Spaces (i)	Revenue & Entertainment (j)	Sport (k)	Seminar Spaces (l)
Primary	1	18	£16,325	£16,425	£14,000	NIL	NIL	NIL	NIL	9	NIL
	1	24	£21,775	£21,900	£18,675	NIL	NIL	NIL	NIL	12	NIL
	2	28	£23,400	£23,525	£20,050	NIL	NIL	NIL	NIL	14	NIL
	2	40	£33,400	£33,600	£28,650	NIL	NIL	NIL	NIL	20	NIL
Secondary	2	36	£30,075	£30,250	£25,800	NIL	NIL	NIL	NIL	18	NIL
	2	42	£35,075	£35,275	£30,075	NIL	NIL	NIL	NIL	21	NIL
	3	72	£68,700	£69,100	£58,925	NIL	NIL	NIL	NIL	36	NIL
	3	83	£79,200	£79,650	£67,925	NIL	NIL	NIL	NIL	42	NIL
	4	81	£75,375	£75,800	£64,650	NIL	NIL	NIL	NIL	41	NIL
	4	92	£85,600	£86,075	£73,425	NIL	NIL	NIL	NIL	46	NIL
Higher	3	85	£81,125	£81,575	£69,575	NIL	NIL	NIL	NIL	43	NIL
	3	97	£92,575	£93,100	£79,400	NIL	NIL	NIL	NIL	49	NIL
	4	94	£87,450	£87,950	£75,025	NIL	NIL	NIL	NIL	47	NIL
	4	107	£99,550	£100,125	£85,375	NIL	NIL	NIL	NIL	54	NIL

TABLE 'B' **ANALYSIS OF COSTS**

£ per Square Metre - Based upon Model 3

Elements of the Building (m)	New £ per m2 (n)	Extension £ per m2 (o)	Remodelled £ per m2 (p)
Main Building Work			
* Foundations	109.07	111.00	4.21
* Superstructure	499.88	468.05	368.13
Finishings			
* Floors	54.32	57.13	72.67
* Walls	26.05	27.15	22.35
* Ceilings	27.70	31.94	46.05
Fittings			
* Furniture	51.64	47.98	48.39
* Fittings	2.13	1.98	2.00
* Sanitary Fittings	147.01	167.89	208.28
* Light fittings	23.51	23.25	25.74
Security			
* Protection	2.70	8.34	8.58
* Surveillance	6.44	8.62	6.25
* Communication	3.67	6.23	5.70
FITTING OUT	**345.17** £ per m2	**380.50** £ per m2	**446.00** £ per m2
TOTAL COST	**954.13** £ per m2	**959.55** £ per m2	**818.34** £ per m2

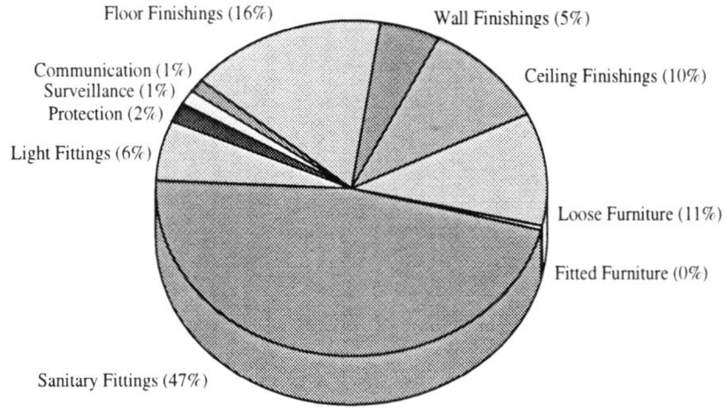

ANALYSIS OF FITTING OUT COSTS

(Costs based upon Table 'B' Column 'p').

LOCKERS

	Width x depth x height	Price £
Steel carcass; coloured doors; heavy duty; epoxy paint finish.		
1 door	300 x 450 x 1800 mm	60 to 80
1 door	300 x 500 x 1800 mm	65 to 85
2 door	300 x 450 x 1800 mm	70 to 95
2 door	300 x 500 x 1800 mm	75 to 100
3 door	300 x 450 x 1800 mm	75 to 105
3 door	300 x 500 x 1800 mm	80 to 110
4 door	300 x 450 x 1800 mm	80 to 110
4 door	300 x 500 x 1800 mm	85 to 115
Steel carcass; coloured doors; ventilated; heavy duty; epoxy paint finish.		
1 door	305 x 305 x 1778 mm	45 to 60
1 door	305 x 457 x 1778 mm	55 to 70
2 door	305 x 305 x 1778 mm	55 to 75
2 door	305 x 457 x 1778 mm	60 to 80
3 door	305 x 305 x 1778 mm	60 to 80
3 door	305 x 457 x 1778 mm	70 to 90
4 door	305 x 305 x 1778 mm	65 to 90
4 door	305 x 457 x 1778 mm	75 to 100
6 door	305 x 305 x 1778 mm	80 to 105
6 door	305 x 457 x 1800 mm	85 to 115
Steel carcass; coloured doors; twin doors full height; ventilated; heavy duty; epoxy paint finish.		
2 door	457 x 457 x 1178 mm	80 to 110
4 door - two person	457 x 457 x 1178 mm	85 to 115
Cube lockers; coloured doors; single doors; ventilated; linking; chromium plated camlocks; epoxy paint finish.		
1 door	305 x 305 x 305 mm	20 to 30
1 door	380 x 380 x 380 mm	25 to 35
1 door	455 x 455 x 455 mm	30 to 40

	Width x depth x height	Price £
Resin board doors; coloured self finish; steel carcass; heavy duty; epoxy paint finish.		
1 door	300 x 450 x 1800 mm	105 to 145
1 door	300 x 500 x 1800 mm	110 to 150
2 door	300 x 450 x 1800 mm	110 to 150
2 door	300 x 500 x 1800 mm	115 to 160
3 door	300 x 450 x 1800 mm	115 to 160
3 door	300 x 500 x 1800 mm	125 to 165
4 door	300 x 450 x 1800 mm	120 to 165
4 door	300 x 500 x 1800 mm	125 to 170
Resin board doors and carcass; coloured self finish; heavy duty.		
1 door	300 x 450 x 1800 mm	140 to 190
1 door	300 x 500 x 1800 mm	145 to 195
2 door	300 x 450 x 1800 mm	145 to 195
2 door	300 x 500 x 1800 mm	150 to 205
3 door	300 x 450 x 1800 mm	165 to 220
3 door	300 x 500 x 1800 mm	170 to 230
4 door	300 x 450 x 1800 mm	180 to 245
4 door	300 x 500 x 1800 mm	190 to 260

LOCKER ACCESSORIES

Assa fittings	
Cam lock	11 to 15
Coin retainer	37 to 50
Coin returner	52 to 70
Spare keys	7 to 9
Master keys	14 to 18
Lowe and Fletcher fittings	
Cam lock	5 to 7
Coin retainer	30 to 40
Coin returner	50 to 67
Spare keys	5 to 7
Master keys	14 to 18

COAT RAILS

		Price £
Stayfast aluminium coat hooks; 300 mm centres; Trespa solid resin rail.		
	price per metre lineal	17 to 23
Wall batten hanger unit; black steel hooks; 300 mm centres; hardwood rail; polished.		
	price per metre lineal	19 to 26
Wall batten hanger unit; black polymide hooks; 300 mm centres; hardwood rail; polished.		
single hooks	price per metre lineal	43 to 58
double hooks	price per metre lineal	55 to 74
Tubular coat rail; steel hooks at 300 mm centres; epoxy paint finish.		
wall hangers	price per metre lineal	45 to 61
extension	price per metre lineal	40 to 55
Tubular coat frame; overhead shelf and hanger; 300 mm centres; heavy duty; epoxy paint finish; battens painted.		
wall hangers	price per metre lineal	86 to 116

SLATTED BENCHES

Extruded polyurethene slatted bench; hygienic, durable, colours; slots onto frame; no screws; steel frame; epoxy

		Price £
Floor support or wall		
Slat bench	1 - 19 mm per metre	80 to 105
Slat bench	20 - 49 mm per metre	70 to 95
Slat bench	50 - 100 mm per metre	65 to 85
Floor support; double island with coat rail		
Slat bench	1 - 19 mm per metre	215 to 290
Slat bench	20 - 49 mm per metre	195 to 260
Slat bench	50 - 100 mm per metre	175 to 235
Changing cubicle bench seats		
Slat bench	1 - 19 mm per metre	60 to 80
Slat bench	20 - 49 mm per metre	55 to 75
Slat bench	50 - 100 mm per metre	50 to 65

WALL BENCHES

	Width x depth x height	Price £
Hardwood single seat; polyurethene finish; steel frame; epoxy paint finish.		
Single seat	762 x 305 x 381 mm	60 to 80
Single seat	915 x 305 x 381 mm	60 to 80
Single seat	1143 x 305 x 381 mm	85 to 115
Hardwood single seat; polyurethene finish; shoe trays; steel frame; epoxy paint finish.		
Key Stage 1 - 2	1500 x 600 x 350 mm	160 to 215
Key Stage 2 - 3	1500 x 600 x 400 mm	160 to 215
Key Stage 3 - 4	1500 x 600 x 450 mm	160 to 215
Hardwood single seat; polyurethene finish; shoe baskets; steel frame; epoxy paint finish.		
Key Stage 1 - 2	1500 x 600 x 350 mm	165 to 220
Key Stage 2 - 3	1500 x 600 x 400 mm	165 to 220
Key Stage 3 - 4	1500 x 600 x 450 mm	165 to 220
Hardwood single seat; polyurethene finish; steel frame; epoxy paint finish.		
Key Stage 1 - 4	1500 x 600 x 350 mm	135 to 180
Hardwood double seat; polyurethene finish; shoe baskets; steel frame; epoxy paint finish.		
Key Stage 1 - 4	1500 x 600 x 350 mm	205 to 275

ISLAND BENCHES

	Width x depth x height	Price £
Hardwood single island seat; polyurethene finish; shoe trays; steel frame; epoxy paint finish.		
Key Stage 1 - 2	1500 x 325 x 350 mm	275 to 375
Key Stage 2 - 3	1500 x 325 x 400 mm	275 to 375
Key Stage 3 - 4	1500 x 325 x 450 mm	275 to 375
Hardwood double island seat; polyurethene finish; steel frame; epoxy paint finish.		
Key Stage 1 - 2	1500 x 600 x 350 mm	305 to 415
Key Stage 2 - 3	1500 x 600 x 400 mm	305 to 415
Key Stage 3 - 4	1500 x 600 x 450 mm	305 to 415
Hardwood double island seat; polyurethene finish; coat hooks; shoe trays; steel frame; epoxy paint finish.		
Key Stage 1 - 2	1500 x 600 x 350 mm	345 to 470
Key Stage 2 - 3	1500 x 600 x 400 mm	345 to 470
Key Stage 3 - 4	1500 x 600 x 450 mm	345 to 470
Combined locker and bench seat; hardwood seat; polyurethene finish; steel carcass/frame; epoxy paint finish.		
Stand/seat	305 x 610 x 762 mm	47 to 63
Stand/seat	305 x 762 x 762 mm	48 to 65
Stand/seat	610 x 610 x 762 mm	61 to 82
Stand/seat	610 x 762 x 762 mm	63 to 85
Stand/seat	915 x 610 x 762 mm	75 to 101
Stand/seat	915 x 762 x 762 mm	78 to 105
Stand/seat	1220 x 610 x 762 mm	89 to 120
Stand/seat	1220 x 762 x 762 mm	93 to 125
Welded tubular steelwork benching; 20 integral hooks; epoxy paint finish; solid ash bench slats; polished.		
Island seat	1500 x 500 x 1700 mm	250 to 340
bench seat (only)	1500 x 244 x 300 mm	85 to 115

A 1 - CLASSROOM AREA ARRANGEMENTS

Model = A typical area arrangement
CR = Classroom

Model 1

(Single bank of three classrooms with covered way)

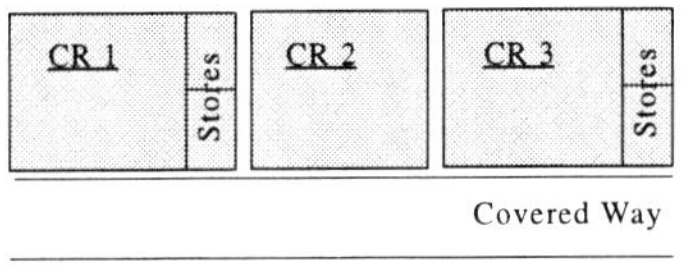

Model 2

(Single bank of five classrooms with covered way)

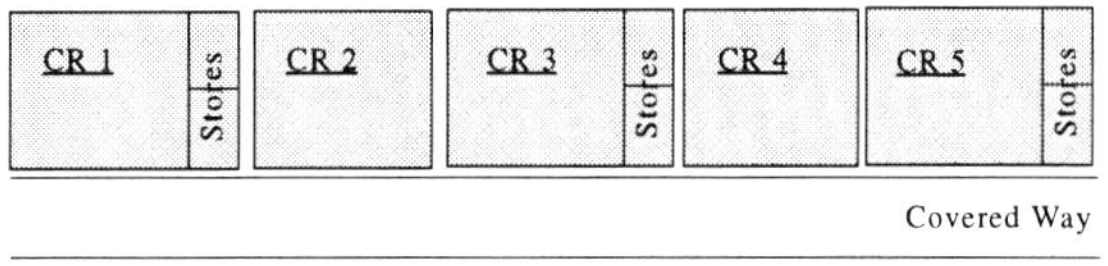

Model 3

(Double bank of six classrooms with corridor)

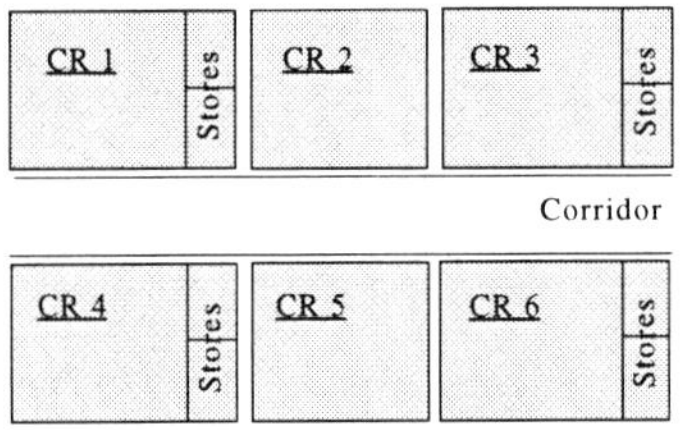

Model 4

(Double bank of ten classrooms with corridor)

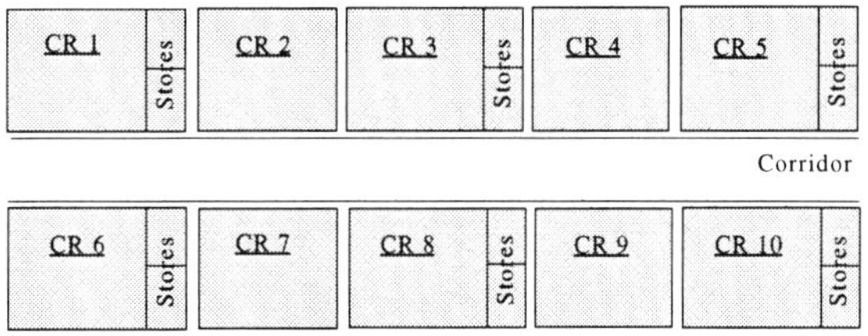

TABLE 'A' CLASSROOMS - SINGLE STOREY

School Type (a)	Model (b)	Area m2 (c)	Total Building Cost (£) New (d)	Total Building Cost (£) Extension (e)	Total Building Cost (£) Re-Modelled (f)	Operational Usage (Nos) Teaching Spaces (g)	School Meals (h)	Assembly Spaces (i)	Revenue & Entertainment (j)	Sport (k)	Seminar Spaces (l)
Primary	1	122	£99,850	£99,850	£99,850	49	NIL	41	49	NIL	61
	1	135	£110,500	£110,500	£110,500	54	NIL	45	54	NIL	68
	2	203	£155,325	£155,325	£155,325	82	NIL	68	82	NIL	102
	2	225	£172,150	£172,150	£172,150	90	NIL	75	90	NIL	113
Secondary	2	349	£267,000	£267,000	£267,000	140	NIL	117	140	NIL	175
	2	349	£267,000	£267,000	£267,000	140	NIL	117	140	NIL	175
	3	435	£309,575	£309,575	£309,575	174	NIL	145	174	NIL	218
	3	450	£320,250	£320,250	£320,250	180	NIL	150	180	NIL	225
	4	725	£477,275	£477,275	£477,275	290	NIL	242	290	NIL	363
	4	750	£493,725	£493,725	£493,725	300	NIL	250	300	NIL	375
Higher	3	450	£320,250	£320,250	£320,250	180	NIL	150	180	NIL	225
	3	460	£327,375	£327,375	£327,375	184	NIL	154	184	NIL	230
	4	750	£493,725	£493,725	£493,725	300	NIL	250	300	NIL	375
	4	767	£504,900	£504,900	£504,900	307	NIL	256	307	NIL	384

TABLE 'B' ANALYSIS OF COSTS

£ per Square Metre - Based upon Model 3

Elements of the Building (m)	New £ per m2 (n)	Extension £ per m2 (o)	Remodelled £ per m2 (p)
Main Building Work			
* Foundations	109.07	111.00	4.21
* Superstructure	315.07	252.66	134.19
Finishings			
* Floors	47.23	49.68	63.19
* Walls	28.66	29.86	24.58
* Ceilings	27.70	31.94	46.05
Fittings			
* Furniture	53.13	49.37	49.79
* Fittings	94.45	87.76	88.51
* Sanitary Fittings	0.00	0.00	0.00
* Light fittings	23.51	23.25	25.74
Security			
* Protection	2.70	8.34	8.58
* Surveillance	6.44	8.62	6.25
* Communication	3.67	6.23	5.70
FITTING OUT	**287.50** £ per m2	**295.04** £ per m2	**318.39** £ per m2
TOTAL COST	**711.65** £ per m2	**658.70** £ per m2	**456.78** £ per m2

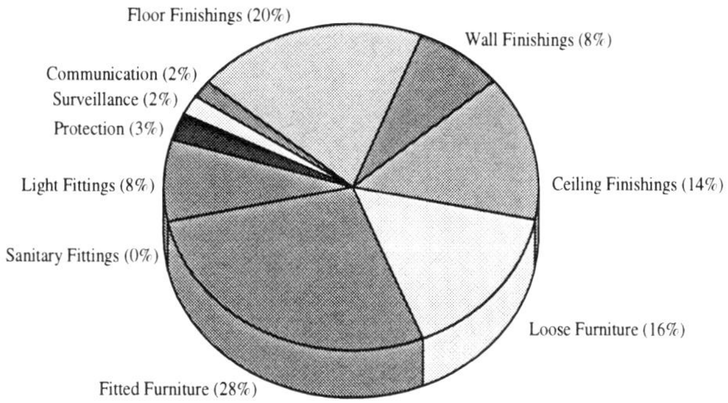

ANALYSIS OF FITTING OUT COSTS

(Costs based upon Table 'B' Column 'p')

A 2 - CLASSOOM AREA ARRANGEMENTS

Model = A typical area arrangement
CR = Classroom

Model 1

(Single bank of six classrooms with covered way)

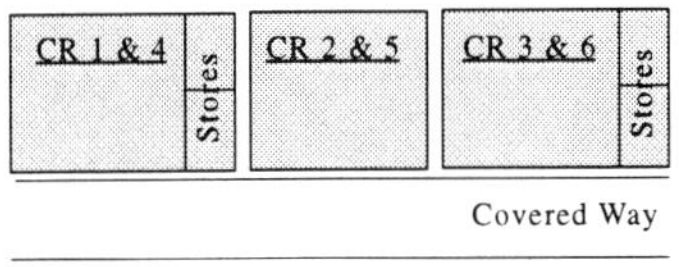

Model 2

(Single bank of ten classrooms with corridor)

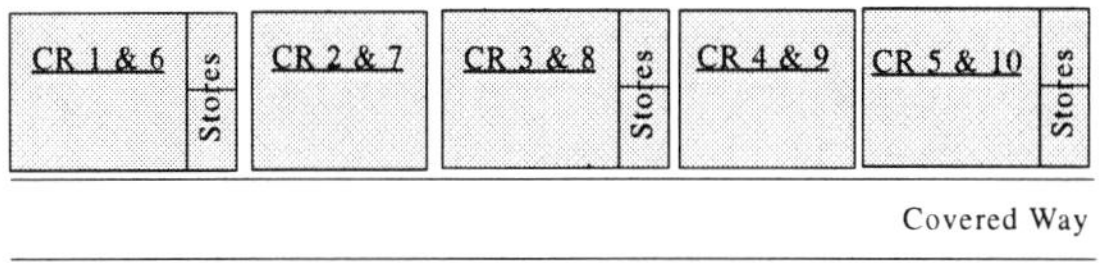

Model 3

(Double bank of twelve classrooms with corridor)

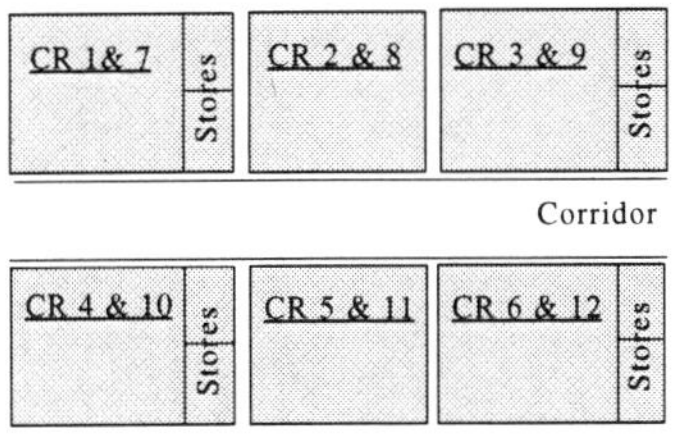

Model 4

(Double bank of twenty classrooms with corridor)

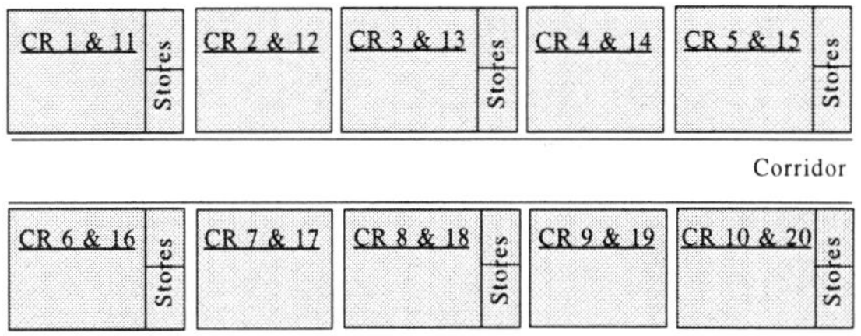

TABLE 'A' CLASSROOMS - TWO STOREY

School Type	Model	Area m2	Total Building Cost (£) New	Extension	Re-Modelled	Operational Usage (Nos) Teaching Spaces	School Meals	Assembly Spaces	Revenue & Entertainment	Sport	Seminar Spaces
(a)	(b)	(c)	(d)	(e)	(f)	(g)	(h)	(i)	(j)	(k)	(l)
Primary	1	229	178,900	164,800	119,600	92	NIL	77	92	NIL	115
	1	254	198,400	182,800	132,600	102	NIL	85	102	NIL	127
	2	381	278,100	256,300	185,900	153	NIL	127	153	NIL	191
	2	424	309,500	285,200	206,900	170	NIL	142	170	NIL	212
Secondary	2	656	478,900	441,200	320,100	263	NIL	219	263	NIL	328
	2	656	478,900	441,200	320,100	263	NIL	219	263	NIL	328
	3	819	556,100	512,400	371,800	328	NIL	273	328	NIL	410
	3	850	577,200	531,800	385,900	340	NIL	284	340	NIL	425
	4	1,364	856,700	789,300	572,700	546	NIL	455	546	NIL	682
	4	1,411	886,200	816,500	592,400	565	NIL	471	565	NIL	706
Higher	3	847	575,200	529,900	384,500	339	NIL	283	339	NIL	424
	3	866	588,100	541,800	393,100	347	NIL	289	347	NIL	433
	4	1,411	886,200	816,500	592,400	565	NIL	471	565	NIL	706
	4	722	453,500	417,800	303,200	289	NIL	241	289	NIL	361

TABLE 'B' ANALYSIS OF COSTS

£ per Square Metre - Based upon Model 3

Elements of the Building (m)	New £ per m2 (n)	Extension £ per m2 (o)	Remodelled £ per m2 (p)
Main Building Work			
* Foundations	57.14	57.60	2.41
* Superstructure	334.33	272.89	133.09
Finishings			
* Floors	47.23	49.68	63.19
* Walls	28.66	29.86	24.58
* Ceilings	27.70	31.94	46.05
Fittings			
* Furniture	53.13	49.37	49.79
* Fittings	94.45	87.76	88.51
* Sanitary Fittings	0.00	0.00	0.00
* Light fittings	23.51	23.25	25.74
Security			
* Protection	2.70	8.34	8.58
* Surveillance	6.44	8.62	6.25
* Communication	3.67	6.23	5.70
FITTING OUT	**287.50** £ per m2	**295.04** £ per m2	**318.39** £ per m2
TOTAL COST	**678.99** £ per m2	**625.53** £ per m2	**453.88** £ per m2

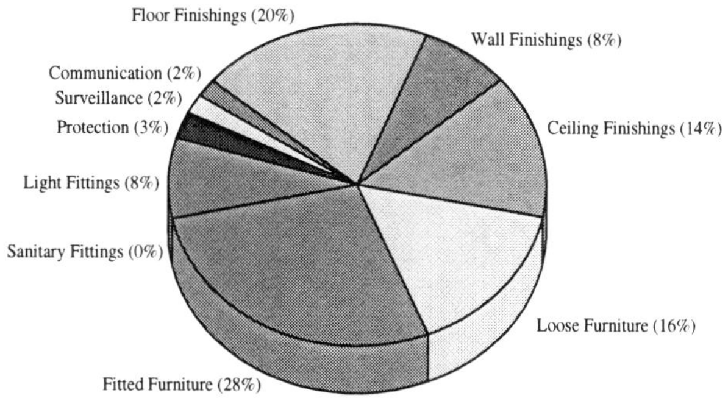

ANALYSIS OF FITTING OUT COSTS

(Costs based upon Table 'B' Column 'p')

CHAIRS

STANDARD; METAL FRAMED

Polypropylene seat; one piece plastic shell; black epoxy steel frame; stacking

	Width x depth x height	Price £
Key Stage 1	350 x 350 x 280 mm	6.4 to 9.4
Key Stage 2	375 x 360 x 320 mm	7 to 10.2
Key Stage 2	410 x 400 x 355 mm	7.2 to 10.8
Key Stage 3	470 x 500 x 390 mm	7.4 to 11.2
Key Stage 3	470 x 500 x 430 mm	7.6 to 11.4
Key Stage 4	470 x 500 x 430 mm	7.6 to 11.4

High impact seat; co-polymer polypropylene; one piece plastic shell; black epoxy steel frame; stacking

	Width x depth x height	Price £
Key Stage 1	364 x 362 x 295 mm	6.8 to 9.2
Key Stage 2	368 x 389 x 335 mm	7.2 to 9.8
Key Stage 2	426 x 440 x 350 mm	7.8 to 10.4
Key Stage 3	430 x 472 x 395 mm	8 to 10.8
Key Stage 3	482 x 506 x 405 mm	8.2 to 11.2
Key Stage 4	486 x 534 x 445 mm	8.4 to 11.4

Polypropylene seat; one piece plastic shell; black epoxy tubular skid frame; stacking

	Width x depth x height	Price £
Key Stage 1	350 x 350 x 280 mm	7.8 to 10.6
Key Stage 2	375 x 360 x 320 mm	8 to 10.8
Key Stage 2	410 x 400 x 355 mm	9 to 12.2
Key Stage 2 - 3	470 x 500 x 390 mm	9.4 to 12.6
Key Stage 3 - 4	470 x 500 x 430 mm	9.4 to 12.6

STANDARD; TIMBER FRAMED

Standard polished wood; plywood seat and back; laminated beech frame; linked; stacking

	Width x depth x height	Price £
Key Stage 1 - 2	355 x 290 x 320 mm	19 to 25.8
Key Stage 2	360 x 320 x 355 mm	20.2 to 27.4
Key Stage 2 - 3	410 x 390 x 390 mm	21.4 to 28.8
Key Stage 3 - 4	410 x 390 x 445 mm	22 to 29.6

Standard polished wood; plywood seat and back; upholstered seat; laminated beech frame; linked; stacking

	Width x depth x height	Price £
Key Stage 1 - 2	355 x 290 x 320 mm	25.2 to 34
Key Stage 2	360 x 320 x 355 mm	26.6 to 36
Key Stage 2 - 3	410 x 390 x 390 mm	28.2 to 38
Key Stage 3 - 4	410 x 390 x 445 mm	28.8 to 39

EASY CHAIRS

Upholstered easy chair; fire resistant; epoxy welded square tubular frame.

	Width x depth x height	Price £
Key Stage 1	380 x 320 x 290 mm	25.2 to 34.2
Key Stage 1	770 x 320 x 290 mm	48.4 to 65.4
Key Stage 1 - 2	440 x 370 x 350 mm	37.4 to 50.6
Key Stage 2	510 x 480 x 370 mm	45.6 to 61.6
Key Stage 2	520 x 480 x 370 mm	53.2 to 72

STOOLS

PLYWOOD SEAT

Stool; seat; epoxy tubular welded frame; four legs; stacking.

	Width x depth x height	Price £
Key Stage 1	470 mm height	11.2 to 15.2
Key Stage 2	520 mm height	11.2 to 15.2
Key Stage 2	570 mm height	11.4 to 15.4
Key Stage 3	620 mm height	11.6 to 15.6
Key Stage 4	670 mm height	11.6 to 15.8

POLYPROPYLENE SEAT

Stool; one piece shell; epoxy tubular welded frame; four legs; skid base; stacking.

	Width x depth x height	Price £
Key Stage 2 - 3	610 mm height	15.4 to 20.8
Key Stage 4	686 mm height	15.8 to 21.2

LECTURE CHAIRS

CHAIRS

Black or chrome frame; clip-on MDF tablet; upholstered seat and back; optional storage undertray.

	Width x depth x height	Price £
Black Frame		
Key Stage 3 - 4	525 x 570 x 465 mm	106 to 143
Chrome Frame		
Key Stage 3 - 4	525 x 570 x 465 mm	118 to 159

Samsonite neutral steel frame; permanent left or right hand tablet; finished in melamine; upholstered seat and back.

	Width x depth x height	Price £
Black Frame		
Key Stage 3 - 4	460 x 560 x 465 mm	56 to 76

TABLES

STACKING

Melamine top; bullnosed edging; steel framed legs; crush bend stacking; epoxy paint finish.

	Length x width x height	Price £
Rectangle		
Key Stage 1	1000 x 500 x 500 mm	25.2 to 34.1
Key Stage 1 - 2	1000 x 500 x 550 mm	25.2 to 34.1
Key Stage 2	1000 x 500 x 600 mm	25.2 to 34.1
Key Stage 2 - 3	1100 x 550 x 650 mm	29.2 to 39.5
Key Stage 3 - 4	1100 x 550 x 700 mm	29.3 to 39.7
Square		
Key Stage 1	1000 x 1000 x 500 mm	41.2 to 55.7
Key Stage 1 - 2	1000 x 1000 x 550 mm	42.2 to 57.1
Key Stage 2	1000 x 1000 x 600 mm	43.3 to 58.5
Key Stage 2 - 3	1000 x 1000 x 650 mm	44.4 to 60.1
Key Stage 3 - 4	1000 x 1000 x 700 mm	45.5 to 61.5
Trapezoidal		
Key Stage 1	1000 x 500 x 500 mm	25.5 to 34.4
Key Stage 1 - 2	1000 x 500 x 550 mm	25.5 to 34.4
Key Stage 2	1000 x 500 x 600 mm	25.5 to 34.4
Key Stage 2 - 3	1100 x 550 x 650 mm	29.2 to 39.5
Key Stage 3 - 4	1100 x 550 x 700 mm	29.7 to 40.1
Circular		
Key Stage 1	1000 dia x 500 mm	42.6 to 57.6
Key Stage 1 - 2	1000 dia x 550 mm	43.7 to 59.1
Key Stage 2	1000 dia x 600 mm	44.7 to 60.5
Examination; folding; melamine top; black tubular steel legs; epoxy paint finish.		
Key Stage 3 - 4	600 x 450 x 700 mm	36.9 to 49.9
Examination; melamine top; epoxy square steel legs; epoxy paint finish.		
Key Stage 3 - 4	600 x 550 x 720 mm	17.6 to 23.7

TABLES

NON STACKING

Melamine top; bullnosed edging; steel framed legs; epoxy paint finsih.

	Length x width x height	Price £
Square		
Key Stage 1	1100 x 1100 x 500 mm	48.4 to 65.4
Key Stage 1 - 2	1100 x 1100 x 550 mm	49.6 to 67.2
Key Stage 2	1100 x 1100 x 600 mm	50.8 to 68.8
Key Stage 2 - 3	1100 x 1100 x 650 mm	52.2 to 70.6
Key Stage 3 - 4	1100 x 1100 x 700 mm	53.4 to 72.4
Circular		
Key Stage 1	550 dia x 500 mm	26.2 to 35.6
Key Stage 1 - 2	550 dia x 550 mm	26.2 to 35.6
Key Stage 1	750 dia x 500 mm	36.4 to 49.2
Key Stage 1 - 2	750 dia x 550 mm	37.2 to 50.4
Key Stage 2	750 dia x 600 mm	38.2 to 51.6
Key Stage 2 - 3	750 dia x 650 mm	39.4 to 53.2
Key Stage 3 - 4	750 dia x 700 mm	40.2 to 54.4
Key Stage 1	1000 dia x 500 mm	43.8 to 59.2
Key Stage 1 - 2	1000 dia x 550 mm	43.8 to 59.2
Key Stage 2	1000 dia x 600 mm	43.8 to 59.2
Key Stage 2 - 3	1000 dia x 650 mm	43.8 to 59.2
Key Stage 3 - 4	1000 dia x 700 mm	43.8 to 59.2
Melamine top; colour range; polyurethene edging; steel framed legs; epoxy paint finish.		
Rectangular		
Key Stage 1	1200 x 600 x 500 mm	42 to 56.8
Key Stage 1 - 2	1200 x 600 x 550 mm	43 to 58.2
Key Stage 2	1200 x 600 x 600 mm	44.1 to 59.6
Key Stage 2 - 3	1200 x 600 x 650 mm	45.2 to 61.1
Key Stage 3 - 4	1200 x 600 x 700 mm	46.2 to 62.5
Trapezoidal		
Key Stage 1 - 2	1000 x trapzl x 550 mm	41.3 to 55.9
Key Stage 2	1000 x trapzl x 600 mm	42.3 to 57.3
Key Stage 2 - 3	1100 x trapzl x 650 mm	43.4 to 58.7
Key Stage 3 - 4	1100 x trapzl x 700 mm	44.5 to 60.1

DESKS

LOCKER DESKS

	Length x width x height	Price £
MDF carcass and locking lid; steel cantilever frame; epoxy paint finish; full width hinges.		
Key Stage 3 - 4	560 x 530 x 700 mm	46.4 to 62.8
Solid beech carcass and locking lid; steel cantilever frame epoxy paint finish.		
Key Stage 3 - 4	560 x 530 x 700 mm	72 to 97.4
MDF carcass, Oak veneer locking lid; hardwood frame; polished.		
Key Stage 1 - 4	610 x 445 x 700 mm	79 to 106.8
Stepped; deep back; shallow front storage; solid beech carcass; locking lid; steel edge; hardwood frame; polished.		
Key Stage 1 - 4	610 x 445 x 700 mm	105.3 to 142.4
Postura; Co-polymer polypropylene tilt action desk top; heavy duty steel frame; epoxy paint finish.		
Key Stage 1	615 x 595 x 485 mm	44.1 to 59.6
Key Stage 1 - 2	615 x 595 x 545 mm	44.1 to 59.6
Key Stage 2	615 x 595 x 570 mm	45.9 to 62.1
Key Stage 2 - 3	615 x 595 x 650 mm	49.6 to 67.1
Key Stage 3 - 4	615 x 595 x 720 mm	49.6 to 67.1

STORAGE - OPEN

Solid beech carcass; polished; static or mobile.

	Width x depth x height	Price £
Static 1 shelf	1040 x 496 x 590 mm	80 to 110
Mobile 1 shelf	1040 x 496 x 650 mm	95 to 125
Static 1 shelf	1040 x 496 x 750 mm	90 to 120
Mobile 1 shelf	1040 x 496 x 810 mm	100 to 135
Static 2 shelves	1040 x 496 x 590 mm	95 to 130
Mobile 2 shelves	1040 x 496 x 650 mm	105 to 140
Static 2 shelves	1040 x 496 x 750 mm	95 to 130
Mobile 2 shelves	1040 x 496 x 810 mm	110 to 145
Static 6 Drawers	1040 x 496 x 590 mm	195 to 260
Mobile 6 Drawers	1040 x 496 x 650 mm	205 to 275
Static 6 Drawers	1040 x 496 x 750 mm	225 to 305
Mobile 6 Drawers	1040 x 496 x 810 mm	235 to 320
Static 1 shelf	360 x 496 x 590 mm	55 to 75
Mobile 1 shelf	360 x 496 x 650 mm	70 to 90
Static 1 shelf	360 x 496 x 750 mm	65 to 85
Mobile 1 shelf	360 x 496 x 810 mm	75 to 100

Solid beech carcass; polished; static or mobile.

	Width x depth x height	Price £
3 shelves	1040 x 496 x 1455 mm	150 to 200
4 shelves	1040 x 496 x 1455 mm	160 to 215
3 shelves, 24 trays	1040 x 496 x 1455 mm	195 to 265
48 trays	1040 x 496 x 1455 mm	230 to 310
12 drawers, 8 trays	1040 x 496 x 1455 mm	365 to 495
3 shelves, 8 drawers	1040 x 496 x 1455 mm	290 to 395

STORAGE - CLOSED

Solid beech carcass; polished; static or mobile; locking.

	Width x depth x height	Price £
Static 1 shelf	1040 x 496 x 590 mm	135 to 180
Mobile 1 shelf	1040 x 496 x 650 mm	145 to 195
Static 1 shelf	1040 x 496 x 750 mm	140 to 190
Mobile 1 shelf	1040 x 496 x 810 mm	150 to 205
Static 2 shelves	1040 x 496 x 590 mm	145 to 195
Mobile 2 shelves	1040 x 496 x 650 mm	155 to 210
Static 2 shelves	1040 x 496 x 750 mm	150 to 200
Mobile 2 shelves	1040 x 496 x 810 mm	160 to 215
Static 6 Drawers	1040 x 496 x 590 mm	245 to 330
Mobile 6 Drawers	1040 x 496 x 650 mm	255 to 345
Static 6 Drawers	1040 x 496 x 750 mm	275 to 375
Mobile 6 Drawers	1040 x 496 x 810 mm	290 to 390
Static 1 shelf	360 x 496 x 590 mm	80 to 110
Mobile 1 shelf	360 x 496 x 650 mm	95 to 125
Static 1 shelf	360 x 496 x 750 mm	90 to 120
Mobile 1 shelf	360 x 496 x 810 mm	100 to 135

Solid beech carcass; polished; static or mobile; locking.

	Width x depth x height	Price £
3 shelves	1040 x 496 x 1455 mm	230 to 315
4 shelves	1040 x 496 x 1455 mm	240 to 325
3 shelves, 24 trays	1040 x 496 x 1455 mm	280 to 375
48 trays	1040 x 496 x 1455 mm	310 to 420
12 drawers, 8 trays	1040 x 496 x 1455 mm	455 to 615
3 shelves, 8 drawers	1040 x 496 x 1455 mm	375 to 505

STORAGE - PERSONAL

Epoxy coated steel frame; solid Beech slatting; 10 hook coat rail; 2 high shelves; 1 low shelf; seat perch.

	Width x depth x height	Price £
Base Unit		
Key Stage 1 - 2	1000 x 500 x 1420 mm	125 to 170
Key Stage 2	1000 x 500 x 1570 mm	145 to 195
Key Stage 3 - 4	1000 x 500 x 1820 mm	170 to 230
Add-on Unit		
Key Stage 1 - 2	975 x 500 x 1420 mm	125 to 170
Key Stage 2	975 x 500 x 1570 mm	145 to 195
Key Stage 3 - 4	975 x 500 x 1820 mm	170 to 230
Key Stage 1 - 2	500 x 500 x 1420 mm	125 to 170
Key Stage 2	500 x 500 x 1570 mm	145 to 195
Key Stage 3 - 4	500 x 500 x 1820 mm	170 to 230

Accessories

Corner shelves (3no.)	500 x 500 mm	125 to 170
Coat rail (10 Hook)	959 x 500 mm	145 to 195
Coat rail (5 Hook)	475 x 500 mm	170 to 230

Epoxy coated Grey steel frame; 18 compartment unit.

Bag storage	915 x 305 x 1830 mm	210 to 280

Single sided bench & coat unit, steel frame, harwood benching, 5 coat hooks & shoe tray.

Key Stage 1 - 2	1500 x 325 x 350 mm	270 to 365
Key Stage 3 - 4	1500 x 325 x 450 mm	270 to 365

PINBOARDS (1)

	Width x depth x height	Price £
Unframed	600 x 900 mm	17 to 22
Unframed	600 x 1200 mm	19 to 26
Unframed	900 x 1200 mm	26 to 34
Unframed	1200 x 1200 mm	31 to 41
Unframed	1200 x 1500 mm	37 to 49
Unframed	1200 x 1800 mm	42 to 57
Unframed	1200 x 2400 mm	54 to 73
Aluminium frame	600 x 900 mm	29 to 39
Aluminium frame	600 x 1200 mm	34 to 46
Aluminium frame	900 x 1200 mm	44 to 59
Aluminium frame	1200 x 1200 mm	52 to 70
Aluminium frame	1200 x 1500 mm	61 to 83
Aluminium frame	1200 x 1800 mm	70 to 94
Aluminium frame	1200 x 2400 mm	89 to 120
Unframed	600 x 900 mm	11 to 15
Unframed	600 x 1200 mm	12 to 16
Unframed	900 x 1200 mm	16 to 21
Unframed	1200 x 1200 mm	19 to 26
Unframed	1200 x 1500 mm	22 to 29
Unframed	1200 x 1800 mm	30 to 40
Unframed	1200 x 2400 mm	38 to 51
Aluminium frame	600 x 900 mm	16 to 21
Aluminium frame	600 x 1200 mm	18 to 24
Aluminium frame	900 x 1200 mm	22 to 30
Aluminium frame	1200 x 1200 mm	26 to 35
Aluminium frame	1200 x 1500 mm	29 to 39
Aluminium frame	1200 x 1800 mm	38 to 52
Aluminium frame	1200 x 2400 mm	48 to 64
Wood frame	600 x 900 mm	18 to 24
Wood frame	600 x 1200 mm	21 to 28
Wood frame	900 x 1200 mm	26 to 35
Wood frame	1200 x 1200 mm	30 to 40
Wood frame	1200 x 1500 mm	34 to 46
Wood frame	1200 x 1800 mm	44 to 60
Wood frame	1200 x 2400 mm	55 to 75

PINBOARDS (2)

	Width x depth x height	Price £
Aluminium frame	600 x 900 mm	24 to 32
Aluminium frame	900 x 1200 mm	35 to 47
Aluminium frame	1200 x 1200 mm	42 to 56
Aluminium frame	1200 x 1500 mm	47 to 63
Aluminium frame	1200 x 1800 mm	53 to 71
Aluminium frame	1200 x 2400 mm	69 to 93
Aluminium frame	600 x 900 mm	16 to 22
Aluminium frame	900 x 1200 mm	24 to 32
Aluminium frame	1200 x 1200 mm	27 to 37
Aluminium frame	1200 x 1500 mm	33 to 45
Aluminium frame	1200 x 1800 mm	43 to 57
Aluminium frame	1200 x 2400 mm	53 to 71
Aluminium frame	600 x 600 mm	16 to 21
Aluminium frame	600 x 900 mm	18 to 24
Aluminium frame	600 x 1200 mm	21 to 28
Aluminium frame	900 x 1200 mm	24 to 33
Aluminium frame	1200 x 1200 mm	29 to 39
Aluminium frame	1200 x 1500 mm	36 to 48
Aluminium frame, Hessian pinboard & HPL whiteboard combined, with central divider.		
White/pinboard	600 x 900 mm	43 to 59
White/pinboard	900 x 1200 mm	60 to 81
White/pinboard	1200 x 1800 mm	87 to 118

NOTICE BOARDS

	Width x depth x height	Price £
Felt covered; grooved.		
Open boards	450 x 600 mm	64 to 87
Open boards	600 x 900 mm	73 to 98
Open boards	1200 x 900 mm	85 to 114
Poster frame	A4	36 to 48
Poster frame	A3	45 to 60
Poster frame	A2	62 to 84
Poster frame	A1	89 to 120
Accessories		
Single brass		365 to 494
Single black		199 to 269
Double brass		664 to 898
Letters		
Alpha numeric	50mm	20 to 27
Alpha numeric	28mm	20 to 27
Alpha numeric	19mm	20 to 27
Felt Grooved; Gold framed; stand.		
	450 x 600 mm	177 to 240
	600 x 900 mm	207 to 280
Single /double toughened glass doors; lockable; felt covered.		
Wood cabinet	900 x 600 mm	79 to 107
Wood cabinet	900 x 1200 mm	114 to 154
Wood cabinet	1200 x 1200 mm	133 to 180
Wood cabinet	1000 x 1500 mm	152 to 206
Wood cabinet	1200 x 1800 mm	203 to 274
Aluminium frame	900 x 600 mm	127 to 171
Aluminium frame	1200 x 600 mm	150 to 202
Aluminium frame	1200 x 900 mm	177 to 240
Aluminium frame	1200 x 1200 mm	215 to 291
Aluminium frame	1500 x 1200 mm	253 to 342
Aluminium frame	1800 x 1200 mm	291 to 393

BLACKBOARDS

	Width x height	Price £
Standard; 2 year guarantee.		
Aluminium frame	900 x 600 mm	14 to 19
Aluminium frame	1200 x 900 mm	21 to 28
Aluminium frame	1200 x 1200 mm	24 to 33
Aluminium frame	1200 x 1500 mm	31 to 42
Aluminium frame	1200 x 1800 mm	36 to 48
Aluminium frame	1200 x 2400 mm	43 to 57
Wood frame	900 x 600 mm	17 to 23
Wood frame	1200 x 900 mm	24 to 33
Wood frame	1200 x 1200 mm	29 to 39
Wood frame	1200 x 1500 mm	36 to 48
Wood frame	1200 x 1800 mm	39 to 53
Wood frame	1200 x 2400 mm	50 to 68
Single sided laminate; 2 year guarantee.		
Aluminium frame	900 x 600 mm	18 to 24
Aluminium frame	1200 x 900 mm	29 to 39
Aluminium frame	1200 x 1200 mm	34 to 46
Aluminium frame	1200 x 1500 mm	41 to 55
Aluminium frame	1200 x 1800 mm	47 to 63
Aluminium frame	1200 x 2400 mm	58 to 78
Wood frame	900 x 600 mm	21 to 28
Wood frame	1200 x 900 mm	33 to 44
Wood frame	1200 x 1200 mm	38 to 52
Wood frame	1200 x 1500 mm	46 to 62
Wood frame	1200 x 1800 mm	53 to 71
Wood frame	1200 x 2400 mm	65 to 88
Single sided steel; magnetic; 5 year guarantee.		
Aluminium frame	900 x 600 mm	46 to 62
Aluminium frame	1200 x 900 mm	71 to 96
Aluminium frame	1200 x 1200 mm	88 to 119
Aluminium frame	1200 x 1500 mm	102 to 138
Aluminium frame	1200 x 1800 mm	121 to 163
Aluminium frame	1200 x 2400 mm	155 to 210

WHITE BOARDS

	Width x height	Price £
High pressure laminate plastic writing surface.		
Double sided	600 x 900 mm	21 to 28
Double sided	900 x 1200 mm	33 to 44
Double sided	1200 x 1200 mm	38 to 51
Double sided	1200 x 1500 mm	46 to 62
Double sided	1200 x 1800 mm	53 to 71
Double sided	1200 x 2400 mm	72 to 97
Steel surface; vitreous enamel finish; high density chipboard core.		
Single sided magnetic	600 x 900 mm	40 to 55
Single sided magnetic	900 x 1200 mm	55 to 75
Single sided magnetic	1200 x 1200 mm	69 to 93
Single sided magnetic	1200 x 1500 mm	89 to 120
Single sided magnetic	1200 x 1800 mm	97 to 131
Single sided magnetic	1200 x 2400 mm	112 to 152
Double sided magnetic	600 x 900 mm	85 to 114
Double sided magnetic	900 x 1200 mm	115 to 156
Double sided magnetic	1200 x 1200 mm	146 to 197
Double sided magnetic	1200 x 1500 mm	203 to 274
Double sided magnetic	1200 x 1800 mm	255 to 345
Double sided magnetic	1200 x 2400 mm	267 to 360
Accessories		
Self adhesive pen tray	300 mm	3 to 4
Self adhesive pen tray	500 mm	3 to 4
Self adhesive pen tray	900 mm	5 to 6
Aluminium pen tray	610mm	12 to 15
Standard board cleaner		7 to 9
- refill		5 to 7
Magnetic board cleaner		13 to 17
- refill		8 to 10
Mobile stands	900 mm	150 to 205
Mobile stands	1200 mm	170 to 230
Mobile stands	1500 mm	175 to 240
Mobile stands	1800 mm	185 to 250

MOBILE BOARDS

	Width x height	Price £
White Board; double sided steel; stand with castors.		
Non-magnetic	900 x 1200 mm	125 to 169
Non-magnetic	1200 x 900 mm	142 to 192
Non-magnetic	1200 x 1200 mm	162 to 219
Non-magnetic	1500 x 1200 mm	179 to 242
Non-magnetic	1800 x 1200 mm	189 to 256
Magnetic	900 x 1200 mm	196 to 265
Magnetic	1200 x 900 mm	216 to 292
Magnetic	1200 x 1200 mm	243 to 329
Magnetic	1500 x 1200 mm	267 to 360
Magnetic	1800 x 1200 mm	287 to 388
White Board; double sided; pinboard reverse side; stand with castors.		
Whiteboard/pinboard	900 x 1200 mm	189 to 256
Whiteboard/pinboard	1200 x 900 mm	199 to 269
Whiteboard/pinboard	1200 x 1200 mm	219 to 297
Whiteboard/pinboard	1500 x 1200 mm	240 to 324
Whiteboard/pinboard	1800 x 1200 mm	263 to 356
White Board; double sided; black or white board; stand with castors.		
Black/black	900 x 1200 mm	162 to 219
Black/black	1200 x 900 mm	182 to 247
Black/black	1200 x 1200 mm	199 to 269
Black/black	1500 x 1200 mm	223 to 301
Black/black	1800 x 1200 mm	256 to 347
Black/black	900 x 1200 mm	162 to 219
Black/black	1200 x 900 mm	182 to 247
Black/black	1200 x 1200 mm	203 to 274
Black/black	1500 x 1200 mm	230 to 310
Black/black	1800 x 1200 mm	263 to 356

ROLLER BOARDS - Free Standing

	Width x height	Price £
THREE PANEL		
MDF Frame		
Chalk/proj/dry wipe	1200 x 2250 mm	211 to 285
Chalk/proj/dry wipe	1500 x 2250 mm	245 to 331
Chalk/proj/dry wipe	1800 x 2250 mm	283 to 382
Chalk/proj/dry wipe	2100 x 2250 mm	333 to 450
Chalk/proj/dry wipe	2400 x 2250 mm	350 to 473
Chalk/proj/dry wipe	2500 x 2510 mm	378 to 511
Solid wood frame.		
Chalk/proj/dry wipe	1200 x 2250 mm	264 to 356
Chalk/proj/dry wipe	1500 x 2250 mm	306 to 413
Chalk/proj/dry wipe	1800 x 2250 mm	343 to 463
Chalk/proj/dry wipe	2100 x 2250 mm	379 to 513
Chalk/proj/dry wipe	2400 x 2250 mm	421 to 570
Chalk/proj/dry wipe	2500 x 2250 mm	527 to 712
Aluminium frame.		
Chalk/proj/dry wipe	1900 x 2300 mm	346 to 468
Chalk/proj/dry wipe	2500 x 2300 mm	405 to 548
FOUR PANEL		
MDF frame.		
Chalk/proj/dry wipe	1200 x 2250 mm	179 to 242
Chalk/proj/dry wipe	1500 x 2250 mm	203 to 274
Chalk/proj/dry wipe	1800 x 2250 mm	266 to 359
Chalk/proj/dry wipe	2100 x 2250 mm	341 to 462
Chalk/proj/dry wipe	2400 x 2250 mm	287 to 388
Chalk/proj/dry wipe	2500 x 2510 mm	369 to 499
Solid wood frame.		
Chalk/proj/dry wipe	1200 x 2250 mm	311 to 421
Chalk/proj/dry wipe	1500 x 2250 mm	353 to 477
Chalk/proj/dry wipe	1800 x 2250 mm	400 to 542
Chalk/proj/dry wipe	2100 x 2250 mm	443 to 599
Chalk/proj/dry wipe	2400 x 2250 mm	490 to 663
Chalk/proj/dry wipe	2500 x 2510 mm	540 to 731
Aluminium frame.		
Chalk/proj/dry wipe	1900 x 2500 mm	405 to 547
Chalk/proj/dry wipe	2500 x 2500 mm	471 to 637

ROLLER BOARDS - Mobile

	Width x height	Price £
THREE PANEL		
Solid wood frame.		
Black/green/dry wipe	1140 x 2130 mm	348 to 470
Black/green/dry wipe	1200 x 1950 mm	373 to 505
Black/green/dry wipe	1210 x 2340 mm	386 to 522
Black/green/dry wipe	1290 x 1900 mm	402 to 544
Black/green/dry wipe	1500 x 1950 mm	407 to 551
Black/green/dry wipe	1630 x 2340 mm	588 to 795
Black/green/dry wipe	1800 x 1950 mm	449 to 607
MDF frame.		
Black/green/dry wipe	1140 x 2130 mm	296 to 400
Black/green/dry wipe	1200 x 1950 mm	317 to 429
Black/green/dry wipe	1210 x 2340 mm	328 to 443
Black/green/dry wipe	1290 x 1900 mm	342 to 462
Black/green/dry wipe	1500 x 1950 mm	346 to 468
Black/green/dry wipe	1630 x 2340 mm	500 to 676
Black/green/dry wipe	1800 x 1950 mm	382 to 516

ROLLER BOARDS - MINI

Solid wood frame.		
3 panels	730 x 1320 mm	211 to 285
4 panels	730 x 1320 mm	230 to 310

SLIDING BOARDS - Wall Fixing

<u>Double sided chalk boards;
horizontal sliding.</u>

White/white	1200 x 900 mm	98 to 133
White/black	1200 x 900 mm	108 to 146
Black/black	1200 x 900 mm	112 to 151
Pair of wall channels	2440 mm	219 to 297
Pair of wall channels; Inset dry marker.	2440 mm	246 to 333

ROLLER BOARDS - Wall Fixing

	Width x height	Price £
TWO PANEL		
MDF frame.		
Chalk/proj/dry wipe	1200 x 1400 mm	195 to 264
Chalk/proj/dry wipe	1500 x 1400 mm	222 to 300
Chalk/proj/dry wipe	1800 x 1400 mm	248 to 335
Chalk/proj/dry wipe	2100 x 1400 mm	274 to 371
Chalk/proj/dry wipe	2400 x 1400 mm	306 to 413
Chalk/proj/dry wipe	2530 x 1250 mm	343 to 463
Solid wood frame.		
Chalk/proj/dry wipe	1200 x 1400 mm	232 to 314
Chalk/proj/dry wipe	1500 x 1400 mm	264 to 356
Chalk/proj/dry wipe	1800 x 1400 mm	295 to 399
Chalk/proj/dry wipe	2100 x 1400 mm	332 to 449
Chalk/proj/dry wipe	2400 x 1400 mm	364 to 492
Chalk/proj/dry wipe	2500 x 1400 mm	385 to 520
Aluminium frame.		
Chalk/proj/dry wipe	1900 x 1360 mm	272 to 367
Chalk/proj/dry wipe	2530 x 1250 mm	340 to 460
THREE PANEL		
MDF frame.		
Chalk/proj/dry wipe	1200 x 1400 mm	232 to 314
Chalk/proj/dry wipe	1500 x 1400 mm	269 to 364
Chalk/proj/dry wipe	1800 x 1400 mm	300 to 406
Chalk/proj/dry wipe	2100 x 1400 mm	332 to 449
Chalk/proj/dry wipe	2400 x 1400 mm	369 to 499
Chalk/proj/dry wipe	2530 x 1250 mm	364 to 492
Solid wood frame.		
Chalk/proj/dry wipe	1200 x 1400 mm	269 to 364
Chalk/proj/dry wipe	1500 x 1400 mm	306 to 413
Chalk/proj/dry wipe	1800 x 1400 mm	343 to 463
Chalk/proj/dry wipe	2100 x 1400 mm	379 to 513
Chalk/proj/dry wipe	2400 x 1400 mm	421 to 570
Chalk/proj/dry wipe	2530 x 1250 mm	364 to 492
Aluminium frame.		
Chalk/proj/dry wipe	1900 x 1360 mm	313 to 423
Chalk/proj/dry wipe	2530 x 1250 mm	381 to 515

A 3 - STUDY AREA ARRANGEMENTS

Model = A typical area arrangement

Model 1 (Group Study Rooms)

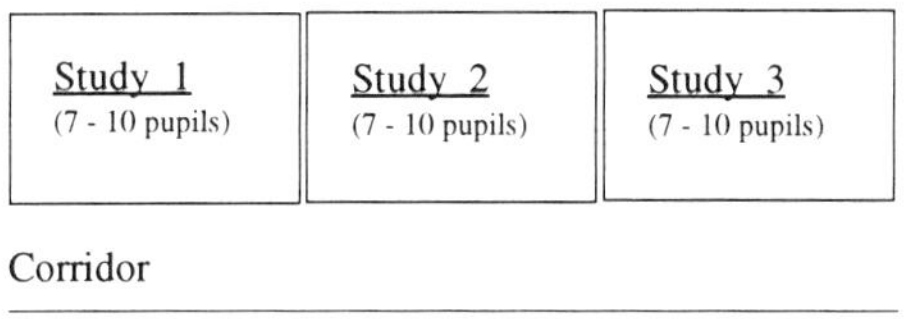

Model 2 (Lecture Rooms)

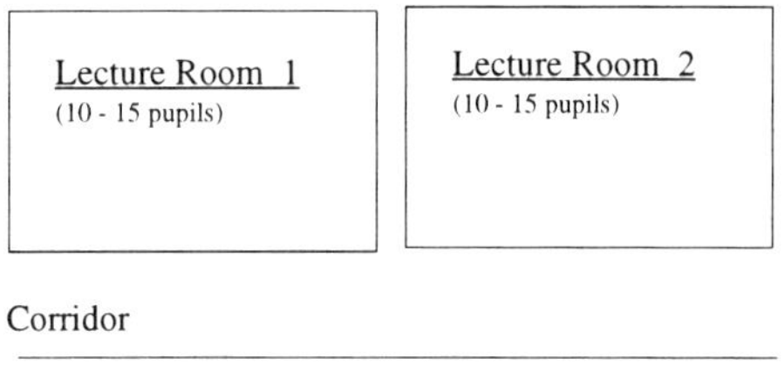

Model 3 (Open Learning Area)

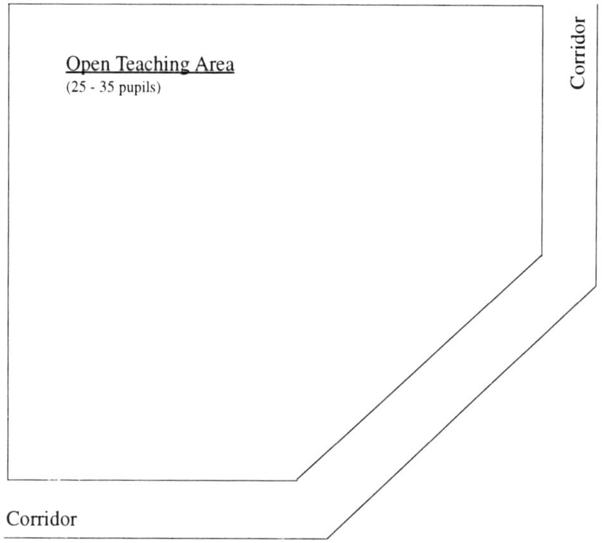

TABLE 'A' STUDY AREAS

School Type (a)	Model (b)	Area m2 (c)	Total Building Cost (£) New (d)	Extension (e)	Re-Modelled (f)	Operational Usage (Nos) Teaching Spaces (g)	School Meals (h)	Assembly Spaces (i)	Revenue & Entertainment (j)	Sport (k)	Seminar Spaces (l)
Primary	0	0	£0	£0	£0	NIL	NIL	NIL	NIL	NIL	NIL
	0	0	£0	£0	£0	NIL	NIL	NIL	NIL	NIL	NIL
	0	0	£0	£0	£0	NIL	NIL	NIL	NIL	NIL	NIL
	0	0	£0	£0	£0	NIL	NIL	NIL	NIL	NIL	NIL
Secondary	1	95	£70,950	£66,275	£45,550	38	48	64	32	NIL	95
	2	107	£77,125	£72,025	£49,500	43	54	72	36	NIL	107
	3	134	£96,575	£90,200	£62,000	54	67	90	45	NIL	134
	0	0	£0	£0	£0	NIL	NIL	NIL	NIL	NIL	NIL
	0	0	£0	£0	£0	NIL	NIL	NIL	NIL	NIL	NIL
	0	0	£0	£0	£0	NIL	NIL	NIL	NIL	NIL	NIL
Higher	1	109	£81,425	£76,025	£52,250	44	55	73	37	NIL	109
	2	123	£88,650	£82,800	£56,900	50	62	82	41	NIL	123
	3	154	£111,000	£103,650	£71,250	62	77	103	52	NIL	154
	0	0	£0	£0	£0	NIL	NIL	NIL	NIL	NIL	NIL

TABLE 'B' ANALYSIS OF COSTS

£ per Square Metre - Based upon Model 3

Elements of the Building (m)	New £ per m2 (n)	Extension £ per m2 (o)	Remodelled £ per m2 (p)
Main Building Work			
* Foundations	109.07	111.00	4.21
* Superstructure	329.24	272.11	136.82
Finishings			
* Floors	47.23	49.68	63.19
* Walls	26.05	27.15	22.35
* Ceilings	27.70	31.94	46.05
Fittings			
* Furniture	64.94	60.34	60.85
* Fittings	106.26	98.73	99.58
* Sanitary Fittings	0.00	0.00	0.00
* Light fittings	23.51	23.25	25.74
Security			
* Protection	2.70	8.34	8.58
* Surveillance	6.44	8.62	6.25
* Communication	3.67	6.23	5.70
FITTING OUT	308.51 £ per m2	314.26 £ per m2	338.28 £ per m2
TOTAL COST	746.83	697.38	479.32

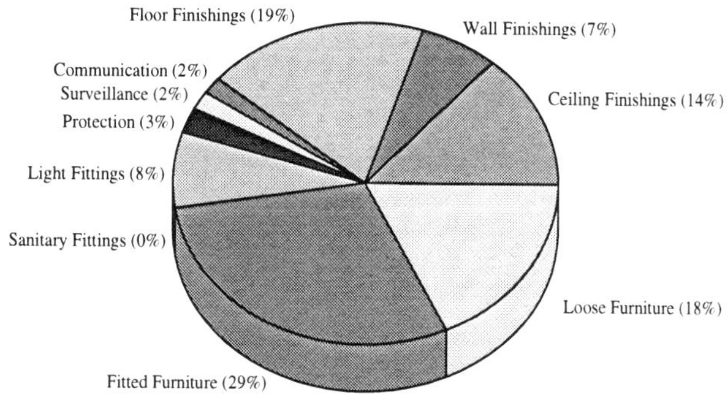

ANALYSIS OF FITTING OUT COSTS

(Costs based upon Table 'B' Column 'p')

WORK STATIONS

	Width x depth x height	Price £
18 mm Work top; beech, oak, or melamine work surface; modesty panels; pinboard backing; perforated side panels; steel frome epoxy paint finish.		
Standard unit		258 to 285
Unit; shelf		273 to 302
Unit; shelf; CPU holder		316 to 349
18 mm Work top; beech, oak, or melamine work surface; modesty panels; pinboard backing; perforated side panels; hardwood frame; polished.		
Table; low divider		
single sided	915 x 610 x 735 mm	255 to 281
double sided	915 x 1220 x 735 mm	317 to 350
Table; short divider		
single sided	915 x 610 x 735 mm	339 to 375
double sided	915 x 1220 x 735 mm	407 to 450
Table; high divider		
single sided	915 x 610 x 735 mm	367 to 406
double sided	915 x 1220 x 735 mm	434 to 480
MFC work top; melamine finish; xorner posts; inset strips; steel frame; epoxy paint finish.		
Standard unit	1000 x 865 x 1270 mm	248 to 274
Coloured strips (4)		26 to 29
Screen; full height; fabric covered; pinboards; accoustic efficient; laminate work surface.		
Single sided	1060 x 800 x 1200 mm	387 to 428
extension piece		290 to 321
Double sided	1060 x 1630 x 1200 mm	678 to 749
extension piece		475 to 525

STUDY CARRELS

	Width x depth x height	Price £
Hardwood worktop; side panels and shelf; solid hardwood lipping; polyurethene finish; steel frame; epoxy paint finish.		
Readers Carrels		
Single sided	732 x 1170 x 480 mm	271 to 300
extension piece		193 to 214
Double sided	1446 x 1170 x 480 mm	443 to 489
extension piece		301 to 332
IT/AV Carrels		
Single sided	732 x 1170 x 480 mm	271 to 300
extension piece		193 to 214
Double sided	1446 x 1170 x 480 mm	443 to 489
extension piece		301 to 332
Modular carrel cluster; hardwood worktop; side panels and shelves; linked; solid hardwood lipping; polyurethene finish; steel frame; epoxy paint finish.		
4 person; 12 dividers		965 to 1067
6 person; 18 dividers		1299 to 1436

SEATS

	Price £
Upholstered seat; polpropylene back and seat; steel frame; epoxy paint finish.	
Side chair	33 to 37
Upholstered seat; solid hardwood back and seat; polished.	
Side chair	77 to 85
Arm chair	80 to 89
Upholstered seat; solid hardwood back and seat; curved arms; polished.	
Side chair	108 to 119
Arm chair	115 to 127
Side chair; seat only	59 to 65
VDU seats; swivel; height adjustment; upholstered; steel frame; epoxy paint finish.	
Basic type	77 to 85
Standard type	93 to 102
High quality	122 to 135

A 4 - NUSERY SCHOOL AREA ARRANGEMENTS

Model 1 (20 - 30 Pupils)

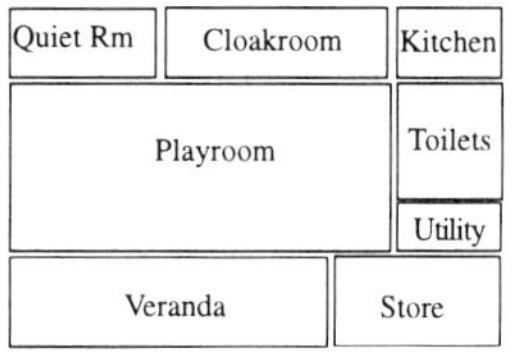

Model 2 (30 - 45 Pupils)

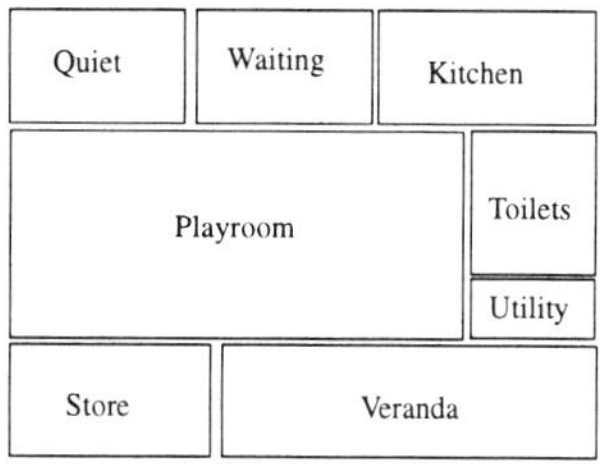

Model 3 (45 - 60 Pupils)

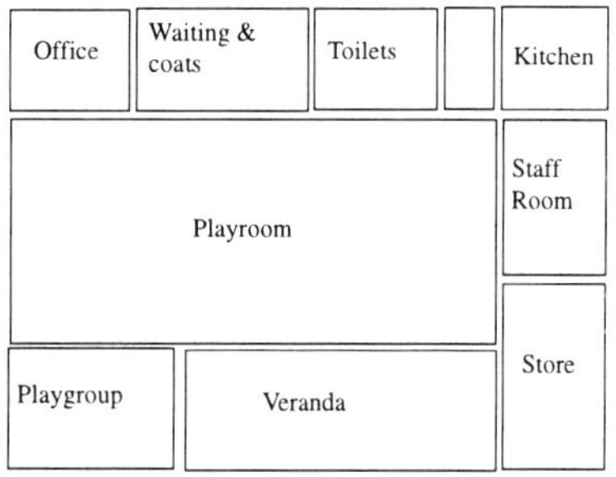

Model 4 (60 - 75 Pupils)

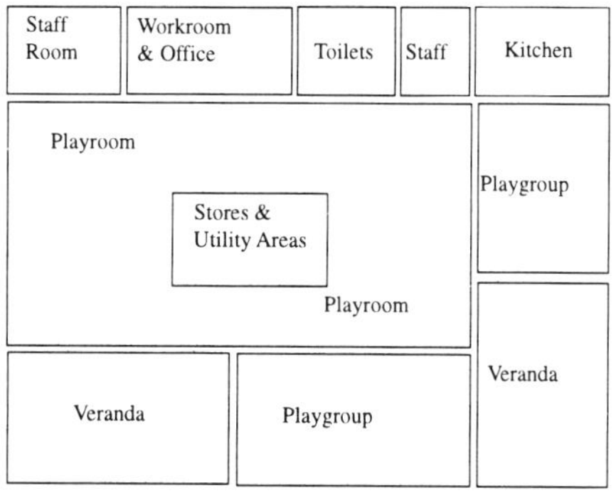

TABLE 'A' NURSERY AND PLAY SCHOOL

School Type (a)	Model (b)	Area m2 (c)	Total Building Cost (£) New (d)	Extension (e)	Re-Modelled (f)	Operational Usage (Nos) Teaching Spaces (g)	School Meals (h)	Assembly Spaces (i)	Revenue & Entertainment (j)	Sport (k)	Seminar Spaces (l)
Primary	1	81	56,700	55,800	30,200	33	41	54	27	17	81
	2	94	68,400	67,300	36,400	38	47	63	32	19	94
	3	136	101,300	99,700	53,900	55	68	91	46	28	136
	4	172	150,400	148,000	80,000	69	86	115	58	35	172
Secondary			NIL	NIL	NIL	NIL	NIL	NIL	NIL	NIL	NIL
			NIL	NIL	NIL	NIL	NIL	NIL	NIL	NIL	NIL
			NIL	NIL	NIL	NIL	NIL	NIL	NIL	NIL	NIL
			NIL	NIL	NIL	NIL	NIL	NIL	NIL	NIL	NIL
			NIL	NIL	NIL	NIL	NIL	NIL	NIL	NIL	NIL
			NIL	NIL	NIL	NIL	NIL	NIL	NIL	NIL	NIL
Higher			NIL	NIL	NIL	NIL	NIL	NIL	NIL	NIL	NIL
			NIL	NIL	NIL	NIL	NIL	NIL	NIL	NIL	NIL
			NIL	NIL	NIL	NIL	NIL	NIL	NIL	NIL	NIL
			NIL	NIL	NIL	NIL	NIL	NIL	NIL	NIL	NIL

TABLE 'B' ANALYSIS OF COSTS

£ per Square Metre - Based upon Model 3

Elements of the Building (m)	New £ per m2 (n)	Extension £ per m2 (o)	Remodelled £ per m2 (p)
Main Building Work			
* Foundations	88.76	82.35	2.36
* Superstructure	312.22	301.42	125.69
Finishings			
* Floors	52.34	52.34	58.36
* Walls	27.58	28.65	22.13
* Ceilings	26.55	26.82	32.88
Fittings			
* Furniture	55.54	55.54	36.66
* Fittings	76.89	76.89	50.75
* Sanitary Fittings	28.65	31.87	12.65
* Light fittings	18.65	20.22	18.64
Security			
* Protection	2.86	3.54	3.54
* Surveillance	7.22	6.55	6.23
* Communication	1.85	2.13	1.98
FITTING OUT	298.13 £ per m2	304.55 £ per m2	243.81 £ per m2
TOTAL COST	699.11 £ per m2	688.32 £ per m2	371.86 £ per m2

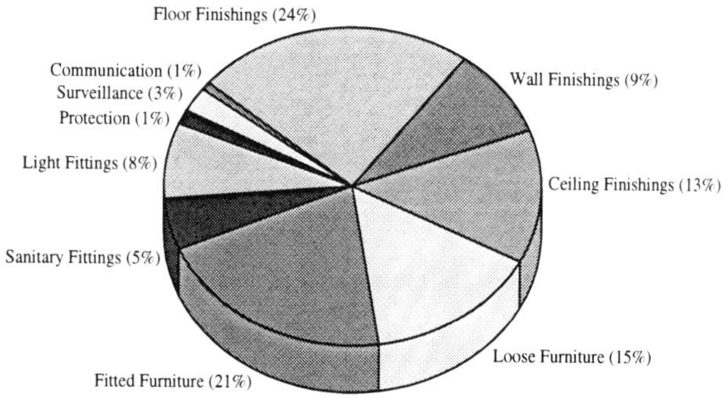

ANALYSIS OF FITTING OUT COSTS

(Costs based upon Table 'B' Column 'p')

CHAIRS

	Width x depth x height	Price £
STACKING		
Polypropylene seat; two piece plastic shell; steel frame; epoxy paint finish.		
Key Stage 1 - Classic	350 x 350 x 280 mm	6.6 to 9
Key Stage 1 - New	350 x 360 x 320 mm	7.4 to 10
Polypropylene seat; two piece plastic shell; fire retardent shell; steel frame; expoxy paint finish.		
Key Stage 1 - Classic	350 x 350 x 280 mm	7.6 to 10.2
Key Stage 1 - New	350 x 360 x 320 mm	8.4 to 11.4
Polypropylene seat; one piece plastic shell; anti static; steel frame; epoxy paint finish.		
Key Stage 1 - Classic	400 x 360 x 280 mm	8.4 to 11.4
Polypropylene seat; two piece plastic shell; coloured; heavy duty; steel frame; epoxy paint finish.		
Key stage 1	350 x 350 x 280 mm	9.2 to 12.6
Postura; one piece injection moulded co-polymer polypropylene back and seat; coloured.		
Key stage 1	364 x 362 x 295 mm	11.2 to 15.2
Polypropylene seat; colours; close stacking; steel frame; skid based; epoxy paint finish.		
Key stage 1	350 x 350 x 280 mm	8.6 to 11.6
Key stage 1 - 2	275 x 360 x 320 mm	8.8 to 12
Traditional; hardwood laminated frame; plywood seat; polished; linking.		
Key stage 1 - 2	355 x 290 x 320 mm	21 to 28.4
Ditto - choice colours	355 x 290 x 320 mm	21.4 to 29
Ditto - upholstered	355 x 290 x 320 mm	27.8 to 37.6

STOOLS & SUITES

	height	Price £
STOOLS		
Plywood seat; polished; stacking; steel frame; epoxy paint finish.		
Key stage 1	470 mm high	13 to 17
Polypropylene seat; stacking; steel tubular frame; epoxy paint finish.		
Key stage 1	430 mm high	16 to 22
Key stage 1 - 2	525 mm high	17 to 22
Polypropylene seat and back rest; stacking; steel frame; epoxy paint finish.		
Key stage 1	510 mm high	18 to 24
Key stage 1 - 2	560 mm high	18 to 24
Solid beech square seat ; stacking; beech legs; polished.		
Key stage 1	510 mm high	25 to 34
Key stage 1 - 2	560 mm high	26 to 34
SUITE FURNITURE		
Upholstered back and seat; laminated birch frame; polished.		
Chair	280 mm high	67 to 90
Sofa	280 mm high	88 to 119
Table	500 mm high	53 to 72
Upholstered back and seat; steel frame; epoxy paint finish.		
Chair	290 mm high	35 to 48
Sofa	290 mm high	67 to 91

BLOCK SEATING

	Width x depth x height	Price £
18 MONTH TO 3 YEARS		
Modular seating; 24kg/m3 foam; FR vinyl covers; Red, Green, Blue or Yellow block colours.		
Sofa	600 x 400 x 170 mm	34 to 45
Chair	300 x 400 x 170 mm	25 to 34
Corner seat	350 x 400 x 170 mm	26 to 35
Square pouffe	400 x 400 x 170 mm	20 to 27
Round pouffe	400 mm diameter	20 to 27
Wedge shaped pouffe	400 mm diameter	20 to 27
3 YEARS TO 6 YEARS		
Modular seating; 24kg/m3 foam; FR vinyl covers; Red, Green, Blue, Yellow or two colour combination.		
Sofa	700 x 480 x 250 mm	51 to 68
Chair	350 x 480 x 250 mm	33 to 44
Corner seat	440 x 480 x 250 mm	39 to 53
Square pouffe	480 x 480 x 250 mm	27 to 37
Rectangular pouffe	600 x 480 x 250 mm	33 to 45
Round pouffe	400 mm diameter	24 to 32
Wedge shaped pouffe	400 mm diameter	30 to 40

WORK STATION

	Width x depth x height	Price £
Mobile; Technology; epoxy coated steel frame (Grey) MDF laminate shelves; complete with plastic trays.		
Key Stage 1 - 2	1200 x 600 x 550 mm	134 to 181

BEAN BAG SEATING

		Price £
100 % Cotton cover; inner bead retaining bag; fire retardEnt; zipped base.		
Beanbag	3 cu ft	19 to 26
Bean chair	3 cu ft	24 to 32
50 % Cotton/ polyester cover; inner bead retaining lining fire retardent; zipped base.		
Beanbag	3 cu ft	18 to 24
Bean chair	3 cu ft	23 to 30
100 % Cotton corduroy cover; inner bead retaining bag; fire retardent; zipped base.		
Beanbag	3 cu ft	19 to 26
Beanbag suite; 50 % Cotton/ polyester cover; inner bead retaining lining; fire retardent; zipped base.		
Beanbag	3 cu ft	19 to 26
Bean chair	3 cu ft	24 to 32
Sofa	6 cu ft	35 to 48

BLOCK MODULES

	Price £
Activity Play, Block Modules; high grade foam with FR; waterproof polyester backed vinyl cover.	
Kit No. 1; Floor area required; 1.8 x 1.2 metres	
Mat (1)	
Cylinder (1)	
Bridge (2)	150 to 205
Kit No. 2; Floor area required; 2.7 x 1.2 metres	
Half circ;e (1)	
U shaped blocks (2)	
Trapezium (1)	235 to 315

PLAY EQUIPMENT

	Width x depth x height	Price £
Kit No. 3; Floor area required; 1.8 x 1.2 metres		
Long cubes (1)		
Mat (1)		
U shaped blocks (2)		190 to 255
Role Play Equipment		
MDF carcass; framed; coloured paint finish.		
Sink unit	400 x 370 x 600 mm	88 to 119
Washing machine	400 x 370 x 600 mm	80 to 108
Fridge	400 x 370 x 600 mm	88 to 119
Cooker	400 x 370 x 600 mm	89 to 120
Ironing board	600 x 170 x 520 mm	17 to 23
Mirror	600 x 460 x 1180 mm	73 to 99
Play theatre seat	770 x 560 x 1200 mm	102 to 138

BOOKCASES

	Width x depth x height	Price £
MDF carcass; coloured paint finish, Red, Blue, Yellow or any colour combination.		
1 shelf, no doors	915 x 300 x 760 mm	102 to 138
4 shelves, no doors	1000 x 330 x 1800 mm	158 to 213
4 shelves, 2 doors	1000 x 330 x 1800 mm	204 to 276
Folding, Mobile; MDF carcass; coloured paint finish, Blue with Red and Yellow shelves.		
2 shelves	915 x 490 x 810 mm	232 to 314

STORAGE

	Width x depth x height	Price £
BOOK TROLLEY		
Beech carcass; MFC laminate finish.		
Display - sloping	1040 x 475 x 910 mm	135 to 183
Display - 2 sided	1040 x 475 x 910 mm	135 to 183
Book bin	1040 x 475 x 590 mm	109 to 147
Book bin	1040 x 475 x 650 mm	109 to 147
Light Oak carcass; laminate finish.		
Display unit	1015 x 450 x 820 mm	190 to 257
Book bin	1015 x 450 x 820 mm	230 to 311
STORAGE - GENERAL		
Beech melamine carcass; MDF doors finished in red, blue or beech, choice of adjustable shelves or trays.		
Static	700 x 496 x 590 mm	142 to 192
Mobile	700 x 496 x 650 mm	148 to 199
Static	700 x 496 x 750 mm	147 to 198
Mobile	700 x 496 x 810 mm	153 to 207
Static	1040 x 496 x 590 mm	158 to 213
Mobile	1040 x 496 x 650 mm	165 to 223
Static	1040 x 496 x 750 mm	164 to 222
Mobile	1040 x 496 x 810 mm	172 to 232
Beech melamine carcass; MDF doors finished in red, blue or beech, adjustable shelves or trays with divider.		
Static	700 x 496 x 590 mm	147 to 198
Mobile	700 x 496 x 650 mm	154 to 208
Static	700 x 496 x 750 mm	152 to 206
Mobile	700 x 496 x 810 mm	161 to 217
Static	1040 x 496 x 590 mm	174 to 235
Mobile	1040 x 496 x 650 mm	181 to 245
Static	1040 x 496 x 750 mm	180 to 243
Mobile	1040 x 496 x 810 mm	188 to 255

STORAGE

	Width x depth x height	Price £
Beech melamine carcass; MDF doors finished in red, blue or beech, combination of adjustable shelves and trays.		
Static	700 x 496 x 590 mm	161 to 218
Mobile	700 x 496 x 650 mm	168 to 227
Static	700 x 496 x 750 mm	171 to 231
Mobile	700 x 496 x 810 mm	178 to 241
Static	1040 x 496 x 590 mm	200 to 270
Mobile	1040 x 496 x 650 mm	207 to 280
Static	1040 x 496 x 750 mm	210 to 284
Mobile	1040 x 496 x 810 mm	219 to 296
Mobile tray units; MDF carcass, light Oak laminate top, rounded edges, complete with trays of any colour		
6 Tray	360 x 495 x 650 mm	46 to 62
12 Tray	700 x 495 x 650 mm	92 to 124
18 Tray	1030 x 495 x 650 mm	139 to 188
Mobile tray units; blue MDF carcass, red MDF top, rounded edges, complete with trays of any colour combination.		
6 Tray	360 x 495 x 650 mm	53 to 71
12 Tray	700 x 495 x 650 mm	98 to 133
18 Tray	1030 x 495 x 650 mm	142 to 192

BOARDS

Drywipe White board - Laminate surface with grid lines and clip-on pen tray, aluminium or plastic frame.		
White board	900 x 600 mm	26 to 34
White board	1200 x 900 mm	40 to 54
White board	1800 x 1200 mm	70 to 94
White board	2400 x 1200 mm	88 to 119
Roller board - Revolving surface on one side, fixed surface on the reverse, wood frame, six colours.		
Chalk surface	740 x 1325 mm	186 to 251
Dry wipe surface	740 x 1325 mm	204 to 276

TEACHING AIDS

	Width x height	Price £
Mobile board; double sided; height adjustable; steel tubular frame finished in red, green, blue or yellow.		
Chalk/chalk	900 x 1150 mm	134 to 182
Dry wipe/dry wipe	900 x 1150 mm	134 to 182
Chalk/dry wipe	900 x 1150 mm	134 to 182
Easel; solid ash frame; two sided; integral pen tray.		
Chalk/chalk	600 x 1130 mm	62 to 83
Dry wipe/dry wipe	600 x 1130 mm	62 to 83
Chalk/dry wipe	600 x 1130 mm	62 to 83

TABLES

	Length x width x height	Price £
NON STACKING		
Work table; red or blue; melamine or natural MDF top; hardwood frame; lacquered.		
Key Stage 1	1220 x 760 x 500 mm	89 to 120
Key Stage 1 - 2	1220 x 760 x 550 mm	89 to 120
Work table; laminated birch top; curved hardwood frame; lacquered.		
Rectangular	800 x 600 x 500 mm	38 to 51
Round	600 dia x 500 mm	38 to 51
Work table; laminated birch frame and top; bullnosed MDF edging.		
Key Stage 1	1000 x 500 x 500 mm	26 to 34

TABLES

	Length x width x height	Price £
STACKING		
General table; MDF teak laminated top; bullnose edge; steel tubular frame; epoxy paint finish.		
Square		
Key Stage 1	1100 x 1100 x 500mm	54 to 72
Key Stage 1 - 2	1100 x 1100 x 550mm	54 to 72
Circular		
Key Stage 1	550 dia x 500 mm	29 to 40
Key Stage 1 - 2	550 dia x 550 mm	29 to 40
Key Stage 1	750 dia x 500 mm	40 to 54
Key Stage 1 - 2	750 dia x 550 mm	40 to 54
Key Stage 1	1000 dia x 500 mm	49 to 66
Key Stage 1 - 2	1000 dia x 550 mm	49 to 66
General table; MDF grey laminate top; colour co-ordinated edges; steel tubular frame; epoxy paint finish.		
Square	600 x 600 x 500 mm	35 to 47
Rectangle	1200 x 600 x 500 mm	52 to 70
Trapezodial	1200 x 520 x 500 mm	50 to 67
General table; MDF laminate top; bullnose edge; crush-bend steel tubular frame; epoxy paint finish.		
Key Stage 1	1000 x 500 x 500 mm	28 to 38
Key Stage 1 - 2	1000 x 500 x 550 mm	28 to 38
LOCKER DESKS		
Standard locker desk; hardwood frame; light oak veneer top; MDF bullnose edge; polished.		
Key Stage 1	610 x 445 x 500 mm	88 to 119
Stepped locker desk; deep back; shallow front; hardwood frame; light oak veneer top; MDF bullnose edge; polished.		
Key Stage 1	610 x 445 x 500 mm	117 to 158

T 5 TECHNOLOGY ROOM AREA ARRANGEMENTS

Model 1

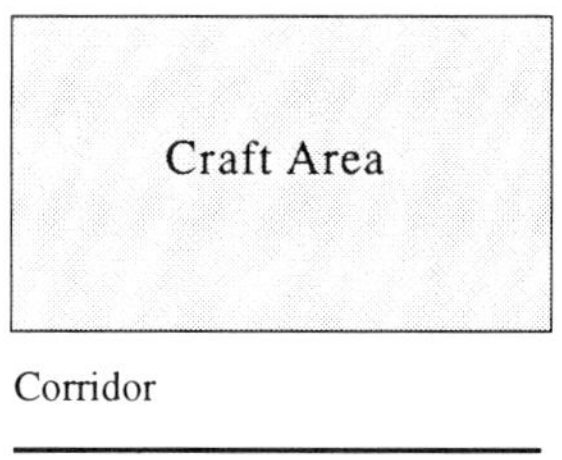

Model 2

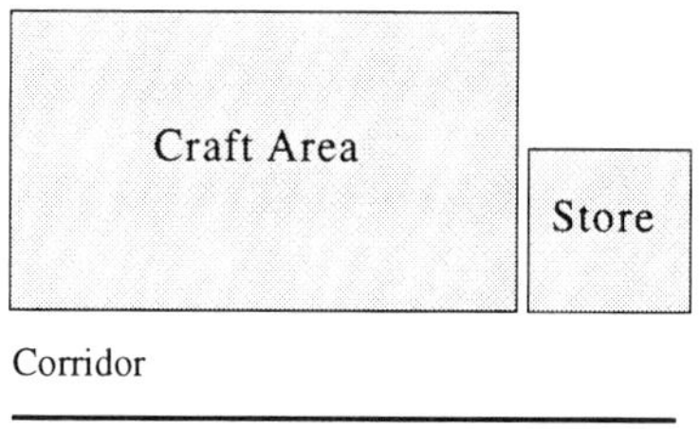

Model 3

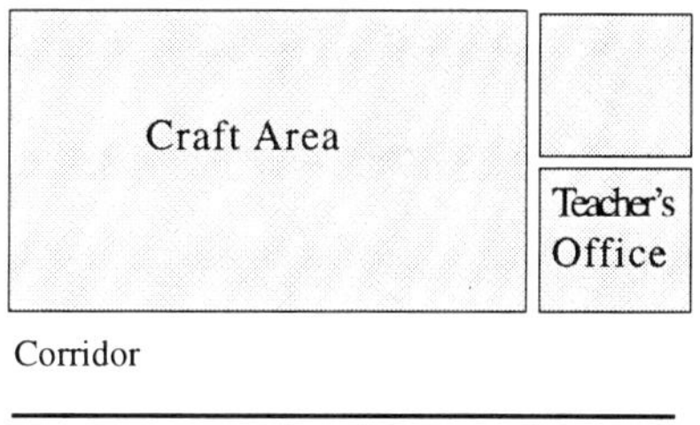

Model 4

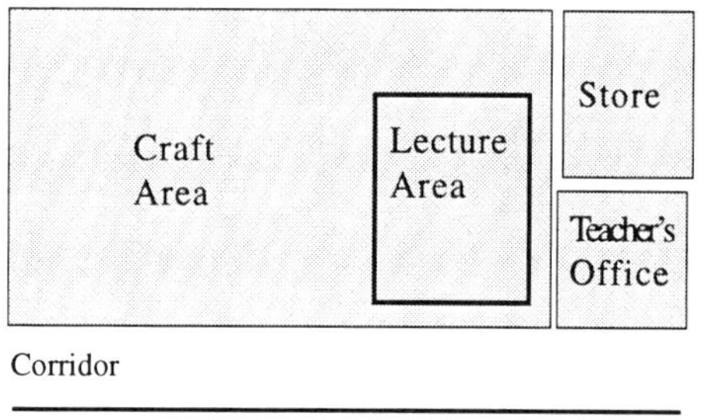

TABLE 'A' **TECHNOLOGY ROOM - SINGLE STOREY**

School Type (a)	Model (b)	Area m2 (c)	Total Building Cost (£) New (d)	Extension (e)	Re-Modelled (f)	Operational Usage (Nos) Teaching Spaces (g)	School Meals (h)	Assembly Spaces (i)	Revenue & Entertainment (j)	Sport (k)	Seminar Spaces (l)
Primary	1	39	28,300	26,450	18,600	16	NIL	26	13	NIL	26
	1	41	29,750	27,800	19,550	17	NIL	28	14	NIL	28
	2	50	36,275	33,900	23,850	20	NIL	34	17	NIL	34
	2	59	42,825	40,000	28,150	24	NIL	40	20	NIL	40
Secondary	2	54	39,200	36,625	25,750	22	NIL	36	18	NIL	36
	2	63	45,725	42,725	30,050	26	NIL	42	21	NIL	42
	3	70	50,625	47,300	33,275	28	NIL	47	24	NIL	47
	3	<u>78</u>	56,600	52,875	37,200	32	NIL	52	26	NIL	52
	4	80	57,950	54,150	38,100	32	NIL	54	27	NIL	54
	4	89	64,750	60,500	42,550	36	NIL	60	30	NIL	60
Higher	3	83	59,950	56,025	39,400	34	NIL	56	28	NIL	56
	3	92	66,750	62,375	43,875	37	NIL	62	31	NIL	62
	4	93	67,575	63,150	44,400	38	NIL	63	32	NIL	63
	4	104	75,200	70,250	49,425	42	NIL	70	35	NIL	70

TABLE 'B' **ANALYSIS OF COSTS**
£ per Square Metre - Based upon Model 3

Elements of the Building (m)	New £ per m2 (n)	Extension £ per m2 (o)	Remodelled £ per m2 (p)
Main Building Work			
* Foundations	109.07	111.00	4.21
* Superstructure	323.97	262.82	146.79
Finishings			
* Floors	47.23	49.68	63.19
* Walls	28.66	29.86	24.58
* Ceilings	27.70	31.94	46.05
Fittings			
* Furniture	53.13	49.37	49.79
* Fittings	94.45	87.76	88.51
* Sanitary Fittings	7.88	8.99	11.16
* Light fittings	23.51	23.25	25.74
Security			
* Protection	2.70	8.34	8.58
* Surveillance	6.44	8.62	6.25
* Communication	3.67	6.23	5.70
FITTING OUT	**295.38** £ per m2	**304.03** £ per m2	**329.55** £ per m2
TOTAL COST	**728.43** £ per m2	**677.86** £ per m2	**480.55** £ per m2

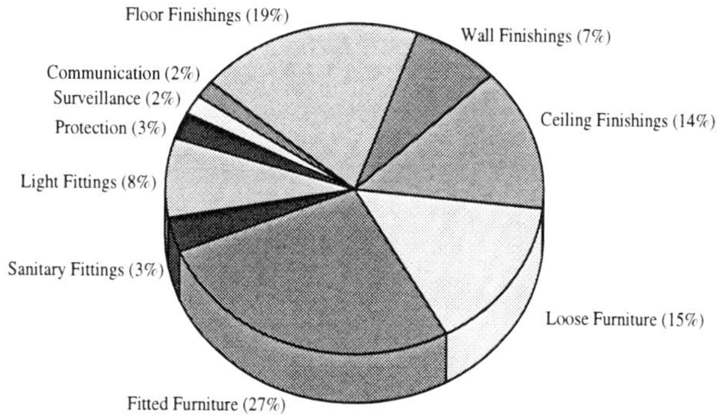

ANALYSIS OF FITTING OUT COSTS
(Costs based upon Table 'B' Column 'p').

WORK UNITS

WALL UNITS

MDF carcass glued and screwed;
steel frames and legs; epoxy paint finish.

	Width x depth x height	Price £
Work surfaces;		
Laminate chipboard	1500 x 600 x 900 mm	105 to 140
Laminate chipboard	1500 x 750 x 900 mm	110 to 150
Solid cored laminate	1500 x 600 x 900 mm	145 to 195
Solid cored laminate	1500 x 750 x 900 mm	160 to 220
Corian	1500 x 600 x 900 mm	220 to 300
Corian	1500 x 750 x 900 mm	245 to 335

MODULAR UNITS

MDF carcass glued and screwed; painted;
optional cupboards under;
steel frames and legs; epoxy paint finish.

Work surfaces;	Octagonal	
Laminate chipboard	1800 dia x 900 mm	450 to 605
Solid cored laminate	1800 dia x 900 mm	610 to 825
Corian	1800 dia x 900 mm	880 to 1190

UNDER BENCH UNITS

MDF carcass; glued and screwed; painted;
plain doors; painted;
aluminium ironmongery; laminate covered shelves.

Single pedestal	500 x 500 x 850 mm	140 to 190
Double pedestal	1000 x 500 x 850 mm	180 to 240
Tray unit	1000 x 500 x 850 mm	190 to 255
Sink unit (excl sink)	1200 x 500 x 850 mm	330 to 445
Sink unit (incl sink)	1200 x 500 x 850 mm	380 to 515

KITCHEN STORAGE UNITS

UNDER BENCH CUPBOARD UNITS

MDF carcass; glued and screwed; laminate internally;
laminate panelled doors;
aluminium ironmongery; laminate covered shelves.

	Width x depth x height	Price £
Single pedestal	300 x 500 x 850 mm	75 to 100
Single pedestal	500 x 500 x 850 mm	80 to 110
Single pedestal	600 x 500 x 850 mm	90 to 120
Double pedestal	900 x 500 x 850 mm	110 to 145
Double pedestal	1000 x 500 x 850 mm	115 to 155
Double pedestal	1200 x 500 x 850 mm	120 to 165

MDF carcass; glued and screwed; laminate internally;
hardwood panelled doors; painted;
aluminium ironmongery; laminate covered shelves.

Single pedestal	300 x 500 x 850 mm	90 to 120
Single pedestal	500 x 500 x 850 mm	95 to 130
Single pedestal	600 x 500 x 850 mm	105 to 140
Double pedestal	900 x 500 x 850 mm	130 to 175
Double pedestal	1000 x 500 x 850 mm	140 to 185
Double pedestal	1200 x 500 x 850 mm	145 to 195

TABLES

Work top; steel frame and legs; epoxy paint finish.

Work surface;		
Laminate chipboard	1200 x 600 x 900 mm	110 to 150
Solid cored laminate	1200 x 600 x 900 mm	135 to 185
Corian	1200 x 600 x 900 mm	245 to 335

WALL CUPBOARD UNITS

MDF carcass; glued and screwed; laminate internally;
laminate panelled doors;
aluminium ironmongery; laminate covered shelves.

	Width x depth x height	Price £
Single	300 x 300 x 570 mm	85 to 115
Single	500 x 300 x 570 mm	90 to 125
Double	900 x 300 x 570 mm	115 to 155
Double	1000 x 300 x 570 mm	120 to 165

MDF carcass; glued and screwed; laminate internally;
hardwood panelled doors; painted;
aluminium ironmongery; laminate covered shelves.

Single	300 x 300 x 570 mm	105 to 140
Single	500 x 300 x 570 mm	110 to 150
Double	900 x 300 x 570 mm	130 to 175
Double	1000 x 300 x 570 mm	135 to 185

MDF carcass; glued and screwed; painted;
glass fronted cupboard doors;
aluminium ironmongery; laminate covered shelves.

Single	300 x 300 x 570 mm	115 to 155
Single	500 x 300 x 570 mm	125 to 165
Double	900 x 300 x 570 mm	145 to 195
Double	1000 x 300 x 570 mm	150 to 200

WORK TABLES

	Length x width x height	Price £
NON STACKING		
22mm Emirite top.		
Work Table	1800 x 900 x 800mm	360 to 490
Work Table	1800 x 1200 x 800mm	450 to 610
Work Table	2400 x 1200 x 800mm	595 to 805
Work Table	3000 x 1200 x 800mm	735 to 995
44mm Emirite top.		
Work Table	1800 x 900 x 800mm	400 to 535
Work Table	1800 x 1200 x 800mm	495 to 670
Work Table	2400 x 1200 x 800mm	665 to 900
Work Table	3000 x 1200 x 800mm	835 to 1130
44mm Emirite top; timber edging.		
Work Table	1800 x 900 x 800mm	475 to 640
Work Table	1800 x 1200 x 800mm	595 to 800
Work Table	2400 x 1200 x 800mm	790 to 1070
Work Table	3000 x 1200 x 800mm	990 to 1340
Flush wood top; not suitable for woodwork.		
Plain Bench	1800 x 900 x 800mm	675 to 915
Plain Bench	1800 x 1200 x 800mm	780 to 1055
Plain Bench	2400 x 1200 x 800mm	915 to 1235
Plain Bench	3000 x 1200 x 800mm	1065 to 1435
Flush wood top; linoleum finish; not suitable for woodwork.		
Plain Bench	1800 x 900 x 800mm	255 to 345
Plain Bench	1800 x 1200 x 800mm	305 to 415
Plain Bench	2400 x 1200 x 800mm	345 to 470
Plain Bench	3000 x 900 x 800mm	335 to 455
Plain Bench	3000 x 1200 x 800mm	385 to 525
Plain Bench	3000 x 1500 x 800mm	440 to 595

WORK BENCHES (1)

	Length x width x height	Price £
SINGLE SIDED		
Adjustable wooden planing tops.		
Key Stage 3 - 4	1500 x 600 mm	325 to 440
Key Stage 3 - 4	1800 x 600 mm	375 to 510
Adjustable planing top; tail vice; pair of steel dogs.		
Key Stage 3 - 4	1500 x 600 mm	580 to 785
Key Stage 3 - 4	1800 x 600 mm	655 to 880
Adjustable planing top; tail vice; pair of steel dogs; double cupboard under; open shelf.		
Key Stage 3 - 4	1500 x 600 mm	835 to 1125
Key Stage 3 - 4	1800 x 600 mm	910 to 1230
Adjustable wooden planing tops; 2 tail vices; surface cramping system; 2 pairs steel dogs.		
Key Stage 3 - 4	1500 x 760 mm	865 to 1170
Cabinet making; tail vice; 4" drawer; pair of steel dogs.		
Key Stage 3 - 4	1500 x 600 mm	665 to 895
Key Stage 3 - 4	1800 x 600 mm	760 to 1025
Teachers bench; beech; well at rear of top; tail vice; pair of steel dogs; cupboard and nest of 4 drawers.		
Instructor	1500 x 600 mm	930 to 1255
Instructor	1800 x 600 mm	1070 to 1450

WORK BENCHES (2)

	Length x width x height	Price £
DUAL SIDED		
Adjustable wooden planing tops.		
Key Stage 1	900 x 600 mm	159 to 215
Key Stage 2 - 3	1200 x 600 mm	218 to 295
Key Stage 3 - 4	1200 x 600 mm	303 to 410
Key Stage 3 - 4	1500 x 760 mm	371 to 502
Adjustable wooden planing tops; two end cupboards.		
Key Stage 3 - 4	1500 x 760 mm	544 to 735
Adjustable wooden planing tops; shelves; two end cupboards.		
Key Stage 3 - 4	1500 x 760 mm	567 to 767
Adjustable wooden planing tops; 2 tail vices; surface cramping system; 2 pairs steel dogs; double cupboard; open shelves.		
Key Stage 3 - 4	1500 x 760 mm	1115 to 1509
TECHNICAL BENCHES		
Bench; linoleum covered top.		
Standard	1200 x 600 x 800 mm	158 to 214
Standard	1200 x 760 x 800 mm	188 to 254
Standard	1800 x 600 x 800 mm	214 to 289
Standard	1800 x 760 x 800 mm	243 to 329
Standard	2400 x 600 x 800 mm	314 to 424
Standard	2400 x 760 x 800 mm	354 to 479
With drawer nest	1800 x 600 x 800 mm	386 to 522
With drawer nest	1800 x 760 x 800 mm	426 to 576
With drawer nest	2400 x 600 x 800 mm	486 to 657
With drawer nest	2400 x 760 x 800 mm	537 to 727
With cupboard unit	1800 x 600 x 800 mm	332 to 449
With cupboard unit	1800 x 760 x 800 mm	372 to 503
With cupboard unit	2400 x 600 x 800 mm	432 to 584
With cupboard unit	2400 x 760 x 800 mm	483 to 654

WORK BENCHES (3)

	Length x width	Price £
CONVERTABLE WOOD & METALWORK BENCHES		
Adjustable wooden planing tops; linoleum finish; steel edging; 2 stops; flaps.		
Dual sided	1500 x 760 mm	557 to 753
Adjustable wooden planing tops; linoleum finish; steel edging; 2 stops; flaps; cupboards under with shelves.		
Dual sided	1500 x 760 mm	737 to 996
Adjustable wooden planing tops; linoleum finish; steel edging; 2 stops; flaps; 2 cupboards under; 4 tool racks.		
Dual sided	1500 x 760 mm	737 to 996
MULTI MEDIA WORK STATION		
Fixed top; stainless steel; steel tubular frame with plated feet; epoxy paint finish.		
Four-way	1200 x 1200 mm	246 to 333
Top plate only	1200 x 1200 mm	82 to 111
Twin cupboards; shelf		237 to 320
Wood vice		20 to 27
Metal vice		27 to 36
COMPACT WELDING BENCH		
20 mm Reversible ground steel top; steel tubular frame with plated feet; epoxy paint finish.		
Bench unit	1200 x 900 x 900 mm	498 to 673

STORAGE

	Width x depth x height	Price £
CUPBOARDS		
Floor units; beech veneered carcase; solid beech rails and lipping; polished; tambour door; locking.		
Unit; no shelves	1800 x 600 x 600 mm	401 to 542
Unit; 3 shelves	1800 x 600 x 600 mm	438 to 592
Pegboard; 6 pegs.		24 to 32
Shelf 600; supports		14 to 19
Unit; no shelves	1800 x 900 x 900 mm	466 to 630
Unit; 3 shelves	1800 x 900 x 900 mm	522 to 706
Pegboard; 9 pegs.		28 to 38
Shelf 900; supports		19 to 26
Unit; no shelves	1800 x 900 x 900 mm	531 to 718
Unit; 3 shelves	1800 x 900 x 900 mm	601 to 813
Pegboard; 12 pegs.		33 to 45
Shelf 1200; supports		24 to 32
TOOL RACK COMPONENTS		
Box; plain top	300 mm	12 to 16
Box; top; pierced.	300 mm	14 to 19
Bit and slotted blocks.	280 x 30 x 20 mm	2 to 3
PEGBOARD AND COMPONENTS		
Single arm		
150 mm long		1.12 to 1.52
225 mm long		1.12 to 1.52
Twin Arm		
225 x 50 mm gap.		2.33 to 3.15
225 x 30 mm gap.		2.5 to 3.38
Terry clips		
12 mm		0.94 to 1.26
18 mm		0.97 to 1.31
25 mm		1.01 to 1.36
38 mm		1.07 to 1.44
Closed twin arm		
225 mm		1.12 to 1.52

RACKING

	Width x depth x height	Price £
LINEAL MATERIALS		
Tubular steel welded frames; 500 mm wide sheet floor trays; bolted connecting member; material retaining drop latch rail.		
Four sectional	2000 x 500 x 2000 mm	397 to 537
Extra 500 mm	500 x 500 x 2000 mm	101 to 136
SHEET MATERIALS		
Tubular steel welded frames; 500 mm wide sheet floor trays; bolted connecting member.		
Four sectional	1800 x 900 x 1000 mm	377 to 510
Extra 500 mm	500 x 500 x 2000 mm	101 to 136

NOTE:-
For prices of standard chairs and tables turn to FURNITURE AND FITTINGS FOR CLASSROOMS see Page 66.

DRAWING BOARDS

	Width x height	Price £
SINGLE UNITS		
A2; laminate surfaces; angled top bench; safety adjustment; steel frame; epoxy paint finish.		
Key Stage 3 - 4	1200 x 725 mm	180 to 240
Key Stage 3 - 4	1500 x 725 mm	205 to 280
Key Stage 3 - 4	1200 x 850 mm	190 to 260
Key Stage 3 - 4	1500 x 850 mm	220 to 300
A2; laminate surfaces; centrally pivoted at top; 4 positional rotation; steel frame; epoxy paint finish.		
Key Stage 3 - 4	1500 x 600 mm	205 to 275
Desks; black or clear track; paralleled motion; rigid frame.		
Key Stage 3 - 4	1500 x 600 mm	205 to 275
DOUBLE UNITS		
A2; laminate surfaces; angled top bench; safety adjustment; steel frame; epoxy paint finish.		
Key Stage 3 - 4	1750 x 725 mm	285 to 385
Key Stage 3 - 4	1750 x 850 mm	300 to 405
A2; laminate surfaces; centrally pivoted at top; 4 positional rotation; steel frame; epoxy paint finish.		
Key Stage 3 - 4	2100 x 600 mm	335 to 455
Desks; laminated surfaces; pull out trays; steel frame; epoxy paint finish.		
2 Person		
2, A2 boards	2000 x 700 mm	325 to 435
1 Person		
1, A2 board	1200 x 700 mm	185 to 250
1, A1 board	1500 x 700 mm	245 to 330

TOOLS (1)

		Price £
VICES - Woodworking		
Light duty	150 mm	23 to 31
Medium duty	175 mm	31 to 41
Heavy duty		
plain screw; dog.	175 mm	64 to 86
plain screw; dog.	225 mm	78 to 105
Quick release		
dog	175 mm	126 to 171
dog	225 mm	171 to 231
dog	263 mm	182 to 246
Portable		
holdfast with collar	263 mm	50 to 67
SAWS - Woodworking		
Tenon		
brass back	250 mm	12 to 16
Dovetail		
brass back	200 mm	13 to 17
Gents		
brass back	150 mm	4 to 6
brass back	200 mm	5 to 6
Bow		
beach frame	300 mm	24 to 32
Jack saw		
	500 mm	6 to 7
	250 mm	5 to 7
Hacksaw		
high leaverable blade	300 mm	6 to 8
tubular frame	300 mm	5 to 7
junior delux; tubular	150 mm	3 to 4
Mitre		
compound	162 mm cutting height	28 to 38
Coping		5 to 7

TOOLS (2)

	Price £
CHISELS - Woodworking	
Turning set; 4 piece; roughing and spindle gouge; screw and scraper.	
per set	42 to 57
Miniature turning set.	
per set	34 to 46
Lino and black carving set; 6 piece.	
per set	21 to 29
Chisel set; 4 piece; polypropylene handles.	
per set	14 to 19
PLANES - Woodworking	
Smoothing No. 4	19 to 26
Block plane No. 220	14 to 19
Spokeshave; fine adjustment.	5 to 7
VICES - Metalworking	
Medium duty; 150 mm; reversible swivel base; plastic jaws.	
per set	28 to 38
Replaceable serrated jaw; anvil face.	
per set	27 to 36
Standard duty; replaceable serrated jaws; 75 mm	
per set	35 to 47
Standard duty; replaceable serrated jaws; 100 mm	
per set	70 to 94
Standard duty; replaceable serrated jaws; 150 mm	
per set	204 to 276
Heavy duty; fitters quick grip.	
per set	248 to 335
Engineers; unbreakable QG; anvil face.	
per set	353 to 478

LABORATORY BENCHES

	Width x depth	Price £
BENCHING		
Wall bench; laminate or hardwood work surface; steel frame; epoxy paint finish		
Hardwood		
Iroko	1500 x 600 mm	135 to 180
Iroko	1800 x 750 mm	150 to 205
Trespa		
Laminate	1500 x 600 mm	155 to 210
Laminate	1500 x 750 mm	180 to 240
Solid mineral filled		
Corian	1500 x 600 mm	245 to 330
Corian	1500 x 750 mm	270 to 370
Modular bench; laminate work surface; steel frame; epoxy paint finish.		
Octagonal		
Iroko	1800 x 1800 mm	550 to 745
Laminate	1800 x 1800 mm	645 to 870
Corian	1800 x 1800 mm	910 to 1230
Under bench unit; MDF carcass; laminate finish.		
Double cupboard	1000 x 600 mm	170 to 230
Tray unit	1000 x 600 mm	180 to 240
Cupboard	1000 x 600 mm	115 to 155
Service pedEstal	1000 x 600 mm	335 to 455
extra for sink unit		60 to 80

EQUIPMENT

	Width x depth	Price £
TABLES		
Tables; laminate work surfaces; steel frame; epoxy paint finish.		
Iroka	1200 x 600 mm	160 to 215
Trespa resin bonded	1200 x 600 mm	160 to 215
Corian	1200 x 600 mm	280 to 380
SINKS		
Stainless steel		190 to 255
Vulcathene drip cup		20 to 30
Corian sink top		280 to 380
Hand wash sink		130 to 170
Eye wash sink		130 to 170
Large wash up sink		420 to 570
FUME CUPBOARDS		
To comply with D*f* EE Design Note No. 29. (Excluding extraction)		1375 to 1860
HAZARDOUS CHEMICALS CUPBOARD		
Complete with underframe All steel with liquid tight shelves. Meets COSHH Regulations.		
Service pedestal	900 x 400 x 900 mm	250 to 335

SERVICES AND STOOLS

	Height	Price £
SERVICE FITTINGS		
Gas tap; single		26 to 34
Gas tap; two way		30 to 40
Gas tap; three way		35 to 45
Gas tap; four way		40 to 50
Single hot or cold water tap		40 to 55
STOOLS		
Stacking; plywood seat; tubular steel legs; epoxy paint finish.		
Key Stage 1	470 mm	14 to 18
Key Stage 2	520 mm	14 to 18
Key Stage 2	570 mm	14 to 19
Key Stage 3	620 mm	14 to 19
Key Stage 4	670 mm	14 to 19
Stacking; solid ash dished seat; tubular steel legs; epoxy paint finish.		
Key Stage 1	510 mm	31 to 41
Key Stage 2	560 mm	31 to 42
Key Stage 2	610 mm	32 to 43
Key Stage 3	660 mm	33 to 44
Stacking; solid square beech seat; tubular steel legs; epoxy paint finish.		
Key Stage 1	510 mm	26 to 34
Key Stage 2	560 mm	27 to 36
Key Stage 2	610 mm	27 to 36
Key Stage 3	660 mm	28 to 38

DESIGNER WORK STATIONS

	Length x width x heigh	Price £
CRAFT TABLES		
Craft; laminate work surface; tray insert; light steel frame; epoxy paint finish.		
	1200 x 600 x 700 mm	60 to 75
Craft; laminate work surface; heavy steel frame; epoxy paint finish.		
	1500 x 800 x 850 mm	160 to 215
Silk screen work surface; reversible; alternative with laminate surface; heavy steel frame; epoxy paint finish.		
	1500 x 1200 x 850mm	370 to 500
Silk screen work surface; laminate edging; 3 mm neoprene top sheet; shelf; storage box; heavy steel cantilever frame; epoxy paint finish.		
	2400 x 1200 x 850mm	530 to 715
Preparation tables; laminate work surface; heavy steel frame; epoxy paint finish.		
Key Stage 3 - 4	1200 x 1200 x 700mm	30 to 40
Key Stage 3 - 4	1800 x 1200 x 700mm	30 to 45
Key Stage 3 - 4	2400 x 1200 x 700mm	35 to 45
Key Stage 4	1200 x 1200 x 850mm	35 to 45
Key Stage 4	1800 x 1200 x 850mm	530 to 715
Key Stage 4	2400 x 1200 x 850mm	530 to 715
Flush top bench 3.2 mm Lino finish; heavy duty epoxy coated steel frame.		
Key Stage 3 - 4	1200 x 600 x 700 mm	255 to 345
Key Stage 3 - 4	1500 x 600 x 700 mm	305 to 415
Key Stage 3 - 4	1800 x 600 x 700 mm	345 to 470
Key Stage 4	1200 x 900 x 700 mm	335 to 455
Key Stage 4	1500 x 900 x 700 mm	385 to 525
Key Stage 4	1800 x 900 x 700 mm	440 to 595

DRAWING BOARDS AND STORAGE

	Width x depth x height	Price £
DRAWING BOARDS		
Desk with board; lightweight steel frame; epoxy paint finish.		
Wood board	1200 x 600 x 700 mm	95 to 130
Laminate board	1200 x 600 x 700 mm	85 to 115
Desk; laminate board; damper safety feature; adjustable angled top; light steel frame; epoxy paint finish.		
Single unit	1200 x 725 x 650 mm	180 to 240
Single unit	1500 x 725 x 650 mm	205 to 280
Single unit	1200 x 850 x 650 mm	190 to 260
Single unit	1500 x 850 x 650 mm	220 to 300
Double unit	1750 x 725 x 650 mm	285 to 385
Double unit	1750 x 850 x 650 mm	300 to 405
T Square and rack		20 to 25
Desk; double; laminate board; adjustable angled top; pull out trays; light steel frame; epoxy paint finish.		
Single unit (A2)	1200 x 620 x 700 mm	185 to 250
Single unit (A1)	1500 x 700 x 700 mm	245 to 330
Double unit (A2)	2000 x 620 x 700 mm	325 to 435
STORAGE		
Plan File chest; eight angle slide drawers; steel carcass; epoxy paint finish.		
Plan Chest		400 to 540
Paper storage cupboard; A2 and A3; two door; six open fronted trays shelves; steel carcass; epoxy paint finish.		
Cupboard		390 to 525
Paper storage cupboard; A2 and A3; two door; six open fronted trays shelves; laminated solid cored carcass; polished.		
Static; 3 shelves	1040 x 700 x 750 mm	205 to 275
Mobile; 3 shelves	1040 x 700 x 810 mm	235 to 315
Static; 6 drawers	1040 x 700 x 750 mm	275 to 370
Mobile; 6 drawers	1040 x 700 x 810 mm	305 to 410

WORK STATIONS AND EASELS

	Width x depth x height	Price £
WORK STATIONS		
Adjustable laminate work surface; heavy steel frame; epoxy paint finish.		
Standard	1000 x 650 x 725 mm	175 to 235
Extra for back screen		60 to 80
under shelf		20 to 25
Paint tray		15 to 20
Laminate work surface; adjustable; track parallel motion; heavy steel frame; epoxy paint finish.		
Standard	1650 x 1050 x 725mm	250 to 335
Mobile; lightbox workstation; MDF laminated work surface; protected edges; fused recessed glass panel; cantilever steel frame; epoxy paint finish.		
Small	760 x 600 x 725 mm	300 to 405
Large	1510 x 810 x 725 mm	340 to 460
Air filter trolley; steel frame; epoxy paint finish.		
Trolley	760 x 760 x 800 mm	300 to 405
Air filter booth	A2 paper	410 to 550
EASELS		
Aluminium frame, black, telescopic locking legs, rubber feet, max height 63".		
Key Stage 2 - 4		35 to 45
Steel frame; black; telescopic quick release locking legs; rotating canvas; rubber feet; max height 82".		
Key Stage 2 - 4		50 to 65
Wooden "A" frame; metal ratchet for adjustable height; tilting action forward or backwards; max height 51".		
Key Stage 2 - 4		75 to 105

CRAFT EQUIPMENT

	Price £
BANDING WHEELS	
Steel construction; frame; epoxy paint finish; fly wheel.	
Floor standing	65 to 90
Table mounted	30 to 40
POTTERY WHEEL	
Steel construction; frame; epoxy paint finish; fly wheel.	
Electrically operated	740 to 1005
Manuallly operated	415 to 560
KILNS	
Steel casing; vitreous finish; fireclay lining; steel frame and stand; epoxy paint finish; complies with COSHH standards.	
Front loading	1865 to 2520
Top loading	935 to 1260

TABLES AND STORAGE UNITS

	Length x width x height	Price £
CRAFT TABLES		
Craft; laminate work surface; tray insert; light steel frame; epoxy paint finish.		
	1200 x 600 x 700 mm	60 to 75
Craft; laminate work surface; heavy steel frame; epoxy paint finish.		
	1500 x 800 x 850 mm	160 to 215
STANDARD - NON STACKING		
Melamine top; bullnosed edging; steel framed legs; epoxy paint finsih.		
Square		
Key Stage 2	1100 x 1100 x 600mm	54 to 72
Key Stage 2 - 3	1100 x 1100 x 650mm	54 to 72
Key Stage 3 - 4	1100 x 1100 x 700mm	54 to 72
Rectangular		
Key Stage 2	1200 x 600 x 600 mm	47 to 63
Key Stage 2 - 3	1200 x 600 x 650 mm	47 to 63
Key Stage 3 - 4	1200 x 600 x 700 mm	47 to 63
STORAGE UNITS - OPEN FRONT		
MDF carcass; laminated; hardwood edging; polished; static or mobile.		
2 shelves	1040 x 496 x 750mm	70 to 94
6 drawers	1040 x 496 x 750mm	163 to 221
3 shelves	1040 x 496 x 1455mm	108 to 146
4 shelves	1040 x 496 x 1455mm	114 to 154
STORAGE UNITS - WITH CUPBOARD DOORS		
MDF carcass; laminated; hardwood edging; polished; static or mobile.		
2 shelves	1040 x 496 x 750mm	121 to 164
6 drawers	1040 x 496 x 750mm	228 to 308
3 shelves	1040 x 496 x 1455mm	190 to 258
4 shelves	1040 x 496 x 1455mm	197 to 266

CHAIRS AND STOOLS

	Width x depth x height	Price £
STANDARD; METAL FRAMED		
High impact seat; co-polymer polypropylene; one piece plastic shell; black epoxy steel frame; stacking.		
Key Stage 2	426 x 440 x 350 mm	8.5 to 11.5
Key Stage 3	430 x 472 x 395 mm	8.8 to 11.9
Key Stage 3	482 x 506 x 405 mm	9.1 to 12.4
Key Stage 4	486 x 534 x 445 mm	9.1 to 12.4
Standard polished wood; plywood seat and back; laminated beech frame; linked; stacking		
Key Stage 2	360 x 320 x 355 mm	22.3 to 30.1
Key Stage 2 - 3	410 x 390 x 390 mm	23.5 to 31.8
Key Stage 3 - 4	410 x 390 x 445 mm	24.2 to 32.7
Lecture chair; black or chrome frame; clip-on MDF tablet; upholstered seat and back; optional storage undertray.		
Black frame		
Key Stage 3 - 4	525 x 570 x 465 mm	16.4 to 157.4
Chrome Frame		
Key Stage 3 - 4	525 x 570 x 465 mm	29.8 to 175.6
Lecture chair; samsonite neutral steel frame; permanent left or right hand tablet; finished in Oak melamine; upholstered seat and back.		
Black Frame		
Key Stage 3 - 4	460 x 560 x 465 mm	61.6 to 83.2
STOOLS		
Stool; plywood seat; tubular welded frame; four legs; epoxy paint finish; stacking.		
Key Stage 2	570 mm height	12.6 to 17
Key Stage 3	620 mm height	12.8 to 17.2
Key Stage 4	670 mm height	12.8 to 17.4
Stool; polypropylene seat; tubular welded frame; four legs; epoxy paint finish; stacking.		
Key Stage 2 - 3	610 mm height	17 to 23
Key Stage 4	686 mm height	17.2 to 23.4

T 11 - MUSIC SUITES AREA ARRANGEMENTS

NOTE: Areas given for models apply to Secondary Schools

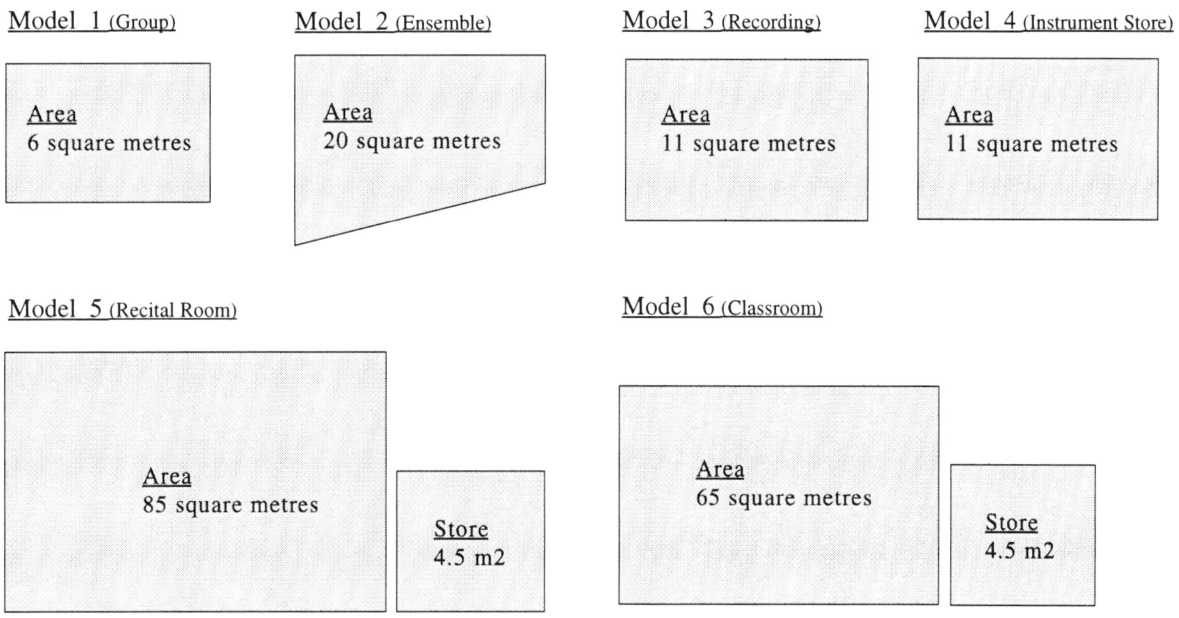

TABLE 'A' **MUSIC SUITES**

School Type (a)	Model (b)	Area m2 (c)	Total Building Cost (£) New (d)	Extension (e)	Re-Modelled (f)	Operational Usage (Nos) Teaching Spaces (g)	School Meals (h)	Assembly Spaces (i)	Revenue & Entertainment (j)	Sport (k)	Seminar Spaces (l)
Primary	0	0	£0	£0	£0	NIL	NIL	NIL	NIL	NIL	NIL
	0	0	£0	£0	£0	NIL	NIL	NIL	NIL	NIL	NIL
	0	0	£0	£0	£0	NIL	NIL	NIL	NIL	NIL	NIL
	0	0	£0	£0	£0	NIL	NIL	NIL	NIL	NIL	NIL
Secondary	1	6	£4,500	£4,200	£2,950	3	NIL	4	2	NIL	6
	2	20	£19,825	£18,525	£12,975	8	NIL	14	7	NIL	20
	3	11	£10,275	£9,625	£6,750	5	NIL	8	4	NIL	11
	4	11	£7,625	£7,125	£5,000	5	NIL	8	4	NIL	11
	5	89	£76,500	£71,475	£50,075	36	NIL	60	30	NIL	89
	6	69	£50,275	£47,000	£32,925	28	NIL	46	23	NIL	69
Higher	3	13	£12,150	£11,350	£7,975	6	NIL	9	5	NIL	13
	4	11	£7,625	£7,125	£5,000	5	NIL	8	4	NIL	11
	5	100	£85,950	£80,325	£56,275	40	NIL	67	34	NIL	100
	6	80	£58,300	£54,475	£38,175	32	NIL	54	27	NIL	80

TABLE 'B' **ANALYSIS OF COSTS**

£ per Square Metre - Based upon Model 3

Elements of the Building (m)	New £ per m2 (n)	Extension £ per m2 (o)	Remodelled £ per m2 (p)
Main Building Work			
* Foundations	109.07	111.00	4.21
* Superstructure	317.78	261.17	132.84
Finishings			
* Floors	54.32	57.13	72.67
* Walls	26.05	27.15	22.35
* Ceilings	27.70	31.94	46.05
Fittings			
* Furniture	100.36	93.25	94.05
* Fittings	75.56	70.21	70.81
* Sanitary Fittings	0.00	0.00	0.00
* Light fittings	23.51	23.25	25.74
Security			
* Protection	2.70	8.34	8.58
* Surveillance	6.44	8.62	6.25
* Communication	3.67	6.23	5.70
FITTING OUT	**320.32** £ per m2	**326.10** £ per m2	**352.18** £ per m2
TOTAL COST	**747.19** £ per m2	**698.27** £ per m2	**489.23** £ per m2

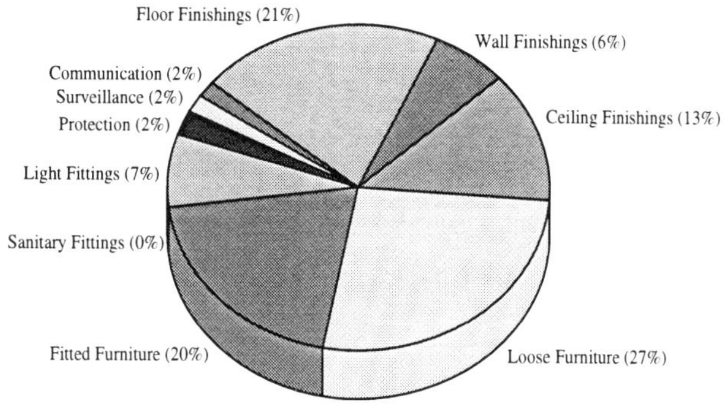

ANALYSIS OF FITTING OUT COSTS

(Costs based upon Table 'B' Column 'p').

MUSICIANS' CHAIRS

	Seat height	Price £
CHAIRS - STACKING		
Beginner's chair; posture design; polypropylene back and seat; steel tubular frame; epoxy paint finish.		
Key Stage 2 - 3	410 mm	42 to 56
Key Stage 3 - 4	460 mm	46 to 62
Student's chair; posture design; polypropylene back and seat; steel tubular frame; epoxy paint finish.		
Key Stage 2 - 3	410 mm	50 to 68
Key Stage 3 - 4	440 mm	53 to 72
Key Stage 3 - 4	470 mm	56 to 75
Musician's chair; postured design; upholstered; textile; steel tubular frame; epoxy paint finish.		
Key Stage 2 - 3	410 mm	108 to 146
Key Stage 3 - 4	440 mm	113 to 153
Key Stage 3 - 4	470 mm	118 to 160
Symphony musician's chair; adjustable design; upholstered back and seat; steel tubular frame; epoxy paint finish.		
Key Stage 3 - 4	470 mm	187 to 253
Key Stage 3 - 4	500 mm	193 to 261
Cellist's chair; extra forward posture design; upholstered back and seat; steel tubular frame; epoxy paint finish.		
Key Stage 3 - 4	470 mm	54 to 73
STOOLS - STACKING		
Plywood seat; tubular welded frame; four legs; epoxy paint finish.		
Key Stage 1	470 mm height	11 to 15
Key Stage 2	520 mm height	12 to 16
Key Stage 2	570 mm height	12 to 16
Key Stage 3	620 mm height	12 to 16
Key Stage 4	670 mm height	12 to 16
Polypropylene one piece seat; steel tubular welded frame; four legs; skid base; epoxy paint finish.		
Key Stage 2 - 3	610 mm height	16 to 21
Key Stage 4	686 mm height	16 to 21

	Seat height	Price £
CHAIRS - FOLDING		
Chair; upholstered; textile; steel tubular frame; epoxy paint finish.		
Key Stage 3 - 4	420 mm	48 to 65
Key Stage 3 - 4	470 mm	50 to 68
Chair; durable nylon back and seat; steel tubular frame; epoxy paint finish.		
Key Stage 3 - 4	420 mm	34 to 45
Key Stage 3 - 4	470 mm	36 to 49
CONDUCTOR'S CHAIR		
Conductor's chair; swivel; adjustable; upholstered; textile; steel tubular frame with foot rest; epoxy paint finish.		
Key Stage 4 - Adult	420 mm	310 to 420
STORAGE - INSTRUMENT		
Locking cabinet, polypropylene shelves; 9 small and 1 large compartments; to accommodate clarinets; flutes; piccolos; oboes; trumpets; Cornets or Alto Saxophones.		
Wood/metal	700 x 740 x 2170 mm	800 to 1085
Locking cabinet, polypropylene shelves; 10 compartments; to accommodate; trombones; bassoons; bass clarinets; Violins; violas or saxophones.		
Wood/metal	700 x 1000 x 2170mm	760 to 1030
Locking cabinet, polypropylene shelves; 3 medium sized compartments; to accommodate; french horns; alto horns or snare drums.		
Wood/metal	700 x 740 x 2170 mm	590 to 800

STANDS AND PODIUM

	Stand height	Price £
MUSIC STAND		
Bravo; adjustable; lightweight; polymer frame with one piece base; 340 x 520 mm music top.		
Key stage 2 - 4	610 x 1160 mm	40 to 55
Rough neck; adjustable; lightweight; polymer frame; one piece base; 340 x 510 mm music top.		
Key stage 2 - 4	610 x 1160 mm	50 to 70
Instrument conductor's; adjustable; lightweight; polymer frame with one piece base; 340 x 520 mm pebble top.		
Key stage 3 - 4	520 x 1160 mm	320 to 430
Choral conductor's; adjustable; lightweight; polymer frame with one piece base; 340 x 520 mm pebble top.		
Key stage 3 - 4	390 x 1160 mm	255 to 345
PODIUM		
Folding; hardboard; steel edging; carpet.		
Key stage 3 - 4	760 x 760 x 200 mm	145 to 195
Base unit; black laminate wood; carpet; built in handle.		
Key stage 3 - 4	1090 x 970 x 150 mm	210 to 285
Upper unit; black laminate wood sides; carpet; built in handle; detachable safety rail.		
Key stage 3 - 4	810 x 970 x 150 mm	30 to 40
Double unit; black laminate wood sides; carpet; built in handle; detachable safety rail.		
Key stage 3 - 4	1090 x 970 x 300 mm	435 to 590
Instrumental conductor's system; podium; black laminate wood sides; carpet; platform; safety rail; chair and shelf.		
Key stage 3 - 4		975 to 1315

T 12 - LABORATORY AREA ARRANGEMENTS

Model = A typical area arrangement
Lab = Laboratory

Model 1

(Single bank two laboratories with covered way)

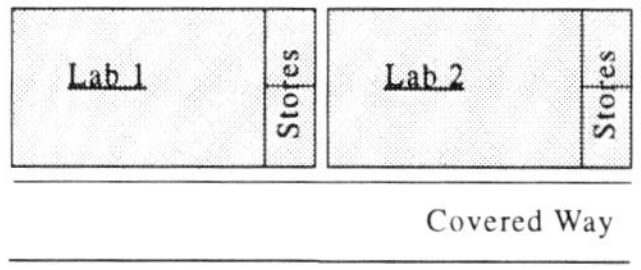

Model 2

(Single bank three laboratories with covered way)

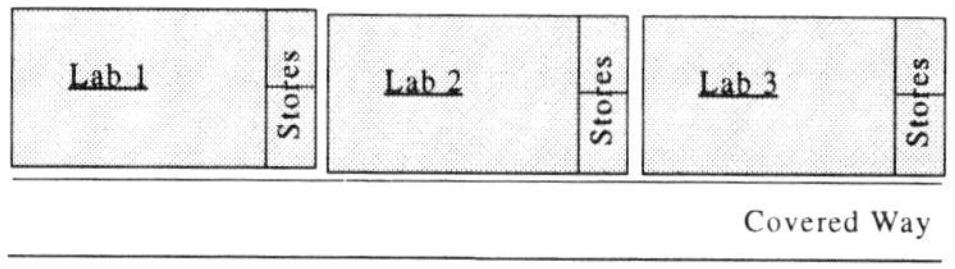

Model 3

(Double bank four laboratories with corridor)

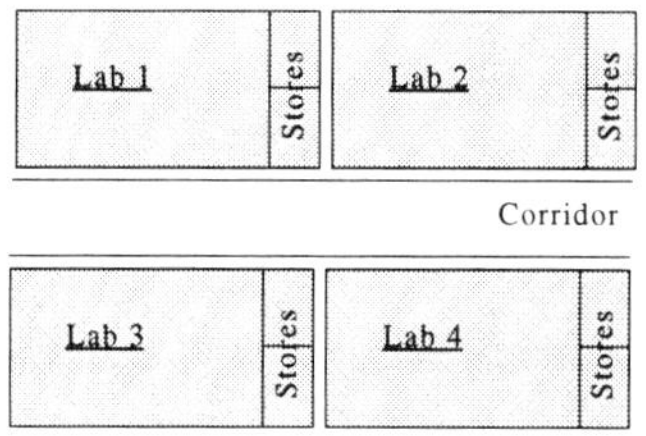

Model 4

(Double bank six laboratories with corridor)

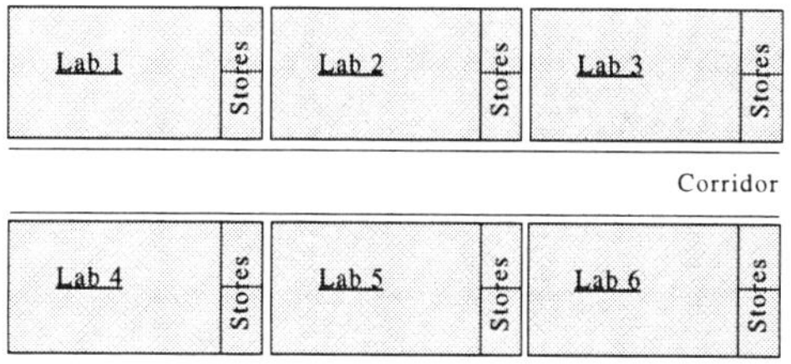

TABLE 'A' LABORATORY - SINGLE STOREY

School Type (a)	Model (b)	Area m2 (c)	Total Building Cost (£) New (d)	Extension (e)	Re-Modelled (f)	Operational Usage (Nos) Teaching Spaces (g)	School Meals (h)	Assembly Spaces (i)	Revenue & Entertainment (j)	Sport (k)	Seminar Spaces (l)
Primary	1	65	£49,150	£45,725	£33,125	26	NIL	33	26	NIL	44
	1	73	£55,200	£51,350	£37,200	30	NIL	37	30	NIL	49
	2	97	£72,950	£67,900	£49,175	39	NIL	49	39	NIL	65
	2	109	£81,975	£76,300	£55,250	44	NIL	55	44	NIL	73
Secondary	2	97	£72,950	£67,900	£49,175	39	NIL	49	39	NIL	65
	2	109	£81,975	£76,300	£55,250	44	NIL	55	44	NIL	73
	3	129	£99,000	£92,125	£66,725	52	NIL	65	52	NIL	86
	3	145	£111,275	£103,550	£75,000	58	NIL	73	58	NIL	97
	4	161	£120,475	£112,100	£81,175	65	NIL	81	65	NIL	108
	4	181	£135,450	£126,025	£91,275	73	NIL	91	73	NIL	121
Higher	3	137	£105,150	£97,850	£70,850	55	NIL	69	55	NIL	92
	3	153	£117,425	£109,275	£79,125	62	NIL	77	62	NIL	102
	4	171	£127,950	£119,075	£86,225	69	NIL	86	69	NIL	114
	4	191	£142,925	£133,000	£96,300	77	NIL	96	77	NIL	128

TABLE 'B' ANALYSIS OF COSTS

£ per Square Metre - Based upon Model 3

Elements of the Building (m)	New £ per m2 (n)	Extension £ per m2 (o)	Remodelled £ per m2 (p)
Main Building Work			
* Foundations	109.08	111.01	4.21
* Superstructure	323.97	262.83	146.80
Finishings			
* Floors	47.23	49.68	63.19
* Walls	28.66	29.86	24.58
* Ceilings	27.70	31.94	46.05
Fittings			
* Furniture	88.55	82.28	82.98
* Fittings	98.00	91.05	91.84
* Sanitary Fittings	7.88	8.99	11.16
* Light fittings	23.51	23.25	25.74
Security			
* Protection	2.70	8.34	8.58
* Surveillance	6.44	8.62	6.25
* Communication	3.67	6.23	5.70
FITTING OUT	**334.35** £ per m2	**340.24** £ per m2	**366.07** £ per m2
TOTAL COST	**767.41** £ per m2	**714.08** £ per m2	**517.07** £ per m2

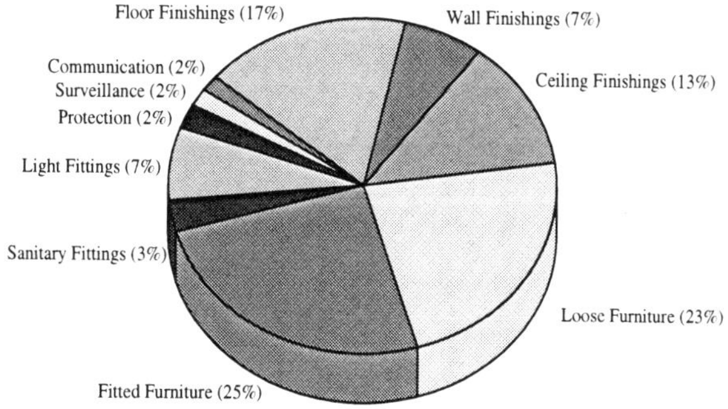

ANALYSIS OF FITTING OUT COSTS

(Costs based upon Table 'B' Column 'p').

T 13 - LABORATORY AREA ARRANGEMENTS

Model = A typical area arrangement
Lab = Laboratory

Model 1

(Single bank four laboratories with covered way)

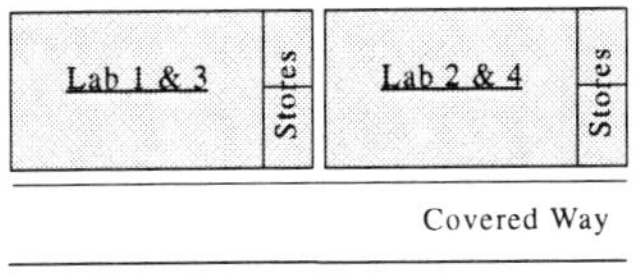

Model 2

(Single bank six laboratories with covered way)

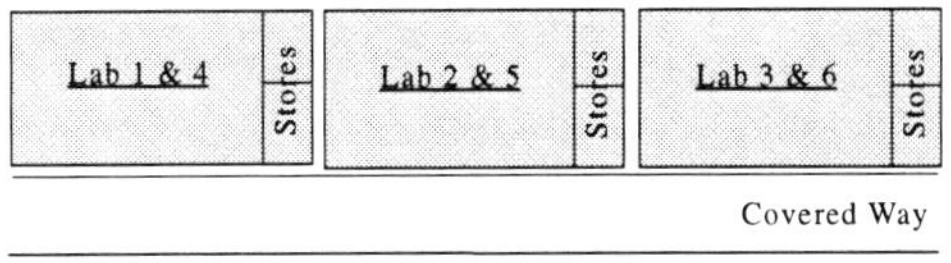

Model 3

(Double bank eight laboratories with corridor)

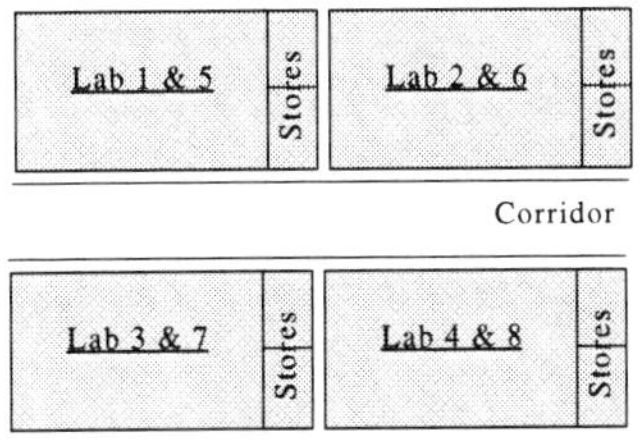

Model 4

(Double bank twelve laboratories with corridor)

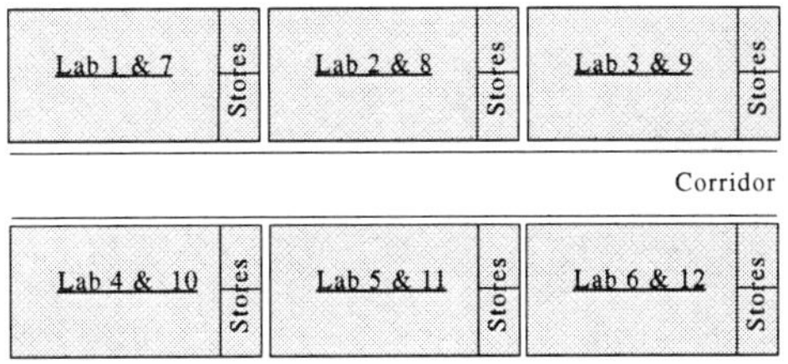

TABLE 'A' LABORATORY - TWO STOREY

School Type (a)	Model (b)	Area m2 (c)	Total Building Cost (£) New (d)	Extension (e)	Re-Modelled (f)	Operational Usage (Nos) Teaching Spaces (g)	School Meals (h)	Assembly Spaces (i)	Revenue & Entertainment (j)	Sport (k)	Seminar Spaces (l)
Primary	1	127	91,750	85,025	64,200	51	NIL	64	51	NIL	85
	1	142	103,025	95,475	72,100	57	NIL	72	57	NIL	95
	2	189	136,200	126,225	95,325	76	NIL	95	76	NIL	127
	2	213	153,050	141,850	107,125	86	NIL	107	86	NIL	142
Secondary	2	189	136,200	126,225	95,325	76	NIL	95	76	NIL	127
	2	213	153,050	141,850	107,125	86	NIL	107	86	NIL	142
	3	252	184,825	171,300	129,350	101	NIL	126	101	NIL	168
	3	283	207,750	192,550	145,400	114	NIL	142	114	NIL	189
	4	314	224,925	208,450	157,400	126	NIL	157	126	NIL	210
	4	353	252,850	234,325	176,950	142	NIL	177	142	NIL	236
Higher	3	267	196,300	181,925	137,375	107	NIL	134	107	NIL	179
	3	298	219,225	203,150	153,425	120	NIL	150	120	NIL	199
	4	333	238,875	221,375	167,175	134	NIL	167	134	NIL	223
	4	372	266,825	247,275	186,725	149	NIL	187	149	NIL	249

TABLE 'B' ANALYSIS OF COSTS
£ per Square Metre - Based upon Model 3

Elements of the Building (m)	New £ per m2 (n)	Extension £ per m2 (o)	Remodelled £ per m2 (p)
Main Building Work			
* Foundations	57.14	57.60	2.41
* Superstructure	343.24	283.06	145.70
Finishings			
* Floors	47.23	49.68	63.19
* Walls	28.66	29.86	24.58
* Ceilings	27.70	31.94	46.05
Fittings			
* Furniture	88.55	82.28	82.98
* Fittings	98.00	91.05	91.84
* Sanitary Fittings	7.88	8.99	11.16
* Light fittings	23.51	23.25	25.74
Security			
* Protection	2.70	8.34	8.58
* Surveillance	6.44	8.62	6.25
* Communication	3.67	6.23	5.70
FITTING OUT	334.35 £ per m2	340.24 £ per m2	366.07 £ per m2
TOTAL COST	734.74 £ per m2	680.91 £ per m2	514.17 £ per m2

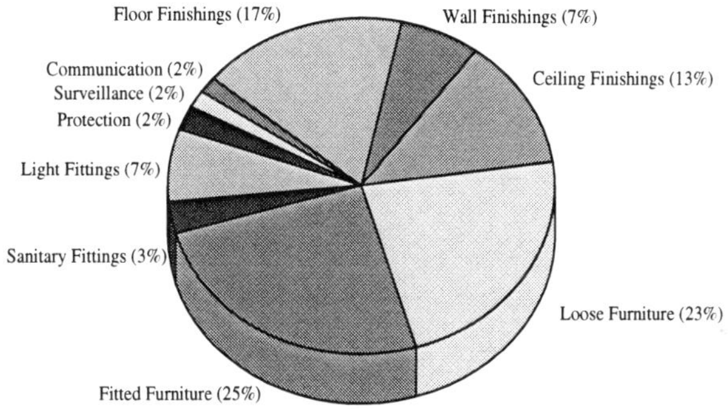

ANALYSIS OF FITTING OUT COSTS
(Costs based upon Table 'B' Column 'p').

WORK UNITS

	Width x depth x height	Price £
WALL UNITS		
MDF carcass glued and screwed; steel frame and legs; epoxy paint finish.		
Work surfaces;		
Laminate chipboard	1500 x 600 x 900 mm	105 to 140
Laminate chipboard	1500 x 750 x 900 mm	110 to 150
Iroko - hardwood	1500 x 600 x 900 mm	120 to 160
Iroko - hardwood	1500 x 750 x 900 mm	135 to 185
Solid cored laminate	1500 x 600 x 900 mm	145 to 195
Solid cored laminate	1500 x 750 x 900 mm	160 to 220
Corian	1500 x 600 x 900 mm	220 to 300
Corian	1500 x 750 x 900 mm	245 to 335
MODULAR ISLAND UNITS		
MDF carcass glued and screwed; painted; optional cupboards under; steel frames and legs; epoxy paint finish.		
Work surfaces;	Octagonal	
Laminate chipboard	1800 x 1800 x 900mm	450 to 605
Iroko - hardwood	1800 x 1800 x 900mm	515 to 695
Solid cored laminate	1800 x 1800 x 900mm	610 to 825
Corian	1800 x 1800 x 900mm	880 to 1190
UNDER BENCH UNITS		
MDF carcase; glued and screwed; painted; plain doors; painted; aluminium ironmongery; laminate covered shelves.		
Single pedestal	500 x 500 x 850 mm	140 to 190
Tray unit	1000 x 500 x 850 mm	190 to 255
Sink unit (excl sink)	1200 x 500 x 850 mm	330 to 445

SINK TOPS AND SPECIAL CUPBOARDS

	Width x depth x height	Price £
TABLES		
Softwood frames and legs; glued and screwed; painted.		
Laminate chipboard	1200 x 600 x 900 mm	110 to 150
Iroko - hardwood	1200 x 600 x 900 mm	110 to 150
Solid cored laminate	1200 x 600 x 900 mm	135 to 185
Corian	1200 x 600 x 900 mm	245 to 335
SINK TOPS AND DRAINERS		
Sink top; integral drainer; waste fittings.		
Stainless steel drainer		
single	500 x 500 x mm	170 to 230
double	1000 x 500 x mm	310 to 420
Vulcan drip cup		20 to 25
Corian drainer		
single	500 x 500 x mm	245 to 335
Hand wash up		110 to 150
Eye wash		110 to 150
Double drainer		380 to 515
FUME CUPBOARD		
To comply with D*f* EE Design Note 29; excluding extract system.		
Single unit		1265 to 1710
HAZARDOUS CHEMICAL CUPBOARD		
Steel construction with under frame; three point locking door; complies with COSHH standards.		
Single unit	900 x 400 x 900 mm	210 to 280

SERVICE FITTINGS

	Height	Price £
SERVICE FITTINGS		
Gas tap; single		26 to 34
Gas tap; two way		30 to 40
Gas tap; three way		35 to 45
Gas tap; four way		40 to 50
Single hot or cold water tap		40 to 55
STOOLS		
Stacking; plywood seat; tubular steel legs; epoxy paint finish.		
Key Stage 1	470 mm	14 to 18
Key Stage 2	520 mm	14 to 18
Key Stage 2	570 mm	14 to 19
Key Stage 3	620 mm	14 to 19
Key Stage 4	670 mm	14 to 19
Stacking; solid ash dished seat; tubular steel legs; epoxy paint finish.		
Key Stage 1	510 mm	31 to 41
Key Stage 2	560 mm	31 to 42
Key Stage 2	610 mm	32 to 43
Key Stage 3	660 mm	33 to 44
Stacking; solid square beech seat; tubular steel legs; epoxy paint finish.		
Key Stage 1	510 mm	26 to 34
Key Stage 2	560 mm	27 to 36
Key Stage 2	610 mm	27 to 36
Key Stage 3	660 mm	28 to 38

COMPUTER WORK STATIONS

	Width x depth x height	Price £
DESKS		
MDF laminated worktop; sliding key board; two MDF laminated shelves; steel frame; epoxy paint finish.		
Key Stage 3 - 4	700 x 520 x 755 mm	100 to 125
Key Stage 3 - 4	700 x 520 x 780 mm	100 to 125
Key Stage 2 - 4	700 x 520 x vary mm	150 to 175
Mobile; locking castors; MDF laminated worktop; adjustable; sliding key board; two MDF laminated shelves; steel frame; epoxy paint finish.		
Key Stage 3 - 4	625 x 490 x 720 mm	125 to 150
Key Stage 3 - 4	625 x 490 x 755 mm	175 to 225
Side extension		30 to 41
Mouse shelf		27 to 37
Paper catcher		18 to 24
MDF laminated worktop; sliding keyboard; cable management system; adjustable; two MDF laminated shelves; steel frame; epoxy paint finish.		
Key Stage 3 - 4	1210 x 810 x 720 mm	225 to 300
Key Stage 3 - 4	1510 x 810 x 720 mm	250 to 325
Key Stage 3 - 4	1810 x 810 x 720 mm	275 to 350
MDF laminated worktop; sliding keyboard; cable management system; adjustable; two MDF laminated shelves; cupboard under; steel frame; epoxy paint finish.		
Key Stage 3 - 4	1200 x 700 x 720 mm	375 to 500
Upper shelf		38 to 51
Power unit		28 to 38
75 mm Rubber wheel		33 to 45
Font winding		56 to 76
SUNDRIES		
Power block;	2 metres long	15 to 20
Cable riser	725 mm high	25 to 30
Cable riser	850 mm high	25 to 35

	Width x depth x height	Price £
DESKS		
Mobile; rubber wheels; post formed laminated worktop; adjustable; sliding key board; removable handles; two post formed laminated shelves; steel frame; epoxy paint finish.		
Full octagon	1800 x 1800 x 720mm	680 to 925
Half octagon	1800 x 900 x 720mm	440 to 600
Corner unit	900 x 900 x 720mm	330 to 450
Link unit	1000 x 600 x 720mm	465 to 625
Link unit	1500 x 600 x 720mm	640 to 875
Cantilever wall fixing type; adjustable; MDF laminated worktop; sliding key board; steel frame; epoxy paint finish.		
Work station	1210 x 610 x 720 mm	130 to 175
Work station	1210 x 810 x 720 mm	150 to 200
Work station	1510 x 610 x 720 mm	145 to 200
Work station	1510 x 810 x 720 mm	165 to 225
Work station	1810 x 610 x 720 mm	165 to 225
Work station	1810 x 810 x 720 mm	200 to 275
Radius corner table	800 mm dia	105 to 150
Radius corner table	1000 mm dia	125 to 175
Square corner table	600 x 600 x 720 mm	80 to 125
Square corner table	650 x 650 x 720 mm	90 to 125
Shelves	1510 mm long	185 to 250
Trays	1510 mm long	285 to 400
Mobile; MDF laminated worktop; two tier; projecting upper shelf; sliding key board; steel frame; epoxy paint finish; three work heights - 560, 660 & 760 mm.		
Trolley	850 x 650 x 660 mm	95 to 125
Trolley	1000 x 650 x 660 mm	100 to 150
Trolley	1200 x 650 x 660 mm	105 to 150
Shelf		10 to 25
Paper feed tray		15 to 25
Keyboard storage		15 to 25
Tidy basket		10 to 25

WORK BENCHES AND TROLLEYS

	Width x depth x height	Price £
WORK BENCHES		
Wall bench; MDF laminated work surface; steel frame; epoxy paint finish.		
Wall Bench; straight	700 x 725 mm	95 to 125
Wall Bench; straight	700 x 850 mm	95 to 130
Octagonal projection table; MDF laminated work surface; steel frame; epoxy paint finish.		
Project table	2000 x 1500 x 720mm	330 to 445
Project table	2000 x 1500 x 850mm	335 to 455
PRINT TROLLEYS		
Printer stand; dot matrix; paper baskets; steel frame; epoxy paint finish.		
Stand	632 x 666 x 670 mm	90 to 120
Printer stand; laser; mid size; adjustable; paper baskets; CD and diskette storage; steel frame; epoxy paint finish.		
Stand	490 x 400 x 660 mm	195 to 265
Printer stand; heavy duty; adjustable; paper baskets; steel frame; epoxy paint finish.		
Stand	700 x 500 x 675 mm	200 to 270
Printer stand; large laser; 6 paper shelves; binder storage; steel frame; epoxy paint finish.		
Stand	490 x 400 x 660 mm	205 to 280
Printer stand; laser; heavy duty; storage cupboard under; steel frame; epoxy paint finish.		
Stand	600 x 550 x 650 mm	260 to 355

S 1 - GYMNASIUM AREA ARRANGEMENTS

Model 1

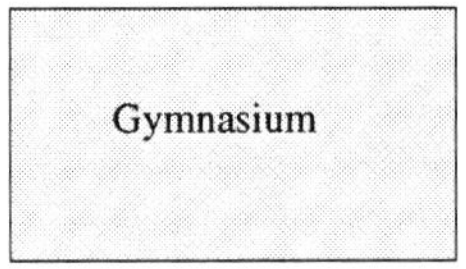

Model 2

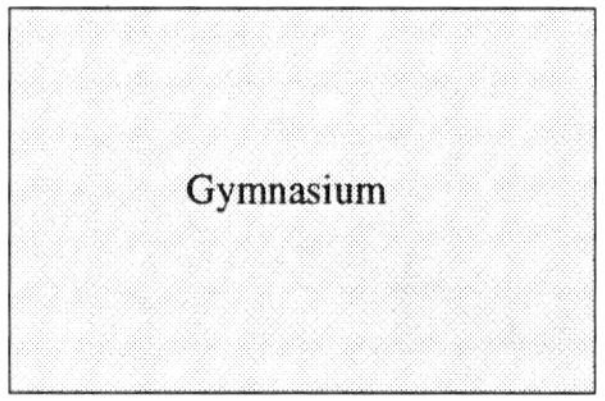

Model 3

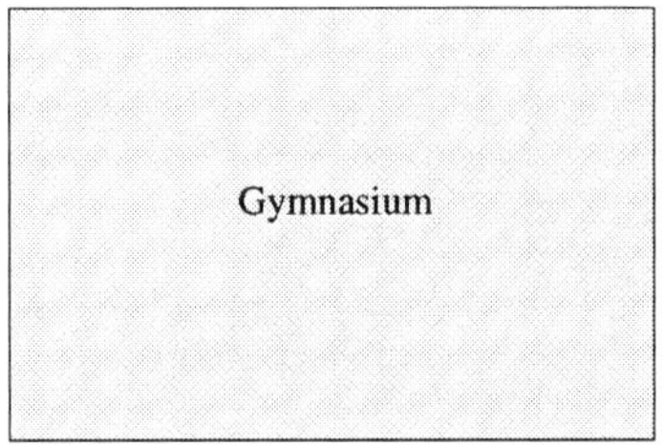

Model 4

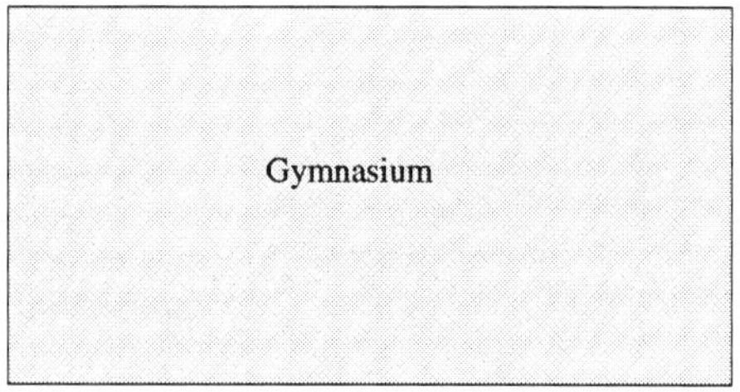

TABLE 'A' GYMNASIUM

School Type (a)	Model (b)	Area m2 (c)	Total Building Cost (£) New (d)	Extension (e)	Re-Modelled (f)	Operational Usage (Nos) Teaching Spaces (g)	School Meals (h)	Assembly Spaces (i)	Revenue & Entertainment (j)	Sport (k)	Seminar Spaces (l)
Primary	1	100	£63,150	£58,500	£34,375	40	NIL	67	34	20	50
	1	170	£107,350	£99,425	£58,450	68	NIL	114	57	34	85
	2	240	£155,625	£144,150	£84,750	96	NIL	160	80	48	120
	2	310	£201,025	£186,175	£109,450	124	NIL	207	104	62	155
Secondary	2	380	£246,400	£228,225	£134,175	152	NIL	254	127	76	190
	2	450	£291,800	£270,275	£158,875	180	NIL	300	150	90	225
	3	520	£354,925	£328,750	£193,250	208	NIL	347	174	104	260
	3	590	£402,700	£373,000	£219,275	236	NIL	394	197	118	295
	4	660	£484,275	£448,550	£263,675	264	NIL	440	220	132	330
	4	730	£535,625	£496,100	£291,625	292	NIL	487	244	146	365
Higher	3	800	£546,050	£505,750	£297,300	320	NIL	534	267	160	400
	3	870	£593,825	£550,000	£323,325	348	NIL	580	290	174	435
	4	940	£689,725	£638,825	£375,525	376	NIL	627	314	188	470
	4	1,010	£741,075	£686,400	£403,500	404	NIL	674	337	202	505

TABLE 'B' ANALYSIS OF COSTS
£ per Square Metre - Based upon Model 3

Elements of the Building (m)	New £ per m2 (n)	Extension £ per m2 (o)	Remodelled £ per m2 (p)
Main Building Work			
* Foundations	140.76	143.26	5.41
* Superstructure	345.74	276.68	129.32
Finishings			
* Floors	59.04	62.10	78.99
* Walls	35.17	36.65	30.17
* Ceilings	27.70	31.94	46.05
Fittings			
* Furniture	35.42	32.91	33.19
* Fittings	2.36	2.19	2.21
* Sanitary Fittings	0.00	0.00	0.00
* Light fittings	23.51	23.25	25.74
Security			
* Protection	2.70	8.34	8.58
* Surveillance	6.44	8.62	6.25
* Communication	3.67	6.23	5.70
FITTING OUT	**196.02** £ per m2	**212.23** £ per m2	**236.88** £ per m2
TOTAL COST	**682.54** £ per m2	**632.17** £ per m2	**371.61** £ per m2

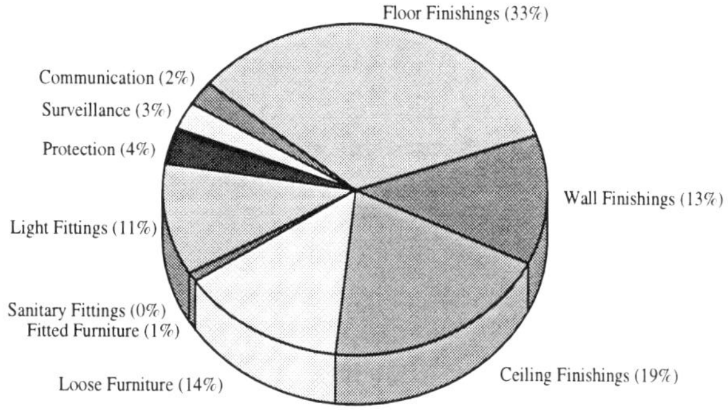

ANALYSIS OF FITTING OUT COSTS
(Costs based upon Table 'B' Column 'p').

S 2 - SQUASH COURT AREA ARRANGEMENTS

Model 1

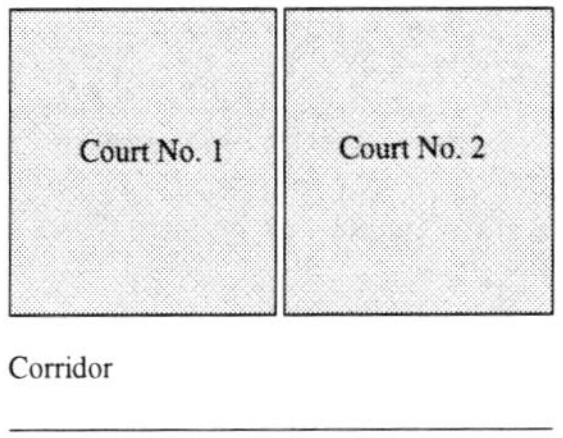

Model 2

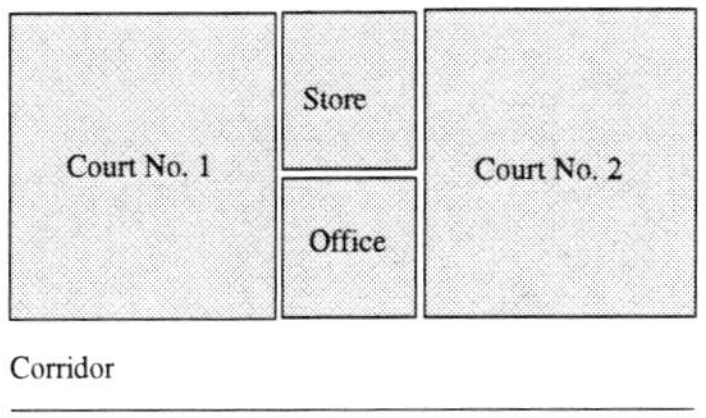

Model 3

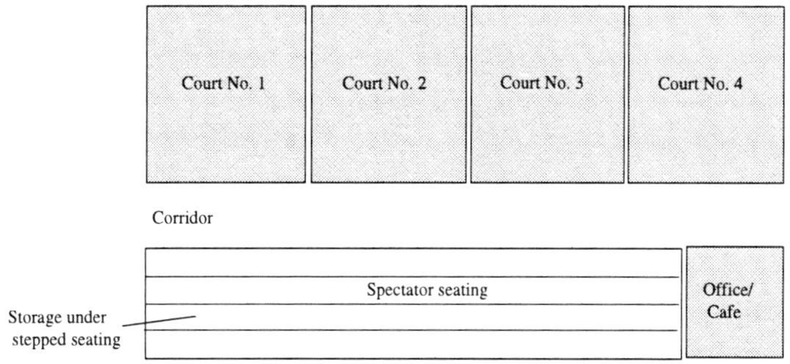

Squash Court Dimensions

Width	6.4 metres
Length	9.7 metres
Height (min)	5 metres

TABLE 'A' SQUASH COURTS

School Type (a)	Model (b)	Area m2 (c)	Total Building Cost (£) New (d)	Extension (e)	Re-Modelled (f)	Operational Usage (Nos) Teaching Spaces (g)	School Meals (h)	Assembly Spaces (i)	Revenue & Entertainment (j)	Sport (k)	Seminar Spaces (l)
Primary	0	0	£0	£0	£0	NIL	NIL	NIL	NIL	NIL	NIL
	0	0	£0	£0	£0	NIL	NIL	NIL	NIL	NIL	NIL
	0	0	£0	£0	£0	NIL	NIL	NIL	NIL	NIL	NIL
	0	0	£0	£0	£0	NIL	NIL	NIL	NIL	NIL	NIL
Secondary	1	128	£76,700	£68,700	£41,750	NIL	NIL	26	4	4	NIL
	2	162	£101,925	£91,275	£55,500	NIL	NIL	33	5	5	NIL
	3	396	£272,875	£244,375	£148,550	NIL	NIL	80	12	12	NIL
	0	0	£0	£0	£0	NIL	NIL	NIL	NIL	NIL	NIL
	0	0	£0	£0	£0	NIL	NIL	NIL	NIL	NIL	NIL
	0	0	£0	£0	£0	NIL	NIL	NIL	NIL	NIL	NIL
Higher	1	128	£76,700	£68,700	£41,750	NIL	NIL	26	4	4	NIL
	2	162	£101,925	£91,275	£55,500	NIL	NIL	33	5	5	NIL
	3	396	£272,875	£244,375	£148,550	NIL	NIL	80	12	12	NIL
	0	0	£0	£0	£0	NIL	NIL	NIL	NIL	NIL	NIL

TABLE 'B' ANALYSIS OF COSTS
£ per Square Metre - Based upon Model 3

Elements of the Building (m)	New £ per m2 (n)	Extension £ per m2 (o)	Remodelled £ per m2 (p)
Main Building Work			
* Foundations	140.76	143.26	5.41
* Superstructure	300.06	227.21	120.47
Finishings			
* Floors	65.80	64.75	88.73
* Walls	25.97	25.55	21.76
* Ceilings	23.16	22.79	36.87
Fittings			
* Furniture	3.54	3.29	3.32
* Fittings	3.54	3.29	3.32
* Sanitary Fittings	0.00	0.00	0.00
* Light fittings	23.51	23.25	25.74
Security			
* Protection	2.70	8.34	8.58
* Surveillance	6.44	8.62	6.25
* Communication	3.67	6.23	5.70
FITTING OUT	**158.34** £ per m2	**166.11** £ per m2	**200.28** £ per m2
TOTAL COST	**599.17** £ per m2	**536.58** £ per m2	**326.16** £ per m2

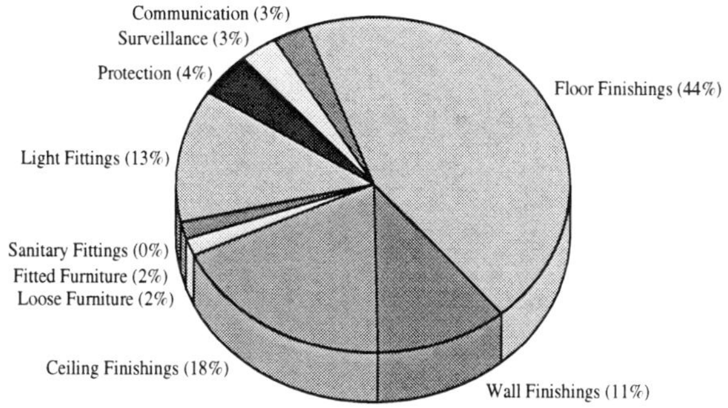

ANALYSIS OF FITTING OUT COSTS
(Costs based upon Table 'B' Column 'p').

S 3 - LEARNER SWIMMING POOL AREA ARRANGEMENT

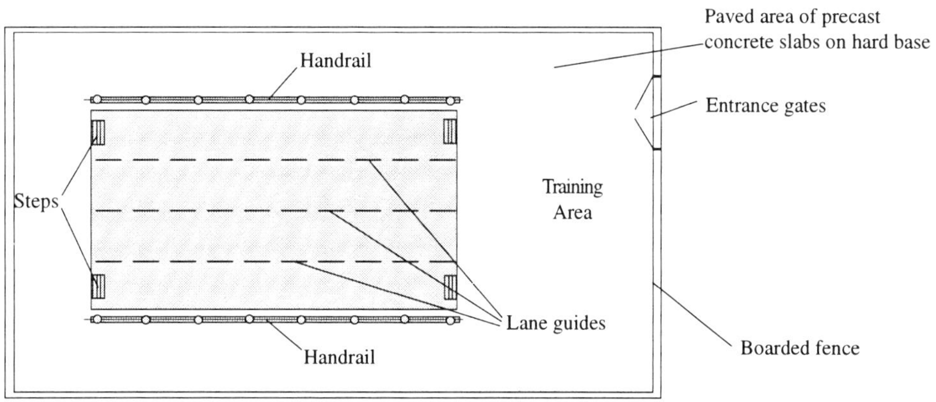

NOTE: Pool areas are rounded up to whole numbers.
Total costs are rounded up to the nearest £500.

NOTE;- All swimming pools have to comply with the Health and Safety Regulations and Building and Planning Regulations which should be strictly observed.

NOTE;- The cost of swimming pools may vary according to the location, size and nature of the site and also care should be taken when using these guides to check the cost of incoming mains services. Costs are based upon information supplied by Purley Pools (UK) Ltd., PO Box 132, Station Approach, Purley, Surrey, Tel; 0181 668 1322

SWIMMING POOLS - LEARNER

Swimmer	Size of Pool		Area	External Pool; Open Air; 4 Hr				External Pool; Covered; 4 Hr Turnover				Internal Pool; Building; 4 Hr Turnover			
Number of Pupils nr	Length m	Width m	Pool only m2	Construction of pool.	Equipment and fittings.	Paved surround and	TOTAL COST	Construction of pool.	Equipment and fittings.	Enclosure only; no changing	TOTAL COST	Construction of pool.	Equipment and fittings.	Building only; no changing	TOTAL COST
20	7.5	2.5	19	£ 6,950	£ 7,775	£10,575	**£ 25,500**	£ 6,950	£ 9,003	£42,315	**£ 58,500**	£ 8,450	£ 9,925	£57,428	**£ 76,000**
25	7.5	5.0	38	£ 10,725	£ 9,725	£13,150	**£ 34,000**	£ 10,725	£ 11,261	£58,590	**£ 81,000**	£ 13,050	£ 12,400	£79,515	**£ 105,000**
25	10.0	5.0	50	£ 12,600	£ 11,025	£14,425	**£ 38,500**	£ 12,600	£ 11,852	£68,040	**£ 92,500**	£ 15,325	£ 14,075	£92,340	**£ 122,000**
30	10.0	7.5	75	£ 17,700	£ 13,625	£17,000	**£ 48,500**	£ 17,700	£ 15,749	£86,940	**£ 120,500**	£ 21,525	£ 17,725	£117,990	**£ 157,500**
25	12.5	5.0	63	£ 15,450	£ 12,325	£15,725	**£ 43,500**	£ 15,450	£ 14,244	£77,490	**£ 107,500**	£ 18,775	£ 16,025	£105,165	**£ 140,000**
30	12.5	7.5	94	£ 21,675	£ 15,575	£18,300	**£ 56,000**	£ 21,675	£ 18,006	£99,015	**£ 139,000**	£ 26,350	£ 20,500	£134,378	**£ 181,500**
30	15.0	5.0	75	£ 18,300	£ 13,625	£17,000	**£ 49,000**	£ 18,300	£ 15,749	£86,940	**£ 121,000**	£ 22,250	£ 18,075	£117,990	**£ 158,500**
35	15.0	7.5	113	£ 25,650	£ 17,525	£19,575	**£ 63,000**	£ 25,650	£ 18,839	£111,090	**£ 156,000**	£ 31,175	£ 23,325	£150,765	**£ 205,500**
35	20.0	7.5	150	£ 33,600	£ 21,175	£22,150	**£ 77,000**	£ 33,600	£ 24,483	£135,240	**£ 193,500**	£ 40,825	£ 28,275	£183,540	**£ 253,000**
40	20.0	10.0	200	£ 43,200	£ 26,025	£24,725	**£ 94,000**	£ 43,200	£ 27,977	£164,640	**£ 236,000**	£ 52,500	£ 35,025	£223,440	**£ 311,000**

Swimmer	Size of Pool		Area	External Pool; Open Air; 2 Hr				External Pool; Covered; 2 Hr Turnover				Internal Pool; Building; 2 Hr Turnover			
Number of Pupils	Length m	Width m	Pool only m2	Construction of pool.	Equipment and fittings.	Paved surround and	TOTAL COST	Construction of pool.	Equipment and fittings.	Enclosure only; no changing	TOTAL COST	Construction of pool.	Equipment and fittings.	Building only; no changing	TOTAL COST
30	7.5	2.5	19	£ 6,950	£ 9,175	£10,575	**£ 27,000**	£ 6,950	£ 10,624	£42,315	**£ 60,000**	£ 8,450	£ 11,712	£57,428	**£ 78,000**
35	7.5	5.0	38	£ 10,725	£ 11,476	£13,150	**£ 35,500**	£ 10,725	£ 13,288	£58,590	**£ 83,000**	£ 13,050	£ 14,632	£79,515	**£ 107,500**
40	10.0	5.0	50	£ 12,600	£ 13,010	£14,425	**£ 40,500**	£ 12,600	£ 13,985	£68,040	**£ 95,000**	£ 15,325	£ 16,609	£92,340	**£ 124,500**
40	10.0	7.5	75	£ 17,700	£ 16,078	£17,000	**£ 51,000**	£ 17,700	£ 18,584	£86,940	**£ 123,500**	£ 21,525	£ 20,916	£117,990	**£ 160,500**
40	12.5	5.0	63	£ 15,450	£ 15,406	£15,725	**£ 47,000**	£ 15,450	£ 17,805	£77,490	**£ 111,000**	£ 18,775	£ 20,031	£105,165	**£ 144,000**
45	12.5	7.5	94	£ 21,675	£ 19,469	£18,300	**£ 59,500**	£ 21,675	£ 22,508	£99,015	**£ 143,500**	£ 26,350	£ 25,625	£134,378	**£ 186,500**
40	15.0	5.0	75	£ 18,300	£ 17,713	£17,000	**£ 53,500**	£ 18,300	£ 20,473	£86,940	**£ 126,000**	£ 22,250	£ 23,498	£117,990	**£ 164,000**
50	15.0	7.5	113	£ 25,650	£ 22,783	£19,575	**£ 68,500**	£ 25,650	£ 24,491	£111,090	**£ 161,500**	£ 31,175	£ 30,323	£150,765	**£ 212,500**
55	20.0	7.5	150	£ 33,600	£ 27,528	£22,150	**£ 83,500**	£ 33,600	£ 31,828	£135,240	**£ 201,000**	£ 40,825	£ 36,758	£183,540	**£ 261,500**
60	20.0	10.0	200	£ 43,200	£ 39,038	£24,725	**£ 107,000**	£ 43,200	£ 41,965	£164,640	**£ 250,000**	£ 52,500	£ 52,538	£223,440	**£ 328,500**

S 4 - SCHOOL AND SOCIAL USAGE SWIMMING POOLAREA ARRANGEMENTS

Model 1

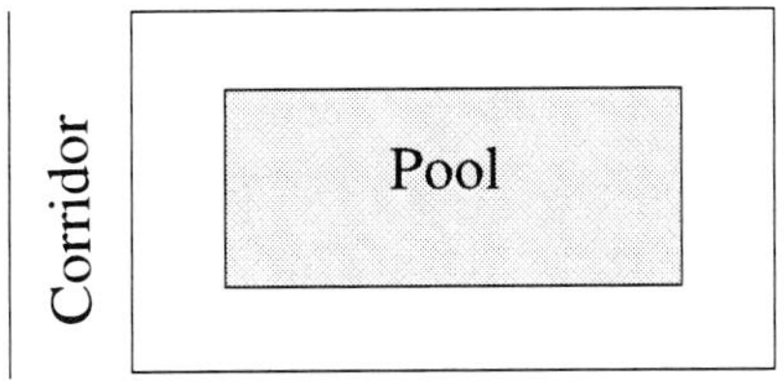

Model 2

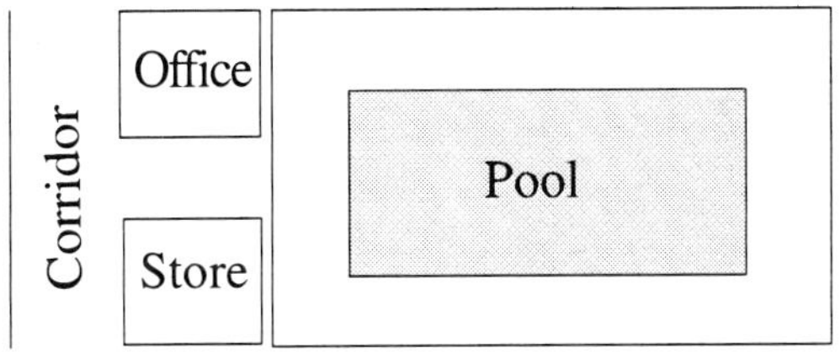

Model 3

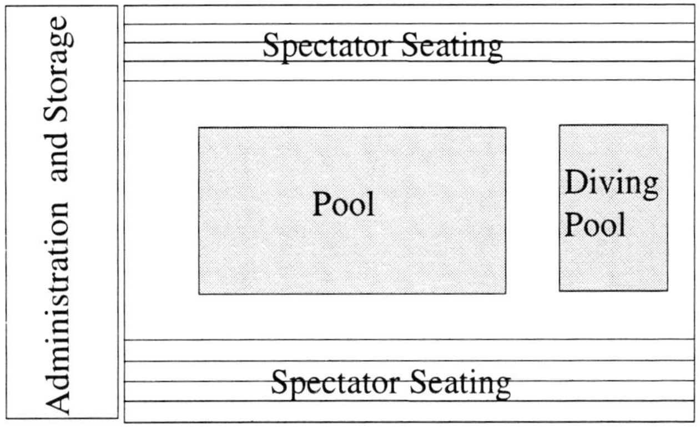

NOTE;- All swimming pools have to comply with the Health and Safety Regulations and Building and Planning Regulations which should be strictly observed.

NOTE: Pool areas are rounded up to whole numbers.
Total costs are rounded up to the nearest £500.

NOTE; All pools require the use of alternative showers and changing accommodation.

NOTE;- The cost of swimming pools may vary according to the location, size and nature of the site and also care should be taken when using these guides to check the cost of incoming mains services.

TABLE 'A' **SWIMMING POOL - SCHOOL AND SOCIAL USE**

	Swimmer	Size of Pool		Area	External Pool; Open Air; 4 Hr				External Pool; Covered; 4 Hr Turnover				Internal Pool; Building; 4 Hr Turnover			
	Number of Pupils	Length m	Width m	Pool only m2	Construction of pool.	Equipment and fittings.	Paved surround and	TOTAL COST	Construction of pool.	Equipment and fittings.	Enclosure only; no changing	TOTAL COST	Construction of pool.	Equipment and fittings.	Building only; no changing	TOTAL COST
Model 1	225	20.0	10.0	200	£ 77,775	£ 73,219	£48,300	**£ 199,500**	£ 77,775	£ 78,719	£188,160	**£ 345,000**	£ 146,475	£ 100,700	£294,000	**£ 541,500**
	275	20.0	12.5	250	£ 87,150	£ 87,519	£53,350	**£ 228,500**	£ 87,150	£ 94,119	£221,760	**£ 403,500**	£ 164,145	£ 120,350	£346,500	**£ 631,000**
	275	25.0	10.0	250	£ 80,100	£ 151,594	£53,350	**£ 285,500**	£ 80,100	£ 162,963	£221,760	**£ 465,000**	£ 160,586	£ 223,625	£346,500	**£ 731,000**
Model 2	350	25.0	12.5	313	£ 97,875	£ 105,394	£58,375	**£ 262,000**	£ 97,875	£ 113,300	£261,360	**£ 473,000**	£ 196,226	£ 163,375	£408,375	**£ 768,000**
	375	33.0	10.0	330	£ 104,588	£ 110,413	£61,400	**£ 276,500**	£ 104,588	£ 118,731	£275,520	**£ 499,000**	£ 209,674	£ 171,150	£430,500	**£ 811,500**
	450	33.0	12.5	413	£ 127,763	£ 133,994	£66,425	**£ 328,500**	£ 127,763	£ 144,100	£324,720	**£ 597,000**	£ 271,688	£ 181,900	£507,375	**£ 961,000**
	550	33.0	15.0	495	£ 150,938	£ 157,644	£71,450	**£ 380,500**	£ 150,938	£ 169,469	£373,920	**£ 694,500**	£ 320,950	£ 248,300	£584,250	**£1,153,500**
Model 3	675	50.0	12.5	625	£ 191,250	£ 267,743	£83,525	**£ 543,000**	£ 191,250	£ 287,824	£459,360	**£ 938,500**	£ 406,656	£ 428,400	£717,750	**£1,553,000**
	800	50.0	15.0	750	£ 225,900	£ 218,556	£88,550	**£ 533,500**	£ 225,900	£ 234,919	£528,960	**£ 990,000**	£ 507,779	£ 360,625	£826,500	**£1,695,000**
	950	50.0	17.5	875	£ 260,550	£ 313,113	£93,600	**£ 667,500**	£ 260,550	£ 336,597	£598,560	**£1,196,000**	£ 585,664	£ 547,950	£935,250	**£2,069,000**

WALL BARS AND CLIMBING FRAMES

	Width x Height	£ per set
FIXED WALL BARS		
Alloy bars;		
1 panel	838 x 2692 mm	245 to 335
2 panels	1676 x 2692 mm	355 to 480
3 panels	2514 x 2692 mm	455 to 620
4 panels	3352 x 2692 mm	550 to 745
Elliptical timber bars;		
1 panel	838 x 2692 mm	340 to 455
2 panels	1676 x 2692 mm	440 to 590
3 panels	2514 x 2692 mm	545 to 735
4 panels	3352 x 2692 mm	640 to 870
CLIMBING FRAMES		
Junior; 3 way	2450 mm high	1015 to 1370
	2750 mm high	1095 to 1485
	3050 mm high	1180 to 1595
Junior classic; 3 way	2927 mm high	1225 to 1655
Senior	3600 mm high	1305 to 1770
Senior Classic	3040 mm high	1390 to 1880
CLIMBING ROPES		
Trackway; 6 ropes	6000 mm long	295 to 400
Frame; 5 ropes	3660 x 3600 mm	295 to 400
Ropes; replacement	87mm circumference	25 to 30
Ropes; replacement	113mm circumference	30 to 40

MATTING

	Width x Length	£ per set
GYM MATS		
Standard;		
19 mm thick	1220 x 915 mm	20 to 30
	1830 x 1220 mm	35 to 50
	2000 x 1000 mm	35 to 50
25 mm thick	1220 x 915 mm	25 to 30
	1830 x 1220 mm	40 to 55
	2000 x 1000 mm	40 to 50
32 mm thick	1830 x 1220 mm	45 to 60
	2000 x 1000 mm	40 to 55
AGILITY MATS		
50 mm thick	1820 x 1220 mm	45 to 65
	2440 x 1220 mm	65 to 85
	3030 x 1220 mm	80 to 105
	3660 x 1220 mm	95 to 130
	4580 x 1220 mm	115 to 155
	6100 x 1220 mm	165 to 225
CRASH MATS		
100 mm thick	1830 x 1220 mm	90 to 120
200 mm thick	1830 x 1220 mm	105 to 140
200 mm thick	2440 x 1220 mm	130 to 175
305 mm thick	2440 x 1220 mm	170 to 225
305 mm thick	3050 x 1220 mm	225 to 300
JUDO MATS		
40 mm thick		
Club; standard	2000 x 1000 mm	55 to 75
Club; celular	2000 x 1000 mm	70 to 90
50 mm thick		
Competition; standard	2000 x 1000 mm	65 to 85
Competition; cellular	2000 x 1000 mm	70 to 95
Trolley		
heavy duty		125 to 165
vertical storage type		115 to 155

EXCERCISE EQUIPMENT

		£ per set
MOVEMENT TABLES		
Primary	900 dia x 760 mm hi	105 to 140
Junior	900 dia x 813 mm hi	110 to 145
Senior	1200 dia x 813mm hi	145 to 190
AGILITY STOOLS		
Primary	760 mm high	75 to 100
Junior	915 mm high	80 to 105
Senior	1006 mm high	85 to 110
TRESTLES		
	915 mm high	60 to 80
	1200 mm high	75 to 95
	1500 mm high	80 to 110
	1830 mm high	90 to 120
NESTING TABLES		
	305 mm high	45 to 60
	610 mm high	50 to 65
	760 mm high	55 to 70
	1060 mm high	75 to 100
PADDED BENCHES		
	1830 mm long	85 to 115
	2140 mm long	90 to 120
	2440 mm long	105 to 140
	2740 mm long	105 to 145
AGILITY LADDERS		
	1830 mm long	65 to 90
	2130 mm long	65 to 90
	2440 mm long	75 to 100
	2740 mm long	80 to 110
BALANCE BENCHES		
Hooks one end	1830 mm long	50 to 70
Hooks one end	2660 mm long	50 to 70
Hooks both ends	2660 mm long	55 to 75
Hooks both ends	3350 mm long	60 to 80
Hooks one end	3350 mm long	115 to 150

EXCERCISE EQUIPMENT

		£ per set
AGILITY POLES		
	1830 mm long	50 to 70
	2130 mm long	50 to 70
	2440 mm long	55 to 75
	2740 mm long	60 to 80
PLANKS		
Standard		
Hooks one end	1830 mm long	50 to 65
Hooks both ends	1830 mm long	65 to 90
Hooks one end	2130 mm long	55 to 70
Hooks both ends	2130 mm long	70 to 90
Hooks one end	2440 mm long	55 to 75
Hooks both ends	2440 mm long	70 to 95
Hooks one end	2660 mm long	85 to 110
Padded		
Hooks one end	1830 mm long	85 to 110
Hooks both ends	1830 mm long	100 to 135
Hooks one end	2130 mm long	85 to 115
Hooks both ends	2130 mm long	100 to 135
STORMING PLANKS		
	1830 mm long	60 to 80
	2130 mm long	65 to 90
	2440 mm long	70 to 95
SLIDES		
	1830 mm long	80 to 110
	2130 mm long	85 to 115
	2440 mm long	90 to 120
	2740 mm long	95 to 130
SPRINGBOARDS		
Training		90 to 120
School type		155 to 210
Competition type		110 to 150

		£ per set
BALANCING BEAMS		
Training		135 to 185
School type		500 to 675
Competition type		640 to 860
POMMEL HORSE		
Square neck type		570 to 775
Drop neck type		590 to 795
wheels		40 to 50
VAULTING HORSE		
Vaulting horse		520 to 705
wheels		40 to 50
VAULTING BUCK		
Vaulting buck		330 to 445
wheels		40 to 50
VAULTING BOXES		
5 sections	1010 mm high	450 to 605
5 sections	1270 mm high	495 to 665
Open wooden bar box on wheels		465 to 630
TRAMPETTE		
frame 1140 mm square		
Solid nylon		205 to 280
25 mm web bed		235 to 320
13 mm web bed		245 to 335

INDOOR SPORTS

	£ per set
BADMINTON	
Nets	
Nylon; 20'	10
Nylon; 24'	10 to 15
Posts	
Tubular post	95 to 125
Triple purpose post	140 to 185
Wall racks	25 to 30
BASKET BALL	
Balls	20 to 25
Goals	
Wheel away	170 to 230
Wall fixing	200 to 270
External use	230 to 310
Backboards	55 to 70

OUTDOOR SPORTS

		£ per set
CRICKET		
Balls		8.5 to 20
Bats		12.75 to 40
Indoor wickets complete		
one wicket		295 to 400
two wickets		380 to 515
FOOTBALL		
Steel posts fitted with nylon nets and hooks		
Junior		330 to 450
Senior		345 to 465
Five a side posts		
	8' x 4'	175 to 235
	12' x 4'	190 to 260
	16' x 4'	205 to 275
Nets - white polythene		
Medium		35 to 45
Heavy		35 to 45
Extra heavy		40 to 55
Super quality		45 to 65
RUGBY		
Steel posts; Epoxy paint finish;		
24' uprights		425 to 570
33' uprights		545 to 740
36' uprights		995 to 1345
Post protectors		115 to 155
Scrummage machine		520 to 700
HOCKEY		
Steel posts; tubular net supports; Epoxy paint finish;		
6" backboards		365 to 490
18" backboards		530 to 715

ATHLETICS

	£ per set
JUMPING	
High jump	64 to 86
Pole vault	169 to 228
Long jump board and pit	73 to 98
International Area kits	
High jump	2740.00
Pole vault	5610.00
HURDLES	
Junior	33 to 44
Senior	36 to 48
Competitive	51 to 69
Starter blocks	25 to 33

		£ per set
DISCUS		
Regulation wood; steel rim		
Girls'	0.75 kg	6 to 8
Women/ Colts'	1 kg	7 to 10
Boys'	1.25 kg	7 to 10
Youths'	1.50 kg	8 to 11
Junior mens'	1.75 kg	9 to 12
Senior mens'	2 kg	9 to 12
Discus circle; all steel; six segments		
Senior mens'		69 to 93
HAMMER		
Competition; Cast	4 kg	14 to 19
Competition; Cast	5 kg	16 to 22
Competition; Cast	6.25 kg	18 to 24
Competition; Cast	7.25 kg	20 to 26
Hammer handle		4 to 5
Hammer wire		3 to 4
JAVELIN		
Competition; steel		
Women/ Colts'	400 GM	49 to 66
Boys'	500 GM	50 to 67
Youths'	600 GM	53 to 71
Junior mens'	700 GM	58 to 78
Senior mens'	800 GM	63 to 85
SHOT		
Juniors'	3.25 kg	8 to 11
Womens'	4 kg	9 to 13
Youths'	5 kg	11 to 15
Junior mens'	6.25 kg	12 to 17
Senior mens'	7.26 kg	14 to 18
Shot circle; all steel; six segments		
Senior mens'		64 to 87
Shot arc		80 to 108

Chapter 3

CYCLICAL MAINTENANCE for EDUCATIONAL PREMISES

THE COST BASE

The Prices, Costs Analyses and Unit Rates for Measured Building Work are all based upon the cost of wages and prices of materials ruling during the first quarter of 1999 and within the Home Counties.

MAINTENANCE CYCLES

A Brief Maintenance Review	Each year	3rd year	7th year
Day to day repairs	x	x	x
Rainwater installation	x		
Replacing broken glass	x		
Overhauling doors and locks	x		
Checking all fire escapes	x		
Checking all alarm systems	x		
Furniture repairs	x		
Inspecting electrical installation	x		
Replacing light fittings and bulbs	x		
Cleaning and maintaining boilers	x		
Repairing pipes and radiators	x		
Cleaning sanitary fittings	x		
Repairing taps & replacing fittings	x		
Flushing out drains	x		
Groundsman's work	x		
External decorations		x	
Roof inspection and repairs		x	
Replacing worn doors and windows			x
Replacing worn floor finishings		x	
Replacing broken furniture		x	
Overhauling kitchen equipment		x	
Upgrading for health and safety		x	
Repair and repointing walls			x
Polishing gymnasium floors			x
Internal re-decorations			x
Boiler cleaning and repair			x
Replacing defective windows			x
Testing electrical installation			x
Testing and upgrading security alarms			x
Replacing defective fittings & wiring			x
Upgrading and replacing furniture			x
Replacing defective sanitary fittings			x
Flushing out and repairing drains			x
Overhauling gates and fencing			x
Relaying defective pavings			x

THE BASIS OF MANAGING MAINTENANCE WORK

A Commercial Approach.
No amount of paperwork can eliminate maintenance problems - dripping taps, leaking roofs, broken windows, blocked drains, faulty locks and electrical faults, but in an increasingly changing management conscious environment a more systematic and commercial approach can pay good dividends.

And it should also be remembered that those repetitive and irritating maintenance problems constantly facing the school manager provide, when properly organised, a real and valuable source of work and income for the Construction Industry.

The Cyclical Maintenance Programme
Head teachers will already be aware that there is a natural cyclical programme associated with maintenance work. It is neither possible nor usually desirable to carry out all of these tasks at one time and much, such as redecoration work, can be phased out over a three or seven year cycle.

The Manager's objective should therefore be to assemble the whole of the anticipated maintenance programme into one or a series of competitive packages which are both attractive and profitable for the contractors and ensure that the premises are always properly maintained and in a safe condition.

Benchmarks and Elemental Yardsticks.
The Maintenance Cycles list on this page shows an example of a logical and systematic analysis of a school's maintenance and operating requirements. The analysis on page 11 provides a guide to their costs and together these will be usefull when considering contractor's tenders or, for the private investors, when considering taking responsibility for these aspects of the school's activities.

The Condition Survey
The first step is to commission an experienced professional architect or surveyor to provide a survey setting out the extent of the existing accommodation and facilities available and their current condition. The cost of the maintenance and repair work should then be assessed as accurately as possible, phased and entered into the benchmark analysis shown on page 265. This will help in determining the viability of each aspect of the school premises and show the salient factors which would influence whether they should be retained or replaced.

This 'Condition Survey' is fundamental to the maintenance programme and the essence of any successful facilities management contract which must show;

i) the extent and use of the existing premises;

ii) the present condition of each element of the premises;

iii) urgent repair work;

iv) the performance standards and Health and Safety requirements;

v) all necessary security requirements.

With this survey to hand, an attempt can be made to programme the maintenance work and divide it into suitable work packages, look at the overall costs and invite tenders and quotations for the urgent repair work.

The programme will result in better standards, reduce costs and often develop opportunities to make favourable financial agreements with third parties.

THE MANAGEMENT PACKAGE

The Growth of Facilities Management
This has enabled the Construction Industry to offer Public Utilities, businesses and institutions such as schools, universities and hospitals an opportunity to reduce their operating and maintenance costs and management's involvement while at the same time improving the condition and performance of their premises.

The contracts may run for a number of years and may relate to particular aspects of the work or to the whole of the maintenance programme. It is usually better to ensure that the packages of work are as comprehensive as possible so that the contractors' bids can then reflect the savings which can be achieved by managing the work on a professional basis. But many small firms and manufacturer's are progressing towards this form of term contract, offering excellent guarantees and service contracts when selling their products.

The Benefits of Professional Management.
The following benefits are normally provided when entering into a proper Facilities Management contract:

- security and warranties;
- single point of contact / call out;
- 24 hour service;
- optimisation of resources;
- co-ordinated field support;
- increased flexibility;
- regulation monitoring and update;
- professional support;
- contractor vetting;
- health and safety procedures.

THE MANAGEMENT CONTRACTOR

Selection of the Contractor.
The selection of suitable contractors to be invited to tender should be approached carefully. It is important to observe all pre-tendering principles previously explained and to ensure that the contractors fully understand their commitment and the precise terms and conditions of such an agreement because of its extended term.

It is also important to ensure that the level and extent of the services required and methods of payment are clear to both parties so that the contractor's performance and maintenance standards can be assessed during the term of the contract against his contractual obligations.

The Maintenance Contract.
The maintenance contract in addition to normal clauses related to the conditions and specification, must set out as a minimum, the following:

- the extent of each element of the work and standard of work;
- the operating standards to be achieved and call out procedures;
- the method to be used for calculating the cost and value of the work;
- arrangements for payment.

Periodic Inspection of the Premises
Ideally it should also make provision for the joint inspection of the premises by a representative of the school and the contractor at specified intervals to ensure that the performance standards set out in the contract are being maintained. Defects should be required to be made good within a specified period and damages should be written into the contract should the contractor fail to meet his contractual obligations.

CONDITIONS AND SPECIFICATIONS

Specification of the Works.
This needs to be written by an architect or surveyor experienced in assessing the condition of buildings and the work required to meet the operating and performance standards required as well as the standard of workmanship. For example the external walls may be badly pointed and the specification of the remedial work and the time span over which the contractor would be called upon to carry out this work needs to be carefully thought through and programmed in a practical way that enables a contractor to meet this requirement. Equally urgent problems with lighting, power and sanitation need immediate attention and the operating procedures need to be carefully specified and monitored.

The Target Cost for Maintenance Work.
The analysis shown on page 11 can be very useful when considering alternative uses or users and when the running costs, related to the annual budget, can be allocated or charged to other parties perhaps having partial use of the premises.

This benchmark and the elemental yardsticks are important and should be reviewed from year to year to reflect actual performance, the cost of management contracts into which the school has entered and the savings and improved state of the premises.

Improvements in the Condition of the Premises.
Responsible maintenance contractors should always improve the standard of their work and the condition of the premises since this is perhaps the best way in the long term of reducing their costs and repetitive maintenance problems. The contract should seek to recover a part of this cost benefit for the school's future overall budget arrangements.

REF	ELEMENT	QUANTITY	UNIT	RATE	PLANNED EXPENDITURE							TOTAL OF ALL COSTS	AVERAGE ANNUAL COSTS
					1st	2nd	3rd	4th	5th	6th	7th		
	ADMINISTRATION												
1	Facilities Management	1	annual	£ 0.00	£ 0	£ 0	£ 0	£ 0	£ 0	£ 0	£ 0	£ 0	£ 0
2	Accountant	1	annual	£ 3,000.00	£ 3,000	£ 3,000	£ 3,000	£ 3,000	£ 3,000	£ 3,000	£ 3,000	£ 21,000	£ 3,000
3	Caretaker	1	annual	£ 12,000.00	£ 12,000	£ 12,000	£ 12,000	£ 12,000	£ 12,000	£ 12,000	£ 12,000	£ 84,000	£ 12,000
4	Security Agreement	1	annual	£ 650.00	£ 650	£ 650	£ 650	£ 650	£ 650	£ 650	£ 650	£ 4,550	£ 650
5	Library	1	annual	£ 7,500.00	£ 7,500	£ 7,500	£ 7,500	£ 7,500	£ 7,500	£ 7,500	£ 7,500	£ 52,500	£ 7,500
6	Technological (Service Agreement)	1	annual	£ 2,500.00	£ 2,500	£ 2,500	£ 2,500	£ 2,500	£ 2,500	£ 2,500	£ 2,500	£ 17,500	£ 2,500
7	Refectory	1	annual	£ 15,000.00	£ 15,000	£ 15,000	£ 22,500	£ 15,000	£ 15,000	£ 22,500	£ 15,000	£ 120,000	£ 17,143
8	Activities	1	annual	£ 7,000.00	£ 7,000	£ 7,000	£ 10,500	£ 7,000	£ 7,000	£ 10,500	£ 7,000	£ 56,000	£ 8,000
	CLEANING AND GROUNDSMAN												
9	Cleaning - General	6155	m2	£ 2.50	£ 15,388	£ 15,388	£ 19,234	£ 15,388	£ 15,388	£ 19,234	£ 15,388	£ 115,406	£ 16,487
10	Cleaning - Toilets	645	m2	£ 5.50	£ 3,548	£ 3,548	£ 3,548	£ 3,548	£ 3,548	£ 3,548	£ 3,548	£ 24,833	£ 3,548
11	Cleaning - Windows	12	months	£ 185.00	£ 2,220	£ 2,220	£ 2,220	£ 2,220	£ 2,220	£ 2,220	£ 2,220	£ 15,540	£ 2,220
12	Cleaning - External	12	months	£ 45.00	£ 540	£ 540	£ 540	£ 540	£ 540	£ 540	£ 540	£ 3,780	£ 540
13	Grass Cutting (26 per annum)	9500	m2	£ 0.01	£ 2,470	£ 2,470	£ 2,470	£ 2,470	£ 2,470	£ 2,470	£ 2,470	£ 17,290	£ 2,470
14	Site Markings	1	annual	£ 300.00	£ 300	£ 300	£ 450	£ 300	£ 300	£ 450	£ 300	£ 2,400	£ 343
15	Gardening	350	m2	£ 6.00	£ 2,100	£ 2,100	£ 2,100	£ 2,100	£ 2,100	£ 2,100	£ 2,100	£ 14,700	£ 2,100
16	Maintenance and Repairs	1	annual	£ 1,500.00	£ 1,500	£ 1,500	£ 3,000	£ 1,500	£ 1,500	£ 1,500	£ 4,500	£ 15,000	£ 2,143
	SERVICE AGREEMENTS												
17	W.C.'s, Seats and Fittings	1	annual	£ 380.00	£ 380	£ 380	£ 380	£ 380	£ 380	£ 380	£ 380	£ 2,660	£ 380
18	Urinal Stalls and Fittings	1	annual	£ 220.00	£ 220	£ 220	£ 220	£ 220	£ 220	£ 220	£ 220	£ 1,540	£ 220
19	Wash Basins and Taps	1	annual	£ 350.00	£ 350	£ 350	£ 350	£ 350	£ 350	£ 350	£ 350	£ 2,450	£ 350
20	Kitchen Equipment	1	annual	£ 165.00	£ 165	£ 165	£ 165	£ 165	£ 165	£ 165	£ 165	£ 1,155	£ 165
21	Electricity Charges	1	annual	£ 6,500.00	£ 6,500	£ 6,500	£ 6,500	£ 6,500	£ 6,500	£ 6,500	£ 6,500	£ 45,500	£ 6,500
22	Renewing Fittings and Bulbs	1	annual	£ 645.00	£ 645	£ 645	£ 3,000	£ 645	£ 645	£ 645	£ 3,000	£ 9,225	£ 1,318
23	Security Equipment	1	annual	£ 600.00	£ 600	£ 600	£ 600	£ 600	£ 600	£ 600	£ 2,500	£ 6,100	£ 871
24	Communication Equipment	1	annual	£ 150.00	£ 150	£ 150	£ 2,000	£ 150	£ 150	£ 150	£ 4,500	£ 7,250	£ 1,036
25	Oil, Gas or Coal	1	annual	£ 4,500.00	£ 4,500	£ 4,500	£ 4,500	£ 4,500	£ 4,500	£ 4,500	£ 4,500	£ 31,500	£ 4,500
26	Boiler Cleaning and Maintenance	1	annual	£ 400.00	£ 400	£ 400	£ 400	£ 400	£ 400	£ 400	£ 6,000	£ 8,400	£ 1,200
27	Repairs and Improvements	1	annual	£ 500.00	£ 500	£ 500	£ 500	£ 500	£ 500	£ 500	£ 500	£ 3,500	£ 500
28	Ventilation	1	annual	£ 180.00	£ 180	£ 180	£ 2,000	£ 180	£ 180	£ 180	£ 3,000	£ 5,900	£ 843
29	Drainage - Cleaning	1	annual	£ 200.00	£ 200	£ 200	£ 200	£ 200	£ 200	£ 200	£ 200	£ 1,400	£ 200
30	Drainage - Improvements	1	repairs	£ 0.00	£ 0	£ 0	£ 2,000	£ 0	£ 0	£ 0	£ 3,000	£ 5,000	£ 714
31	Roads and Paths	1	repairs	£ 0.00	£ 0	£ 0	£ 3,000	£ 0	£ 0	£ 0	£ 5,000	£ 8,000	£ 1,143
32	Fencing	1	annual	£ 120.00	£ 120	£ 120	£ 2,000	£ 120	£ 120	£ 120	£ 2,000	£ 4,600	£ 657

REF	ELEMENT	QUANTITY	UNIT	RATE	PLANNED EXPENDITURE							TOTAL OF ALL COSTS	AVERAGE ANNUAL COSTS
					1st	2nd	3rd	4th	5th	6th	7th		
	REDECORATION WORK												
33	Internal - Windows and Doors	1	annual	£ 200	£ 200	£ 200	£ 200	£ 200	£ 200	£ 200	£ 10,000	£ 11,200	£ 1,600
34	Internal - Walls	1	annual	£ 750	£ 750	£ 750	£ 2,000	£ 750	£ 750	£ 750	£ 10,000	£ 15,750	£ 2,250
35	Internal - Ceilings	1	annual	£ 200	£ 200	£ 200	£ 200	£ 200	£ 200	£ 200	£ 5,000	£ 6,200	£ 886
36	External Complete	1	annual	£ 300	£ 300	£ 300	£ 15,000	£ 300	£ 300	£ 300	£ 8,500	£ 25,000	£ 3,571
	FLOOR COVERINGS												
37	Wood Floors	800	m2	£ 3.45	£ 2,760	£ 2,760	£ 5,520	£ 2,760	£ 2,760	£ 2,760	£ 5,520	£ 24,840	£ 3,549
38	Carpets	1200	m2	£ 0.60	£ 720	£ 720	£ 8,100	£ 720	£ 720	£ 1,080	£ 10,800	£ 22,860	£ 3,266
39	Terrazzo	1700	m2	£ 1.20	£ 2,040	£ 2,040	£ 3,060	£ 2,040	£ 2,040	£ 3,060	£ 15,300	£ 29,580	£ 4,226
40	Vinyl	2800	m2	£ 0.25	£ 700	£ 700	£ 18,900	£ 700	£ 700	£ 1,050	£ 25,200	£ 47,950	£ 6,850
	ROOF COVERINGS												
41	Flat	3250	m2	£ 0.35	£ 1,138	£ 1,138	£ 2,500	£ 1,138	£ 1,138	£ 1,138	£ 12,000	£ 20,188	£ 2,884
42	Pitched	3550	m2	£ 0.25	£ 888	£ 888	£ 2,000	£ 888	£ 888	£ 888	£ 5,500	£ 11,938	£ 1,705
43	Roof Lights	56	nr	£ 5.00	£ 280	£ 280	£ 280	£ 280	£ 280	£ 280	£ 1,000	£ 2,680	£ 383
44	Rainwater Pipes and Guttering	1	annual	£ 300.00	£ 300	£ 300	£ 300	£ 300	£ 300	£ 300	£ 300	£ 2,100	£ 300
	DOORS												
45	Door Repairs and Re-glazing	1	annual	£ 650	£ 650	£ 650	£ 650	£ 650	£ 650	£ 650	£ 650	£ 4,550	£ 650
46	Door Renewal	1	renewal	£ 240	£ 240	£ 240	£ 3,000	£ 240	£ 240	£ 240	£ 4,400	£ 8,600	£ 1,229
47	Ironmongery - Clean and Oil	1	annual	£ 150	£ 150	£ 150	£ 300	£ 150	£ 150	£ 150	£ 450	£ 1,500	£ 214
48	Ironmongery - Renew	1	renewal	£ 0	£ 0	£ 0	£ 2,000	£ 0	£ 0	£ 0	£ 2,000	£ 4,000	£ 571
	WINDOWS												
49	Window Repair and Re-glazing	1	annual	£ 650	£ 650	£ 650	£ 650	£ 650	£ 650	£ 650	£ 650	£ 4,550	£ 650
50	Window Renewal	1	renewal	£ 0	£ 0	£ 0	£ 5,000	£ 0	£ 0	£ 0	£ 30,000	£ 35,000	£ 5,000
51	Ironmongery - Clean and Oil	1	annual	£ 75	£ 75	£ 75	£ 75	£ 75	£ 75	£ 75	£ 75	£ 525	£ 75
52	Ironmongery - Renew	1	renewal	£ 0	£ 0	£ 0	£ 2,000	£ 0	£ 0	£ 0	£ 2,000	£ 4,000	£ 571
	FITTINGS												
53	Classroom	1	annual	£ 300	£ 300	£ 300	£ 5,000	£ 300	£ 300	£ 300	£ 10,000	£ 16,500	£ 2,357
54	Library	1	annual	£ 75	£ 75	£ 75	£ 2,500	£ 75	£ 75	£ 75	£ 5,000	£ 7,875	£ 1,125
55	Cloakroom and Toilets	1	annual	£ 300	£ 300	£ 300	£ 300	£ 300	£ 300	£ 300	£ 5,000	£ 6,800	£ 971
56	Gymnasium	1	annual	£ 250	£ 250	£ 250	£ 3,000	£ 250	£ 250	£ 250	£ 5,000	£ 9,250	£ 1,321

NOTE: The costs shown in these schedules are guides based upon an examination of actual expenditure in various schools. A planned analysis should also take into account known items of expenditure and the repairs and alterations required from examination of the Condition Survey.

Chapter 4

UNIT RATES FOR MEASURED WORKS for EDUCATIONAL PREMISES

All Unit Rates are based upon metric measurements and rates of wages and prices of materials ruling at 1 January 1999 and exclude Value Added Tax

See page 12 for LEA Regional Price Variations

UNIT RATES

Unit Rates have been used extensively in the Building Industry for many years for pricing building works. Properly applied they provide a relatively straightforward and reliable basis for preparing budgets, pricing and checking tenders, pricing variations in the extent of the works and comparing the cost of alternative materials.

Construction price books have been published since 1850 containing information about building costs and including unit rates for use by builders, surveyors, architects and engineers and the publishers of this book are now selling the 124th edition of *Spon's Architects' and Builders' Price Book*. *Spon's* is also recognised in the courts and used by experts as a basis for settling disputes about the cost of building work.

This book is focused particularly on the cost and performance of a wide range of both modern and historic educational premises including new building work and re-fitting, extending, repairing and maintaining existing premises. Information, cost data and unit rates are presented in schedule form to allow easy reference and comparison between the cost of alternative forms of construction. It is hoped this will allow ease of access and help head teachers, principals and governors and consultants in the management, development and use of their premises.

The information in this book should be used carefully and it is advisable that the reader gains a general understanding of 'units of measurement', how they are derived, the different types of unit rates, how the different cost elements are incorporated into the unit rates, how they can be varied and the ways that the rates can be applied.

METHOD OF MEASUREMENT

Metric Measurements.
The Unit Rates and Composite Unit Rates are based upon the metric measuring system.

Standard Method of Measurement
This standard convention (trade practice) which provides a uniform and consistent basis for measuring and pricing building works has been devised and agreed by building contractors and the professional institutions within the UK construction industry.

The schedules of rates set out in this book are intended to adhere closely to these conventions. The rules are generally not difficult to follow and mainly reflect natural units of measurement, but care must always be taken to ensure that the rates are applied to the units of measurements indicated.

Abbreviations and Mathematical Symbols
The standard abbreviations indicating the units of measurement are as follows;-

single, pairs or sets	nr, pair or set
linear metre	m
superficial metre	m2
cubic metre	m3
by weight	kilo or tonne (ton)
liquid measure	litre

The standard mathematical symbols are as follows;-

<	less than	+	add or plus
>	greater than	-	deduct or less
=	equal to	x	multiply by
		/	divide by

Trade Order
The schedules of rates are set out in standard trade order as shown on pages 104 and 105.

COMPOSITION OF UNIT RATES

Content of Key Unit Rates and Composite Rates.
The key unit rates and composite rates:

include:
- i) Labour, material, small tools and plant;
- ii) Supervision and preliminaries;
- iii) Builders' overheads and profit.

exclude:
- i) Temporary works such as hoardings;
- ii) Preparation works (unless otherwise stated);
- iii) Professional Fees;
- iv) Value Added Tax.

and are based upon:
- i) Rates of wages and prices of materials and plant ruling at 1 January 1999;
- ii) Regional price levels for 'Outer London';
- iii) Tender Price Index - 329, see page 107;
- iv) Allowances for the builders' preliminaries, overhead charges and profit as follows;

Type of Work	Prelims	O'heads	Profit
i) New Work	6%	7.5%	3.5%
ii) Refurb work	8%	9 %	4.5%
iii) Repairs	incl	12.5%	7 %
iv) Maintenance	incl	5 %	10 %

CONTRACT VALUES

The contract values for projects appropriate for the application of the unit rates are as follows:

Type of Work	Contract value
i) New premises	>£100,000 <£1,000,000
ii) Extensions	>£100,000 <£1,000,000
iii) Refurbishment work	>£100,000 <£1,000,000
iv) Repairs & maintenance	<£ 20,000

BUILDING PRICE INDEX

YEAR	QUARTER	INDEX
1970	1st Quarter	277
1997	1st Quarter	289
	2nd Quarter	286
	3rd Quarter	289
	4th Quarter	301
1998	1st Quarter	311
	2nd Quarter	319
	3rd Quarter	325
	4th Quarter	328
1999	1st Quarter	329
	2nd Quarter *	333
	3rd Quarter *	339
	4th Quarter *	341
2000	1st Quarter *	343
	2nd Quarter *	346
	3rd Quarter *	352
	4th Quarter *	345
2001	1st Quarter *	349
	2nd Quarter *	355
	3rd Quarter *	361
	4th Quarter *	366

NOTE;- The Building Price Indices above reflect a reasonable guide to the level of tenders received in the industry taking account both of the levels of inflation related to the cost of both labour and materials and the effect of demand upon the prices submitted in competition by Builders. The indices marked with an asterisk are suggested to be a reasonable forecast of future tender prices.

RESTORATION WORKS

The cost of restoration work can be expensive and vary considerably according to the age and standard of the work required. The following are a guide to the additional cost of making good work over the key unit rates set out in this Chapter.

Old buildings	multiply by 1.25
Listed buildings	multiply by 1.85
Fine art work	multiply by 2.35

UNDERSTANDING THE RATES

Types of Rates and Cost Indicators

The following types of unit rates and cost indicators are included in the schedules:

i) **Key Unit Rates**
Each trade section provides standard schedules of rates covering the cost of the main items of work (referred to as *Key Unit Rates*).

ii) **Composite Unit Rates**
Each trade section also provides *Composite Rates* (covering groups of items) and these rates are combined with the graphic illustrations in the pictorial glossaries.

iii) **Variable Cost Factors**
Each trade section provides schedules and graphs which facilitate the adjustment of the key rates to allow for alternative specifications of materials and workmanship, changes in the form of construction and prices of materials.

iv) **Variable Contract Values**
Where the proposed contract value falls outside the cost bands set out on the previous page, Fig No.4.3 provides a bar chart showing the percentage adjustments necessary to convert standard key rates appropriate for repairs, small works and major contracts.

v) **Regional Variations in Prices**
Adjustments to take account of these factors can be made by reference to Page 12.

vi) **Inflation and Competitive Conditions**
Building prices will vary to take account of inflation and the level of competition will depend upon the level of demand in the industry at the time - see table of indices in previous column.

USING THE RATES

Understanding the Schedules of Unit Rates

The schedules of unit rates have been carefully structured to allow costs to be assessed with different degrees of accuracy as shown on page 109 and summarized as follows;-

'A' - Guide Prices
The key unit rates and composite rates in the schedules as shown provide the guide to costs and comparisons between costs when considering the use of alternative materials and forms of construction.

'B' - Average Prices
The key unit rates and Composite rates in the schedules can be adjusted to provide more accurate average unit rates, to take account of more detailed specifications and the actual cost of the materials, by making the appropriate adjustments in accordance with the factors and constants set out in each trade section.

'C' - Accurate Prices
These may be required when close estimates of cost are necessary. They can be obtained by adjusting the rates as 'B' above and adjusting the allowances for preliminaries, profit and overheads according to the anticipated value and type as shown in Fig No. 4.1.

'D' - Accurate Competitive Prices
These prices are obtained in accordance with 'C' above and then by making further adjustments to take account of regional variations and inflation as shown.

'E' - Composite Guide Prices
These are obtained by referring to the Pictorial Glossary at the start of each trade section.

ANALYSIS OF KEY UNIT RATES

Unit rates are based upon the cost of work per unit of measurement (viz. 1 metre superficial) and include the cost of labour, materials, plant and equipment necessary to carry out the work.
Page 106 shows a list of units of measurement.
Labour costs are based upon the time taken to do the work and are calculated from 'constants' (built up from historical experience). Surveyors and estimators usually adust standard constants according to their own judgement of the time required to carry out an operation.
Plant costs are included in the unit rates, but they may be priced separately by builders and included in the cost of their 'preliminaries'. See Page 106, viz; excavators, cement mixers.
Material costs are based upon their purchase price plus the cost of waste, delivery and storage.
Overheads and Profit are included in the unit rates but may be priced separately by builders and included in the cost of their 'preliminaries'.

Fig. No. 4.1 Typical Analysis of a Unit Rate

Common brickwork; no frogs; rate per m2; 65 mm metric commons; gauged mortar (1:3);
One brick wall; 215 mm thick

Labour	Constant	Rate £	Waste	Total	
Craftsman	1.67	7.91		13.22	
Labourer	0.67	6.43		4.30	
				17.52	17.52
Materials					
Bricks	120	135	5%	17.01	
Mortar	0.12	38.25	10%	5.05	
				22.06	22.06
Plant			Fuel		
Mortar Mill	0.21	13.45	6.00	4.08	
				4.08	4.08
Nett Prime Costs					43.66
ADD Overheads and Profit (17%)					7.42
TOTAL RATE (per m2)					£51.09

ANALYSIS OF LABOUR RATES

Labour is generally the most significant cost factor in any project and forms a substantial part of its overall cost. It is also subject to economic pressures. The typical standard rate analysis set out below (upon which the key unit rates are based) for payment of wages to craftsmen and general operatives, is often overtaken by the 'going rate' for self employed workers, the hire of 'contract labour' or the general shortage of skilled labour, leaving builders and clients vulnerable to 'market forces'. Fig No. 4.2 sets out the analysis of normal trade rates, but professional help must be sought if arguments arise about labour costs.
Daywork is a term applied within the Industry to labour employed on minor alteration work during the course of a building project. Schedules of 'Daywork Rates' are published by the Royal Institution of Chartered Surveyors, but these should not be used as a basis for negotiation of the cost of building works.

Fig. No. 4.2 Typical Analysis of a Labour Rate

Typical Labour Rate Analysis

	Weeks	Rate	Craftsman £	Rate	Labourer £
Wages	47.80	188.37	9,004.09	156.00	7,456.80
Lost time	0.90	188.37	169.53	156.00	140.40
Overtime	1.74	188.37	327.76	156.00	271.44
Sub Total			9,501.38		7,868.64
Bonuses		30%	2,850.41	25%	1,967.16
Sick Pay			-		-
Training Levy	1	year	24.51		20.30
Public Holidays	1.60	188.37	301.39	156.00	249.60
Employer's holiday contribution	47	19.65	923.55	19.65	923.55
N.I.	47.8	14.36	686.41	11.89	568.34
Severance Pay		1.5%	171.56	1.5%	144.46
Sub Total			14,459.21		11,742.05
Insurances		2.00%	289.18	2.00%	234.84
Cost per Annum			14,748.39		11,976.89
Hourly Rate			£ 7.91		£ 6.43

THE CONTRACT VALUES

Building costs are divided between the net or 'prime' cost of carrying out the work and the additional costs incurred by the builder including the 'preliminaries' such as supervision, insurances, storage of materials and temporary services, and the contribution the builder may wish to apply in respect of his 'overheads and profit'. The size and nature of a project has a direct effect on the cost of preliminaries, overheads and profit. For example complex works may demand more site supervision, and in the case of larger contracts, the builder's anticipated contribution towards his overheads may be reduced as a percentage of the prime cost. The key unit rates in this book include an assessment of the additional cost of preliminaries, overheads and profit as set out on Page 106 so that they indicate the 'total' cost of the work. However the key unit rates can be varied if required by adjusting the percentage rates or by reference to Fig No. 4.3 below.

Fig. No. 4.3 Adjustment of Contract Values

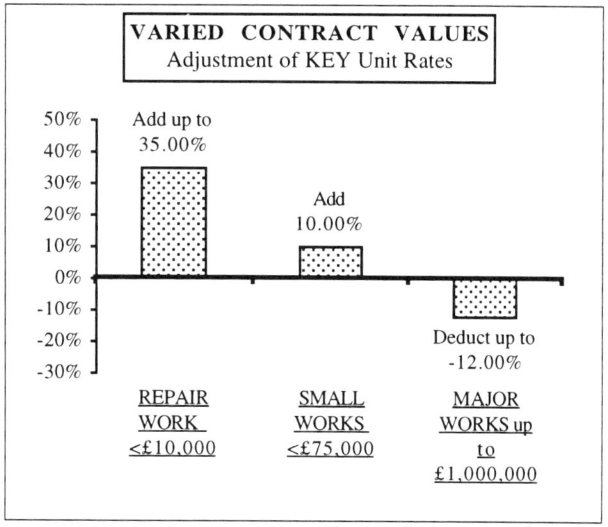

METHOD OF CALCULATING PRICES

A' GUIDE PRICES

Key Rate	x	Quantity	=	**TOTAL**

B' AVERAGE PRICES

Key Rate (or Composite Rate)	x	Specification Factor	x	Material Price	x	Quantity	=	**TOTAL**

C' ACCURATE PRICES

Key Rate (or Composite Rate)	x	Specification Factor	x	Material Price	x	Regional Factor	x	Quantity	=	**TOTAL**

D' ACCURATE COMPETITIVE PRICES

Key Rate (or Composite Rate)	x	Specification Factor	x	Material Price	x	Regional Factor	x	Size of Contract	x	Inflationary Factor	x	Quantity	=	**TOTAL**

E' COMPOSITE GUIDE PRICES

Composite Rate from Pictorial Glossary	x	Regional Factor	x	Quantity	=	**TOTAL**

All Unit Rates are based upon metric measurements and rates of wages and prices of materials ruling at 1 January 1999 and exclude Value Added Tax

PERFORMANCE STANDARDS.

Value for Money
Sourcing cheaper prices is an expedient that does not always provide a satisfactory solution. This chapter provides a guide to the complex assessments and decisions necessary to achieve a reasonable sense of 'value for money' when investing in new buildings or undertaking the refurbishment of existing premises.

It is equally important to consider the life and relative performance properties of the materials and methods of construction to be used in the project.

The Performance Icons
In order to provide a *quick informative guide* to the performance of various materials and methods we have assembled a series of 'Icons' based upon the recommendation of the Manufacturers and Trade Associations which are attached, where appropriate, to the key rates contained in this chapter to provide a guide to the performance of the materials selected as follows:

i) the performance qualities of materials specified;

ii) their economic performance over their normal lifespan based on 'cost in use'
(Graded from 1 to 100 in units of 10) and;

iii) a guide to the guarantees and warranties that are available for the materials and workmanship involved in their installation.

Best use of Materials and Methods
At the start of each trade section, where appropriate, we have included a *quick and informative guide* to help in the selection of materials and methods of construction for specified areas of accommodation to be provided in Schools or Universities, taking account of both cost and performance and based upon Manufacturers' recommendations.

	Antistatic	Materials which provide insulation against a build up of static electric current.
	Chemical Resistant	Materials that resist damage from chemicals for use in areas such as laboratories.
	Colour Fastness	Materials with pigments and dies resistant to fading and water damage.
	Designer Finish	Materials used to enhance appearances, often obtained without additional cost.
	Durability	Hard wearing materials for use in areas such as corridors or multimaterial workshops.
	Dust free	Sealed materials such as granolithic or terrazzo used to avoid the creation of dust.
	Easy cleaning	Reduces time and labour and the cost of cleaning as well as maintenance.
	Fire Resistant	Materials which resist the spread of flame as required by the building regulations.
	Flexible	Resilient materials with good wearing qualities also used to cover poor surfaces.
	Heavy Duty Areas	Materials which resist abrasion due to light industrial and heavy pedestrian use.
	Hygienic	Materials which meet hygienic standards nessary in cooking, dining and medical areas.
	Non slip	Safety materials which resist slipping generally in toilets, showers and on staircases.
	Seamless	Sheet materials with welded joints which prevent water seepage in wet areas.
	Steam Cleaning	Materials which withstand steam cleaning.

MAINTENANCE INDICES

The following Icons are also attached, where appropriate, to the schedules of key rates in this chapter to provide a guide to the maintenance performance standards which should be looked for when selecting materials for the proposed works. The Icons show on a scale of 1 to 100 in units of 10, the economic value of the materials during their life cycle and related to their maintenance and upkeep, the higher the value reflecting the best quality.

Maintenance Icon
showing period and guarantee which should be provided

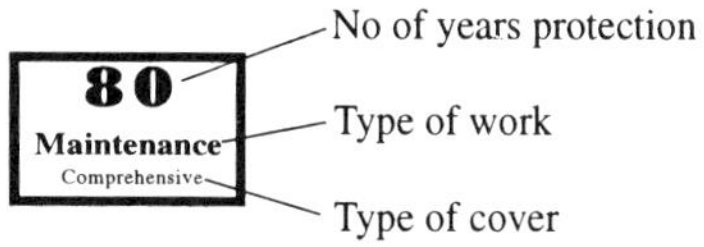

WARRANTY AGREEMENTS

Generally
Guarantees and warranties for all building work should always be obtained wherever possible to provide protection against their failure or early replacement and building disputes. The following Icons are attached, where appropriate, to the schedules of key rates in this chapter to provide a guide to the type and length of warranties available from manufacturers and specialists. Guarantees and warranties are generally available from most suppliers and specialists covering both the materials used and the workmanship entailed in its installation, but they should be read with great care and not accepted on their face value.

Materials only / **Materials and Workmanship**

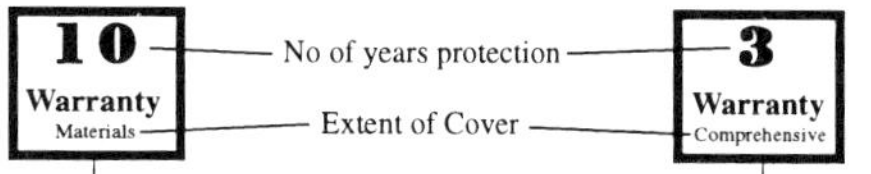

Typical cover provided	
Roofs	10 - 30 years
Floors	5 - 10 years
Windows	5 - 15 years

Typical cover provided	
Roofs	3 - 7 years
Floors	1 - 5 years
Windows	5 - 7 years

Construction of the Graphs.
The graph lines reflect variations in the unit cost of building work depending upon changes in the purchase price of the constituent materials.

i) Axis "x - x" Shows the purchase price of the main materials to be used in the work.
ii) Axis "y - y" Shows the revised unit rate for the selected building work.

Reading the Graphs.

1. Select the purchase price of the building materials on axis x - x;
2. Follow the vertical co-ordinate to the intersection with the graph line related to the specification of the work;
3. From the intersection with the graph line follow the horizontal co-ordinate to the left hand vertical axis y - y;
4. At the intersection of the horizontal co-ordinate with axis y - y, read the revised unit rate;
5. Adjust all other key unit rates within the section on a proportional basis.

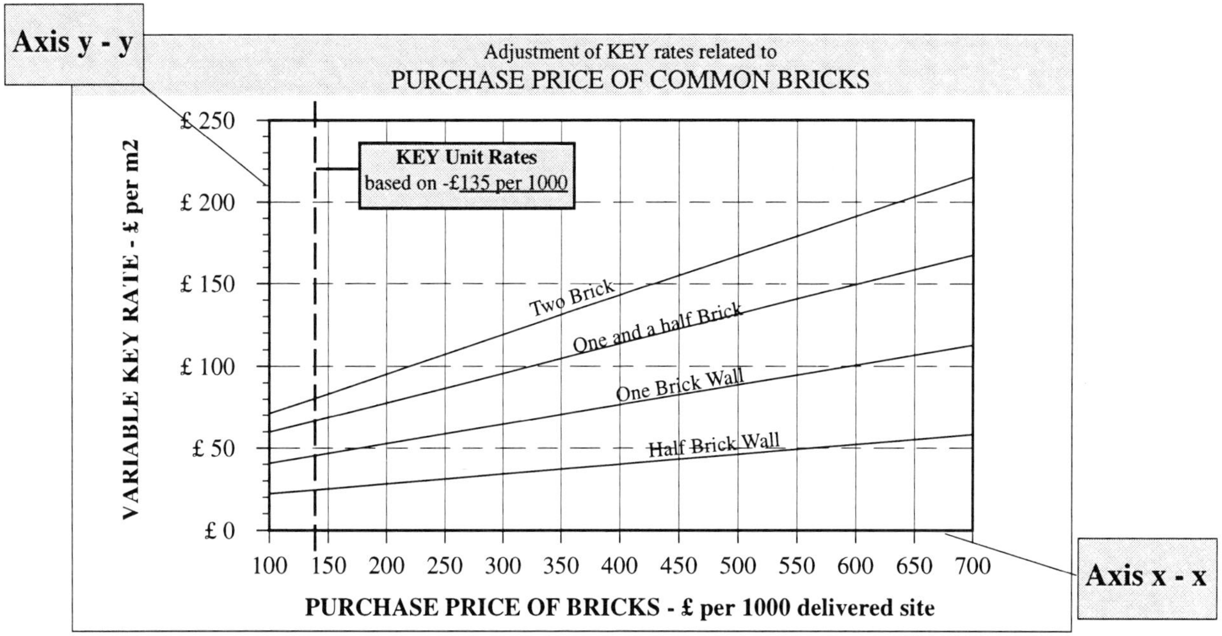

Example 1 HOW TO ADJUST; Key Unit Rates
to take account of the actual cost of materials to obtain
Adjusted Unit Rates:

The adjusted key unit rates using the graph:

Following steps 1 to 5 set out above;
Purchase price of common bricks = £300.00 per 1000;
Unit Rate, see page 123; one brick wall = £43.25 per m2;
Adjusted key unit rate from graph = £60.00 per m2

To adjust the rates following the unit rate selected:

The percentage increase in the key unit rate
(£60.00/£43.25)*100/1 = 139%
To adjust the following rates;

Plain walls	£45.41 * 139%	=	£63.11 per m2
Dwarf walls	£46.09 * 139%	=	£64.07 per m2
Skin of hollow wall	£46.55 * 139%	=	£64.70 per m2

Example 2 HOW TO ADJUST; Key Rates
to allow for
Inflation:

To adjust key unit rates to allow for inflation;

From Indices in schedule on page 107;

i) Current indices 1st Quarter 1999	=	329
ii) Contract to start 3rd Quarter 1999	=	339
To adjust multiply by (1+ (339-329/329)	=	1.0304

To adjust the unit rate selected:

Adjusted rates from the previous column;

Plain walls	£63.11 * 1.0304 = £65.03 per m2
Dwarf walls	£64.07 * 1.0304 = £66.02 per m2
Skin of hollow wall	£64.70 * 1.0304 = £66.67 per m2

Example 3 HOW TO ADJUST; New Key Rates
to obtain
Sub-Contractor's prices:

To calculate sub-contract prices;

From table on page 106;

i) Omit main contractor's overheads	=	7.5%
ii) Omit main contractor's profit	=	3.5%

Total % reduction = 11.0% multiply by 89%

To adjust the unit rate selected:

Adjusted rates from the previous column;

Plain walls	£65.03 * 89% = £57.88 per m2
Dwarf walls	£66.02 * 89% = £58.76 per m2
Skin of hollow wall	£66.67 * 89% = £59.34 per m2

TEMPORARY HOARDINGS

- Prepare working areas; erect secured units;
- Maintain during contract work; repairs;
- Clear away at completion; licence fees.

	Unit	New	Refurb
Hoarding; 18 mm external plywood; softwood posts			
2.4 metres high	m	52.15	59.97
ADD for the following;			
Pedestrian door	nr	98.50	113.28
Works access gates	pair	231.78	266.54
Decoration one side	m2	8.46	9.73
Protected walkway	m	28.97	33.32
Temporary Protective Roofs			
Scaffolding and waterproof covering to;			
Single storey building	m2	13.34	13.91
Two storey building	m2	16.12	16.80
Vertical weather proofing to perimeter			
Single storey building	m	7.17	7.47
Two storey building	m	10.56	11.01
Dustproof Screens; 2.4 m high; sw framing			
Coverings;			
Polythene sheeting;	m2	20.64	21.52
Plywood; 6mm	m2	25.01	26.07
ADD for the following;			
Access door	nr	72.25	75.33
Decoration one side	m2	8.11	8.46
Dustproof & soundproof screen; 2.4 m high; sw framing			
Covering both sides; fibreglass insulation			
Hardboard; 3 mm	m2	29.93	31.20
Plywood; 6mm	m2	36.26	37.81
ADD for the following;			
Access door	nr	104.76	109.22
Decoration one side	m2	11.77	12.27
Temporary Chestnut Fencing			
Fencing; 1.2 m high;	m	7.50	7.82

DEMOLITION WORK

- Pull down to ground level; break up floor slab;
- Strip out services; disconnect supplies;
- Clear away debris; disposal fees by Contractor.

	Unit	New	Refurb
Buildings - timber floor and frame			
One storey	m2	9.49	10.91
Buildings - brick or stone built; timber floors			
One storey	m2	13.20	15.18
Two storey	m2	12.89	14.82
Buildings - steel framed; timber floors			
One storey	m2	16.92	19.46
Two storey	m2	16.23	18.66
Add to rate for each extra floor	m2	0.93	1.07
Buildings - concrete floors and frame			
One storey	m2	21.38	24.59
Two storey	m2	20.58	23.67
Add to rate for each extra floor	m2	1.19	1.37
Foundations - break up existing (up to 1 metre deep)			
Insitu concrete			
Foundations	m3	61.01	70.16
Beds			
100 mm average thick	m2	5.91	6.80
150 mm average thick	m2	9.79	11.26
Reinforced concrete			
Foundations	m3	79.31	91.20
Beds			
100 mm average thick	m2	7.69	8.85
150 mm average thick	m2	12.73	14.64
Brick or stone walls; cement mortar			
Half brick thick	m2	8.91	10.25
One brick thick	m2	15.52	17.85
One and a half brick thick	m2	22.21	25.55
over 2 bricks thick	m3	30.05	34.55

PULLING DOWN WALLS

- Pull down including finishings; protection;
- Cut back finishings to perimeter as necessary;
- Clear away debris; disposal fees by Contractor.

	Unit	New	Refurb
Concrete walls			
100 mm thick	m2	9.47	10.89
150 mm thick	m2	14.18	16.31
200 mm thick	m2	16.56	19.04
100 mm thick - reinforced	m2	11.97	13.77
150 mm thick - reinforced	m2	17.94	20.63
200 mm thick - reinforced	m2	20.94	24.08
Brick walls; gauged mortar			
Half brick thick	m2	9.73	11.19
One brick thick	m2	16.94	19.48
One and a half brick thick	m2	24.26	27.90
over 2 bricks thick	m3	81.33	93.53
ADD if in cement mortar	%		
Blockwork walls - lightweight blocks; gauged mortar			
75 mm thick	m2	6.59	7.58
100 mm thick	m2	7.78	8.95
150 mm thick	m2	9.75	11.21
Blockwork walls - high density blocks; cement mortar			
75 mm thick	m2	9.56	10.99
100 mm thick	m2	11.28	12.97
150 mm thick	m2	14.14	16.26
200 mm thick	m2	21.35	24.56
Timber framed partitions - plasterboard both sides			
75 mm thick	m2	5.61	6.45
100 mm thick	m2	7.17	8.24
Metal framed partitions - plasterboard both sides			
75 mm thick	m2	5.21	5.99
100 mm thick	m2	6.43	7.40

BREAKING OUT - FLOORS

- Concrete; break out; cut reinforcement;
- Timber; take up boarding; pull down ceiling;
- Finishings; break up; clean surfaces for new;
- Clear away debris; disposal fees by Contractor.

	Unit	New	Refurb
Suspended floors			
Concrete; floor			
floor			
100 mm thick	m2	11.49	13.21
150 mm thick	m2	15.77	18.13
landing			
100 mm thick	m2	14.36	16.51
150 mm thick	m2	19.71	22.66
staircase; 20 steps 1 m wide	nr	4.23	4.86
Boarded & joisted; plaster soffit			
floor			
200 mm thick	m2	11.74	13.50
250 mm thick	m2	13.75	15.82
landing			
150 mm thick	m2	10.06	11.57
staircase			
flight; 20 steps 1 m wide	nr	5.03	5.79
Existing floor finishings			
Insitu			
Screeds	m2	5.53	6.36
Granolithic finish	m2	7.04	8.10
Terrazzo	m2	8.94	10.28
Asphalt	m2	5.10	5.86
Tile or Sheet			
Tiling; clay	m2	8.94	10.28
Tiling; plastic	m2	0.85	0.97
Sheet; plastic	m2	0.85	0.97
Carpets (loose)	m2	1.02	1.17
Carpets (fixed)	m2	2.21	2.55
Timber			
Floor boarding; denailing	m2	9.73	11.19
Wood block	m2	8.39	9.64

BREAKING OUT - FINISHINGS

- Insitu finishings; hack off; clean and form key;
- Ceilings; pull down; denail joists if necessary;
- Timber; pull down complete with grounds;
- Clear away debris; disposal fees by Contractor.

	Unit	New	Refurb
Hack off existing wall finishings			
Insitu;			
Plasterwork	m2	2.07	2.38
Cement rendering	m2	3.62	4.17
Roughcast	m2	3.24	3.73
Moulded cornice	m	1.27	1.46
Tiling;			
Glazed tiling	m2	4.70	5.40
Glazed tiling and screed	m2	5.18	5.96
Panelling;			
Lincrusta dado lining	m2	2.99	3.44
Timber panelling	m2	5.49	6.31
Skirtings			
Insitu			
100 mm high	m	2.01	2.31
150 mm high	m	2.38	2.74
Timber			
100 mm high	m	0.77	0.88
150 mm high	m	0.91	1.05
Dado or picture rails			
Insitu			
75 mm high	m	1.65	1.89
Timber			
75 mm high	m	0.67	0.77
Pull down existing ceiling finishings			
Insitu; Lath and plaster			
up to 3 m high	m2	6.36	7.31
over 3 m high	m2	8.55	9.84
beam casing; 600 mm girth	m	6.54	7.52
Insitu; Plasterboard and finish			
up to 3 m high	m2	3.85	4.43
over 3 m high	m2	5.18	5.96
beam casing; 600 mm girth	m	3.96	4.56

REPAIRING DOORS AND WINDOWS

- Doors; ease and repair; overhaul ironmongery;
- Windows; ease and repair; overhaul ironmongery;
- Glazing; replace broken glass; make good putty;
- Ironmongery; replace if required; Clear away debris.

	Unit	New	Refurb
Existing doors			
Dismantle; take out complete;			
Single	nr	12.21	14.04
Pairs	nr	18.58	21.37
Overhaul; ease and adjust;			
Single	nr	32.94	37.88
Pairs	nr	42.70	49.10
Existing windows			
Dismantle; take out complete;			
timber or metal			
size 900 x 1050 mm	nr	18.07	20.78
size 1220 x 1650 mm	nr	25.89	29.77
Overhaul; ease and adjust;			
Sash windows; timber			
size 900 x 1050 mm	nr	34.16	39.28
size 1220 x 1650 mm	nr	42.70	49.10
Casement windows; timber			
size 900 x 1050 mm	nr	28.06	32.27
size 1220 x 1650 mm	nr	31.72	36.47
Ironmongery; replace existing with new			
Locks; door			
Mortice lock (£6.50)	nr	18.85	21.67
Ditto; 5 lever (£17.00)	nr	31.11	35.77
Handle; door			
Knob furniture (£4.50)	pair	11.59	13.33
Lever handles (£9.00)	pair	13.72	15.78
Hinges			
100 mm steel (£1.75)	nr	7.62	8.77
Overhead door closure; (£65.00)	nr	103.69	119.24
Windows			
Catches; brass (£4.50)	nr	14.03	16.13
Weights; sash cord (£11.00)	nr	19.09	21.95

CALCULATING THE COSTS

Example; How to Use the Tables

Cutting a New Door Opening in Historic Building

OPENING - Select rate according to size; 1500 mm wide; door with fanlight		234.79
Making Good Finishings		
FACE 'A' - Select finishing; Facing Bricks		
Select rate according to size		61.11
ADD - for historic restoration	30%	18.33
FACE 'B' - Select finishing; Plastered Walls		
Select rate according to size		41.33
ADD - for historic restoration	30%	12.40
TOTAL COST		367.95

Note;- Alternatively this method can be used for window openings using Page 115.

RESTORATION IN HISTORIC BUILDINGS

Extra costs if work is in a Listed Historic Building

- Facing brickwork - ADD 65% for special facings, strings or quoins;
- Rubble walling - ADD 85% for dressed quoins or strings;
- Ashlar stone facing - ADD 85% for raised quoins or strings;
- Plastered walls - ADD 35% for decorative moulded panelling;
- Wood panelling walls - ADD 65% for special feature moulded panels.

Note; See page 107 for general guidelines for work in old buildings and fine art work.

DOOR OPENINGS

Cost of Cutting New Openings

Width mm	Single door 900	Single door 1050	Pair doors 1500	Pair doors 2000
Height	**Hollow wall - 275 mm thick**			
normal	136.10	155.17	224.31	280.68
with fanlight	143.57	163.39	234.79	293.67
high opening	148.74	169.08	242.05	302.67
	Hollow wall - 387 mm thick			
normal	167.35	191.62	299.42	376.58
with fanlight	176.50	201.79	312.70	393.29
high opening	182.82	208.84	321.89	404.87
	Solid wall - 104 mm thick			
normal	65.34	75.21	108.22	136.96
with fanlight	68.64	78.89	113.03	143.03
high opening	70.93	81.44	116.36	147.23
	Solid wall - 215 mm thick			
normal	106.15	122.15	173.20	217.43
with fanlight	112.14	128.80	181.82	228.23
high opening	116.29	133.41	187.79	235.72
	Solid wall - 319 mm thick			
normal	150.63	165.41	248.74	312.16
with fanlight	158.89	174.17	260.76	327.32
high opening	164.62	180.24	269.08	337.81
	Solid wall - 450 mm thick			
normal	187.24	216.15	308.34	386.31
with fanlight	197.87	227.98	323.80	405.78
high opening	205.22	236.18	334.49	419.27
	Stud framed partition - 125 mm thick			
normal	54.49	62.31	89.09	113.09
with fanlight	57.32	65.44	93.10	118.09

Making Good Finishings

(Cost for each wall face)

Width mm	Single door 900	Single door 1050	Pair doors 1500	Pair doors 2000
Height	**A - Ordinary brickwork pointed.**			
normal	9.66	10.63	13.52	16.73
with fanlight	9.91	10.87	13.76	16.98
high opening	10.08	11.04	13.93	17.15
	B - Facing brickwork pointed			
normal	36.14	37.90	43.18	49.05
with fanlight	38.93	40.69	45.97	51.84
high opening	40.85	42.61	47.90	53.76
	C - Stonework, coursed rubble			
normal	48.39	50.60	57.23	64.59
with fanlight	52.27	54.48	61.11	68.47
high opening	54.95	57.16	63.79	71.15
	D - Stonework, ashlar			
normal	94.02	101.03	122.04	145.38
with fanlight	99.66	106.67	127.68	151.02
high opening	103.57	110.57	131.58	154.92
	E - Plastered walls			
normal	31.76	33.51	38.75	44.58
with fanlight	34.33	36.08	41.33	47.16
high opening	36.11	37.86	43.11	48.94
	F - Glazed tiled walls			
normal	104.60	108.01	118.21	129.55
with fanlight	114.47	117.87	128.08	139.42
high opening	121.30	124.71	134.91	146.25
	G - Wood panelled walls			
normal	119.57	126.49	147.23	170.29
with fanlight	128.03	134.94	155.69	178.75
high opening	133.88	140.80	161.55	184.60

WINDOW OPENINGS

Cost of Cutting New Openings

Height	Width of Windows (mm) 900	1050	1500	2000
	Hollow wall - 275 mm thick			
600	90.72	104.97	159.68	200.01
900	99.33	114.45	171.77	215.00
1450	115.12	131.84	193.93	242.48
	Hollow wall - 387 mm thick			
600	116.45	134.98	225.60	283.65
900	127.00	146.72	240.91	302.94
1450	146.33	168.24	268.99	338.30
	Solid wall - 104 mm thick			
600	39.05	45.51	68.30	85.70
900	42.86	49.76	73.86	92.70
1450	49.85	57.55	84.04	105.54
	Solid wall - 215 mm thick			
600	67.42	78.90	116.36	145.48
900	74.34	86.57	126.30	157.95
1450	87.02	100.64	144.54	180.82
	Solid wall - 319 mm thick			
600	102.38	119.84	178.15	222.96
900	111.91	130.46	192.02	240.44
1450	129.39	149.93	217.45	272.50
	Solid wall - 450 mm thick			
600	127.78	149.94	221.91	277.40
900	140.04	163.60	239.74	299.87
1450	162.52	188.63	272.43	341.07
	Stud framed partition - 125 mm thick			
600	29.89	34.60	52.04	65.66
900	33.15	38.20	56.67	71.43
1450	39.12	44.80	65.15	76.18

Making Good Finishings

(Cost for each wall face)

Height	Width of Windows (mm) 900	1050	1500	2000
	A - Ordinary brickwork pointed.			
600	12.28	13.57	17.44	21.74
900	12.52	13.81	17.68	21.99
1450	12.69	13.98	17.85	22.16
	B - Facing brickwork pointed			
600	42.68	45.25	52.98	61.57
900	45.46	48.04	55.77	64.36
1450	47.39	49.96	57.69	66.28
	C - Stonework, coursed rubble			
600	58.06	61.48	71.73	83.12
900	61.94	65.35	75.61	87.00
1450	64.62	68.04	78.29	89.69
	D - Stonework, ashlar			
600	106.44	114.99	140.65	169.17
900	112.07	120.63	146.29	174.81
1450	115.98	124.53	150.20	178.71
	E - Plastered walls			
600	42.80	45.93	55.31	65.74
900	45.37	48.50	57.89	68.32
1450	47.15	50.28	59.67	70.10
	F - Glazed tiled walls			
600	102.58	105.73	115.18	125.67
900	112.45	115.60	125.04	135.54
1450	119.28	122.43	131.88	142.37
	G - Wood panelled walls			
600	119.99	126.96	147.86	171.09
900	128.45	135.41	156.32	179.55
1450	134.30	141.27	162.18	185.40

SPECIFICATION HEADINGS

Cutting Openings - Existing Walls

- Insert needles over openings exceeding 1.2 metre wide; wedge up and pin;
- Erect shoring to face of wall over opening where necessary;
- Cut away brickwork or stonework;
- Cut back wall finishings on both sides of opening for a minimum width of 400 mm;
- Build in new lintel over new opening;
- Provide cavity gutter over opening where necessary;
- Face up jambs ready for new finishings;
- Doors; form threshold as existing; new water bar;
- Windows; form cill; insert damp proof course;
- Remove shoring; clear away debris.

Making Good - Existing Finishings

- Piece in facing bricks or stonework in gauged mortar to head, jambs and cill where exposed;
- point joints to match existing;
- Form arch over in common bricks, facing bricks or stonework to match existing; point joints to match existing;
- Render reveals and face of wall internally and externally as necessary; make good;
- Make good wall tiling to reveal and face of wall to match existing as necessary;
- Piece out wood, tile or rendered skirting to both jambs as existing; make good up to existing;
- Doors; Make good threshold; lay new floor covering to match existing;
- Windows; form brick, tile or stone cill to match existing; build in ends; point all round.

CALCULATING THE COSTS

Example: How to Use the Tables

Filling in Window Opening

OPENING - Select rate according to size; 1500 mm wide; door with fanlight		126.63
Making Good Finishings		
FACE 'A' - Select finishing; Facing Bricks		
Select rate according to size		68.65
ADD - for historic restoration	0%	0.00
FACE 'B' - Select finishing; Plastered Walls		
Select rate according to size		66.03
ADD - for historic restoration	0%	0.00
TOTAL GUIDE PRICE		261.31

RESTORATION IN HISTORIC BUILDINGS

Extra costs if work is in a Listed Historic Building

- Facing Brickwork - ADD 65% for special facings, strings or quoins;
- Rubble walling - ADD 85% for dressed quoins or strings;
- Ashlar stone facing - ADD 85% for raised quoins or strings;
- Plastered walls - ADD 35% for decorative moulded panelling;
- Wood panelling walls - ADD 65% for special feature moulded panels.

Note: See page 107 for general guidelines for work in old buildings and fine art work.

DOOR OPENINGS

Cost of Filling In New Openings

Width mm	Single door 900	Single door 1050	Pair doors 1500	Pair doors 2000
Height	**Hollow wall - 275 mm thick**			
normal	151.15	172.24	229.08	290.40
with fanlight	166.67	189.62	252.05	319.56
high opening	177.42	201.65	267.95	339.76
	Hollow wall - 387 mm thick			
normal	205.42	233.85	312.74	398.54
with fanlight	226.96	258.04	344.89	439.54
high opening	241.86	274.79	367.15	467.92
	Solid wall - 104 mm thick			
normal	81.98	94.44	125.43	158.01
with fanlight	89.74	103.13	136.91	172.59
high opening	95.11	109.15	144.86	182.69
	Solid wall - 215 mm thick			
normal	136.25	156.06	209.09	266.15
with fanlight	150.02	171.56	229.75	292.56
high opening	159.56	182.29	244.06	310.85
	Solid wall - 319 mm thick			
normal	192.48	208.73	292.81	372.80
with fanlight	212.60	230.24	322.82	411.04
high opening	226.53	245.14	343.59	437.51
	Solid wall - 450 mm thick			
normal	245.46	278.49	371.20	472.35
with fanlight	271.65	307.86	410.08	521.81
high opening	289.79	328.19	437.00	556.04
	Stud framed partition - 125 mm thick			
normal	58.82	68.36	90.57	113.40
with fanlight	63.92	74.06	98.09	122.93
high opening	67.45	78.01	103.29	129.53

Making Good Finishings

(Cost for each wall face)

Height	900	1050	1500	2000
	A - Ordinary brickwork pointed.			
normal	3.40	3.84	5.16	6.63
with fanlight	3.79	4.28	5.74	7.37
high opening	4.06	4.58	6.14	8.13
	B - Facing brickwork pointed			
normal	45.83	51.99	70.45	90.96
with fanlight	50.33	57.05	77.20	99.59
high opening	53.45	60.55	81.87	105.56
	C - Stonework, coursed rubble			
normal	58.34	65.99	88.96	114.47
with fanlight	64.60	73.04	98.35	126.48
high opening	68.94	77.92	104.86	134.79
	D - Stonework, ashlar			
normal	130.40	149.12	205.30	267.71
with fanlight	139.51	159.37	218.97	285.18
hign opening	145.82	166.47	228.43	297.27
	E - Plastered walls			
normal	46.39	52.75	71.83	93.03
with fanlight	50.55	57.43	78.06	101.00
high opening	53.43	60.67	82.38	106.51
	F - Glazed tiled walls			
normal	152.67	172.85	233.40	300.67
with fanlight	168.62	190.79	257.32	331.23
high opening	179.66	203.21	273.88	352.39
	G - Wood panelled walls			
normal	150.50	171.08	232.79	301.37
with fanlight	164.17	186.45	253.29	327.56
high opening	173.63	197.10	267.49	345.70

WINDOW OPENINGS

Cost of Filling In New Openings

Height	Width of Windows 900	1050	1500	2000
Hollow wall - 275 mm thick				
600	64.01	74.64	100.13	126.60
900	81.91	94.69	126.63	160.26
1450	114.74	131.46	175.21	221.96
Hollow wall - 387 mm thick				
600	84.50	98.03	132.21	168.33
900	109.35	125.94	169.30	215.64
1450	154.90	177.10	237.31	302.36
Solid wall - 104 mm thick				
600	38.41	45.64	60.95	76.11
900	47.36	55.67	74.20	92.94
1450	63.78	74.06	98.49	123.79
Solid wall - 215 mm thick				
600	58.90	69.03	93.03	117.84
900	74.79	86.92	116.88	148.32
1450	103.93	119.70	160.60	204.19
Solid wall - 319 mm thick				
600	79.49	87.90	124.33	158.07
900	102.70	112.73	158.95	202.19
1450	145.27	158.25	222.42	283.08
Solid wall - 450 mm thick				
600	98.35	113.58	152.86	194.65
900	128.58	147.46	197.72	251.71
1450	183.99	209.59	279.97	356.32
Stud framed partition - 125 mm thick				
600	30.20	36.35	48.37	59.88
900	36.08	42.92	57.04	70.87
1450	46.87	54.99	72.94	91.03

Making Good Finishings

(Cost for each wall face)

Height	Width of Windows 900	1050	1500	2000
A - Ordinary brickwork pointed.				
600	3.62	4.09	5.51	7.10
900	4.00	4.53	6.10	7.84
1450	4.27	4.83	6.50	8.35
B - Facing brickwork pointed				
600	40.70	46.00	61.90	79.56
900	45.20	51.06	68.65	88.18
1450	48.32	54.57	73.32	94.15
C - Stonework, coursed rubble				
600	53.25	60.05	80.47	103.16
900	59.51	67.10	89.87	115.16
1450	63.85	71.98	96.37	123.48
D - Stonework, ashlar				
600	115.23	131.43	180.02	234.01
900	124.34	141.68	193.69	251.47
1450	130.65	148.77	203.15	263.56
E - Plastered walls				
600	39.17	44.32	59.79	76.98
900	43.33	49.00	66.03	84.95
1450	46.20	52.24	70.34	90.46
F - Glazed tiled walls				
600	148.07	167.49	225.73	290.45
900	164.02	185.43	249.65	321.01
1450	175.06	197.85	266.21	342.17
G - Wood panelled walls				
600	137.31	155.68	210.80	272.04
900	150.98	171.06	231.30	298.24
1450	160.44	181.70	245.49	316.37

SPECIFICATION HEADINGS

Filling in Openings - Existing Walls

- Take out and clear away old door or window complete with frame and architrave;
- Cut back defective wall finishings on both sides for a minimum width of 200 mm and prepare for new;
- Cut out defective brickwork, stonework or blockwork and piece in with new;
- Provide pockets in existing for bonding with new filling, or provide firfix bonding strip;
- Provide damp course over where necessary;
- Fill in opening with brick, block or stonework in gauged mortar to thickness of existing wall;
- Face up both sides of wall to match existing;
- Piece in skirtings and dado rails;
- Remove shoring; clear away debris.

Making Good - Existing Finishings

- Plasterwork; dubb out as necessary to allow new finishings to be finished flush;
- Cut back edges in straight lines to existing solid wall and clean off dust;
- Plaster; cover filling to match existing and make good all round;
- Facing brickwork; face up to match existing and make good all round;
- Wall tiling; cover infilling to match existing as necessary and bond into existing;
- Piece out wall panelling to match existing;
- Piece out wood, tile or rendered skirting to both sides of infilling; make good up to existing.

SPECIFICATION OF SUB - SOIL

NOTE;- ADD to composite rates and excavation unit rates for the following ground conditions. (See BS8004: 1986)

Type of Ground	Load Bearing Capacity/ m2	Cost Rating
♦ Loose sand or soil	< 75 kN	+ 12 %
♦ Firm sand or soil	> 100 < 300 kN	+ 8 %
♦ Clay -firm	> 75 < 300 kN	+ 6.5 %
♦ Clay - heavy firm	> 300 < 600 kN	+ 5 %
♦ Compacted soil	< 300 kN	0 %
♦ Gravel	> 200 < 600 kN	- 5 %
♦ Soft rock	> 2000 < 4000 kN	- 7.5%
♦ Hard rock	< 10000 kN	+ 12 %

GLOSSARY

Auger A hand or machine driven tool for digging circular holes for piles.

Backfilling Excavated materials returned as filling around foundation walls, well rammed and consolidated.

Cart away Load into lorries and clear away from site including tipping charge.

Damp proof A layer of impervious material laid on course walls or floor slabs to prevent rising damp.

Heave The swelling of clay caused by an increase in its water content.

Foundations to External Hollow walls

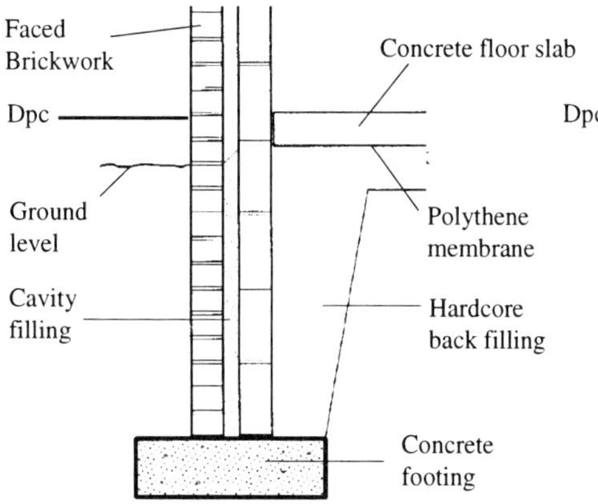

Strip Foundation
External Walls
£90.00 per metre

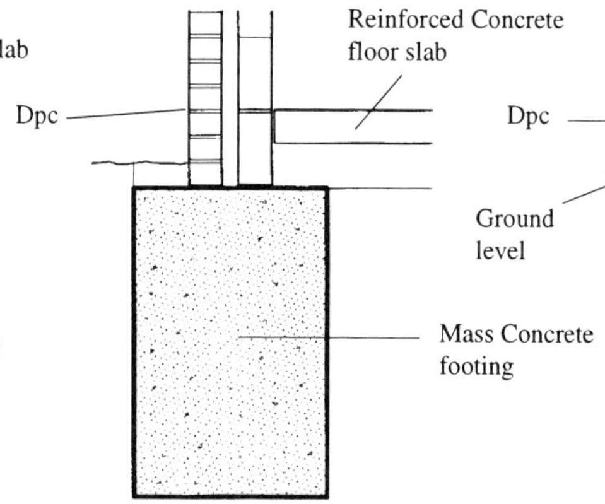

Ground Beam
External Walls
£100.00 per metre

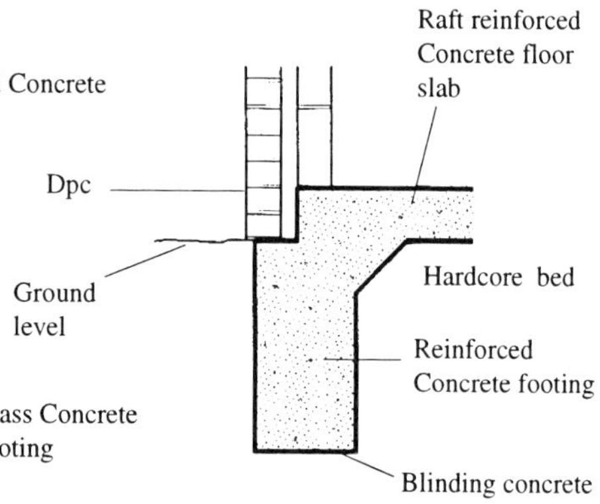

Toe Ground Beam
External Walls
£75.00 per metre

Foundations to Internal walls

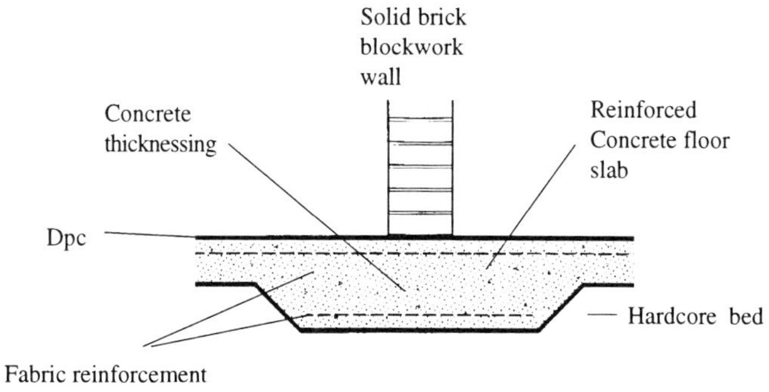

Thicknessing to Floor Slab
Internal Walls
£20.00 per metre

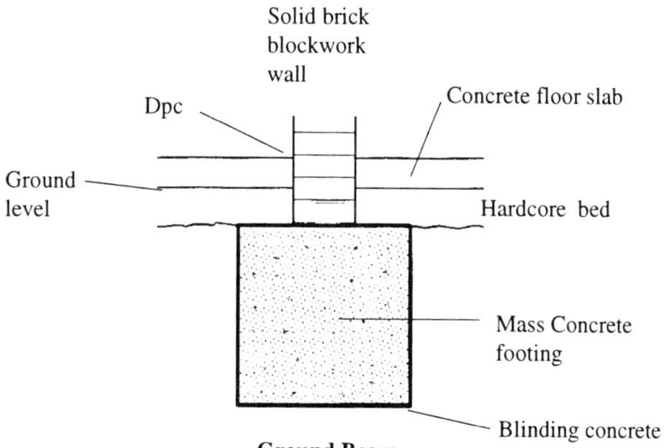

Ground Beam
Internal Walls
£35.00 per metre

EXCAVATION - by Hand

- Setting out; temporary profiles; hand digging;
- Load into barrows; wheel.

	Unit	New	Refurb
Digging - by hand			
Top soil; deposit in spoil heaps			
150 mm (average) deep	m2	1.57	1.72
Excavate to reduce site levels			
< 1 metre deep	m3	10.99	12.09
>1 metre < 2 metre deep	m3	12.81	14.09
Excavate for pits			
< 2 metres deep	m3	15.62	17.18
> 2 metres < 4 metres deep	m3	23.75	26.12
Excavate trenches			
< 2 metres deep	m3	16.86	18.55
> 2 metres < 4 metres deep	m3	20.62	22.68
ADD Extra for cost of;-			
digging below water table	m3	2.19	2.41
breaking up hard rock	m3	36.51	40.16
working next Main Highway	m3	6.25	6.87
Disposal - Surplus excavated material			
Back fill into trenches	m3	6.71	7.38
Wheel and deposit in temporary spoil heaps			
< 50 metres	m3	9.06	9.97
>50 < 100 metres	m3	13.12	14.43
Dig from spoil heaps and spread and level <200 mm deep	m3	11.00	12.10
Load and transport off site			
< 5 km	m3	19.81	21.79
> 5 < 10 km	m3	26.47	29.12
Temporary support to earthwork			
Depth of excavation			
< 1 metre	m3	2.84	3.13
> 1 < 2 metre	m3	3.54	3.90

EXCAVATION - by Machine

- Setting out; temporary profiles; machine digging;
- Load into dumpers; wheel; deposit in spoil heaps;
- Temporary earth supports; remove groundwater.

	Unit	New	Refurb
Digging - by machine			
Top soil; deposit in spoil heaps			
150 mm (average) deep	m2	0.97	1.06
Excavate to reduce site levels			
< 1 metre deep	m3	1.20	1.32
>1 metre < 2 metre deep	m3	1.03	1.13
Excavate for pits			
< 2 metres deep	m3	4.81	5.29
> 2 metres < 4 metres deep	m3	5.07	5.57
Excavate trenches			
< 2 metres deep	m3	3.93	4.32
> 2 metres < 4 metres deep	m3	4.37	4.81
ADD Extra for cost of;-			
digging below water table	m3	2.07	2.28
breaking up hard rock	m3	24.80	27.28
working next Main Highway	m3	13.01	14.31
Disposal - Surplus excavated material			
Back fill into trenches	m3	4.90	5.38
Wheel and deposit in temporary spoil heaps			
< 50 metres	m3	0.99	1.09
>50 < 100 metres	m3	1.77	1.95
Dig from spoil heaps and spread and level <200 mm deep	m3	2.55	2.81
Load and transport off site			
< 1 km	m3	17.66	19.42
> 1 < 5 km	m3	20.03	22.03
Temporary support to earthwork			
Depth of excavation			
< 1 metre	m3	2.84	3.13
> 1 < 2 metres	m3	3.54	3.90

FILLING

- Crushed stone; clean; all in aggregate;
- Broken brick or stone; 50 mm mesh;
- Reclaimed hardcore; clean; 50 mm mesh.

	Unit	New	Refurb
Crushed stone - by hand			
Filling > 300 mm deep	m3	34.62	38.08
Beds			
100 mm thick	m2	4.33	4.76
150 mm thick	m2	5.97	6.57
Broken brick or stone - by hand			
Filling > 300 mm deep	m3	23.04	25.34
Beds			
100 mm thick	m2	2.88	3.17
150 mm thick	m2	3.97	4.37
Reclaimed materials - by hand			
Filling > 300 mm deep	m3	16.27	17.90
Beds			
100 mm thick	m2	2.03	2.24
150 mm thick	m2	2.81	3.09
Imported Soil - by machine			
Filling > 300 mm deep	m3	13.85	15.23
Crushed stone - by machine			
Filling > 300 mm deep	m3	29.72	32.69
Beds			
100 mm thick	m2	3.71	4.09
150 mm thick	m2	5.13	5.64
Broken brick or stone - by machine			
Filling > 300 mm deep	m3	18.20	20.02
Beds			
100 mm thick	m2	2.28	2.50
150 mm thick	m2	3.14	3.45
Reclaimed materials - by machine			
Filling > 300 mm deep	m3	11.73	12.90
Beds			
100 mm thick	m2	1.47	1.61
150 mm thick	m2	2.02	2.23

CONCRETE

- Ready mix; dense aggregate; 25 N/mm2; delivered;
- Transporting across site; filling into formwork;
- Compacting around reinforcement; curing.

	Unit	New	Refurb
Insitu concrete; without any steel reinforcement			
Foundations			
Strip footings	m3	79.82	87.81
Ground beams	m3	91.00	100.10
Column bases	m3	87.81	96.59
Beds			
100 mm thick	m2	8.18	9.00
150 mm thick	m2	12.27	13.50
Steps	m3	95.79	105.37
Insitu concrete; reinforced with steelwork			
Foundations			
Strip footings	m3	85.81	94.39
Ground beams	m3	97.82	107.61
Column bases	m3	94.39	103.83
Beds			
100 mm thick	m2	8.80	9.68
150 mm thick	m2	13.19	14.51
Frame			
Beams	m3	101.55	111.70
Columns	m3	107.98	118.77
Casing to steel work	m3	112.00	123.20
Walls			
150 mm thick	m2	15.24	16.77
200 mm thick	m2	19.54	21.49
Suspended floors			
150 mm thick	m2	14.13	15.54
200 mm thick	m2	18.84	20.72
Staircases			
steps	m3	117.63	129.40
landings 100 mm thick	m2	12.35	13.59

NOTE;- Unit Rates for concrete work do not include for the cost of any reinforcement or formwork.

FORMWORK

- Temporary timber moulds for casting concrete;
- Formwork to face of concrete; watertight;
- Struts, collars & spacers; striking and clearance.

	Unit	New	Refurb
Formwork; to faces of concrete without reinforcement			
Foundations			
Footings & ground beams	m2	21.89	24.08
Column bases	m2	22.98	25.28
Edge of beds;			
100 mm thick	m	2.74	3.01
150 mm thick	m	4.10	4.51
Formwork; to faces of concrete with reinforcement			
Foundations			
Footings & ground beams	m2	26.82	29.50
Column bases	m2	28.16	30.97
Edge of beds			
100 mm thick	m	3.35	3.69
150 mm thick	m	5.03	5.53
Frame			
Soffit of suspended floors			
< 3.0 metres high	m2	23.20	25.52
< 3.0 metres high; sloping	m2	25.31	27.84
Beams	m2	29.65	32.62
Columns	m2	27.44	30.18
Columns; circular	m2	40.99	45.09
Walls	m2	25.03	27.53
Staircases			
Sloping soffit	m2	27.84	30.62
Outer edge of steps	m2	33.41	36.75
Riser 150 mm high	m	5.22	5.74
ADD; Extra cost for;-			
Formwork > 3.0 metres high		1.60	1.76
Special finishings; rubbing down		15%	15%
Formwork left in at completion		25%	25%
Forming nibs or rebates		5.12	5.64
Mortices		1.92	2.11

STEEL REINFORCEMENT

- Mild steel; annealed tying wire; spacers;
- Rods; cut and bent; labelled; hooked ends;
- Mesh; lapped one full mesh at all joints.

	Unit	New	Refurb
Plain round steel bars			
Main bars; columns and beams			
12 mm	kg	0.80	0.88
16 mm	kg	0.77	0.84
20 mm	kg	0.74	0.82
25 mm	kg	0.73	0.80
32 mm	kg	0.71	0.78
40 mm	kg	0.77	0.85
Links and stirrups			
6 mm	kg	0.86	0.95
8 mm	kg	0.88	0.96
10 mm	kg	0.91	1.00
Steel fabric reinforcement			
In beds and suspended slabs			
B283; 100 x 200 mm; 3.73 kg	m2	3.44	3.78
B385; 100 x 200 mm; 4.53 kg	m2	4.01	4.41
B503; 100 x 200 mm; 5.93 kg	m2	4.99	5.49
B785; 100 x 200 mm; 8.14 kg	m2	6.41	7.05
B1131 100 x 200 mm; 10.90 kg	m2	8.15	8.97
Wrapping to beams and columns			
D98; 200 x 200 mm; 1.54 kg	m2	3.82	4.20
A193; 200 x 200 mm; 3.02 kg	m2	4.84	5.32
A142; 200 x 200 mm; 2.22 kg	m2	4.35	4.78
ADD; Extra costs of materials and labour			
Materials			
Deformed steel bars		-10%	-8%
High yield reinforcement		5%	5%
Stainless steel reinforcement		above rates times 12	
Labour			
Cutting and bending on site		17%	25%
Raking cutting fabric	m	1.68	1.85
Circular cutting fabric	m	2.60	2.85

SUNDRIES

	Unit	New	Refurb
Special Finishing to insitu floors			
Extra cost for;-			
Trowelling surfaces smooth			
level	m2	0.99	1.09
falls or falls and cross falls	m2	1.10	1.21
Power float finish			
level	m2	2.02	2.22
falls or falls and cross falls	m2	2.14	2.35
Surface hardener	m2	4.75	5.22
Bush hammering; machine operated			
Surfaces > 300 mm wide	m2	7.29	8.02
Surfaces < 300 mm wide	m2	7.76	8.53
Waterproof Membranes			
2000 Gauge polythene	m2	1.88	2.07
RIW Waterproofing Liquid;	m2	5.20	5.72

REPAIRS AND RESTORATION

	Unit	New	Refurb
Cleaning concrete surfaces			
Jet pressure spray with diluted additive, remove cement stains	m2	-	3.77
Remove existing paint	m2	-	4.82
Repair surfaces of concrete; cut out defective areas; wire brush exposed reinforcement; apply rust preventative liquid; prime concrete; make good surface with filling compound.			
Surfaces > 5 m2			
> 300 mm wide	m2	-	19.99
< 300 mm wide	m2	-	26.58
Surfaces < 5 m2; isolated areas			
> 300 mm wide	m2	-	34.97
< 300 mm wide	m2	-	46.52

NOTE ;- These rates exclude the cost of scaffolding

SUSPENDED FLOOR SYSTEMS

- Proprietary design by specialist; complete;
- Concrete; pots or planks; reinforcement;

Unit rates per m2

Maximum Span feet	Maximum Span metre	Floor Loading (kN per m2) 1.50	3.00	7.50	10.00
Prestressed concrete planks; lightweight blocks					
12.50	4.26	18.89	19.35	20.27	20.73
15.00	5.11	19.36	19.84	20.78	21.25
17.50	5.96	19.85	20.33	21.30	21.78
20.00	6.82	20.34	20.84	21.83	22.33
22.50	7.67	20.85	21.36	22.38	22.89
25.00	8.52	21.37	21.89	22.94	23.46
Prestressed precast concrete planks (unit rate per m2)					
15	3.61	21.70	22.23	23.29	23.82
20	4.82	22.24	22.79	23.87	24.41
25	6.02	22.80	23.36	24.47	25.02
30	7.22	23.37	23.94	25.08	25.65
35	8.43	23.95	24.54	25.71	26.29
40	9.63	24.55	25.15	26.35	26.95
Dow Mac Double Tee Beams; 50 mm screed					
20	6.82	28.30	28.99	30.37	31.06
30	10.22	29.00	29.71	31.13	31.83
40	13.63	29.73	30.45	31.90	32.63
50	17.04	30.47	31.22	32.70	33.44
60	20.45	31.23	32.00	33.52	34.28
70	23.86	32.01	32.80	34.36	35.14
Dow Mac Double Tee Beams; 75 mm screed					
20	6.82	32.39	33.18	34.76	35.55
30	10.22	33.20	34.01	35.62	36.43
40	13.63	34.03	34.86	36.51	37.34
50	17.04	34.88	35.73	37.43	38.28
60	20.45	35.75	36.62	38.36	39.24
70	23.86	36.64	37.54	39.32	40.22

PRECAST UNITS

- Dense aggregate concrete; grade 32 N/mm2;
- Moulds; surface fair finishings where required;
- All necessary reinforcement; hopisting and fixing.

Unit Rate each lineal

Maximum Span feet	metre	Square section (width x height) 102 x 152	102 x 229	152 x 229	Boot shaped 225 x 152	225 x 229
Lintels; precast concrete						
3.0	1.02	3.47	3.56	3.73	3.81	3.90
4.0	1.36	3.53	3.61	3.78	3.87	3.96
5.0	1.70	3.58	3.67	3.84	3.93	4.01
6.0	2.04	3.63	3.72	3.90	3.99	4.07
7.0	2.39	3.69	3.78	3.96	4.05	4.14
8.0	2.73	3.74	3.83	4.02	4.11	4.20
9.0	3.07	3.80	3.89	4.08	4.17	4.26
Lintels; prestressed concrete						
3.0	1.02	4.54	4.65	4.87	4.99	5.10
4.0	1.36	4.61	4.72	4.95	5.06	5.17
5.0	1.70	4.68	4.79	5.02	5.14	5.25
6.0	2.04	4.75	4.87	5.10	5.21	5.33
7.0	2.39	4.82	4.94	5.17	5.29	5.41
8.0	2.73	4.89	5.01	5.25	5.37	5.49
9.0	3.07	4.97	5.09	5.33	5.45	5.57

COPING STONES

Unit Rate per metre

Section	Splayed			Saddleback	
Width / Thickness	177 x 75	304 x 75	425 x 125	315 x 425	425 x 125
Coping; precast concrete					
	9.03	9.25	9.69	9.91	10.13
Coping; reconstructed stone					
	13.31	13.63	14.28	14.60	14.93

TYPES OF BRICKS

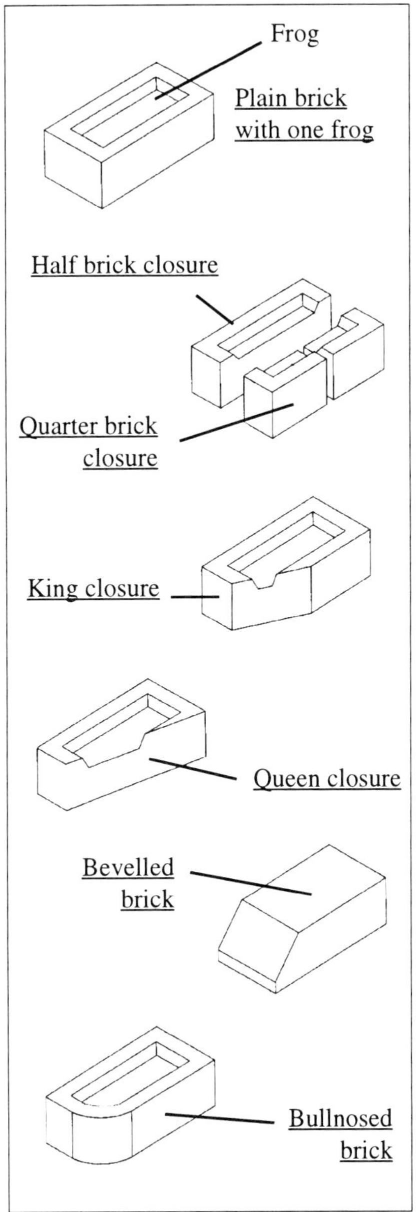

TYPES OF BRICK BONDING

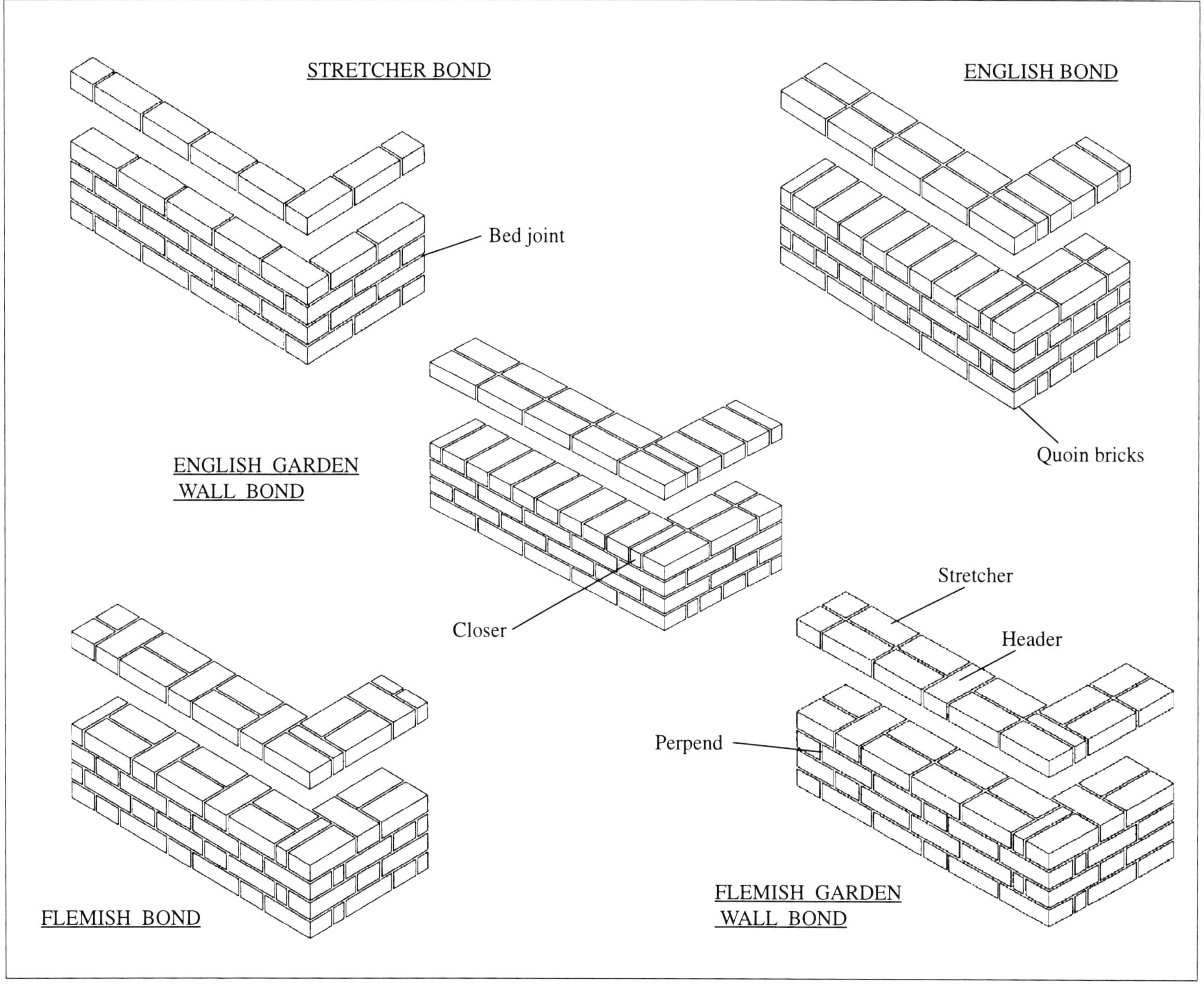

Adjustment of KEY rates related to
PURCHASE PRICE OF COMMON BRICKS

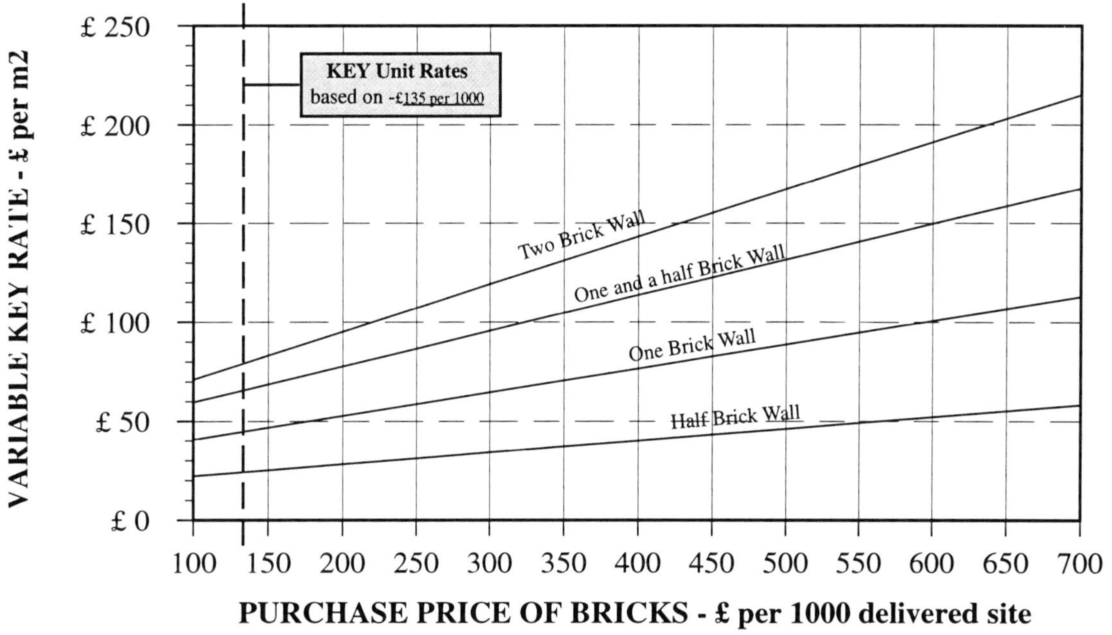

ADJUSTMENTS FOR USE OF OTHER MATERIALS

Extra cost of building in Imperial Bricks

Adjust the key rates for brickwork if built in Imperial bricks;		Brick thickness 68 mm	73 mm
half brick wall	m2	-1.50%	-1.85%
one brick wall	m2	-1.60%	-1.93%
one and a half brick wall	m2	-1.65%	-2.05%

Extra cost of mortar for bricks with two frogs

Add to the key rates for brickwork if the common bricks have two frogs;		One Frog	Two Frogs
half brick wall	m2	incl	0.39
one brick wall	m2	incl	0.77
one and a half brick wall	m2	incl	1.15

Variation in cost of different mortar mixes

Adjust the key rates for brickwork for the following alternative mixes;	Mix 1:1:6	Mix 1:2:9	Mix 1:3
gauged mortar	incl	1.25%	-
cement mortar	-	-	1.65%
coloured mortar	1.45%	1.45%	1.50%
refractory mortar	3.50%	3.50%	3.50%
proprietary brand mortar	3.25%	3.25%	3.25%
sulphate resistant cement	4.50%	-	5.20%

NOTE; The extra cost of special mortar for pointing is set out under the Section "Price Adjustments" for facing brickwork. See Page 127

65 mm COMMON BRICKS

- price **£135.00** per 1000 delivered to site;
- metric size 215 x 102.5 x 65 mm;
- bedded in gauged mortar (1:1:6) ;
- joints 9.5 mm thick; one frog.

	Unit	New	Refurb
Half Brick (HB); 102.5 mm thick			
Foundation wall	m2	23.36	26.28
Plain wall	m2	24.53	27.59
Dwarf wall	m2	24.89	28.00
Skin of cavity wall	m2	25.14	28.28
Extra cost where built;-			
circular on plan	m2	5.43	6.11
as backing or thicknessing	m2	22.19	24.96
working overhand	m2	1.88	2.12
One brick wall (1B); 215 mm thick			
Foundation wall	m2	43.25	48.66
Plain wall	m2	45.41	51.09
Dwarf wall	m2	46.09	51.86
Skin of cavity wall	m2	46.55	52.37
Extra cost where built;-			
circular on plan	m2	8.75	9.84
as backing or thicknessing	m2	21.19	23.84
working overhand	m2	2.80	3.15
One and a half brick wall (1.5B); 317.5 mm thick			
Foundation wall	m2	62.71	70.55
Plain wall	m2	65.85	74.08
Dwarf wall	m2	66.84	75.19
Skin of cavity wall	m2	67.50	75.93
Extra cost where built;-			
circular on plan	m2	12.69	14.27
as backing or thicknessing	m2	30.73	34.57
working overhand	m2	4.06	4.56

Properties *Management*

90 Maintenance Comprehensive

Thickness	Brick Bond	No Wall in Common bricks	Wall faced ONE side Commons	Facings	Bed Joint	Perpend	Wall faced BOTH sides Commons	Facings	Bed Joint	Perpend	Mortar Volume (1 frog) m3
65 mm Metric Bricks; size 215 x 102.5 x 65 mm											
HB	Stretcher	60	0	60	13	4	0	60	26	8	0.024
	English	60	0	75	13	6	0	75	26	12	0.024
	Flemish	60	0	80	13	5	0	80	26	10	0.024
	Garden Wall	60	0	75	13	5	0	75	26	10	0.024
1B	Stretcher	120	60	60	13	4	0	120	26	8	0.058
	English	120	30	90	13	6	0	120	26	12	0.058
	Flemish	120	40	80	13	5	0	120	26	10	0.058
	Garden Wall	120	45	75	13	5	0	120	26	10	0.058
1.5B	Stretcher	180	120	60	13	4	60	120	26	8	0.091
	English	180	90	90	13	6	30	150	26	12	0.091
	Flemish	180	100	80	13	5	30	150	26	10	0.091
	Garden Wall	180	105	75	13	5	30	150	26	10	0.091
2B	Stretcher	240	180	60	13	4	120	120	26	8	0.125
	English	240	150	90	13	6	60	180	26	12	0.125
	Flemish	240	160	80	13	5	80	160	26	10	0.125
	Garden Wall	240	165	75	13	5	90	150	26	10	0.125
68 mm (2 5/8th") **Imperial;** size 215 x 102.5 x 68 mm											
HB	Stretcher	56	0	56	13	4	0	56	26	8	0.023
	English	56	0	69	13	6	0	69	26	12	0.023
	Flemish	56	0	74	13	5	0	74	26	10	0.023
	Garden Wall	56	0	69	13	5	0	69	26	10	0.023
1B	Stretcher	111	55	56	13	4	0	111	26	8	0.056
	English	111	28	83	13	6	0	111	26	12	0.056
	Flemish	111	37	74	13	5	0	111	26	10	0.056
	Garden Wall	111	42	69	13	5	0	111	26	10	0.056
1.5B	Stretcher	167	111	56	13	4	56	111	26	8	0.089
	English	167	84	83	13	6	28	139	26	12	0.089
	Flemish	167	93	74	13	5	28	139	26	10	0.089
	Garden Wall	167	98	69	13	5	28	139	26	10	0.089
2B	Stretcher	222	166	56	13	4	111	111	26	8	0.122
	English	222	139	83	13	6	55	167	26	12	0.122
	Flemish	222	148	74	13	5	74	148	26	10	0.122
	Garden Wall	222	153	69	13	5	83	139	26	10	0.122

Wall Thickness	Brick Bond	No Wall in Common bricks	Wall faced ONE side Commons	Facings	Bed Joint	Perpend	Wall faced BOTH sides Commons	Facings	Bed Joint	Perpend	Mortar Volume (1 frog)
51 mm (2") **Imperial;** size 215 x 102.5 x 51 mm											
HB	Stretcher	71	0	71	17	4	0	71	34	8	0.029
	English	71	0	89	17	5	0	89	34	10	0.029
	Flemish	71	0	95	17	5	0	95	34	10	0.029
	Garden Wall	71	0	89	17	5	0	89	34	10	0.029
1B	Stretcher	143	72	71	17	4	0	143	34	8	0.067
	English	143	36	107	17	5	0	143	34	10	0.067
	Flemish	143	48	95	17	5	0	143	34	10	0.067
	Garden Wall	143	54	89	17	5	0	143	34	10	0.067
1.5B	Stretcher	214	143	71	17	4	71	143	34	8	0.106
	English	214	107	107	17	5	35	179	34	10	0.106
	Flemish	214	119	95	17	5	35	179	34	10	0.106
	Garden Wall	214	125	89	17	5	35	179	34	10	0.106
2B	Stretcher	286	215	71	17	4	143	143	34	8	0.144
	English	286	179	107	17	5	72	214	34	10	0.144
	Flemish	286	191	95	17	5	96	190	34	10	0.144
	Garden Wall	286	197	89	17	5	107	179	34	10	0.144
73 mm (2 7/8") **Imperial;** size 215 x 102.5 x 73 mm											
HB	Stretcher	52	0	52	12	4	0	52	24	8	0.022
	English	52	0	65	12	6	0	65	24	12	0.022
	Flemish	52	0	70	12	5	0	70	24	10	0.022
	Garden Wall	52	0	65	12	5	0	65	24	10	0.022
1B	Stretcher	104	52	52	12	4	0	104	24	8	0.054
	English	104	26	78	12	6	0	104	24	12	0.054
	Flemish	104	34	70	12	5	0	104	24	10	0.054
	Garden Wall	104	39	65	12	5	0	104	24	10	0.054
1.5B	Stretcher	157	105	52	12	4	53	104	24	8	0.086
	English	157	79	78	12	6	27	130	24	12	0.086
	Flemish	157	87	70	12	5	27	130	24	10	0.086
	Garden Wall	157	92	65	12	5	27	130	24	10	0.086
2B	Stretcher	209	157	52	12	4	105	104	24	8	0.118
	English	209	131	78	12	6	52	157	24	12	0.118
	Flemish	209	139	70	12	5	70	139	24	10	0.118
	Garden Wall	209	144	65	12	5	79	130	24	10	0.118

SCAFFOLDING

	Unit	New	Refurb
Lump Sum installation charge			
Charges for installation and dismantling at completion.			
up to 20 metres long on plan			
up to 10 metre high;	m2	4.69	5.39
10 to 20 metres high;	m2	5.86	6.74
over 20 metres long on plan			
up to 10 metre high;	m2	4.34	4.99
10 to 20 metres high;	m2	5.42	6.23
Cantilevered supports for fans	m	2.03	2.34
Protection; plastic sheeting	m2	1.05	1.21
Weekly hire charges			
ADD for Weekly hire charges;			
staging up to 15 metre high;	m2	0.39	0.45
staging 15 to 30 metres high;	m2	0.49	0.56
Cantilevered supports for fans	m	1.75	2.02
Work platform; 1 metre wide;	m	1.63	1.87
Protective Fans			
Corrugated sheeting	m	8.74	10.05
Scaffold boarding	m	10.41	11.97
Tarpaulines	m	8.55	9.83
Towers			
1.2 x 1.2 x 3 metres high	nr	28.60	32.89
1.8 x 1.8 x 3 metres high	nr	42.90	49.33
1.8 x 1.8 x 5 metres high	nr	48.07	55.28
Platform for MINOR works			
Trestles and steps	nr	9.44	10.85

PRICING GUIDE

- Key Rates for brickwork do not include the cost of scaffolding or towers.
- ADD the appropriate rate from the above table to adjust your key rates to include for scaffolding or trestles.

CAVITY CONSTRUCTION

	Unit	New	Refurb
Forming Cavities in Walls			
Form cavity between brick or block skins			
50 mm wide including ties	m2	0.90	1.02
75 mm wide including ties	m2	0.97	1.09
Cavity Insulation			
Expanded polystyrene insulation board			
50 mm thick; between wall ties	m2	6.69	7.53
75 mm thick; between wall ties	m2	8.72	9.82
Sealing Cavity at openings			
50 mm wide	m	4.05	4.56
50 mm wide including dpc	m	5.91	6.65
75 mm wide including dpc	m	6.18	6.96

LINTELS & REINFORCEMENT

	Unit	New	Refurb
Concrete Lintels			
Precast concrete; reinforced.			
102 x 150 mm	m	13.55	15.25
215 x 225 mm	m	19.81	22.29
Prestressed plate concrete; reinforced.			
100 x 70 mm	m	7.30	8.21
150 x 70 mm	m	9.50	10.69
215 x 70 mm	m	12.98	14.60
Combined Metal Boot Lintels			
202 x 143 mm high	m	24.34	27.38
202 x 219 mm high	m	32.87	36.98
Brick Reinforcement			
Brick Reinforcement			
60 mm wide; GBF100B	m	0.68	0.76
100 mm wide; GBF160B	m	0.84	0.94
160 mm wide; GBF60B	m	1.07	1.21

DAMP PROOF COURSES

	Unit	New	Refurb
Vertical			
Synthapruf bituminous paint; wall surfaces.			
two coats	m2	6.02	6.77
three coats	m2	8.29	9.33
Horizontal			
Lead cored bituminous; fibre based.			
over 300 mm wide	m2	17.72	19.93
75 mm wide	m	1.59	1.79
102.5 mm wide	m	2.18	2.45
225 mm wide	m	4.78	5.38
Two courses slates; cement mortar (1:3); pointing.			
over 300 mm wide	m	27.13	30.52
75 mm wide	m	2.44	2.75
102.5 mm wide	m	3.34	3.75
225 mm wide	m	7.32	8.24
Cavity Gutters			
Lead cored bituminous; fibre based.			
over 300 mm girth	m2	20.03	22.53
225 mm girth	m	5.05	5.68
Lead; Grade 3; welted joints.			
over 300 mm girth	m2	66.29	74.58
225 mm girth	m	16.71	18.79

PRICING GUIDE

- Key Unit Rates are intended to provide a general guide to the cost of the work including the cost of all labour, materials, and plant, but not scaffolding or trestles;
- The Pictorial Glossary on page 122 also indicates, where appropriate, composite key rates including scaffolding and trestles;
- Adjustments of the key rates can be made by reference to the guides set out on pages 106 to 111.

POINTING

Flush
£1.83 per m2

Flush, hollowed
£2.24 per m2

Recessed
£1.65 per m2

Weather Struck
£1.45 per m2

Keyed
£1.15 per m2

Raked, pointed
£5.08 per m2

Raked, tuck pointed
£8.88 per m2

Raked bastard tuck pointed
£11.46 per m2

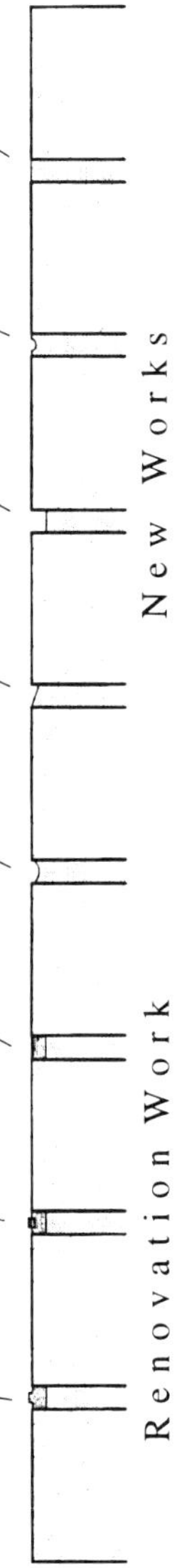

CILLS

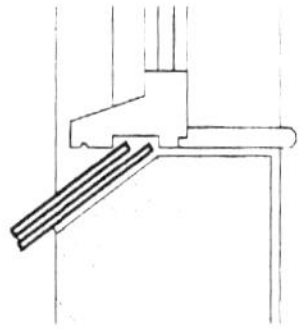

Plain Roofing tiles;
£ 9.99 per metre

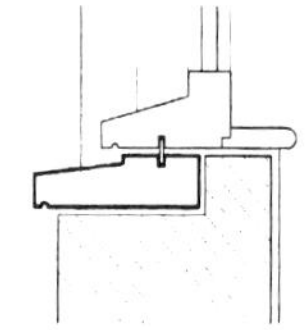

Stone cill; 250 x 65 mm;
£22.39 per metre

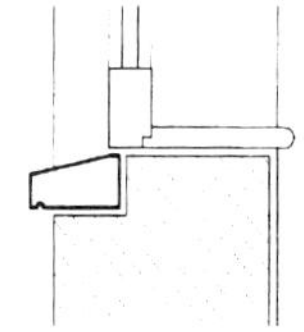

Concrete cill; 135 x 65 mm;
£ 9.28 per metre

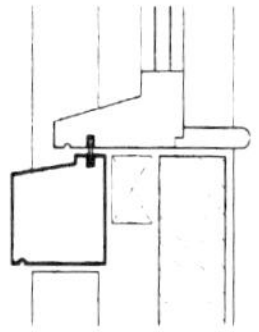

Stone cill; 135 x 140 mm;
£26.79 per metre

COPINGS

Brick on edge; tile creasing
£17.42 per metre

Bullnosed brick on edge;
£21.79 per metre

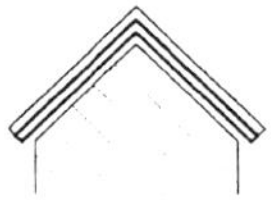

Tiled coping;
£20.02 per metre

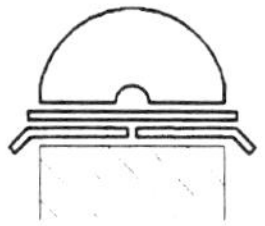

Segmental brick on edge;
£24.77 per metre

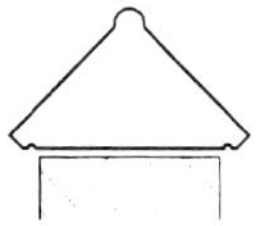

Weathered stone;
£22.71 per metre

Terra cotta;
£35.59 per metre

ARCHES

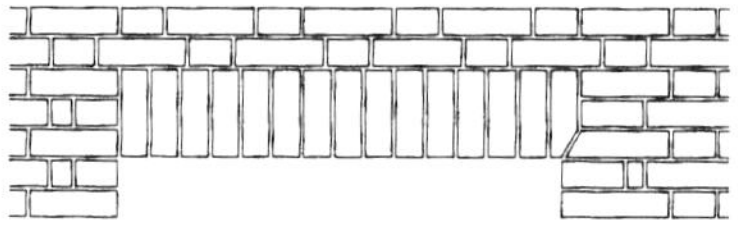

Brick on End;
£11.42 per metre

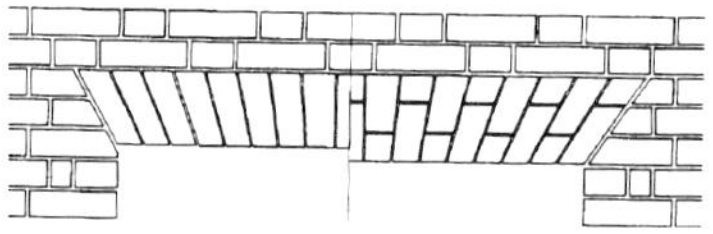

Gauged;
£45.79 per metre

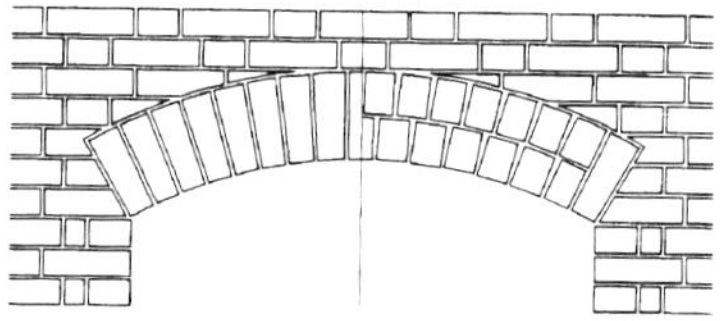

Segmental;
£35.19 per metre

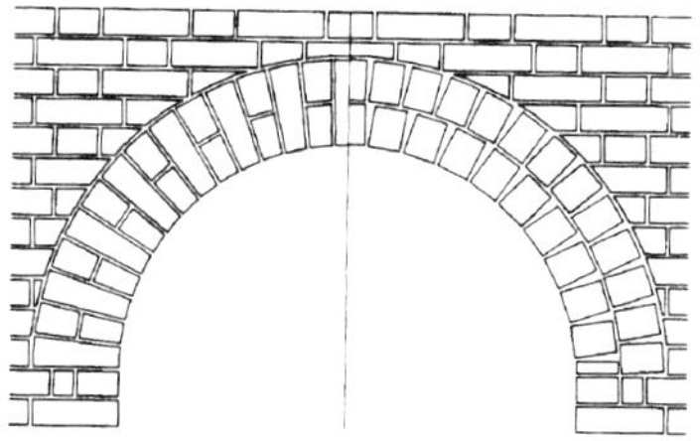

Semi-circular;
£31.11 per metre

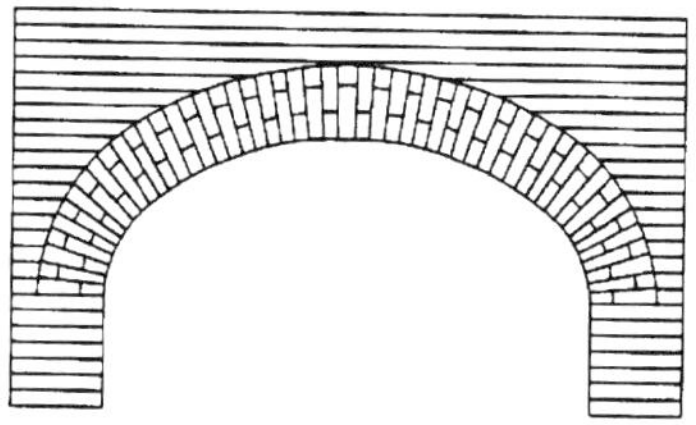

Elliptical;
£85.80 per metre

NOTE; See Page No. 129 for cost of repair work

Adjustment of KEY rates related

PURCHASE PRICE OF FACING BRICKS

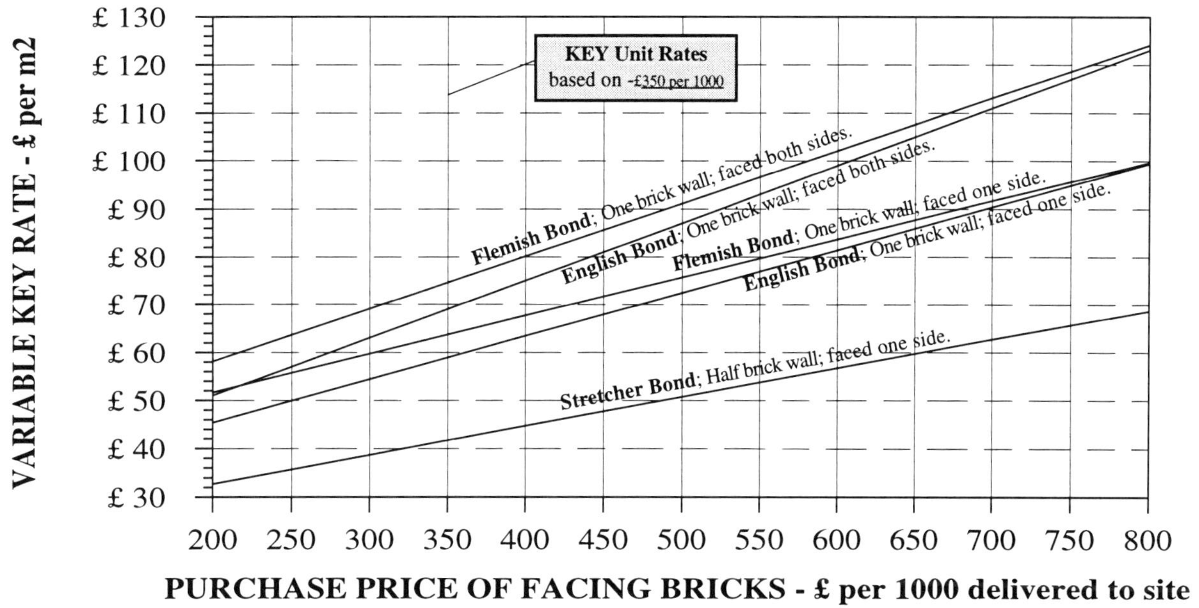

ADJUSTMENTS FOR USE OF OTHER MATERIALS AND METHODS

	Unit	New	Refurb
Imperial Bricks			
DEDUCT from key rates for facework		Brick Thickness	
if built in Imperial sized bricks		68mm	73mm
Half Brick Wall	m2	-1.04	-1.17
One Brick Wall	m2	-1.65	-1.85
One and a Half Brick Wall	m2	-2.54	-2.85
Decorative Features			
Extra for oversailing and pointing			
Single course 32 mm	m	2.01	2.26
Two courses 64 mm overall	m	4.21	4.73
Extra for bevelled plinth headers			
Single course; pointing	m	7.87	8.85
Bull's eye openings; half brick wall			
600 mm diameter	nr	91.49	102.93
1200 mm diameter	nr	146.38	164.68

Pointing the Brickwork Joints

	Unit	New	Refurb
Pointing; in the general building mortar			
Flush	m2	1.65	1.85
Flush, hollowed	m2	2.01	2.26
Recessed (raked)	m2	1.49	1.67
Struck (Weatherstruck)	m2	incl	incl
Keyed	m2	1.04	1.17
Raking and pointing with selected mortar			
Gauged Mortar (1:1:6); tinted			
Raked and pointed	m2	4.57	5.15
Raked; tuck pointed	m2	7.99	8.99
Raked; bastard tuck pointed	m2	10.31	11.60

65 mm FACING BRICKWORK

- price **£350.00** per 1000 delivered to site;
- metric size 215 x 102.5 x 65 mm;
- bedded in gauged mortar (1:1:6);
- struck joints 9.5 mm thick; no frog.

	Unit	New	Refurb
Stretcher Bond			
Half brick - faced ONE side	m2	42.24	47.51
Half brick - faced BOTH sides	m2	56.28	63.32
Extra cost built circular on plan	m2	9.80	11.02
English Bond			
Wall faced ONE side			
Half brick (HB) (with snap headers	m	41.20	46.35
One brick (1B)	m	59.63	67.09
One and a half brick (1.5B)	m	81.60	91.80
Extra cost built circular on plan	m2	9.96	11.20
Wall faced BOTH sides			
Half brick (HB) (with snap headers	m	46.53	52.35
One brick (!B)	m	69.86	78.59
One and a half brick (1.5B)	m	92.80	104.39
Extra cost built circular on plan	m2	13.20	14.85
Flemish Bond			
Wall faced ONE side			
Half brick (HB) (with snap headers	m	44.54	50.10
One brick (1B)	m	64.47	72.53
One and a half brick (1.5B)	m	88.22	99.24
Extra cost built circular on plan	m2	10.76	12.11
Wall faced BOTH sides			
Half brick (HB) (with snap headers	m	50.30	56.59
One brick (1B)	m	75.52	84.96
One and a half brick (1.5B)	m	100.32	112.86
Extra cost built circular on plan	m2	14.27	16.06

Properties *Management*

BRICKWORK - LABOURS

	Unit	New	Refurb
Sundry Labours			
Finish face of common brickwork fair and point with a flush joint.			
Walls; flush joint.	m2	1.37	1.54
Ditto circular on plan.	m2	2.48	2.79
EXTRA over key rates for;-			
Fair cutting	m	2.84	3.20
Ditto circular on elevation	m	4.04	4.54
Bed wood frame or plate in mortar			
Wood plate	m	2.15	2.41
Wood frame; point one side	m	2.75	3.09
Wedge and pin new brickwork to existing soffit with one course of slates; cement mortar (1:3).			
Half brick (HB)	m	7.19	8.09
One brick (1B)	m	11.06	12.44
One and a half brick (1.5B)	m	14.71	16.55
Bonding new and existing brickwork			
Block bond END of new brickwork to existing including cutting pockets 3 courses high in existing wall.			
Half brick (HB)	m	8.57	9.64
One brick (1B)	m	15.58	17.52
One and a half brick (1.5B)	m	22.71	25.55
Block bond FACE new brick backing to existing including cutting pockets 3 courses high in existing wall.			
Wall; 4 pockets per m2	m2	27.91	31.40
Toothing and bonding END of new brick wall to existing including cutting pockets alternative courses			
Half brick (HB)	m	10.71	12.04
One brick (1B)	m	19.47	21.91
One and a half brick (1.5B)	m	28.39	31.94

CHIMNEY POTS

	Unit	New	Refurb
Chimney Pots - Terracotta			
Pots; bedded in gauged mortar (1:1:6)			
Beaded; square base; round top.			
185 to 150 mm; 300 mm long	nr	31.51	35.45
225 to 200 mm; 300 mm long	nr	35.95	40.44
Anti down draught.			
240 to 150 mm; 600 mm long	nr	50.64	56.97
240 to 175 mm; 600 mm long	nr	57.28	64.44
Ornamental; Crown type.			
250 to 195 mm; 600 mm long	nr	79.68	89.64
290 to 290 mm; 625 mm long	nr	91.51	102.95
Flaunch around chimney pots			
Average 50 mm thick	m2	9.56	10.76

FLUE LINERS

	Unit	New	Refurb
Terracotta beaded flue liners			
Liner; 150 mm diameter	m	23.28	33.17
Liner; 225 mm diameter	m	48.77	69.49
Liner; 250 mm diameter	m	56.90	81.09
Fire brick lining; 102.5 mm thick			
Flue; 220 x 220 mm in clear	m	114.02	162.48
Flue; 347 x 347 mm in clear	m	180.42	257.10
Gas appliance flue system			
Precast refractory concrete block flue system; refractory mortar joints; access, exit and offset units; terminals.			
115 System	m	48.67	69.36
150 System	m	64.90	92.48

AIR BRICKS

	Unit	New	Replace
Terracotta air bricks			
225 x 65 mm	nr	2.79	3.14
225 x 140 mm	nr	3.39	3.81
Cast iron air bricks			
225 x 65 mm	nr	3.63	4.08
225 x 140 mm	nr	4.40	4.95
Openings for air bricks			
Openings through brickwork or blockwork for air brick; line with slates; gauged mortar (1:1:6)			
Half brick wall	nr	6.86	7.72
One brick wall	nr	9.91	11.15
1 1/2 brick wall	nr	12.96	14.58

FIREPLACES

	Unit	New	Refurb
Fire backs			
Solid; bedded in refractory mortar			
400 mm	nr	18.54	23.64
450 mm	nr	25.50	32.51
Ditto cut out for boiler;			
400 mm	nr	20.45	26.08
450 mm	nr	35.75	45.58
Frets and stools			
Black vitreous enamel			
400 mm	nr	34.77	39.11
450 mm	nr	46.36	52.15
Lustre finish			
400 mm	nr	40.56	45.63
450 mm	nr	52.15	58.67

WALLS

	Unit	Repairs
Making good defective walls - isolated areas < 2 m2		
Cut out area defective brickwall and re-build in gauged mortar.		
Half brick wall (HB)	m2	85.12
One brick wall (1B)	m2	159.71
ADD; metric faced brickwork	m2	25.08
ADD; imperial faced brickwork	m2	30.49
Making good structural cracks		
Cut out crack in brickwall and re-build in gauged mortar (1:1:6)		
Half brick wall (HB)	m	34.26
One brick wall (1B)	m	64.28
ADD; metric faced brickwork	m2	10.10
ADD; imperial faced brickwork	m2	12.27

CILLS

	Unit	New	Refurb
New cills; gauged mortar; flush pointing			
Plain roofing tiles; two courses	m	10.11	11.37
Plain brick on edge; splayed	m	8.08	9.09
Bullnose brick on edge; splayed	m	23.38	26.30
Concrete; 135 x 65 mm	m	9.40	10.57
Stone; 250 x 65 mm	m	22.66	25.50
Stone; 135 x 140 mm	m	27.12	30.51
Replace cills; cut out existing; short lengths			
Plain roofing tiles; two courses	m		16.31
Plain brick on edge; splayed	m		11.97
Bullnose brick on edge; splayed	m		30.85
Concrete; 135 x 65 mm	m		12.83
Stone; 250 x 65 mm	m		30.39
Stone; 135 x 140 mm	m		35.23

COPINGS

	Unit	New	Repairs
New copings; gauged mortar; flush pointing			
Brick on edge with tile creasing	m	17.64	19.84
Double bullnosed brick on edge	m	22.06	24.82
Segmental brick on edge	m	25.08	28.21
Apex tiling	m	20.26	22.79
Weathered stone	m	22.99	25.86
Terracotta	m	36.03	40.53
Replace copings; cut out existing; short lengths			
Brick on edge with tile creasing	m		24.24
Bullnosed brick on edge	m		27.81
Segmental brick on edge	m		31.27
Apex tiling	m		26.24
Weathered stone	m		29.24
Terracotta	m		44.76

ARCHES

	Unit	New	Repairs
New arches; gauged mortar; flush pointing			
Flat; half brick; bearer bar	m	15.19	17.09
Flat; one brick; bearer bar	m	14.46	16.26
Gauged; one brick; bearer bar	m	57.94	65.19
Segmental; one brick	m	44.53	50.09
Semi-circular; one brick	m	39.37	44.29
Elliptical; one brick	m	108.57	122.14
Replace arches; cut out existing			
Flat; half brick; bearer bar	m		40.15
Flat; one brick; bearer bar	m		38.22
Gauged; one brick; bearer bar	m		153.19
Segmental; one brick	m		117.71
Semi-circular; one brick	m		104.07
Elliptical; one brick	m		287.03

WALL SURFACES

	Unit	New	Refurb
Cleaning existing brick surfaces			
Jet pressure spray with diluted additive; remove cement stains;	m2	-	2.41
Remove existing paint; rake out joints brickwork and re-point.	m2	-	5.36
Removing existing rendering			
Sparrow peck face of existing brickwork to form key for new rendering.	m2	-	2.15
Hack off existing rendering and rake out joints brickwork to form key for new.			
wall plaster.	m2	-	2.22
cement rendering	m2	-	3.87
Rake out joints existing brickwork			
Rake out joints existing brickwork;			
lime mortar	m2	-	5.41
cement mortar	m2	-	10.73

Re-point faced brickwork

	rates per square metre		
Mortars	(1:3)	(1:1:6)	(1:2:9)
Flush	1.82	1.76	1.77
Flush, hollowed	2.23	2.15	2.16
Recessed (Raked)	1.65	1.59	1.60
Struck (Weather struck)	1.51	1.46	1.47
Keyed	1.15	1.11	1.11
Tuck pointed	8.84	8.54	8.58
Bastard tuck pointed	11.40	11.02	11.07
ADD to key rates pointing for the following;	rates per square metre		
work in small areas			
tinted mortar	2.12	2.05	2.06
stone dust aggregate	4.25	4.11	4.13

GLOSSARY

Accelerator	A liquid product added to new concrete or building mortar to speed up the setting time.
Beam filling	Brickwork or blockwork built between floor or ceiling joists.
Bond	The arrangement of bricks or blocks to provide maximum strength and good appearance where exposed.
Wall ties	Galvanised or stainless steel ties made of twisted wire or flat steel and built into walls of cavity construction to tie both leaves together.
Cavity flashing	A metal tray fixed across a cavity to drain water away from the inner leaf of a hollow external wall.
Cement Fillet	A triangular section of mortar set at the junction of horizontal and vertical surfaces.
Compo	Building mortar comprising cement, hydrated lime and sand used for building bricks or blockwork.
Damp proof course	A layer of impervious material laid horizontally or vertically to prevent the ingress of moisture.
Efflorescence	The deposit of crystallized salt on the surface of brick or block walls due to the evaporation of moisture when drying out.
Lintel	A concrete, timber or steel structural member set over an opening to carry the load of the brick, block or stone work over.

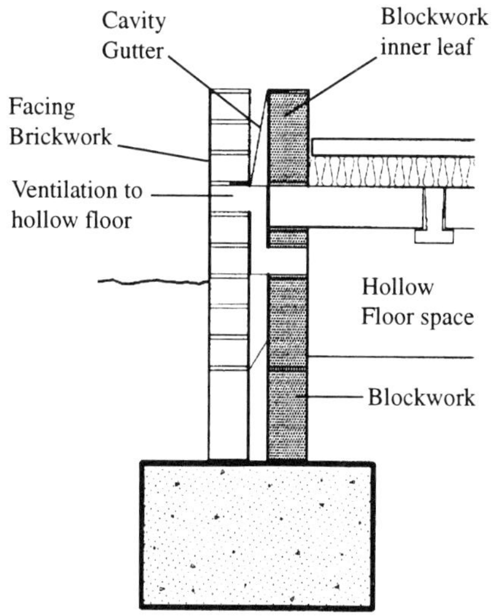

Foundation to External Wall
External facing brickwork
Unfilled cavity
Internal blockwork
£90.00 per linear metre.

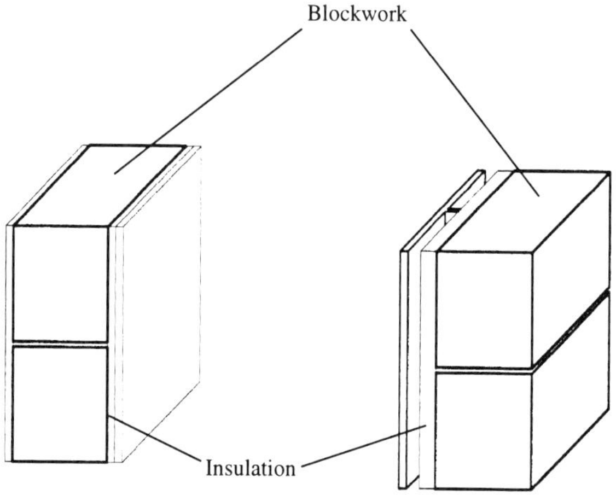

Solid Wall
External Rendering
Blockwork wall
Insulate dry Lining
£46.00 per square metre

Solid Wall
External insulated cladding
Blockwork wall
Internal fair face
£57.00 per square metre

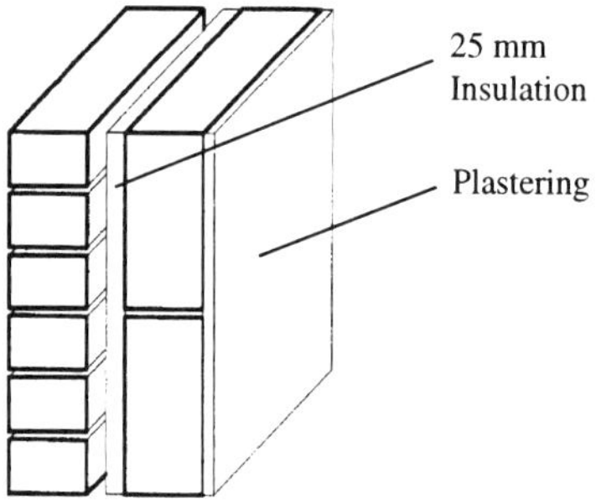

External Cavity Wall
External facing brickwork
Partial fill insulated cavity
Internal blockwork, plastered
£74.00 per square metre

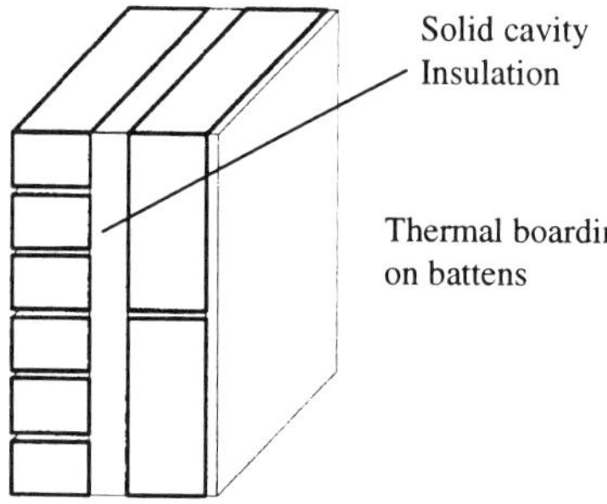

External Cavity Wall
External facing brickwork
Full fill insulated cavity
Internal blockwork, plastered
£78.00 per square metre

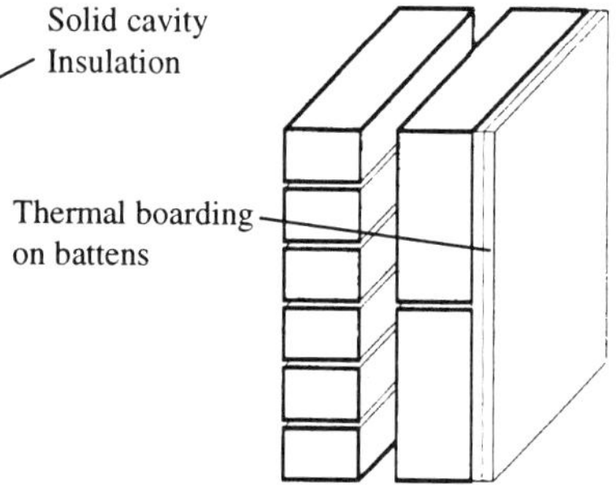

External Cavity Wall
External facing brickwork
Unfilled cavity
Internal insulated dry lining
£80.00 per square metre

LIGHTWEIGHT

- Cost price 100 mm thick - Standard; **£5.86** m2;
- Cost price 100 mm thick - Smooth; **£10.25** m2;
- Size 440 x 215 mm; strength 3.5 N/mm2;
- Bedded in gauged mortar (1:1:6); 9.5 mm joints.

	Unit	New	Refurb
Plain Walls - Standard			
75 mm thick	m2	12.77	14.37
100 mm thick	m2	16.04	18.04
140 mm thick	m2	20.37	22.92
215 mm thick	m2	27.92	31.41
ADD for extra cost where;-			
finished fair face; each side	m2	0.68	0.77
built as pier or casing	m2	0.75	0.85
built circular on plan	m2	5.03	5.66
Plain Walls - Smooth Finish			
100 mm thick	m2	24.23	27.26
140 mm thick	m2	31.99	35.99
150 mm thick	m2	33.59	37.78
215 mm thick	m2	45.37	51.04
ADD for extra cost where;-			
finished fair face; each side	m2	0.28	0.32
built as pier or casing	m2	0.74	0.83
built circular on plan	m2	5.30	5.96
Labours on blockwork			
ADD for extra cost of labours;-			
Block bond end to new brickwork			
75 mm thick	m	3.20	3.60
100 mm thick	m	4.67	5.25
140 mm thick	m	7.39	8.31
215 mm thick	m	10.63	11.96
Ditto to existing brickwork			
75 mm thick	m	6.34	7.13
100 mm thick	m	9.25	10.40
140 mm thick	m	14.63	16.45
215 mm thick	m	21.04	23.67

HEAVYWEIGHT

- Cost price 100 mm thick - Standard; **£6.45** m2;
- Cost price 100 mm thick - Coloured; **£15.50** m2;
- Size 440 x 215 mm; strength 3.5 N/mm2;
- Bedded in gauged mortar (1:1:6); 9.5 mm joints.

	Unit	New	Refurb
Plain Walls - 7.00 N/mm2			
75 mm thick	m2	18.72	21.06
100 mm thick	m2	19.46	21.89
140 mm thick	m2	25.47	28.65
215 mm thick	m2	36.14	40.65
ADD for extra cost where;-			
finished fair face; each side	m2	0.87	0.98
built as pier or casing	m2	1.60	1.80
built circular on plan	m2	5.53	6.22
Plain Walls - Coloured blocks and mortar			
100 mm thick	m2	39.81	44.79
140 mm thick	m2	55.31	62.23
190 mm thick	m2	73.98	83.22
215 mm thick	m2	78.08	87.84
ADD for extra cost where;-			
finished fair face; each side	m2	0.28	0.32
built as pier or casing	m2	0.74	0.83
built circular on plan	m2	5.30	5.96
Labours on blockwork			
ADD for extra cost of labours;-			
Block bond end to new brickwork			
75 mm thick	m	5.13	5.77
100 mm thick	m	7.27	8.17
140 mm thick	m	10.97	12.34
215 mm thick	m	12.38	13.93
Ditto to existing brickwork			
75 mm thick	m	10.16	11.43
100 mm thick	m	14.39	16.19
140 mm thick	m	21.72	24.43
215 mm thick	m	24.51	27.57

HOLLOW

- Cost price 100 mm thick - Standard; **£7.55** m2;
- Cost price 100 mm thick - Coloured; **£14.75** m2;
- Size 440 x 215 mm; strength 3.5 N/mm2;
- Bedded in gauged mortar (1:1:6); 9.5 mm joints.

	Unit	New	Refurb
Plain Walls - 7.00 N/mm2			
100 mm thick	m2	22.70	25.54
140 mm thick	m2	27.60	31.05
190 mm thick	m2	35.68	40.14
215 mm thick	m2	34.47	38.78
ADD for extra cost where;-			
finished fair face; each side	m2	0.97	1.09
built as pier or casing	m2	1.82	2.05
built circular on plan	m2	6.12	6.88
Plain Walls - Coloured blocks and mortar			
100 mm thick	m2	35.18	39.58
140 mm thick	m2	43.88	49.36
190 mm thick	m2	55.16	62.06
215 mm thick	m2	59.73	67.19
ADD for extra cost where;-			
finished fair face; each side	m2	0.35	0.39
built as pier or casing	m2	0.82	0.92
built circular on plan	m2	5.70	6.42
Labours on blockwork			
ADD for extra cost of labours;-			
Block bond end to new brickwork			
75 mm thick	m	3.72	4.19
100 mm thick	m	5.27	5.93
140 mm thick	m	7.95	8.95
215 mm thick	m	8.97	10.10
Ditto to existing brickwork			
75 mm thick	m	7.37	8.29
100 mm thick	m	10.43	11.73
140 mm thick	m	15.75	17.72
215 mm thick	m	17.77	19.99

FLINT

- Cost price - Standard; **£85.00** per ton delivered site;
- Natural stone to pass 75 to 100 mm rings;
- Bedded in gauged mortar (1:1:6); grouting solid;
- Faced walls; pointed with a neat flush joint.

	Unit	New	Refurb
Walls; uncoursed			
300 mm thick;			
faced one side	m2	79.48	89.41
faced both sides	m2	85.06	95.69
400 mm thick;			
faced one side	m2	103.85	116.83
faced both sides	m2	109.42	123.10
ADD to rates for;			
Levelling for dpc or copings	m2	6.20	6.97
Ditto on rake	m2	7.82	8.80
Walling; coursed			
300 mm thick;			
faced one side	m2	80.46	90.52
faced both sides	m2	86.03	96.79
400 mm thick;			
faced one side	m2	105.16	118.30
faced both sides	m2	110.73	124.57
ADD to rates for;			
Finishing on rake for copings	m2	8.05	9.06
Stone finishings; dressed			
Walls			
Corner stones	m	20.86	23.47
Windows and Door openings			
Heads	m	28.97	32.59
Jambs	m	24.34	27.38
Cills	m	37.08	41.72
Thresholds	m	20.86	23.47
Lacing course; two tiles			
300 mm wide	m	9.85	11.08
400 mm wide	m	13.13	14.78

UNCOURSED RANDOM RUBBLE

- Cost price - Standard; **£95.00** per ton delivered site;
- Natural stone to pass 75 to 100 mm rings;
- Bedded in gauged mortar (1:1:6); grouting solid;
- Faced walls; pointed with a neat flush joint.

	Unit	New	Refurb
Rubble walling; laid dry			
Walls; natural faces both sides			
275 mm thick	m2	61.78	69.50
450 mm thick	m2	90.75	102.10
ADD to rates for copings			
275 mm thick	m	11.01	12.39
450 mm thick	m	15.12	17.01
Rubble walling; in gauged mortar (1:2:9)			
Walls; faced and pointed one side			
275 mm thick	m2	74.18	83.45
450 mm thick	m2	109.96	123.71
550 mm thick	m2	130.41	146.71
Walls; faced and pointed both sides			
275 mm thick	m2	78.63	88.45
450 mm thick	m2	114.41	128.71
550 mm thick	m2	134.85	151.71
Piers and chimneys; faced and pointed one side			
275 mm thick	m2	85.31	95.97
450 mm thick	m2	126.45	142.26
550 mm thick	m2	149.97	168.71
ADD to rates for;			
Levelling; dpc or copings	m2	6.94	7.81
Ditto on rake	m2	9.37	10.54
Coursed random coping			
275 x 200 (average) mm high	m	15.72	17.68
450 x 200 (average) mm high	m	20.64	23.22
550 x 200 (average) mm high	m	28.05	31.56
Arches; formed of rough stones			
Flat	m	27.65	31.10
Segmental	m	47.39	53.32

COURSED RANDOM RUBBLE

- Cost price - Standard; **£95.00** per ton delivered site;
- Natural stone to pass 75 to 100 mm rings;
- Bedded in gauged mortar (1:1:6); grouting solid;
- Faced walls; pointed with a neat flush joint.

	Unit	New	Refurb
Walls			
150 mm thick	m2	67.76	76.23
225 mm thick	m2	98.45	110.75
300 mm thick	m2	129.13	145.28
ADD to rates for;			
Facing and pointing; each side	m2	6.37	7.17
Levelling; dpc or copings	m2	6.06	6.82
Coping			
150 x 100 mm (average)	m	13.19	14.84
225 x 100 mm (average)	m	15.72	17.69
300 x 100 mm (average)	m	18.25	20.53
Arches			
Flat	m	21.44	24.12
Segmental	m	41.72	46.93

Example **1** **HOW TO ADJUST; Key Unit Rates**
to take account of the actual cost of materials to obtain
Flint wall; 300 mm thick.

The adjusted key unit rates using the graph:

Cost of flint included in Key Rates = £ 85.00 per ton
Actual purchase price of flint = £115.00 per ton
Unit Rate, see page 132; flint wall = £ 79.48 per m2
Adjusted key unit rate from graph = £100.00 per m2

To adjust the rates following the unit rate selected:

The percentage increase in the key unit rate
(£100.00/£ 79.48)*100/1 = 126%

To adjust the following rates;

300 mm wall fcd o/s £ 79.48 * 126% = £100.14 per m2
Ditto fcd b/s £ 85.06 * 126% = £107.17 per m2

SQUARED RANDOM RUBBLE

- Cost price - Standard; **£95.00** per ton delivered site;
- Natural stone to pass 75 to 100 mm rings;
- Bedded in gauged mortar (1:1:6); grouting solid;
- Faced walls; pointed with a neat flush joint.

	Unit	New	Refurb
Walls			
150 mm thick	m2	72.50	81.56
225 mm thick	m2	105.26	118.42
300 mm thick	m2	138.02	155.27
ADD to rates for;			
Facing; pointing each side	m2	7.33	8.25
Levelling; dpc or copings	m2	3.94	4.43
Coping			
150 x 100 mm (average)	m	15.17	17.07
225 x 100 mm (average)	m	18.08	20.34
300 x 100 mm (average)	m	20.99	23.61
Arches			
Flat	m	31.09	34.97
Segmental	m	60.49	68.05

REGULAR COURSED RUBBLE

- Cost price - Standard; **£95.00** per ton delivered site;
- Natural stone to pass 75 to 100 mm rings;
- Bedded in gauged mortar (1:1:6); grouting solid.
- faced walls; pointed with a neat flush joint.

	Unit	New	Refurb
Walls			
150 mm thick	m2	95.69	107.65
225 mm thick	m2	127.20	143.10
300 mm thick	m2	163.38	183.80
ADD to rates for;			
Facing; pointing each side	m2	11.98	13.48
Levelling; dpc or copings	m2	6.44	7.24
Coping			
150 x 100 mm (average)	m	18.37	20.66
225 x 100 mm (average)	m	21.89	24.62
300 x 100 mm (average)	m	25.41	28.59
Arches			
Flat	m	50.80	57.15
Segmental	m	98.86	111.22

Adjustment of KEY rates related

PURCHASE PRICE OF STONE

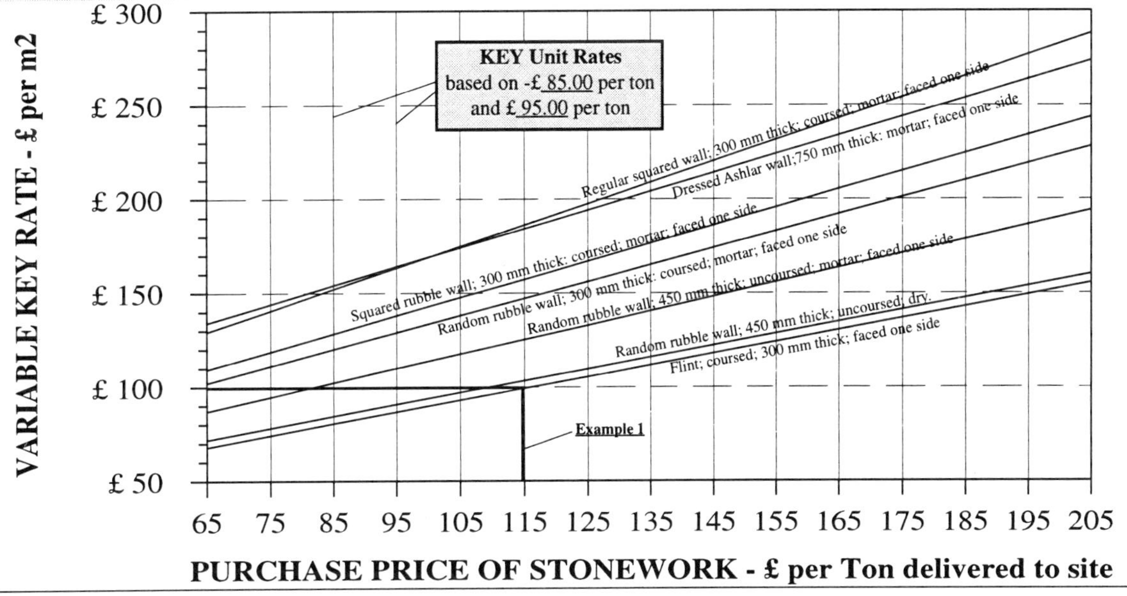

DRESSED ASHLAR

- Cost price - Standard; **£130.00** per ton delivered site;
- Natural stone to pass 75 to 100 mm rings;
- Bedded in gauged mortar (1:1:6); grouting solid;
- Faced walls; pointed with a neat flush joint.

	Unit	New	Refurb
Facework			
Cladding; bedded against backing			
50 mm thick	m2	156.45	176.00
63 mm thick	m2	197.01	221.63
75 mm thick	m2	203.96	229.46
100 mm thick	m2	225.98	254.23
ADD to rates for;			
Fair returns			
50 mm wide	m	0.93	1.04
63 or 75 mm wide	m	1.33	1.50
100 mm wide	m	2.03	2.28
Raking cutting			
50 mm wide	m	14.72	16.56
63 or 75 mm wide	m	17.67	19.88
100 mm wide	m	22.66	25.49
Dressed Features			
Copings; splayed and throated; angles and ends			
300 x 50 mm	m	108.66	122.24
350 x 75 mm	m	149.03	167.66
400 x 100 mm	m	176.67	198.75
450 x 100 mm	m	193.58	217.78
Band Courses; moulded; angles and ends			
150 x 75 mm	m	102.85	115.71
200 x 100 mm	m	129.85	146.08
250 x 150 mm	m	170.06	191.32
350 x 200 mm	m	283.92	319.42
Surrounds to openings; moulded			
75 x 100 mm	m	65.82	74.05
100 x 100 mm	m	74.19	83.47
200 x 175 mm	m	157.61	177.31
300 x 225 mm	m	200.36	225.40

GLOSSARY

Built up Roofing	Two or three layers of bituminous felt bonded together with bitumen.
Cheek	Side to a dormer window
Dormer window	A window projecting through a sloping roof.
Flashing	An impervious cover at the junction of a flat or sloping roof, usually in lead, zinc or leadcored bituminous felt to prevent the ingress of rain water.
Flaunching	Weathered cement mortar filling around chimney pots.
Gable end	Triangular shaped wall at the end of a pitched roof with two sloping faces.
Hipped roof	A roof with four sloping sides.
Mansard roof	A roof with two sloping faces each side at different pitches.
Mono pitched roof	A roof with only one sloping face.
Purlin	An horizontal structural member supporting rafters at mid span.
Rafter	The sloping structural members supporting the roof coverings.
Soffit board	Horizontal boarding to the soffit of over hanging eaves to a roof.

SECTIONS THROUGH FLAT ROOFS

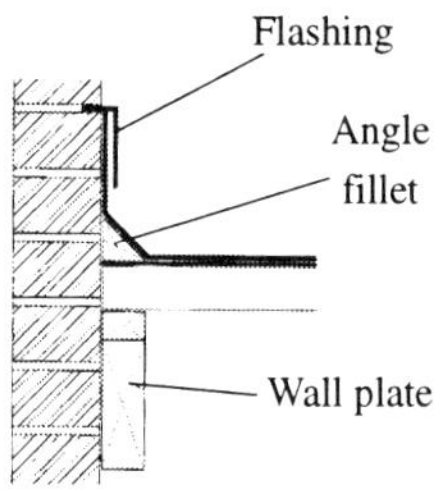

Abutment with wall
£26.50 per metre

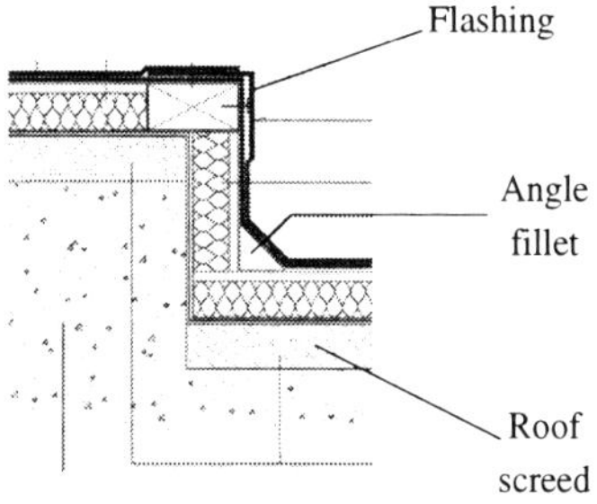

Step in flat roof
£35.50 per metre

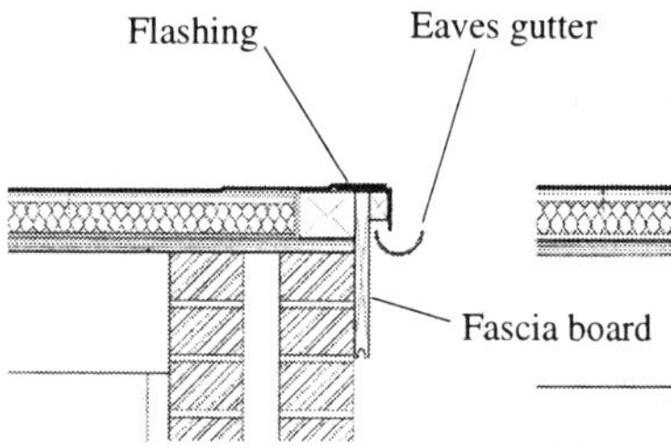

Verge with gutter
£33.50 per metre

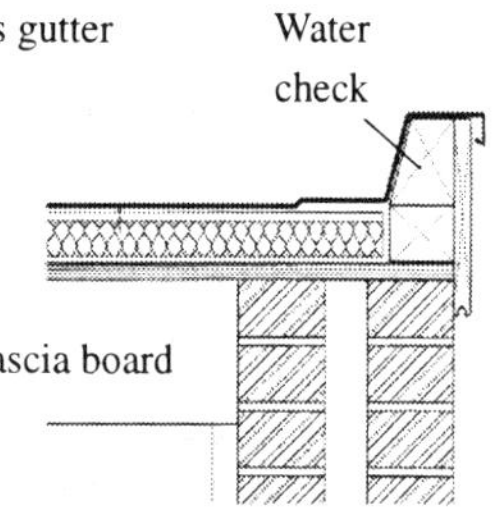

Verge with water check
£32.50 per metre

SECTIONS THROUGH SLOPING ROOFS

Top edge flashing
£20.50 per metre

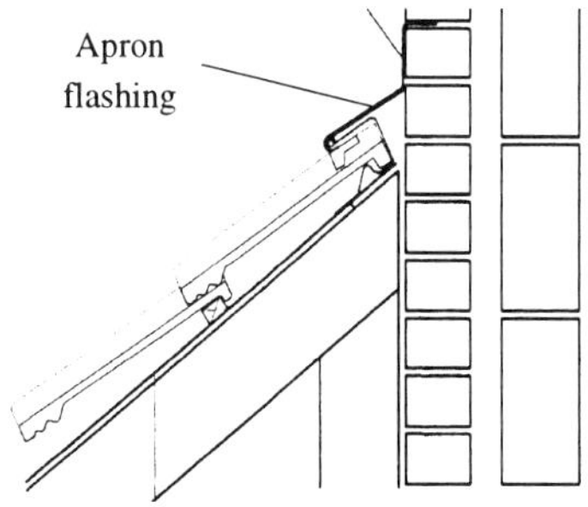

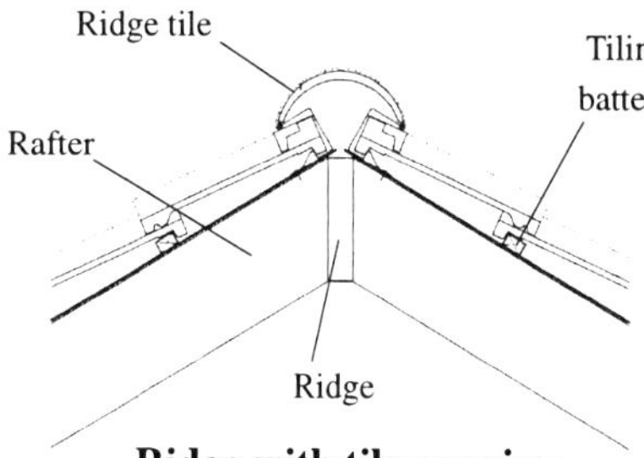

Ridge with tile capping
£26.50 per metre

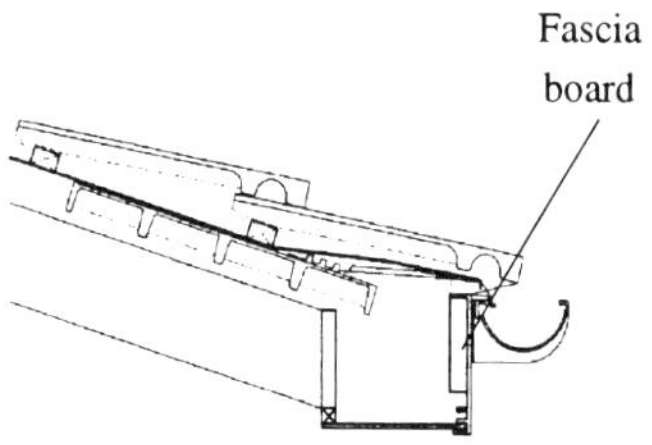

Ventilated Eaves with gutter
£32.50 per metre

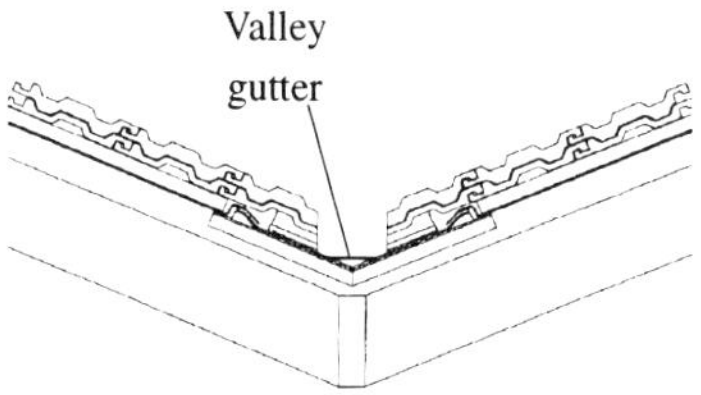

Valley Gutter
£32.00 per metre

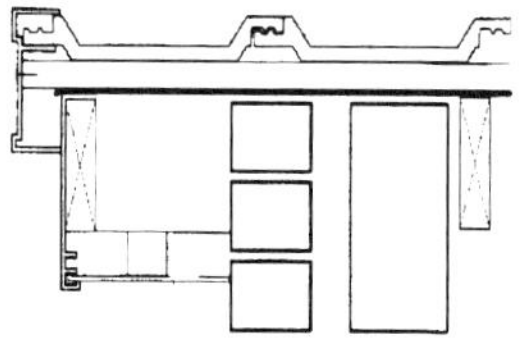

Ventilated Verge with waterstop
£39.00 per metre

SHEET METAL

- Aluminium; Copper; Lead; Steel; Zinc:
- Head laps; 150 mm: Side laps; one corrugation;
- Fixings; Copper or approved by the manufacturer;
- Flashings; Standard as Code of Practice.

	Unit	New	Refurb
Aluminium; 0.90 mm thick			
Pitched roof; 10° to 50°	m2	44.00	47.30
Vertical	m2	47.37	50.92
Extra; curved work;	m2	16.39	17.62
Extra; flashings; average rate	m	14.75	15.86
Copper; 23 Swg			
Pitched roof; 10° to 50°	m2	84.92	91.29
Vertical	m2	89.47	96.18
Extra; curved work;	m2	20.36	21.89
Extra; flashings; average rate	m	25.67	27.60
Lead; Code 5			
Pitched roof; 10° to 50°	m2	102.67	110.37
Vertical	m2	108.66	116.81
Extra; curved work;	m2	21.64	23.26
Extra; flashings; average rate	m	34.69	37.29
Steel; .50 mm			
Mill finish; coloured; fixings:			
Roof; sloping	m2	26.85	28.86
Walls; vertical	m2	28.09	30.20
Extra; flashings; average rate	m	8.29	8.91
PVC coated finish; fixings:			
Roof; sloping	m2	21.12	22.71
Walls; vertical	m2	23.17	24.91
Extra; flashings; average rate	m	10.69	11.49
Zinc; 23 Swg			
Pitched roof; 10° to 50°	m2	53.23	57.22
Dormer windows	m2	62.80	67.51
Extra; curved work;	m2	16.39	17.62
Extra; flashings; average rate	m	14.75	15.86

BUILT UP FELT

- Felt; glass fibre based; bonding with bitumen;
- Polyester; 2 layers as manufacturer's instructions;
- Elastomeric; 2 layers as last described;
- Hiload Elastomeric; 2 layers as last described.

	Unit	New	Refurb
Felt; Three layers to BS 747			
Roof; flat or sloping	m2	10.27	11.04
Extra for;			
White stone spar chippings	m2	3.01	3.24
300 x 300 mm tiles	m2	36.66	39.41
Skirtings; <200 mm;	m	7.98	8.58
Skirtings; >200 mm<400 mm;	m	9.48	10.19
Lining to gutter; 500 mm girth;	m	13.62	14.64
Polyester; Two layers; High performance			
Roof; flat or sloping	m2	14.15	15.22
Extra for;			
White stone spar chippings	m2	3.01	3.24
Skirtings; <200 mm;	m	8.45	9.09
Skirtings; >200 mm<400 mm;	m	10.14	10.91
Lining to gutter; 500 mm girth;	m	17.34	18.64
Elastomeric; Two layers			
Roof; flat or sloping	m2	13.72	14.75
Extra for;			
White stone spar chippings	m2	3.01	3.24
Skirtings; <200 mm;	m	8.31	8.94
Skirtings; >200 mm<400 mm;	m	9.93	10.67
Lining to gutter; 500 mm girth;	m	16.89	18.16
Hyload Elastomeric			
Roof; flat or sloping	m2	22.34	24.02
Extra for;			
White stone spar chippings	m2	3.01	3.24
Skirtings; <200 mm;	m	21.72	23.35
Skirtings; >200 mm<400 mm;	m	29.32	31.52
Lining to gutter; 500 mm girth;	m	44.32	47.64

PROPRIETARY SYSTEMS

- Laid as described in Manufacturer's specification;
- Vapour check; 32 mm polystyrene insulation;
- Underfelt; two layers of felt as specified;
- Finished; white spar stone chippings.

	Unit	New	Refurb
Bituminous felt flat roofing system			
0.70 Decking; galvanised steel			
46 mm deep; 2.96 m span	m2	33.93	36.47
60 mm deep; 3.74 m span	m	35.03	37.66
0.90 Decking; aluminium			
35 mm deep; 1.79 m span	m2	37.54	40.36
60 mm deep; 2.34 m span	m	39.12	42.05
Bituminous high performance felt flat roofing system			
0.70 Decking; galvanised steel			
46 mm deep; 3.12 m span	m2	36.66	39.41
70 mm deep; 3.74 m span	m	47.33	50.88
0.90 Decking; aluminium			
46 mm deep; 1.79 m span	m2	40.19	43.21
70 mm deep; 2.56 m span	m	51.02	54.84
Polyester flat roofing system			
0.70 Decking; galvanised steel			
46 mm deep; 2.96 m span	m2	37.89	40.73
60 mm deep; 3.74 m span	m	38.99	41.91
0.90 Decking; aluminium			
35 mm deep; 1.79 m span	m2	41.50	44.61
60 mm deep; 2.34 m span	m	43.08	46.31
Elastomeric flat roofing system			
0.70 Decking; galvanised steel			
46 mm deep; 2.96 m span	m2	37.49	40.30
60 mm deep; 3.74 m span	m	38.59	41.49
0.90 Decking; aluminium			
35 mm deep; 1.79 m span	m2	41.10	44.19
60 mm deep; 2.34 m span	m	42.68	45.88

FIBRE CEMENT SLATES

- Cost £235.00 per 1000 delivered site; regular widths;
- Regular courses; 75 mm lap; sarking underfelt;
- Each slate fixed two aluminium alloy nails;
- 50 x 19 mm Softwood battens, treated.

	Unit	New	Refurb
Slates; size 200 x 400 mm			
Sloping; minimum 20°	m2	44.70	48.05
Curved/ convex/ concave	m2	51.40	55.26
Vertical	m2	48.72	52.38
Vertical; curved on plan	m2	56.03	60.23
Extra for eaves	m	10.12	10.88
Extra for verge	m	6.79	7.30
Extra for valley gutter	m	9.19	9.88
Slates; size 250 x 500 mm			
Sloping; minimum 20°	m2	39.57	42.54
Curved/ convex/ concave	m2	45.51	48.92
Vertical	m2	41.97	45.12
Vertical; curved on plan	m2	48.27	51.89
Extra for eaves	m	9.86	10.60
Extra for verge	m	6.62	7.12
Extra for valley gutter	m	8.96	9.63
Slates; size 300 x 600 mm			
Sloping; minimum 20°	m2	33.84	36.38
Vertical	m2	36.26	38.98
Extra for eaves	m	9.62	10.34
Extra for verge	m	6.45	6.94
Extra for valley gutter	m	8.74	9.39
Sundries			
Cutting			
square or raking	m	6.11	6.56
curved	m	8.94	9.61
Ridge tiles; half round; end	nr	35.34	37.99
Hip tiles; half round; ends	nr	52.10	56.00

NATURAL BLUE SLATES

- Cost £325.00 per 1000 delivered site; regular widths;
- Regular courses; 75 mm lap; sarking underfelt;
- Each slate fixed two aluminium alloy nails;
- 44 x 25 mm Softwood battens, treated.

	Unit	New	Refurb
Slates; size 205 x 305 mm			
Sloping; minimum 20°	m2	48.16	51.78
Curved/ convex/ concave	m2	55.39	59.54
Vertical	m2	51.64	55.51
Vertical; curved on plan	m2	59.39	63.84
Extra for eaves	m	18.48	19.87
Extra for verge	m	14.23	15.30
Extra for valley gutter	m	21.38	22.98
Slates; size 255 x 455 mm			
Sloping; minimum 20°	m2	54.18	58.24
Curved/ convex/ concave	m2	62.30	66.98
Vertical	m2	36.25	38.97
Vertical; curved on plan	m2	41.69	44.81
Extra for eaves	m	18.02	19.37
Extra for verge	m	13.88	14.92
Extra for valley gutter	m	20.85	22.41
Slates; size 305 x 610 mm			
Sloping; minimum 20°	m2	63.99	68.79
Vertical	m2	66.07	71.02
Extra for eaves	m	17.57	18.89
Extra for verge	m	13.53	14.54
Extra for valley gutter	m	20.33	21.85
Sundries			
Cutting			
square or raking	m	12.77	13.73
curved	m	18.54	19.93
Ridge tiles; half round; end	nr	29.19	31.38
Hip tiles; half round; ends	nr	55.87	60.06

NATURAL BLUE SLATES

- Cost £345.00 per 1000 delivered site; irregular widths;
- Diminishing courses; sarking underfelt;
- Each slate fixed two aluminium alloy nails;
- 44 x 25 mm Softwood battens, treated.

	Unit	New	Refurb
Slates; size 205 x 305 mm - 305 x 610 mm			
Sloping; minimum 20°	m2	65.02	69.90
Curved/ convex/ concave	m2	74.77	80.38
Vertical	m2	74.77	80.38
Vertical; curved on plan	m2	85.99	92.44
Extra for eaves	m	23.11	24.84
Extra for verge	m	17.79	19.12
Extra for valley gutter	m	63.49	68.25
Slates; size 255 x 560 mm - 305 x 760 mm			
Sloping; minimum 20°	m2	69.96	75.21
Curved/ convex/ concave	m2	80.46	86.49
Vertical	m2	80.46	86.49
Vertical; curved on plan	m2	92.52	99.46
Extra for eaves	m	24.86	26.73
Extra for verge	m	19.14	20.58
Extra for valley gutter	m	68.32	73.44
Slates; size 305 x 760 mm - 760 x 1200 mm			
Sloping; minimum 20°	m2	73.46	78.97
Vertical	m2	84.48	90.81
Extra for eaves	m	26.10	28.06
Extra for verge	m	20.10	21.61
Extra for valley gutter	m	71.73	77.11
Sundries			
Cutting			
square or raking	m	17.24	18.53
curved	m	25.03	26.91
Ridge tiles; half round; end	nr	29.92	32.17
Hip tiles; half round; ends	nr	57.27	61.56

NATURAL GREEN SLATES

- Cost £1,550.00 per 1000 delivered site; regular widths;
- Regular courses; 75 mm lap; sarking underfelt;
- Each slate fixed two aluminium alloy nails;
- 44 x 25 mm Softwood battens, treated.

	Unit	New	Refurb
Slates; size 205 x 305 mm			
Sloping; minimum 20°	m2	90.74	97.55
Curved/ convex/ concave	m2	104.36	112.18
Vertical	m2	97.29	104.59
Vertical; curved on plan	m2	111.89	120.28
Extra for eaves	m	34.83	37.44
Extra for verge	m	26.81	28.82
Extra for valley gutter	m	40.28	43.31
Slates; size 255 x 455 mm			
Sloping; minimum 20°	m2	102.08	109.73
Curved/ convex/ concave	m2	117.39	126.19
Vertical	m2	68.30	73.42
Vertical; curved on plan	m2	78.54	84.43
Extra for eaves	m	33.96	36.50
Extra for verge	m	26.14	28.10
Extra for valley gutter	m	39.28	42.22
Slates; size 305 x 610 mm			
Sloping; minimum 20°	m2	120.57	129.61
Vertical	m2	124.48	133.81
Extra for eaves	m	33.11	35.59
Extra for verge	m	25.49	27.40
Extra for valley gutter	m	38.30	41.17
Sundries			
Cutting			
square or raking	m	24.06	25.87
curved	m	34.94	37.56
Ridge tiles; half round; end	nr	42.33	45.50
Hip tiles; half round; ends	nr	81.02	87.09

NATURAL GREEN SLATES

- Cost £1,825.00 per 1000 delivered site; irregular widths;
- Diminishing courses; sarking underfelt;
- Each slate fixed two aluminium alloy nails;
- 44 x 25 mm Softwood battens, treated.

	Unit	New	Refurb
Slates; size 205 x 305 mm - 305 x 610 mm			
Sloping; minimum 20°	m2	114.27	122.84
Curved/ convex/ concave	m2	131.41	141.27
Vertical	m2	122.52	131.71
Vertical; curved on plan	m2	140.90	151.46
Extra for eaves	m	43.86	47.14
Extra for verge	m	33.76	36.30
Extra for valley gutter	m	50.73	54.53
Slates; size 255 x 560 mm - 305 x 760 mm			
Sloping; minimum 20°	m2	128.54	138.18
Curved/ convex/ concave	m2	147.82	158.91
Vertical	m2	86.01	92.46
Vertical; curved on plan	m2	98.91	106.32
Extra for eaves	m	42.76	45.97
Extra for verge	m	32.92	35.39
Extra for valley gutter	m	49.46	53.17
Slates; size 305 x 760 mm - 760 x 1200 mm			
Sloping; minimum 20°	m2	151.83	163.22
Vertical	m2	156.75	168.51
Extra for eaves	m	41.69	44.82
Extra for verge	m	32.10	34.50
Extra for valley gutter	m	48.22	51.84
Sundries			
Cutting			
square or raking	m	30.30	32.57
curved	m	43.99	47.29
Ridge tiles; half round; end	nr	43.39	46.64
Hip tiles; half round; ends	nr	83.04	89.27

NATURAL COTSWOLD SLATES

- Cost £2,025.00 per 1000 delivered site; irregular widths;
- Diminishing courses; sarking underfelt;
- Each slate fixed two aluminium alloy nails;
- 44 x 38 mm Softwood battens, treated.

	Unit	New	Refurb
Slates; size 205 x 305 mm - 305 x 610 mm			
Sloping; minimum 20°	m2	143.90	154.69
Curved/ convex/ concave	m2	165.48	177.89
Vertical	m2	154.28	165.85
Vertical; curved on plan	m2	177.43	190.73
Extra for eaves	m	55.22	59.37
Extra for verge	m	42.52	45.71
Extra for valley gutter	m	63.88	68.67
Slates; size 255 x 560 mm - 305 x 760 mm			
Sloping; minimum 20°	m2	161.87	174.01
Curved/ convex/ concave	m2	186.15	200.11
Vertical	m2	108.30	116.43
Vertical; curved on plan	m2	124.55	133.89
Extra for eaves	m	53.84	57.88
Extra for verge	m	41.45	44.56
Extra for valley gutter	m	62.28	66.96
Slates; size 305 x 760 mm - 760 x 1200 mm			
Sloping; minimum 20°	m2	191.19	205.53
Vertical	m2	197.39	212.19
Extra for eaves	m	52.50	56.44
Extra for verge	m	40.42	43.45
Extra for valley gutter	m	60.73	65.28
Sundries			
Cutting			
square or raking	m	38.16	41.02
curved	m	55.40	59.55
Ridge tiles; half round; end	nr	43.39	46.64
Hip tiles; half round; ends	nr	83.04	89.27

CONCRETE TILE; PLAIN; Machine Made

- Cost £210.00 per 1000 delivered site; 165 x 265 mm;
- 65 mm Lap; sarking underfelt; well lapped;
- Every tile nailed two nails every fifth course;
- 38 x 19 mm Softwood battens, treated.

	Unit	New	Refurb
Roof coverings; exceeding 40° slope			
Sloping; minimum 40°	m2	39.90	42.89
Curved/ convex/ concave	m2	45.89	49.33
Extra for eaves	m	6.38	6.86
Extra for verge	m	5.94	6.39
Extra for valley gutter	m	28.64	30.79
Extra for hip tiles	m	14.14	15.20
Roof coverings; 40° slope			
Sloping; minimum 40°	m2	41.90	45.04
Curved/ convex/ concave	m2	48.18	51.80
Extra for eaves	m	6.70	7.20
Extra for verge	m	6.24	6.71
Extra for valley gutter	m	30.07	32.33
Extra for hip tiles	m	14.85	15.96
Wall coverings; vertical			
Vertical	m2	39.56	42.53
Vertical; curved on plan	m2	41.54	44.65
Extra for abutment	m	5.67	6.10
Extra for verge	m	6.23	6.70
Extra for vertical angle tiles	m	17.68	19.00
Sundries			
Extra; alternative coloured tiles	m2	3.05	3.28
Extra; ornamental tiles	m2	8.17	8.79
Cutting			
square or raking	m	4.21	4.52
curved	m	7.36	7.92
Ridge tiles; half round; end	nr	12.39	13.32
Hip tiles; half round; ends	nr	14.14	15.20

CLAY TILE; ORNAMENTAL; Machine Made

- Cost £330.00 per 1000 delivered site; 165 x 265 mm;
- 65 mm Lap; sarking underfelt; well lapped;
- Every tile nailed two nails every fifth course;
- 38 x 19 mm Softwood battens, treated.

	Unit	New	Refurb
Roof coverings; exceeding 40° slope			
Sloping; minimum 40°	m2	37.32	40.11
Curved/ convex/ concave	m2	42.91	46.13
Extra for eaves	m	6.12	6.58
Extra for verge	m	5.88	6.32
Extra for valley gutter	m	41.46	44.57
Extra for hip tiles	m	43.72	47.00
Roof coverings; 40° slope			
Sloping; minimum 40°	m2	39.18	42.12
Curved/ convex/ concave	m2	45.06	48.44
Extra for eaves	m	6.43	6.91
Extra for verge	m	6.17	6.64
Extra for valley gutter	m	43.54	46.80
Extra for hip tiles	m	45.91	49.35
Wall coverings; vertical			
Vertical	m2	37.28	40.08
Vertical; curved on plan	m2	39.14	42.08
Extra for abutment	m	5.34	5.74
Extra for verge	m	5.88	6.32
Extra for vertical angle tiles	m	41.46	44.57
Sundries			
Extra; alternative coloured tiles	m2	8.48	9.11
Extra; ornamental tiles	m2	12.71	13.66
Cutting			
square or raking	m	4.45	4.79
curved	m	7.79	8.38
Ridge tiles; half round; end	nr	19.27	20.72
Hip tiles; half round; ends	nr	43.72	47.00

CLAY TILE; PLAIN; Hand Made

- Cost £475.00 per 1000 delivered site; 165 x 265 mm;
- 65 mm Lap; sarking underfelt; well lapped:
- Every tile nailed two nails every fifth course;
- 38 x 19 mm Softwood battens, treated.

	Unit	New	Refurb
Roof coverings; exceeding 40° slope			
Sloping; minimum 40°	m2	59.16	63.60
Curved/ convex/ concave	m2	68.04	73.14
Extra for eaves	m	8.31	8.93
Extra for verge	m	6.16	6.62
Extra for valley gutter	m	42.93	46.15
Extra for hip tiles	m	45.18	48.57
Roof coverings; 40° slope			
Sloping; minimum 40°	m2	62.12	66.78
Curved/ convex/ concave	m2	71.44	76.80
Extra for eaves	m	8.72	9.38
Extra for verge	m	6.47	6.95
Extra for valley gutter	m	45.07	48.45
Extra for hip tiles	m	47.44	51.00
Wall coverings; vertical			
Vertical	m2	56.58	60.82
Vertical; curved on plan	m2	59.41	63.86
Extra for abutment	m	5.93	6.37
Extra for verge	m	6.16	6.62
Extra for vertical angle tiles	m	51.27	55.12
Sundries			
Extra; alternative coloured tiles	m2		
Extra; ornamental tiles	m2	5.16	5.55
Cutting			
square or raking	m	4.70	5.05
curved	m	8.22	8.84
Ridge tiles; half round; end	nr	19.04	20.47
Hip tiles; half round; ends	nr	45.18	48.57

CLAY PANTILES

- Cost (as stated) per 1000 delivered site; regular widths;
- Regular courses; 75 mm lap; sarking underfelt;
- Each slate fixed two aluminium alloy nails;
- 38 x 25 mm Softwood battens, treated.

	Unit	New	Refurb
Pantiles (£587.00 per 1000)			
Sloping; minimum 22.5°	m2	23.37	25.13
Sloping; exceeding 45°	m2	26.88	28.89
Extra for alternative colours	m2	0.61	0.66
Extra for eaves	m	8.72	9.38
Extra for verge	m	12.53	13.47
Extra for valley gutter	m	5.62	6.05
Double Roman (£722.00 per 1000)			
Sloping; minimum 22.5°	m2	25.43	27.34
Sloping; exceeding 45°	m2	29.25	31.44
Extra for alternative colours	m2	0.61	0.66
Extra for eaves	m	8.72	9.38
Extra for verge	m	12.53	13.47
Extra for valley gutter	m	5.62	6.05
Interlocking (£666.00 per 1000)			
Sloping; minimum 22.5°	m2	24.10	25.91
Sloping; exceeding 45°	m2	27.72	29.80
Extra for alternative colours	m2	0.61	0.66
Extra for eaves	m	8.72	9.38
Extra for verge	m	12.53	13.47
Extra for valley gutter	m	5.62	6.05
Sundries			
Cutting			
square or raking	m	5.49	5.90
curved	m	9.61	10.33
holes	nr	2.93	3.15
Ridge tiles; cutting both sides	m	26.41	28.39
Hip tiles; cutting both sides	m	31.58	33.95

INTERLOCKING CONCRETE TILES

- Cost (as stated) per 1000 delivered site; regular widths;
- Regular courses; 75 mm lap; sarking underfelt;
- Each slate fixed two aluminium alloy nails;
- 38 x 25 mm Softwood battens, treated.

	Unit	New	Refurb
Ludlow (£580.00 per 1000)			
Sloping; minimum 22.5°	m2	17.82	19.16
Sloping; exceeding 45°	m2	20.50	22.03
Extra for alternative colours	m2	included	
Extra for eaves	m	1.59	1.70
Extra for verge	m	4.90	5.27
Extra for valley gutter	m	24.43	26.27
Anglia (£514.00 per 1000)			
Sloping; minimum 22.5°	m2	15.83	17.02
Sloping; exceeding 45°	m2	18.21	19.57
Extra for alternative colours	m2	included	
Extra for eaves	m	1.59	1.70
Extra for verge	m	4.90	5.27
Extra for valley gutter	m	24.43	26.27
County (£635.00 per 1000)			
Sloping; minimum 22.5°	m2	19.24	20.68
Sloping; exceeding 45°	m2	22.12	23.78
Extra for alternative colours	m2	included	
Extra for eaves	m	1.59	1.70
Extra for verge	m	4.90	5.27
Extra for valley gutter	m	24.43	26.27
Sundries			
Cutting			
square or raking	m	5.49	5.90
curved	m	9.61	10.33
holes	nr	2.93	3.15
Ridge tiles; cutting both sides	m	17.22	18.52
Hip tiles; cutting both sides	m	20.59	22.14

SUNDRY TILES

- Reconstructed stone slates; 75 mm lap;
- Red Cedar Shingles; treated; 125 mm gauge;
- Costs (as stated) d/d site; sarking underfelt;
- 38/38 x 25 mm Softwood battens, treated.

	Unit	New	Refurb
Hardrow Reconstructed stone (£650.00 per 1000)			
Sloping; 457 x 305 mm	m2	34.59	37.19
Sloping; 457 x 457 mm	m2	31.98	34.38
Extra for eaves	m2	8.43	9.06
Extra for verge	m	6.66	7.16
Extra for valley gutter	m	12.52	13.46
Ridge tile	m	34.54	37.13
Hip tile	m	30.96	33.28
Reconstructed stone random ((£18.50 per m2)			
Sloping; 300 - 550 mm long	m2	43.62	46.89
Extra for eaves	m2	10.07	10.83
Extra for verge	m	7.15	7.68
Extra for valley gutter	m	14.84	15.96
Ridge tile	m	39.34	42.29
Hip tile	m	41.74	44.87
Red sawn cedar shingles (£12.00 per m2)			
Sloping roof coverings	m2	40.32	43.35
Extra for eaves	m2	5.38	5.78
Extra for verge	m	2.83	3.04
Extra for valley	m	7.36	7.91
Preformed ridge or hip	m	16.39	17.62
Flashings			
Nuralite			
150 mm girth	m	10.73	11.53
250 mm girth	m	15.29	16.44
Lead; Code 5			
150 mm girth	m	28.04	30.14
250 mm girth	m	42.20	45.36

PIPEWORK		uPVC Coloured; self finish			ALUMINIUM Powder coat finish			CAST IRON Paint every 5 years			STEEL Paint every 5 years			MAINTENANCE	
Rainwater Pipes and Fittings		New	Refurb	Repair	New	Refurb	Repair	New	Refurb	Repair	New	Refurb	Repair	Paint one coat oil paint	Flush out
63.5 mm diameter down pipe	m	7.06	7.94	10.24	16.26	18.29	23.57	25.20	28.35	36.54	10.24	11.52	14.84	2.07	0.35
Extra for;															
shoe	nr	4.64	5.22	6.72	8.74	9.84	12.68	18.50	20.81	26.82	6.72	7.57	9.75	included in above rates	
bend	nr	7.07	7.96	10.26	11.15	12.55	16.17	15.01	16.89	21.77	10.26	11.54	14.87		
junction, single branch	nr	10.38	11.67	15.05	15.27	17.18	22.15	25.25	28.41	36.61	15.05	16.93	21.82		
offset pipe	nr	9.50	10.68	13.77	20.16	22.68	29.23	22.92	25.78	33.23	13.77	15.49	19.97		
access pipe	nr	16.19	18.22	23.48	17.96	20.20	26.04	44.25	49.79	64.17	23.48	26.42	34.05		
75.5 mm diameter	m	7.47	8.41	10.84	18.49	20.80	26.81	25.72	28.93	37.29	10.84	12.19	15.72	2.27	0.35
Extra for;															
shoe	nr	4.87	5.48	7.06	11.48	12.91	16.64	18.90	21.26	27.41	7.06	7.95	10.24	included in above rates	
bend	nr	6.27	7.05	9.09	13.29	14.95	19.27	23.76	26.73	34.45	9.09	10.23	13.18		
junction, single branch	nr	11.06	12.45	16.04	17.58	19.78	25.49	27.70	31.16	40.16	16.04	18.05	23.26		
offset pipe	nr	7.94	8.93	11.51	22.67	25.50	32.87	23.82	26.80	34.54	11.51	12.95	16.70		
access pipe	nr	17.42	19.60	25.27	19.43	21.86	28.18	47.22	53.12	68.47	25.27	28.42	36.63		
100 mm diameter	m	10.95	12.32	15.87	28.49	32.05	41.31	33.02	37.15	47.88	15.87	17.86	23.02	3.32	0.35
Extra for;															
shoe	nr	8.16	9.18	11.83	13.01	14.63	18.86	23.97	26.96	34.75	11.83	13.31	17.16	included in above rates	
bend	nr	11.50	12.94	16.68	16.74	18.83	24.27	22.65	25.48	32.84	16.68	18.77	24.19		
junction, single branch	nr	16.13	18.14	23.39	21.51	24.19	31.18	31.67	35.63	45.93	23.39	26.31	33.91		
offset pipe	nr	17.72	19.94	25.70	23.64	26.60	34.28	38.48	43.29	55.79	25.70	28.91	37.26		
access pipe	nr	19.76	22.23	28.65	21.25	23.90	30.81	49.88	56.11	72.32	28.65	32.23	41.54		
Roof outlets															
Flat		Not available									Not available			Not required	
75.5 mm diameter outlet	nr				79.53	89.47	115.32	98.65	110.98	143.04					
100 mm diameter outlet	nr				101.29	113.96	146.88	106.55	119.87	154.50					
Domed															
75.5 mm diameter outlet	nr				82.24	92.52	119.24	118.30	133.09	171.54					
100 mm diameter outlet	nr				99.48	111.92	144.25	126.24	142.03	183.05					

EAVES GUTTERS		uPVC Coloured; self finish			ALUMINIUM Powder coat finish			CAST IRON Paint every 5 years			STEEL Paint every 5 years			MAINTENANCE		
Eaves Gutters and Fittings		New	Refurb	Repair	New	Refurb	Repair	New	Refurb	Repair	New	Refurb	Repair	Paint one coat oil Paint	Flush Out	One Coat Bitumen
		(112 mm diameter only)														
Half Round; 100 mm diameter	m	7.29	8.21	10.58	16.82	18.92	24.38	17.72	19.94	25.70	Not available			2.35	0.42	1.56
Extra for;-																
stopped ends	nr	2.85	3.21	4.13	6.98	7.86	10.12	5.69	6.40	8.25				included in above rates		
outlet	nr	4.96	5.58	7.19	9.00	10.13	13.06	10.39	11.69	15.07						
angle	nr	5.95	6.69	8.62	8.12	9.14	11.78	11.92	13.41	17.28						
		(170 mm diameter only)														
Half Round; 125 mm diameter	m	12.72	14.31	18.45	20.61	23.19	29.89	20.64	23.22	29.92	Not available			3.22	0.42	1.95
Extra for;-																
stopped ends	nr	5.13	5.77	7.44	9.25	10.41	13.41	6.76	7.61	9.81				included in above rates		
outlet	nr	8.10	9.11	11.74	11.34	12.75	16.44	13.23	14.88	19.18						
angle	nr	11.98	13.48	17.38	10.65	11.98	15.44	13.69	15.40	19.86						
		(112 mm diameter only)														
Ogee; 100 mm diameter	m	7.29	8.21	10.58	20.74	23.33	30.07	19.41	21.83	28.14	Not available			2.35	0.42	1.56
Extra for;-																
stopped ends	nr	2.85	3.21	4.13	5.29	5.95	7.66	6.88	7.74	9.97				included in above rates		
outlet	nr	4.96	5.58	7.19	9.91	11.15	14.37	11.04	12.42	16.00						
angle	nr	5.95	6.69	8.62	7.79	8.76	11.29	12.49	14.05	18.11						
Ogee; 125 mm diameter	m	Not available			24.80	27.89	35.95	22.19	24.97	32.18	Not available			3.22	0.42	1.95
Extra for;-																
stopped ends	nr				5.99	6.73	8.68	8.28	9.31	12.00				included in above rates		
outlet	nr				11.34	12.75	16.44	12.89	14.50	18.69						
angle	nr				9.90	11.13	14.35	14.51	16.32	21.04						
											(200 x 100 mm overall only)					
Box; 114 mm wide	m	7.98	8.98	11.57	24.80	27.89	35.95	22.19	24.97	32.18	29.80	33.52	43.20	3.45	0.36	2.15
Extra for;-																
stopped ends	nr	3.07	3.45	4.45	5.99	6.73	8.68	8.28	9.31	12.00	14.22	16.00	20.63			
outlet	nr	5.56	6.25	8.06	11.34	12.75	16.44	12.89	14.50	18.69	24.89	28.00	36.09			
angle	nr	6.39	7.19	9.26	9.90	11.13	14.35	14.51	16.32	21.04	26.78	30.12	38.83			

GLOSSARY

Basic Timber Prices

Softwood

General structural timber (GS grade)	£ 205	per m3
Special structural timber (SS grade)	£ 218	per m3
Selected for 1st and 2nd fixings	£ 270	per m3

Hardwood

Afromosia	£ 845	per m3
Beech	£ 945	per m3
Columbian Pine	£ 715	per m3
Ebony	£ 3,150	per m3
Idigbo	£ 795	per m3
Iroko	£ 770	per m3
Mahogany; African	£ 765	per m3
Mahogany; South American	£ 915	per m3
Maple	£ 1,650	per m3
Meranti	£ 845	per m3
Oak; English	£ 2,100	per m3
Oak; American	£ 1,950	per m3
Oak; Japanese	£ 2,350	per m3
Sapele	£ 760	per m3
Teak	£ 1,880	per m3
Utile	£ 1,560	per m3
Walnut	£ 1,230	per m3

Timber Treatments

Wood Preservatives;		
brush coats	£ 2.25	per m2
pressure impregnated	£25 to 45	per m3
Fire Protective Treatments		
flame proof treatment	£85 to 95	per m3

STRUCTURAL TIMBERWORK

- SOFTWOOD; Sawn; Cost £**200.00** per m3;
- Standard lengths; delivered to site;
- Tanalised; fixing - connectors or nails.

	Unit	New	Refurb
Floors			
50 x 100 mm	m	2.62	2.95
50 x 125 mm	m	3.00	3.37
50 x 150 mm	m	3.49	3.93
75 x 100 mm	m	3.63	4.09
75 x 125 mm	m	3.82	4.30
75 x 150 mm	m	4.37	4.91
75 x 200 mm	m	5.50	6.19
Partitions			
38 x 75 mm	m	1.96	2.21
38 x 100 mm	m	2.52	2.83
50 x 75 mm	m	2.39	2.69
50 x 100 mm	m	2.89	3.25
75 x 75 mm	m	3.21	3.61
100 x 100 mm	m	5.13	5.77
Roofs - Flat			
38 x 75 mm	m	1.51	1.70
38 x 100 mm	m	1.70	1.92
50 x 75 mm	m	2.28	2.56
50 x 100 mm	m	2.75	3.10
50 x 150 mm	m	3.58	4.02
75 x 100 mm	m	3.38	3.80
Roofs - Pitched			
38 x 75 mm	m	1.53	1.72
38 x 100 mm	m	2.29	2.58
50 x 75 mm	m	2.39	2.69
50 x 100 mm	m	2.89	3.25
50 x 150 mm	m	3.41	3.84
75 x 100 mm	m	3.56	4.00
75 x 150 mm	m	4.59	5.16

WOOD FRAMES

- SOFTWOOD; Wrot; Cost £**320.00** per m3;
- Standard lengths; delivered to site;
- Framing; fixing - nails and screws.

	Unit	New	Refurb
Purpose Made Frames and Linings			
Jambs and heads			
32 x 63 mm	m	4.06	4.57
32 x 100 mm	m	4.89	5.51
32 x 150 mm	m	5.95	6.70
38 x 75 mm	m	4.88	5.49
38 x 100 mm	m	5.54	6.23
38 x 115 mm	m	6.27	7.05
38 x 150 mm	m	7.30	8.21
50 x 100 mm	m	6.89	7.75
50 x 125 mm	m	7.92	8.91
63 x 88 mm	m	7.70	8.66
63 x 100 mm	m	8.11	9.12
63 x 125 mm	m	9.42	10.60
75 x 100 mm	m	9.11	10.25
75 x 125 mm	m	11.08	12.46
75 x 150 mm	m	12.65	14.23
100 x 100 mm	m	11.78	13.25
100 x 150 mm	m	16.12	18.13
Mullions and transoms			
32 x 63 mm	m	6.82	7.68
32 x 100 mm	m	7.85	8.83
32 x 150 mm	m	9.07	10.21
38 x 75 mm	m	7.93	8.92
38 x 100 mm	m	8.78	9.88
38 x 115 mm	m	9.47	10.66
38 x 150 mm	m	10.67	12.01
50 x 100 mm	m	10.14	11.41
50 x 125 mm	m	11.57	13.02
63 x 88 mm	m	11.22	12.62
63 x 100 mm	m	11.71	13.18
75 x 100 mm	m	13.03	14.66

WOOD FRAMES

- HARDWOOD; Wrot; Cost **£765.00** per m3;
- Standard lengths; delivered to site;
- Framing; fixing - screws; pelleted.

		New	Refurb
Purpose Made Frames and Linings			
Jambs and heads			
32 x 63 mm	m	7.24	8.15
32 x 100 mm	m	8.97	10.10
32 x 150 mm	m	11.08	12.47
38 x 75 mm	m	8.72	9.81
38 x 100 mm	m	10.27	11.55
38 x 115 mm	m	11.64	13.10
38 x 150 mm	m	13.47	15.15
50 x 100 mm	m	12.89	14.50
50 x 125 mm	m	15.17	17.07
63 x 88 mm	m	14.16	15.93
63 x 100 mm	m	15.49	17.42
63 x 125 mm	m	18.32	20.61
75 x 100 mm	m	17.64	19.85
75 x 125 mm	m	21.67	24.38
75 x 150 mm	m	25.09	28.23
100 x 100 mm	m	22.72	25.56
100 x 150 mm	m	32.04	36.05
Mullions and transoms			
32 x 63 mm	m	11.71	13.18
32 x 100 mm	m	13.86	15.60
32 x 150 mm	m	16.38	18.43
38 x 75 mm	m	13.71	15.43
38 x 100 mm	m	15.64	17.59
38 x 115 mm	m	17.09	19.22
38 x 150 mm	m	19.16	21.56
50 x 100 mm	m	18.45	20.76
50 x 125 mm	m	21.46	24.14
63 x 88 mm	m	20.08	22.59
63 x 100 mm	m	21.73	24.44
75 x 100 mm	m	24.51	27.57

WOOD FINISHING

- SOFTWOOD; Wrot; Cost **£235.00** per m3;
- Standard lengths; delivered to site;
- Untreated; fixing with wire nails.

		New	Refurb
Eaves and Fascia boarding			
25 x 150 mm	m	2.61	2.94
32 x 150 mm	m	2.86	3.22
32 x 200 mm	m	3.40	2.93
Small mouldings; Quadrants and scotia			
19 mm	m	1.72	1.93
25 mm	m	2.31	2.00
Plain; Skirtings and cover fillets			
15 x 25 mm	m	1.06	1.19
15 x 50 mm	m	1.50	1.68
19 x 50 mm	m	1.73	1.95
19 x 75 mm	m	1.93	2.17
25 x 50 mm	m	1.96	2.21
25 x 75 mm	m	2.13	2.39
25 x 100 mm	m	2.38	2.68
38 x 100 mm	m	3.10	3.48
38 x 150 mm	m	4.02	4.53
38 x 200 mm	m	5.03	5.66
Moulded; Architraves, picture rails and window boards			
18 x 50 mm	m	1.87	2.11
18 x 75 mm	m	2.07	2.32
25 x 50 mm	m	2.06	2.31
25 x 75 mm	m	2.27	2.55
25 x 100 mm	m	2.49	2.81
25 x 150 mm	m	3.12	3.51
25 x 200 mm	m	3.59	4.03
38 x 75 mm	m	2.61	2.93
38 x 100 mm	m	2.87	3.23
38 x 150 mm	m	3.59	4.03
38 x 200 mm	m	4.12	4.64

WOOD FINISHING

- HARDWOOD; Wrot; Cost **£765.00** per m3;
- Standard lengths; delivered to site;
- Untreated; fixing with wire nails.

		New	Refurb
Eaves and Fascia boarding			
25 x 150 mm	m	7.88	8.86
32 x 150 mm	m	8.64	9.72
32 x 200 mm	m	10.26	8.85
Small mouldings; Quadrants and scotia			
19 mm	m	5.18	5.83
25 mm	m	6.99	6.03
Plain; Skirtings and cover fillets			
15 x 25 mm	m	3.20	3.60
15 x 50 mm	m	4.52	5.09
19 x 50 mm	m	5.23	5.89
19 x 75 mm	m	5.82	6.54
25 x 50 mm	m	5.92	6.66
25 x 75 mm	m	6.42	7.23
25 x 100 mm	m	7.19	8.09
38 x 100 mm	m	9.35	10.52
38 x 150 mm	m	12.15	13.67
38 x 200 mm	m	15.19	17.09
Moulded; Architraves, picture rails and window boards			
18 x 50 mm	m	5.66	6.36
18 x 75 mm	m	6.24	7.02
25 x 50 mm	m	6.21	6.99
25 x 75 mm	m	6.85	7.70
25 x 100 mm	m	7.53	8.47
25 x 150 mm	m	9.41	10.59
25 x 200 mm	m	10.83	12.18
38 x 75 mm	m	7.87	8.86
38 x 100 mm	m	8.66	9.74
38 x 150 mm	m	10.83	12.18
38 x 200 mm	m	12.45	14.01

TIMBER SIZE	FLOORS AND FLAT ROOFS						PARTITIONS						PITCHED ROOFS					
	Softwood				Hardwood		Softwood				Hardwood		Softwood				Hardwood	
	Untreated		Tanalised		Untreated		Untreated		Tanalised		Untreated		Untreated		Tanalised		Untreated	
Nominal	Sawn	Wrot	Sawn	Wrot	Sawn	Wrot	Sawn	Wrot	Sawn	Wrot	Sawn	Wrot	Sawn	Wrot	Sawn	Wrot	Sawn	Wrot
13 x 25	0.66	0.77	0.71	0.82	2.60	2.95	0.71	0.83	0.76	0.88	2.80	3.18	0.68	0.79	0.72	0.84	2.67	3.03
13 x 38	0.70	0.84	0.75	0.90	2.76	3.22	0.76	0.90	0.81	0.96	2.97	3.47	0.72	0.86	0.77	0.92	2.83	3.31
13 x 50	0.74	0.91	0.79	0.96	2.90	3.46	0.80	0.97	0.85	1.04	3.12	3.73	0.76	0.93	0.81	0.99	2.98	3.55
19 x 50	0.81	0.99	0.86	1.05	3.18	3.77	0.87	1.06	0.93	1.13	3.43	4.06	0.83	1.01	0.89	1.08	3.27	3.87
19 x 75	0.92	1.15	0.98	1.23	3.63	4.39	0.99	1.24	1.06	1.32	3.90	4.73	0.95	1.18	1.01	1.26	3.72	4.51
19 x 100	1.04	1.31	1.10	1.40	4.07	5.00	1.12	1.41	1.19	1.50	4.38	5.38	1.06	1.35	1.13	1.43	4.18	5.13
25 x 25	0.83	0.96	0.89	1.02	3.26	3.69	0.90	1.03	0.95	1.10	3.51	3.97	0.85	0.98	0.91	1.05	3.35	3.78
25 x 38	0.91	1.06	0.97	1.13	3.57	4.08	0.98	1.14	1.04	1.22	3.84	4.39	0.93	1.09	0.99	1.16	3.66	4.19
25 x 50	0.98	1.16	1.04	1.23	3.85	4.44	1.05	1.25	1.12	1.33	4.14	4.78	1.01	1.19	1.07	1.26	3.95	4.56
25 x 75	1.13	1.35	1.20	1.44	4.43	5.19	1.21	1.46	1.29	1.55	4.77	5.58	1.16	1.39	1.23	1.48	4.54	5.32
25 x 100	1.38	1.64	1.46	1.75	5.40	6.29	1.48	1.77	1.58	1.88	5.81	6.77	1.41	1.68	1.50	1.79	5.54	6.45
25 x 125	1.52	1.83	1.62	1.95	5.98	7.02	1.64	1.97	1.75	2.10	6.43	7.56	1.56	1.88	1.67	2.00	6.13	7.21
25 x 150	1.77	2.12	1.89	2.26	6.95	8.12	1.90	2.28	2.03	2.43	7.47	8.73	1.82	2.17	1.93	2.31	7.13	8.33
25 x 175	1.92	2.31	2.04	2.46	7.53	8.84	2.06	2.48	2.20	2.65	8.10	9.52	1.97	2.37	2.10	2.52	7.72	9.07
25 x 200	2.07	2.50	2.20	2.66	8.11	9.57	2.22	2.69	2.37	2.86	8.73	10.30	2.12	2.56	2.26	2.73	8.32	9.82
32 x 25	0.97	1.11	1.03	1.18	3.81	4.27	1.04	1.19	1.11	1.27	4.10	4.59	1.00	1.13	1.06	1.21	3.91	4.38
32 x 50	1.16	1.34	1.24	1.43	4.56	5.17	1.25	1.45	1.33	1.54	4.90	5.57	1.19	1.38	1.27	1.47	4.67	5.31
32 x 75	1.35	1.58	1.44	1.68	5.30	6.07	1.45	1.70	1.55	1.81	5.71	6.53	1.39	1.62	1.48	1.73	5.44	6.23
32 x 100	1.64	1.91	1.75	2.03	6.43	7.33	1.76	2.05	1.88	2.18	6.92	7.89	1.68	1.96	1.79	2.08	6.60	7.52
32 x 150	2.02	2.37	2.15	2.53	7.93	9.11	2.17	2.55	2.31	2.72	8.53	9.80	2.07	2.43	2.21	2.59	8.13	9.35
32 x 200	2.40	2.83	2.56	3.02	9.42	10.88	2.58	3.05	2.75	3.25	10.13	11.70	2.46	2.91	2.62	3.10	9.66	11.16
38 x 25	1.01	1.16	1.08	1.23	3.97	4.47	1.09	1.25	1.16	1.33	4.27	4.81	1.04	1.19	1.11	1.27	4.07	4.58
38 x 38	1.23	1.40	1.31	1.49	4.83	5.39	1.32	1.50	1.41	1.60	5.19	5.80	1.26	1.43	1.34	1.53	4.95	5.53
38 x 50	1.44	1.62	1.53	1.73	5.65	6.27	1.55	1.75	1.65	1.86	6.08	6.74	1.48	1.67	1.57	1.77	5.79	6.43
38 x 75	1.67	1.90	1.78	2.02	6.55	7.32	1.80	2.04	1.91	2.18	7.05	7.88	1.71	1.95	1.83	2.07	6.73	7.51
38 x 100	2.10	2.36	2.23	2.52	8.23	9.12	2.26	2.54	2.40	2.71	8.85	9.81	2.15	2.42	2.29	2.58	8.44	9.36
38 x 125	2.43	2.73	2.58	2.91	9.52	10.54	2.61	2.94	2.78	3.13	10.24	11.34	2.49	2.80	2.65	2.98	9.77	10.82
38 x 150	2.76	3.10	2.94	3.30	10.81	11.97	2.97	3.33	3.16	3.55	11.63	12.88	2.83	3.18	3.01	3.39	11.09	12.28
50 x 50	1.85	2.05	1.97	2.18	7.25	7.92	1.99	2.20	2.12	2.34	7.80	8.52	1.90	2.10	2.02	2.23	7.44	8.12
50 x 75	2.14	2.38	2.28	2.53	8.39	9.19	2.30	2.56	2.45	2.72	9.03	9.89	2.19	2.44	2.34	2.60	8.61	9.43
50 x 100	2.53	2.80	2.69	2.99	9.91	10.84	2.72	3.02	2.89	3.21	10.66	11.67	2.59	2.88	2.76	3.06	10.17	11.13
50 x 125	2.82	3.13	3.00	3.34	11.05	12.12	3.03	3.37	3.23	3.59	11.89	13.04	2.89	3.21	3.08	3.42	11.33	12.43
50 x 150	3.20	3.56	3.41	3.79	12.57	13.77	3.45	3.83	3.67	4.08	13.52	14.81	3.29	3.65	3.50	3.89	12.89	14.12
50 x 175	3.60	3.99	3.83	4.25	14.11	15.44	3.87	4.29	4.12	4.57	15.18	16.61	3.69	4.09	3.93	4.36	14.48	15.84
50 x 200	4.00	4.43	4.26	4.72	15.69	17.15	4.30	4.77	4.58	5.08	16.88	18.45	4.10	4.55	4.37	4.84	16.10	17.59
50 x 225	4.30	4.78	4.58	5.09	16.89	18.48	4.63	5.14	4.93	5.47	18.17	19.88	4.42	4.90	4.70	5.22	17.33	18.96
50 x 250	4.71	5.22	5.01	5.55	18.47	20.18	5.06	5.61	5.39	5.98	19.87	21.72	4.83	5.35	5.14	5.70	18.95	20.71
75 x 75	2.73	3.01	2.91	3.20	10.73	11.65	2.94	3.24	3.13	3.45	11.54	12.54	2.81	3.09	2.99	3.29	11.01	11.95
75 x 100	3.26	3.57	3.47	3.80	12.77	13.82	3.50	3.84	3.73	4.09	13.75	14.87	3.34	3.66	3.56	3.90	13.11	14.18
75 x 125	3.68	4.03	3.92	4.29	14.44	15.61	3.96	4.34	4.22	4.62	15.53	16.80	3.78	4.13	4.02	4.40	14.81	16.02
75 x 150	4.10	4.49	4.37	4.78	16.10	17.41	4.42	4.83	4.70	5.15	17.32	18.73	4.21	4.61	4.48	4.91	16.52	17.86
75 x 175	4.53	4.95	4.82	5.27	17.76	19.20	4.87	5.33	5.19	5.68	19.11	20.66	4.65	5.08	4.95	5.41	18.22	19.70
75 x 200	5.05	5.51	5.38	5.87	19.81	21.37	5.43	5.93	5.79	6.32	21.31	22.99	5.18	5.65	5.52	6.02	20.32	21.92
75 x 225	5.47	5.97	5.83	6.36	21.47	23.16	5.89	6.43	6.27	6.85	23.10	24.92	5.61	6.13	5.98	6.53	22.03	23.76
75 x 250	5.99	6.53	6.38	6.96	23.51	25.33	6.45	7.03	6.87	7.49	25.30	27.26	6.15	6.70	6.55	7.14	24.13	25.99
100 x 100	4.27	4.61	4.55	4.91	16.76	17.91	4.60	4.96	4.89	5.29	18.03	19.27	4.38	4.73	4.67	5.04	17.19	18.38
100 x 150	5.63	6.04	5.99	6.44	22.07	23.48	6.05	6.50	6.45	6.92	23.75	25.26	5.77	6.20	6.15	6.60	22.64	24.09
100 x 175	6.21	6.66	6.61	7.09	24.35	25.88	6.68	7.17	7.11	7.63	26.20	27.85	6.37	6.83	6.78	7.28	24.98	26.55
100 x 200	7.00	7.49	7.46	7.98	27.47	29.13	7.53	8.06	8.02	8.59	29.56	31.34	7.18	7.69	7.65	8.19	28.18	29.89
100 x 225	8.10	8.62	8.62	9.18	31.76	33.54	8.71	9.28	9.28	9.88	34.17	36.09	8.31	8.85	8.85	9.42	32.59	34.41
100 x 250	8.83	9.40	9.41	10.01	34.65	36.56	9.50	10.11	10.12	10.77	37.28	39.34	9.06	9.64	9.65	10.27	35.55	37.51
100 x 300	10.31	10.95	10.98	11.66	40.43	42.59	11.09	11.78	11.81	12.54	43.51	45.83	10.57	11.23	11.26	11.96	41.48	43.70

LOCATION	SOFTWOOD					PLYWOOD		CHIPBOARD					BLOCKBOARD		MDF		HARDBOARD		
	Sawn	Wrot	Wrot	Wrot	Wrot														
	Square Edge	Square Edge	Tongued and grooved	Tongued, grooved and 'V' jointed	Shiplap weather boards	Birch faced	Marine grade	Standard grade	Moisture resistant	Flooring grade	Flooring grade tongued and grooved	Flooring with 25 mm of insulation	Birch faced	Hardwood veneered	Standard grade	Fire resistant	Standard grade; 3.2 mm thick	White coated; 3.2 mm thick	Oiled; 3.2 mm thick
Floors																			
18 mm thick	8.62	11.27	11.37	12.23	-	17.84	18.76	5.20	6.11	5.69	5.94	8.82	16.83	20.10	14.73	17.59	2.89	-	3.00
19 mm thick	9.10	11.89	12.01	12.91	-	18.84	19.80	5.46	6.42	5.97	6.24	9.11	-	-	-	-	-	-	-
22 mm thick	10.18	12.83	13.12	14.11	-	21.38	22.18	6.23	7.32	6.81	7.12	9.99	-	-	-	-	-	-	-
25 mm thick	10.76	13.14	13.60	14.62	-	22.89	23.47	7.35	8.64	8.03	8.40	11.27	24.46	27.95	21.40	24.45	-	-	-
30 mm thick	11.74	13.66	14.39	15.47	-	25.39	25.63	8.91	10.46	9.73	10.17	13.05	-	-	-	-	-	-	-
Ducts and Covers																			
12 mm thick																			
over 300 mm wide	5.39	7.04	7.11	7.65	-	12.73	13.39	-	-	-	-	-	13.59	15.99	11.89	14.00	-	-	-
<300 mm wide	5.65	7.38	7.37	7.92	-	14.64	15.39	-	-	-	-	-	16.98	19.99	14.86	17.49	-	-	-
19 mm thick																			
over 300 mm wide	8.17	10.67	10.78	11.58	-	19.29	20.28	-	-	-	-	-	19.23	23.37	16.83	20.45	-	-	-
<300 mm wide	8.56	11.18	11.17	12.00	-	22.19	23.33	-	-	-	-	-	24.04	29.21	21.03	25.56	-	-	-
Walls																			
18 mm thick								-	-	-	-	-	-	-	-	-	-	-	-
over 300 mm wide	-	-	-	-	9.42	17.40	18.29	-	-	-	-	-	16.41	19.60	14.36	17.15	-	4.25	-
<300 mm wide	-	-	-	-	12.53	20.62	21.68	-	-	-	-	-	19.45	23.23	17.02	20.32	-	5.02	-
Roofs																			
19 mm thick																			
laid flat	9.10	11.89	12.01	-	-	18.84	19.80	6.17	7.25	6.74	7.05	10.30	17.68	21.11	15.47	18.47	3.10	-	3.22
laid to slope	9.53	12.46	12.44	-	-	19.38	20.35	6.48	7.61	7.08	7.40	10.81	20.33	24.27	17.79	21.24	-	-	-
vertical; to dormers	12.16	15.89	15.93	-	-	21.66	22.77	7.45	8.75	8.14	8.51	12.43	25.41	30.34	22.23	26.55	-	-	-
25 mm thick																			
laid flat	10.76	13.14	13.60	-	-	22.89	23.47	8.31	9.76	9.08	9.49	12.74	25.68	29.34	22.47	25.68	-	-	-
laid to slope	11.20	13.67	14.03	-	-	23.43	24.02	8.72	10.25	9.53	9.96	13.37	29.54	33.75	25.84	29.53	-	-	-
vertical; to dormers	13.22	16.13	-	-	-	26.32	26.99	10.03	11.79	10.96	11.46	15.38	36.92	42.18	32.30	36.91	-	-	-
Eaves and Verge soffites																			
15 mm thick																			
over 300 mm wide	7.18	9.39	9.48	10.10	-	18.23	19.27	-	-	-	-	-	15.56	18.03	13.61	15.78	-	-	-
<300 mm wide	7.53	9.84	9.82	10.46	-	20.96	22.17	-	-	-	-	-	17.89	20.74	15.66	18.14	-	-	-
18 mm thick																			
over 300 mm wide	8.62	11.27	11.37	12.11	-	20.36	21.41	-	-	-	-	-	19.36	23.12	16.94	20.23	-	-	-
<300 mm wide	9.03	11.81	11.79	12.55	-	23.42	24.62	-	-	-	-	-	22.26	26.58	19.48	23.26	-	-	-
Gutters; bottoms and sides																			
19 mm thick																			
over 300 mm wide	16.38	19.99	21.03	-	-	19.74	20.24	-	-	-	-	-	20.33	24.27	17.79	21.24	-	-	-
<300 mm wide	18.83	22.99	24.18	-	-	22.70	23.28	-	-	-	-	-	23.38	27.91	20.45	24.42	-	-	-
25 mm thick																			
over 300 mm wide	18.04	22.02	22.70	-	-	24.74	25.38	-	-	-	-	-	25.21	28.00	22.06	24.50	-	-	-
<300 mm wide	20.75	25.33	26.10	-	-	28.45	29.18	-	-	-	-	-	28.99	32.20	25.37	28.17	-	-	-

GLOSSARY – Pictorial Glossary

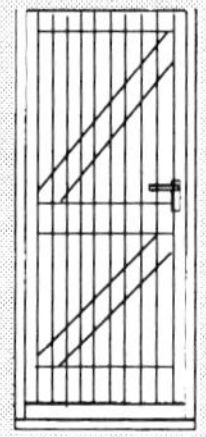

Ledged, framed and braced doo

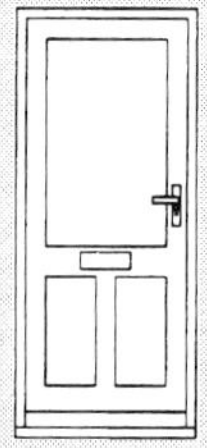

Three panelled door.

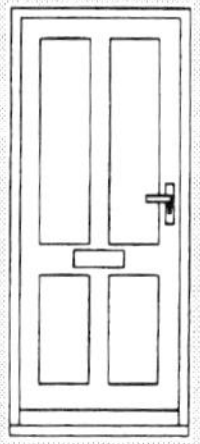

Four panelled doordoor.

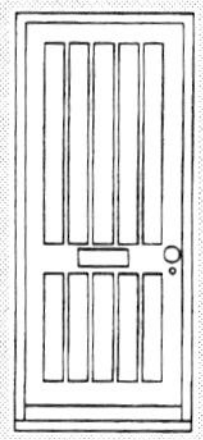

Glazed open panelled.

DOOR TYPE	SW/HW	DOOR: Door	DOOR: Glass, glazing strip, hardwood beads	IRONMONGERY: Hinges	Door handles	Locks	Push plates	Kicking plates	Door closures, Perko type	Bolts	Signs	FRAME: Lining	Frame	Stop	Intumescent strip	Architrave	PAINT: Three coats oil paint	Two coats polyurethene paint, waxed	TOTAL
NOTES; See Manufacturers' catalogues for full range of doors and metric doors (M). Rates for plywood faced flush doors allow for birch (SW) or Sapele (HW) on both sides.	Softwood (SW) Hardwood (HW)																		
Ledged framed and braced																			
44 x 762 x 1981 mm	SW	125.19		6.14	10.60	13.51				15.69			63.40			21.87	31.42		287.82
44 x 838 x 1981 mm	SW	125.19		6.14	10.60	13.51				15.69			65.76			22.25	32.63		291.77
44 x 762 x 1981 mm	HW	162.42		7.63	16.07	13.51				15.69			80.46			45.61	31.42		372.82
44 x 838 x 1981 mm	HW	162.42		7.63	16.07	13.51				15.69			80.46			46.72	32.63		375.13
Four panelled door; plywood panels.																			
44 x 762 x 1981 mm	SW	181.57		5.21	9.67	23.45			26.48	14.77	10.23		63.40			21.87	31.42		388.07
44 x 838 x 1981 mm	SW	185.13		5.21	9.67	23.45			26.48	14.77	10.23		65.76			22.25	32.63		395.58
44 x 762 x 1981 mm	HW	330.60		7.63	16.07	24.38	6.65	8.80	35.89	18.68	10.23		138.66			45.61		41.79	684.99
44 x 838 x 1981 mm	HW	330.60		7.63	16.07	24.38	6.65	8.80	35.89	18.68	10.23		142.64			46.72		43.39	691.67
Three panelled door; 1 glazed top panel; 2 plywood bottom panels.																			
44 x 762 x 1981 mm	SW	174.23	45.40	5.21	10.60	23.45			26.48	14.77	10.23		63.40			14.52	31.42		419.71
44 x 838 x 1981 mm	SW	174.23	45.40	5.21	10.60	23.45			26.48	14.77	10.23		65.76			14.76	32.63		423.51
44 x 762 x 1981 mm	HW	191.37	50.40	7.63	16.07	23.45	6.18	8.34	35.89	18.68	10.23		138.66			20.47		41.79	569.15
44 x 838 x 1981 mm	HW	191.37	50.40	7.63	16.07	23.45	6.18	8.34	35.89	18.68	10.23		142.64			20.96		43.39	575.23
Flush, plywood faced; half hour fire door.																			
44 x 762 x 1981 mm	SW	98.54		9.62	16.07	23.45	6.18	8.34	86.70	18.68	10.23		48.79			16.10	31.42		374.12
44 x 838 x 1981 mm	SW	102.07		9.62	16.07	23.45	6.18	8.34	86.70	18.68	10.23		48.79			16.33	32.63		379.09
44 x 826 x 2040 mm (M)	SW	103.86		9.62	16.07	23.45	6.18	8.34	86.70	18.68	10.23		48.79			16.29	32.60		380.81
44 x 762 x 1981 mm	HW	140.21		9.62	16.07	23.45	6.18	8.34	86.70	18.68	10.23		48.79			22.76		41.79	432.82
44 x 838 x 1981 mm	HW	140.21		9.62	16.07	23.45	6.18	8.34	86.70	18.68	10.23		48.79			23.19		43.39	434.85
44 x 826 x 2040 mm (M)	HW	140.21		9.62	16.07	23.45	6.18	8.34	86.70	18.68	10.23		48.79			23.11		43.36	434.74
Solid timber doors and frame; transome and mullions; ten 100 mm wide vertical glazed panels.																			
Pair 44 x 826 x 2040 mm (M)	SW	294.82		22.10	18.39	30.83	10.98	16.21	172.93	31.21	11.15		109.34			19.35	32.60		769.92
Pair 44 x 826 x 2040 mm (M)	HW	380.37		22.10	18.39	30.83	10.98	16.21	172.93	31.21	11.15		151.31			19.35	32.60		897.44

GLOSSARY: Pictorial Glossary	DOOR TYPE: NOTES; See Manufacturers' catalogue for full range of doors and metric doors (M). Rates for plywood faced flush doors allow for birch (SW) or Sapele (HW) on both sides.	Softwood (SW) Hardwood (HW)	DOOR: Door	DOOR: Glass, safety glazing strip, hardwood beads	IRONMONGERY: Hinges	IRONMONGERY: Door handles	IRONMONGERY: Locks	IRONMONGERY: Push plates	IRONMONGERY: Kicking plates	IRONMONGERY: Door closures, Perko type	IRONMONGERY: Bolts	IRONMONGERY: Signs	FRAME: Lining	FRAME: Frame	FRAME: Stop	FRAME: Intumescent strip	FRAME: Architrave	PAINT: Three coats oil paint	PAINT: Two coats polyurethene paint, waxed	TOTAL
Four panelled door. 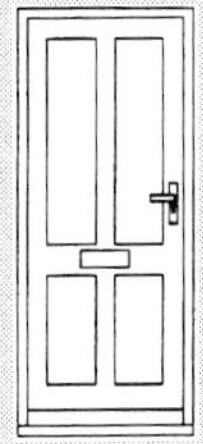	**Four panelled door,** plywood panels																			
	35 x 762 x 1981 mm	SW	152.23		5.48	10.60	11.30					10.23	36.60		8.05		41.24	31.42		307.15
	35 x 838 x 1981 mm	SW	158.99		5.48	10.60	11.30					10.23	37.71		8.05		42.34	32.63		317.32
	35 x 762 x 1981 mm	HW	343.09		6.97	16.07	12.40	6.18	8.34			14.65	83.48		12.23		99.25		41.79	644.44
	35 x 838 x 1981 mm	HW	358.17		6.97	16.07	12.40	6.18	8.34			14.65	84.58		12.23		104.77		43.39	667.75
Three panelled door. 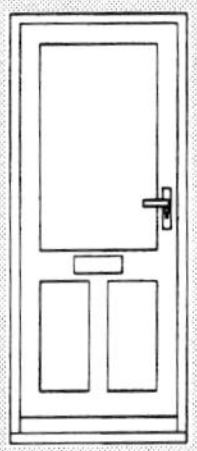	**Three panelled door;** 1 top panel glazed; 2 plywood bottom panels.																			
	35 x 762 x 1981 mm	SW	96.47	61.75	5.48	10.60	11.30					10.23	36.60		8.05		41.24	31.42		313.14
	35 x 838 x 1981 mm	SW	103.02	61.75	5.48	10.60	11.30					10.23	37.71		8.05		42.34	32.63		323.11
	35 x 762 x 1981 mm	HW	318.79	61.75	6.97	16.07	12.40	6.18	8.34			14.65	83.48		12.23		99.25		41.79	681.89
	35 x 838 x 1981 mm	HW	333.88	61.75	6.97	16.07	12.40	6.18	8.34			14.65	84.58		12.23		104.77		43.39	705.21
Flush door. 	**Flush, plywood faced;** hardwood lipping two edges.																			
	35 x 762 x 1981 mm	SW	55.36		5.48	10.60	11.30					10.23	36.60		8.05		40.63	31.42		209.67
	35 x 838 x 1981 mm	SW	57.09		5.48	10.60	11.30					10.23	36.60		8.05		40.63	32.63		212.61
	40 x 762 x 1981 mm	SW	56.87		5.48	10.60	11.30					10.23	36.60		8.05		40.63	31.42		211.18
	40 x 838 x 1981 mm	SW	58.60		5.48	10.60	11.30					10.23	36.60		8.05		40.63	32.63		214.12
	35 x 762 x 1981 mm	HW	66.54		6.97	16.07	12.40	6.18	8.34	32.18		10.23	83.48		12.23		99.25	31.42		385.29
	35 x 838 x 1981 mm	HW	68.18		6.97	16.07	12.40	6.18	8.34	32.18		14.65	83.48		12.23		99.25	32.63		392.55
	40 x 762 x 1981 mm	HW	104.41		6.97	16.07	12.40	6.18	8.34	32.18		14.65	83.48		12.23		99.25	31.42		427.58
	40 x 838 x 1981 mm	HW	107.82		6.97	16.07	12.40	6.18	8.34	32.18		14.65	83.48		12.23		99.25	32.63		432.19
Flush door with vision panel. 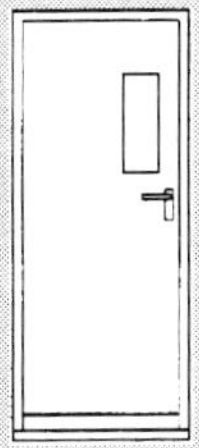	**Flush; plywood faced; half hour fire door;** hardwood lipping two long edges; vision panels size 200 x 450 mm.																			
	44 x 762 x 1981 mm	SW	74.76	36.60	8.38	16.07	23.45			82.99		10.23	43.23		9.19	26.27	41.74	31.42		404.33
	44 x 838 x 1981 mm	SW	77.33	36.60	8.38	16.07	23.45			82.99		10.23	43.23		9.19	26.27	41.74	32.63		408.11
	44 x 762 x 1981 mm	HW	109.29	44.62	8.38	16.07	23.45	6.18	8.34	90.72		10.23	85.69		13.11	29.06	101.45	31.42		578.01
	44 x 838 x 1981 mm	HW	115.77	44.62	8.38	16.07	23.45	6.18	8.34	90.72		10.23	85.69		13.11	29.06	101.45	32.63		585.69
	Pair 44 x 826 x 2040 mm (M)	SW	228.91	83.78	13.52	31.59	28.97	10.05	13.64	165.05	31.21	29.00		103.54		39.13	54.13	32.60		865.13
	Pair 44 x 826 x 2040 mm (M)	HW	346.78	86.57	13.52	38.21	28.97	10.05	13.64	165.05	31.21	29.00		181.51		43.31	117.94	32.60		1,138.4

GLOSSARY

Window References and Symbols

ALMN Aluminium
C Opening casement window
HW Hardwood
M Modular window
N Narrow windows
SW Softwood
S Swept head
V Top hung opening vent

 Top hung

 Bottom hung

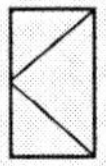 Side hung casement

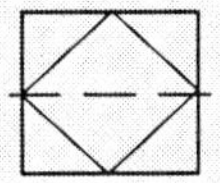 Centre horizontal pivot

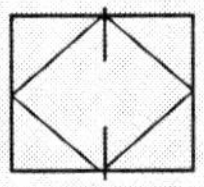 Centre vertical pivot

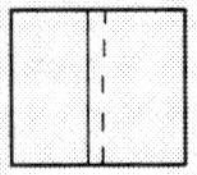 Horizontal sliding

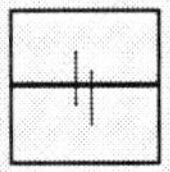 Vertical sliding sash

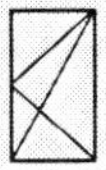 Tilt and turn

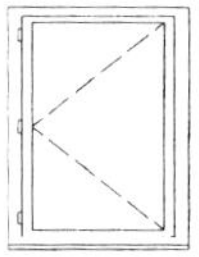

Standard - Width 630 mm

Height mm	Window Reference	SW £	HW £	ALMN £	uPVC £
750	107C	88.75	143.41	108.45	113.19
900	109C	97.46	183.02	134.28	140.39
1050	110C	104.20	188.48	137.63	143.81
1200	112C	110.92	195.99	142.32	148.64
1350	113C	120.09	211.20	152.09	158.85

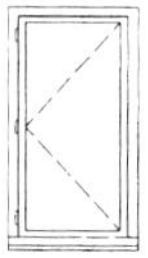

Narrow Module - Width 488 mm

Height mm	Window Reference	SW £	HW £	ALMN £	uPVC £
750	N07C	86.00	133.37	102.13	106.65
900	N09C	91.05	165.98	123.44	129.09
1050	N10C	95.40	171.80	127.08	132.85
1200	N12C	100.75	181.05	132.99	139.00
1350	N13C	107.12	190.91	139.30	145.57

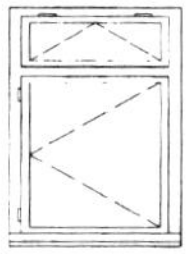

Standard - Width 630 mm

Height mm	Window Reference	SW £	HW £	ALMN £	uPVC £
750	107V	90.37	137.89	104.81	109.34
900	109V	94.51	142.37	107.51	112.06
1050	110V	99.97	181.79	133.22	139.15
1200	112V	105.65	188.27	137.23	143.26
1350	113V	109.97	193.96	140.73	146.83

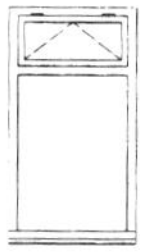

Narrow Module - Width 488 mm

Height mm	Window Reference	SW £	HW £	ALMN £	uPVC £
750	N07V	78.99	120.40	93.59	97.61
900	N09V	84.96	128.93	99.03	103.26
1050	N10V	88.87	162.22	120.77	126.17
1200	N12V	93.94	171.47	126.68	132.32
1350	-	-	-	-	-

Standard - Width 630 mm

Height mm	Window Reference	SW £	HW £	ALMN £	uPVC £
750	-	-	-	-	-
900	-	-	-	-	-
1050	110T	127.88	246.74	176.01	184.42
1200	112T	134.61	254.49	180.86	189.42
1350	113T	141.25	262.08	185.61	194.31

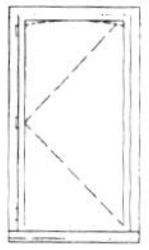

Swept head - Width 630 mm

Height mm	Window Reference	SW £	HW £	ALMN £	uPVC £
750	S107C	103.31	161.82	120.58	126.02
900	S109C	108.53	192.60	140.60	147.08
1050	S110C	113.79	199.70	145.02	151.62
1200	S112C	120.62	210.16	151.66	158.52
1350	S113C	129.89	226.05	161.87	169.20

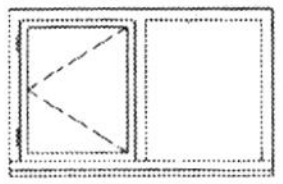

Standard - Width 1200 mm

Height mm	Window Reference	SW £	HW £	ALMN £	uPVC £
750	207C	130.61	198.07	143.24	149.43
900	209C	138.10	249.05	176.29	184.16
1050	210C	148.31	259.70	182.77	190.78
1200	212C	159.15	273.23	191.15	199.41
1350	213C	172.71	293.24	203.79	212.55

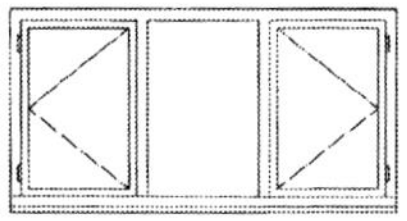

Standard - Width 1700 mm

Height mm	Window Reference	SW £	HW £	ALMN £	uPVC £
750	307CC	191.44	298.09	208.07	217.52
900	309CC	204.43	382.65	262.99	275.30
1050	310CC	216.95	397.72	272.13	284.64
1200	312CC	232.33	424.44	288.95	302.10
1350	313CC	252.61	451.15	305.76	319.56

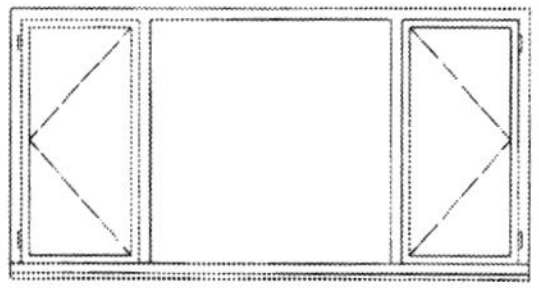

Standard - Width 2339 mm

Height mm	Window Reference	SW £	HW £	ALMN £	uPVC £
750	407CC	238.36	368.91	253.36	264.80
900	409CC	253.01	447.67	304.14	318.08
1050	410CC	268.11	469.36	317.33	331.58
1200	412CC	286.93	491.07	330.53	345.10
1350	-	-	-	-	-

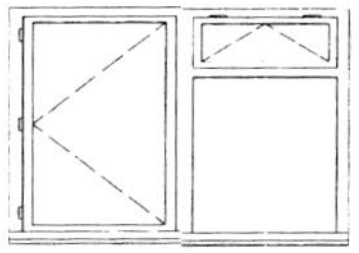

Standard - Width 1200 mm

Height mm	Window Reference	SW £	HW £	ALMN £	uPVC £
750	207CV	154.11	241.54	171.88	179.73
900	209CV	163.22	304.32	212.70	222.68
1050	210CV	171.95	314.29	218.73	228.83
1200	212CV	183.07	326.69	226.36	236.67
1350	213CV	196.24	344.97	237.87	248.61

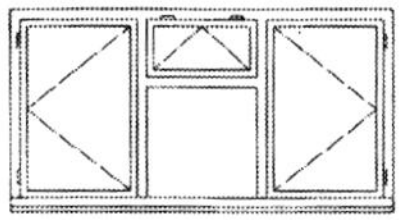

Standard - Width 1700 mm

Height mm	Window Reference	SW £	HW £	ALMN £	uPVC £
750	307CCV	215.02	341.71	236.80	247.92
900	309CCV	227.75	431.36	295.08	309.25
1050	310CCV	240.32	446.47	304.24	318.62
1200	312CCV	256.05	465.64	316.09	330.82
1350	313CCV	276.34	490.62	331.76	347.07

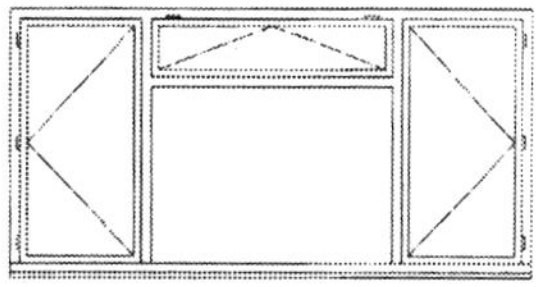

Standard - Width 2339 mm

Height mm	Window Reference	SW £	HW £	ALMN £	uPVC £
750	-	-	-	-	-
900	-	-	-	-	-
1050	410CWC	302.15	534.19	360.03	376.77
1200	412CWC	327.07	554.26	372.15	389.14
1350	413CWC	232.12	584.83	391.19	408.83

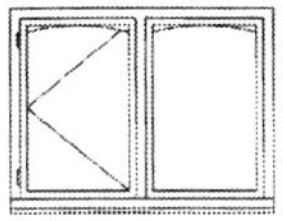

Standard - Width 1200 mm

Height mm	Window Reference	SW £	HW £	ALMN £	uPVC £
750	S207C	168.82	268.76	189.81	198.70
900	S209C	178.78	404.39	278.62	292.43
1050	S210C	188.20	418.45	287.35	301.43
1200	S212C	200.75	439.90	300.94	315.58
1350	S213C	216.00	467.97	318.89	334.34

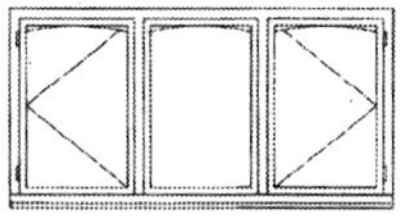

Swept head - Width 1770 mm

Height mm	Window Reference	SW £	HW £	ALMN £	uPVC £
750	S307CC	238.62	385.38	265.57	278.36
900	S309CC	252.47	463.67	316.36	331.77
1050	S310CC	265.86	481.31	327.20	342.90
1200	S312CC	282.95	506.83	343.22	359.53
1350	S313CC	305.00	543.79	366.78	384.13

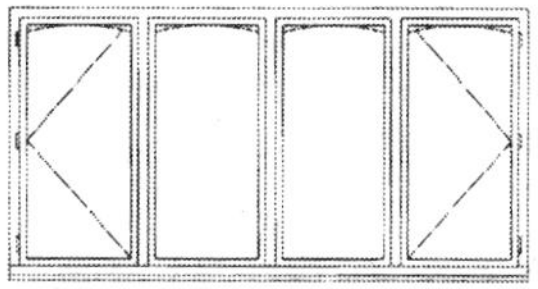

Swept head - Width 2339 mm

Height mm	Window Reference	SW £	HW £	ALMN £	uPVC £
750	-	-	-	-	-
900	S309CMC	274.92	486.12	329.47	344.88
1050	S310CMC	291.94	507.39	342.38	358.09
1200	S312CMC	312.66	536.54	360.48	376.79
1350	S313CMC	338.33	565.70	368.24	395.50

Ref	Size		Price	Mastic	Glazing		Decoration		TOTAL COSTS		
Window Reference	Width	Height	Softwood; Including ironmongery	Seal around window with mastic	Single glazing 4 mm thick	Double glazing 3 / 4 mm sealed units 24 mm thick	Woodwork; Oil paint; two undercoats; one finishing coat.	Woodwork; preservative; two coats.	Softwood window; 4 mm clear glass; Oil paint.	Softwood window; 4 mm clear glass; Wood Preservative.	Extra cost; Double glazing;
109C	630	900	74.24	3.67	11.42	34.66	9.91	7.01	99.24	96.34	23.24
110C	630	1050	77.23	4.03	13.45	40.82	11.56	8.18	106.27	102.89	27.37
112C	630	1200	80.20	4.39	15.48	46.97	13.21	9.35	113.28	109.42	31.50
113C	630	1350	85.62	4.75	17.50	53.13	14.86	10.51	122.74	118.39	35.62
109V	630	900	71.30	3.67	11.42	34.66	9.91	7.01	96.30	93.40	23.24
110V	630	1050	73.00	4.03	13.45	40.82	11.56	8.18	102.05	98.66	27.37
112V	630	1200	74.93	4.39	15.48	46.97	13.21	9.35	108.01	104.15	31.50
113V	630	1350	75.49	4.75	17.50	53.13	14.86	10.51	112.61	108.26	35.62
110T	630	1050	100.91	4.03	13.45	40.82	11.56	8.18	129.95	126.57	27.37
112T	630	1200	103.88	4.39	15.48	46.97	13.21	9.35	136.96	133.10	31.50
113T	630	1350	106.77	4.75	17.50	53.13	14.86	10.51	143.89	139.54	35.62
N09C	488	900	72.82	3.33	8.62	26.18	7.68	5.43	92.45	90.20	17.55
N10C	488	1050	74.22	3.69	10.16	30.82	8.96	6.33	97.02	94.40	20.67
N12C	488	1200	76.63	4.05	11.69	35.47	10.23	7.24	102.61	99.61	23.79
N13C	488	1350	80.06	4.41	13.22	40.12	11.51	8.14	109.20	105.83	26.90
N09V	488	900	66.73	3.33	8.62	26.18	7.68	5.43	86.36	84.12	17.55
N10V	488	1050	67.69	3.69	10.16	30.82	8.96	6.33	90.49	87.87	20.67
N12V	488	1200	69.83	4.05	11.69	35.47	10.23	7.24	95.80	92.80	23.79
S109C	630	900	85.31	3.67	11.42	34.66	9.91	7.01	110.31	107.41	23.24
S110C	630	1050	86.82	4.03	13.45	40.82	11.56	8.18	115.86	112.47	27.37
S112C	630	1200	89.90	4.39	15.48	46.97	13.21	9.35	122.98	119.11	31.50
S113C	630	1350	95.42	4.75	17.50	53.13	14.86	10.51	132.54	128.19	35.62
209C	1200	900	94.87	5.04	22.64	68.73	18.88	13.35	141.43	135.90	46.09
210C	1200	1050	98.08	5.40	26.66	80.93	22.02	15.58	152.16	145.72	54.27
212C	1200	1200	101.94	5.76	30.68	93.13	25.17	17.80	163.55	156.18	62.45
213C	1200	1350	108.50	6.12	34.70	105.33	28.31	20.03	177.64	169.35	70.63
209CV	1200	900	119.98	5.04	22.64	68.73	18.88	13.35	166.54	161.02	46.09
210CV	1200	1050	121.73	5.40	26.66	80.93	22.02	15.58	175.81	169.37	54.27
212CV	1200	1200	125.85	5.76	30.68	93.13	25.17	17.80	187.46	180.09	62.45
213CV	1200	1350	132.04	6.12	34.70	105.33	28.31	20.03	201.17	192.89	70.63
S209C	1200	900	135.54	5.04	22.64	68.73	18.88	13.35	182.10	176.57	46.09
S210C	1200	1050	137.97	5.40	26.66	80.93	22.02	15.58	192.05	185.61	54.27
S212C	1200	1200	143.53	5.76	30.68	93.13	25.17	17.80	205.14	197.77	62.45
S213C	1200	1350	151.79	6.12	34.70	105.33	28.31	20.03	220.93	212.64	70.63
309CC	1700	900	143.63	6.24	32.49	98.62	26.74	18.91	209.10	201.27	66.13
310CC	1700	1050	146.32	6.60	38.26	116.12	31.20	22.07	222.37	213.24	77.86
312CC	1700	1200	151.88	6.96	44.03	133.63	35.65	25.22	238.51	228.08	89.60
313CC	1700	1350	162.32	7.32	49.79	151.13	40.11	28.37	259.54	247.80	101.34
307CCV	1700	750	164.05	5.88	26.72	81.11	22.28	15.76	218.93	212.41	54.39
309CCV	1700	900	166.95	6.24	32.49	98.62	26.74	18.91	232.42	224.60	66.13
310CCV	1700	1050	169.69	6.60	38.26	116.12	31.20	22.07	245.75	236.61	77.86
312CCV	1700	1200	175.59	6.96	44.03	133.63	35.65	25.22	262.23	251.79	89.60
313CCV	1700	1350	186.05	7.32	49.79	151.13	40.11	28.37	283.28	271.54	101.34
S309CC	1700	900	191.67	6.24	32.49	98.62	26.74	18.91	257.14	249.32	66.13
S310CC	1700	1050	195.23	6.60	38.26	116.12	31.20	22.07	271.29	262.16	77.86
S312CC	1700	1200	202.49	6.96	44.03	133.63	35.65	25.22	289.13	278.70	89.60
S313CC	1700	1350	214.71	7.32	49.79	151.13	40.11	28.37	311.93	300.19	101.34
409CC	2339	900	169.77	7.77	45.07	136.81	36.79	26.02	259.41	248.64	91.73
410CC	2339	1050	171.41	8.13	53.07	161.09	42.92	30.36	275.54	262.98	108.02
412CC	2339	1200	176.76	8.49	61.08	185.38	49.06	34.70	295.39	281.03	124.30
S410CC	2339	1050	205.44	8.13	53.07	161.09	42.92	30.36	309.57	297.01	108.02
S412CC	2339	1200	216.91	8.49	61.08	185.38	49.06	34.70	335.53	321.17	124.30
S413CC	2339	1350	108.50	8.85	69.08	209.66	55.19	39.03	241.62	225.46	140.59
S309CM	2339	900	191.67	7.77	45.07	136.81	36.79	26.02	281.31	270.54	91.73
S310CM	2339	1050	195.23	8.13	53.07	161.09	42.92	30.36	299.36	286.80	108.02
S312CM	2339	1200	202.49	8.49	61.08	185.38	49.06	34.70	321.12	306.76	124.30
S313CM	2339	1350	214.71	8.85	69.08	209.66	55.19	39.03	347.82	331.67	140.59

Ref	Size		Price	Mastic	Glazing		Decoration		TOTAL COSTS		
Window Reference	Width	Height	Hardwood; Including ironmongery	Seal around window with mastic	Single glazing 4 mm thick	Double glazing 3 / 4 mm sealed units 24 mm thick	Woodwork; Oil paint; two undercoats; one finishing coat.	Woodwork; preservative; two coats.	Hardwood window; 4 mm clear glass; Oil paint.	Hardwood window; 4 mm clear glass; Wood Preservative.	Extra cost; Double glazing;
107C	630	750	123.94	3.31	9.39	28.51	8.26	5.84	144.91	142.49	19.12
109C	630	900	159.80	3.67	11.42	34.66	9.91	7.01	184.80	181.90	23.24
110C	630	1050	161.51	4.03	13.45	40.82	11.56	8.18	190.56	187.17	27.37
112C	630	1200	165.27	4.39	15.48	46.97	13.21	9.35	198.35	194.48	31.50
113C	630	1350	176.73	4.75	17.50	53.13	14.86	10.51	213.85	209.50	35.62
109V	630	900	119.15	3.67	11.42	34.66	9.91	7.01	144.15	141.25	23.24
110V	630	1050	154.83	4.03	13.45	40.82	11.56	8.18	183.87	180.48	27.37
112V	630	1200	157.55	4.39	15.48	46.97	13.21	9.35	190.62	186.76	31.50
113V	630	1350	159.49	4.75	17.50	53.13	14.86	10.51	196.61	192.26	35.62
110T	630	1050	219.77	4.03	13.45	40.82	11.56	8.18	248.81	245.43	27.37
112T	630	1200	223.77	4.39	15.48	46.97	13.21	9.35	256.85	252.98	31.50
113T	630	1350	227.61	4.75	17.50	53.13	14.86	10.51	264.73	260.38	35.62
N09C	488	900	147.75	3.33	8.62	26.18	7.68	5.43	167.38	165.14	17.55
N10C	488	1050	150.63	3.69	10.16	30.82	8.96	6.33	173.43	170.81	20.67
N12C	488	1200	156.93	4.05	11.69	35.47	10.23	7.24	182.91	179.91	23.79
N13C	488	1350	163.84	4.41	13.22	40.12	11.51	8.14	192.99	189.62	26.90
N09V	488	900	110.70	3.33	8.62	26.18	7.68	5.43	130.33	128.09	17.55
N10V	488	1050	141.05	3.69	10.16	30.82	8.96	6.33	163.85	161.23	20.67
N12V	488	1200	147.35	4.05	11.69	35.47	10.23	7.24	173.33	170.33	23.79
S109C	630	900	169.39	3.67	11.42	34.66	9.91	7.01	194.39	191.49	23.24
S110C	630	1050	172.73	4.03	13.45	40.82	11.56	8.18	201.77	198.38	27.37
S112C	630	1200	179.44	4.39	15.48	46.97	13.21	9.35	212.52	208.65	31.50
S113C	630	1350	191.58	4.75	17.50	53.13	14.86	10.51	228.70	224.35	35.62
207C	1200	750	161.82	4.68	18.62	56.53	15.73	11.13	200.85	196.25	37.91
209C	1200	900	205.81	5.04	22.64	68.73	18.88	13.35	252.37	246.84	46.09
210C	1200	1050	209.47	5.40	26.66	80.93	22.02	15.58	263.56	257.11	54.27
212C	1200	1200	216.02	5.76	30.68	93.13	25.17	17.80	277.63	270.26	62.45
213C	1200	1350	229.03	6.12	34.70	105.33	28.31	20.03	298.17	289.88	70.63
209CV	1200	900	261.08	5.04	22.64	68.73	18.88	13.35	307.64	302.12	46.09
210CV	1200	1050	264.06	5.40	26.66	80.93	22.02	15.58	318.14	311.70	54.27
212CV	1200	1200	269.47	5.76	30.68	93.13	25.17	17.80	331.08	323.72	62.45
213CV	1200	1350	280.77	6.12	34.70	105.33	28.31	20.03	349.91	341.62	70.63
S209C	1200	900	361.15	5.03	22.62	68.65	18.85	13.33	407.66	402.14	46.03
S210C	1200	1050	368.23	5.39	26.63	80.83	21.99	15.56	422.24	415.81	54.20
S212C	1200	1200	382.68	5.75	30.65	93.02	25.14	17.78	444.22	436.86	62.37
S213C	1200	1350	403.76	6.11	34.66	105.20	28.28	20.00	472.82	464.54	70.54
309CC	1700	900	321.85	6.23	32.45	98.49	26.71	18.89	387.24	379.42	66.04
310CC	1700	1050	327.09	6.59	38.21	115.98	31.16	22.04	403.05	393.93	77.77
312CC	1700	1200	343.98	6.95	43.97	133.46	35.61	25.19	430.52	420.09	89.49
313CC	1700	1350	360.87	7.31	49.73	150.95	40.06	28.34	457.97	446.24	101.21
307CCV	1700	750	290.74	5.87	26.69	81.01	22.26	15.74	345.56	339.04	54.32
309CCV	1700	900	370.56	6.23	32.45	98.49	26.71	18.89	435.95	428.14	66.04
310CCV	1700	1050	375.84	6.59	38.21	115.98	31.16	22.04	451.80	442.68	77.77
312CCV	1700	1200	385.18	6.95	43.97	133.46	35.61	25.19	471.71	461.29	89.49
313CCV	1700	1350	400.33	7.31	49.73	150.95	40.06	28.34	497.43	485.71	101.21
S309CC	1700	900	402.87	6.23	32.45	98.49	26.71	18.89	468.26	460.44	66.04
S310CC	1700	1050	410.68	6.59	38.21	115.98	31.16	22.04	486.64	477.52	77.77
S312CC	1700	1200	426.37	6.95	43.97	133.46	35.61	25.19	512.91	502.48	89.49
S313CC	1700	1350	453.50	7.31	49.73	150.95	40.06	28.34	550.60	538.88	101.21
409CC	2339	900	364.43	7.76	45.02	136.64	36.75	25.99	453.95	443.20	91.62
410CC	2339	1050	372.65	8.12	53.01	160.89	42.87	30.32	476.66	464.11	107.88
412CC	2339	1200	380.91	8.48	61.00	185.15	48.99	34.65	499.38	485.04	124.15
S410CC	2339	1050	437.48	8.12	53.01	160.89	42.87	30.32	541.49	528.94	107.88
S412CC	2339	1200	444.10	8.48	61.00	185.15	48.99	34.65	562.57	548.23	124.15
S413CC	2339	1350	461.21	8.84	68.99	209.40	55.12	38.99	594.16	578.03	140.41
S309CM	2339	900	402.87	7.76	45.02	136.64	36.75	25.99	492.40	481.64	91.62
S310CM	2339	1050	410.68	8.12	53.01	160.89	42.87	30.32	514.68	502.14	107.88
S312CM	2339	1200	426.37	8.48	61.00	185.15	48.99	34.65	544.85	530.51	124.15
S313CM	2339	1350	442.08	8.84	68.99	209.40	55.12	38.99	575.03	558.90	140.41

Ref	Size		Price	Mastic	Glazing		Sub Frame		Decoration		TOTAL COSTS			
Window Reference	Width	Height	Insulating cold bridge; Including ironmongery	Seal around window with mastic	Single glazing 4 mm thick	Double glazing 3 / 4 mm sealed units 24 mm thick	Softwood; 75 x 50 mm; internal bead	Hardwood; 75 x 50 mm; internal bead	Woodwork; Oil paint; two undercoats; one finishing coat.	Woodwork; preservative; two coats.	Aluminium window; 4 mm clear glass; No sub-frame.	Aluminium window; 4 mm clear glass; Softwood sub-frame.	Aluminium window; 4 mm clear glass; Hardwood sub-frame.	Extra cost; Double glazing;
109C	630	900	120.27	3.67	11.42	34.66	9.86	14.11	4.60	3.29	135.36	149.81	152.75	23.24
110C	630	1050	121.40	4.03	13.45	40.82	10.82	15.49	5.05	3.61	138.88	154.75	157.97	27.37
112C	630	1200	123.87	4.39	15.48	46.97	11.79	16.87	5.50	3.93	143.74	161.03	164.54	31.50
113C	630	1350	131.42	4.75	17.50	53.13	12.75	18.26	5.95	4.25	153.68	172.38	176.18	35.62
109V	630	900	93.49	3.67	11.42	34.66	9.86	14.11	4.60	3.29	108.58	123.04	125.97	23.24
110V	630	1050	116.99	4.03	13.45	40.82	10.82	15.49	5.05	3.61	134.47	150.34	153.57	27.37
112V	630	1200	118.78	4.39	15.48	46.97	11.79	16.87	5.50	3.93	138.65	155.94	159.45	31.50
113V	630	1350	120.07	4.75	17.50	53.13	12.75	18.26	5.95	4.25	142.32	161.03	164.83	35.62
112T	630	1200	162.41	4.39	15.48	46.97	11.79	16.87	5.50	3.93	182.27	199.56	203.08	31.50
113T	630	1350	164.94	4.75	17.50	53.13	12.75	18.26	5.95	4.25	187.19	205.90	209.70	35.62
N07C	488	750	92.78	2.97	7.09	21.53	7.97	11.41	3.72	2.66	102.85	114.55	116.92	14.44
N09C	488	900	112.33	3.33	8.62	26.18	8.94	12.80	4.17	2.98	124.29	137.40	140.07	17.55
N10C	488	1050	114.22	3.69	10.16	30.82	9.91	14.18	4.62	3.30	128.07	142.60	145.55	20.67
N12C	488	1200	118.38	4.05	11.69	35.47	10.87	15.56	5.07	3.62	134.12	150.06	153.31	23.79
N13C	488	1350	122.93	4.41	13.22	40.12	11.84	16.95	5.53	3.95	140.56	157.92	161.45	26.90
N09V	488	900	87.92	3.33	8.62	26.18	8.94	12.80	4.17	2.98	99.88	112.99	115.66	17.55
N10V	488	1050	107.91	3.69	10.16	30.82	9.91	14.18	4.62	3.30	121.76	136.29	139.24	20.67
N12V	488	1200	112.07	4.05	11.69	35.47	10.87	15.56	5.07	3.62	127.81	143.75	146.99	23.79
S109C	630	900	126.58	3.67	11.42	34.66	9.86	14.11	4.60	3.29	141.68	156.13	159.07	23.24
S110C	630	1050	128.78	4.03	13.45	40.82	10.82	15.49	5.05	3.61	146.26	162.13	165.36	27.37
S112C	630	1200	133.21	4.39	15.48	46.97	11.79	16.87	5.50	3.93	153.07	170.36	173.88	31.50
S113C	630	1350	141.20	4.75	17.50	53.13	12.75	18.26	5.95	4.25	163.46	182.16	185.96	35.62
209C	1200	900	150.58	5.04	22.64	68.73	13.53	19.36	6.31	4.51	178.26	198.10	202.13	46.09
210C	1200	1050	152.99	5.40	26.66	80.93	14.49	20.75	6.76	4.83	185.05	206.31	210.63	54.27
212C	1200	1200	157.30	5.76	30.68	93.13	15.46	22.13	7.21	5.15	193.74	216.42	221.03	62.45
213C	1200	1350	165.87	6.12	34.70	105.33	16.43	23.51	7.67	5.48	206.70	230.79	235.68	70.63
209CV	1200	900	186.99	5.04	22.64	68.73	13.53	19.36	6.31	4.51	214.67	234.51	238.54	46.09
210CV	1200	1050	188.95	5.40	26.66	80.93	14.49	20.75	6.76	4.83	221.01	242.27	246.59	54.27
212CV	1200	1200	192.52	5.76	30.68	93.13	15.46	22.13	7.21	5.15	228.96	251.63	256.24	62.45
213CV	1200	1350	199.96	6.12	34.70	105.33	16.43	23.51	7.67	5.48	240.78	264.87	269.77	70.63

Ref	Size		Price	Mastic	Glazing		Sub Frame		Decoration		TOTAL COSTS			
Window Reference	Width	Height	Insulating cold bridge; Including ironmongery	Seal around window with mastic	Single glazing 4 mm thick	Double glazing 3 / 4 mm sealed units 24 mm thick	Softwood; 75 x 50 mm; internal bead	Hardwood; 75 x 50 mm; internal bead	Woodwork; Oil paint; two undercoats; one finishing coat.	Woodwork; preservative; two coats.	Aluminium window; 4 mm clear glass; No sub-frame.	Aluminium window; 4 mm clear glass; Softwood sub-frame.	Aluminium window; 4 mm clear glass; Hardwood sub-frame.	Extra cost; Double glazing;
S209C	1200	900	252.91	5.04	22.64	68.73	13.53	19.36	6.31	4.51	280.59	300.43	304.47	46.09
S210C	1200	1050	257.57	5.40	26.66	80.93	14.49	20.75	6.76	4.83	289.63	310.89	315.21	54.27
S212C	1200	1200	267.09	5.76	30.68	93.13	15.46	22.13	7.21	5.15	303.54	326.21	330.82	62.45
S213C	1200	1350	280.98	6.12	34.70	105.33	16.43	23.51	7.67	5.48	321.80	345.89	350.79	70.63
309CC	1700	900	227.02	6.24	32.49	98.62	16.75	23.97	7.82	5.58	265.75	290.31	295.30	66.13
310CC	1700	1050	230.47	6.60	38.26	116.12	17.71	25.36	8.27	5.90	275.33	301.31	306.59	77.86
312CC	1700	1200	241.60	6.96	44.03	133.63	18.68	26.74	8.72	6.23	292.58	319.98	325.55	89.60
313CC	1700	1350	252.72	7.32	49.79	151.13	19.65	28.12	9.17	6.55	309.83	338.65	344.50	101.34
307CC	1700	750	206.52	5.88	26.72	81.11	15.78	22.59	7.36	5.26	239.13	262.27	266.98	54.39
309CC	1700	900	259.11	6.24	32.49	98.62	16.75	23.97	7.82	5.58	297.84	322.40	327.39	66.13
310CC	1700	1050	262.59	6.60	38.26	116.12	17.71	25.36	8.27	5.90	307.44	333.42	338.70	77.86
312CC	1700	1200	268.74	6.96	44.03	133.63	18.68	26.74	8.72	6.23	319.72	347.12	352.69	89.60
313CC	1700	1350	278.72	7.32	49.79	151.13	19.65	28.12	9.17	6.55	335.83	364.64	370.50	101.34
S309CC	1700	900	280.39	6.24	32.49	98.62	16.75	23.97	7.82	5.58	319.12	343.69	348.68	66.13
S310CC	1700	1050	285.54	6.60	38.26	116.12	17.71	25.36	8.27	5.90	330.39	356.38	361.66	77.86
S312CC	1700	1200	295.87	6.96	44.03	133.63	18.68	26.74	8.72	6.23	346.86	374.26	379.82	89.60
S313CC	1700	1350	313.74	7.32	49.79	151.13	19.65	28.12	9.17	6.55	370.85	399.67	405.52	101.34
409CC	2339	900	255.07	7.77	45.07	136.81	20.86	29.86	9.74	6.95	307.91	338.51	344.73	91.73
410CC	2339	1050	260.49	8.13	53.07	161.09	21.83	31.25	10.19	7.28	321.69	353.71	360.22	108.02
412CC	2339	1200	265.92	8.49	61.08	185.38	22.80	32.63	10.64	7.60	335.49	368.93	375.72	124.30
S410CC	2339	1050	303.19	8.13	53.07	161.09	21.83	31.25	10.19	7.28	364.40	396.42	402.92	108.02
S412CC	2339	1200	307.55	8.49	61.08	185.38	22.80	32.63	10.64	7.60	377.12	410.55	417.35	124.30
S413CC	2339	1350	318.82	8.85	69.08	209.66	23.76	34.01	11.09	7.92	396.75	431.60	438.68	140.59
S309C	2339	900	280.39	7.77	45.07	136.81	20.86	29.86	9.74	6.95	333.24	363.84	370.06	91.73
S310C	2339	1050	285.54	8.13	53.07	161.09	21.83	31.25	10.19	7.28	346.74	378.76	385.27	108.02
S312C	2339	1200	295.87	8.49	61.08	185.38	22.80	32.63	10.64	7.60	365.44	398.88	405.67	124.30
S313C	2339	1350	295.87	8.85	69.08	209.66	23.76	34.01	11.09	7.92	373.80	408.66	415.74	140.59

Ref	Size		Price	Mastic	Glazing		Sub Frame		Decoration		TOTAL COSTS			
Window Reference	Width	Height	UPVC window; Including ironmongery	Seal around window with mastic	Single glazing 4 mm thick	Double glazing 3 / 4 mm sealed units 24 mm thick	Softwood: 75 x 50 mm; internal bead	Hardwood: 75 x 50 mm; internal bead	Woodwork; Oil paint; two undercoats; one finishing coat.	Woodwork; preservative; two coats.	uPVC window; 4 mm clear glass; No sub-frame.	uPVC window; 4 mm clear glass; Softwood sub-frame.	uPVC window; 4 mm clear glass; Hardwood sub-frame.	Extra cost; Double glazing;
109C	630	900	126.38	3.41	10.61	32.19	9.15	13.10	4.27	3.05	140.39	153.82	156.55	21.59
110C	630	1050	127.58	3.74	12.49	37.91	10.05	14.39	4.69	3.35	143.81	158.55	161.54	25.42
112C	630	1200	130.19	4.08	14.37	43.62	10.95	15.67	5.11	3.65	148.64	164.70	167.96	29.25
113C	630	1350	138.18	4.41	16.25	49.34	11.84	16.95	5.53	3.95	158.85	176.22	179.75	33.08
109V	630	900	98.05	3.41	10.61	32.19	9.15	13.10	4.27	3.05	112.06	125.49	128.22	21.59
110V	630	1050	122.91	3.74	12.49	37.91	10.05	14.39	4.69	3.35	139.15	153.89	156.88	25.42
112V	630	1200	124.81	4.08	14.37	43.62	10.95	15.67	5.11	3.65	143.26	159.31	162.58	29.25
113V	630	1350	126.17	4.41	16.25	49.34	11.84	16.95	5.53	3.95	146.83	164.20	167.73	33.08
110T	630	1050	168.18	3.74	12.49	37.91	10.05	14.39	4.69	3.35	184.42	199.15	202.15	25.42
112T	630	1200	170.97	4.08	14.37	43.62	10.95	15.67	5.11	3.65	189.42	205.47	208.74	29.25
113T	630	1350	173.64	4.41	16.25	49.34	11.84	16.95	5.53	3.95	194.31	211.68	215.21	33.08
N09C	488	900	117.98	3.09	8.01	24.31	8.30	11.88	3.87	2.77	129.09	141.26	143.74	16.30
N10C	488	1050	119.99	3.43	9.43	28.63	9.20	13.17	4.29	3.07	132.85	146.34	149.08	19.19
N12C	488	1200	124.38	3.76	10.85	32.94	10.10	14.45	4.71	3.37	139.00	153.81	156.82	22.09
N13C	488	1350	129.20	4.10	12.27	37.26	11.00	15.74	5.13	3.67	145.57	161.70	164.97	24.98
N09V	488	900	92.16	3.09	8.01	24.31	8.30	11.88	3.87	2.77	103.26	115.44	117.91	16.30
N10V	488	1050	113.31	3.43	9.43	28.63	9.20	13.17	4.29	3.07	126.17	139.66	142.40	19.19
N12V	488	1200	117.71	3.76	10.85	32.94	10.10	14.45	4.71	3.37	132.32	147.13	150.14	22.09
S109C	630	900	133.06	3.41	10.61	32.19	9.15	13.10	4.27	3.05	147.08	160.50	163.23	21.59
S110C	630	1050	135.39	3.74	12.49	37.91	10.05	14.39	4.69	3.35	151.62	166.36	169.36	25.42
S112C	630	1200	140.07	4.08	14.37	43.62	10.95	15.67	5.11	3.65	158.52	174.58	177.84	29.25
S113C	630	1350	148.53	4.41	16.25	49.34	11.84	16.95	5.53	3.95	169.20	186.57	190.10	33.08
209C	1200	900	158.45	4.68	21.03	63.83	12.56	17.98	5.86	4.19	184.16	202.58	206.33	42.80
210C	1200	1050	161.00	5.01	24.76	75.16	13.46	19.27	6.28	4.49	190.78	210.52	214.53	50.40
212C	1200	1200	165.56	5.35	28.50	86.49	14.36	20.55	6.70	4.79	199.41	220.46	224.74	58.00
213C	1200	1350	174.63	5.68	32.23	97.82	15.25	21.83	7.12	5.08	212.55	234.92	239.46	65.59
209CV	1200	900	196.98	4.68	21.03	63.83	12.56	17.98	5.86	4.19	222.68	241.11	244.85	42.80
210CV	1200	1050	199.05	5.01	24.76	75.16	13.46	19.27	6.28	4.49	228.83	248.57	252.58	50.40
212CV	1200	1200	202.82	5.35	28.50	86.49	14.36	20.55	6.70	4.79	236.67	257.72	262.00	58.00
213CV	1200	1350	210.70	5.68	32.23	97.82	15.25	21.83	7.12	5.08	248.61	270.98	275.53	65.59
S209C	1200	900	266.72	4.68	21.03	63.83	12.56	17.98	5.86	4.19	292.43	310.86	314.60	42.80
S210C	1200	1050	271.65	5.01	24.76	75.16	13.46	19.27	6.28	4.49	301.43	321.17	325.18	50.40
S212C	1200	1200	281.73	5.35	28.50	86.49	14.36	20.55	6.70	4.79	315.58	336.63	340.91	58.00
S213C	1200	1350	296.42	5.68	32.23	97.82	15.25	21.83	7.12	5.08	334.34	356.71	361.25	65.59
309CC	1700	900	239.33	5.79	30.17	91.58	15.55	22.26	7.26	5.18	275.30	298.11	302.74	61.41
310CC	1700	1050	242.98	6.13	35.53	107.84	16.45	23.55	7.68	5.48	284.64	308.76	313.67	72.31
312CC	1700	1200	254.76	6.46	40.89	124.10	17.35	24.83	8.10	5.78	302.10	327.55	332.72	83.21
313CC	1700	1350	266.52	6.80	46.24	140.35	18.25	26.12	8.51	6.08	319.56	346.32	351.76	94.11
307CCV	1700	750	217.65	5.46	24.82	75.32	14.66	20.98	6.84	4.89	247.92	269.42	273.79	50.51
309CCV	1700	900	273.28	5.79	30.17	91.58	15.55	22.26	7.26	5.18	309.25	332.06	336.70	61.41
310CCV	1700	1050	276.96	6.13	35.53	107.84	16.45	23.55	7.68	5.48	318.62	342.75	347.65	72.31
312CCV	1700	1200	283.47	6.46	40.89	124.10	17.35	24.83	8.10	5.78	330.82	356.26	361.44	83.21
313CCV	1700	1350	294.03	6.80	46.24	140.35	18.25	26.12	8.51	6.08	347.07	373.83	379.27	94.11
S309CC	1700	900	295.80	5.79	30.17	91.58	15.55	22.26	7.26	5.18	331.77	354.58	359.22	61.41
S310CC	1700	1050	301.25	6.13	35.53	107.84	16.45	23.55	7.68	5.48	342.90	367.03	371.93	72.31
S312CC	1700	1200	312.18	6.46	40.89	124.10	17.35	24.83	8.10	5.78	359.53	384.97	390.15	83.21
S313CC	1700	1350	331.09	6.80	46.24	140.35	18.25	26.12	8.51	6.08	384.13	410.89	416.32	94.11
409CC	2339	900	269.01	7.22	41.86	127.05	19.38	27.73	9.04	6.46	318.08	346.50	352.27	85.19
410CC	2339	1050	274.74	7.55	49.29	149.60	20.27	29.02	9.46	6.76	331.58	361.32	367.36	100.31
412CC	2339	1200	280.49	7.89	56.72	172.15	21.17	30.30	9.88	7.06	345.10	376.15	382.46	115.44
S410CC	2339	1050	319.93	7.55	49.29	149.60	20.27	29.02	9.46	6.76	376.77	406.50	412.54	100.31
S412CC	2339	1200	324.54	7.89	56.72	172.15	21.17	30.30	9.88	7.06	389.14	420.19	426.50	115.44
S413CC	2339	1350	336.46	8.22	64.15	194.71	22.07	31.59	10.30	7.36	408.83	441.20	447.78	130.56
S309CM	2339	900	295.80	7.22	41.86	127.05	19.38	27.73	9.04	6.46	344.88	373.30	379.07	85.19
S310CM	2339	1050	301.25	7.55	49.29	149.60	20.27	29.02	9.46	6.76	358.09	387.82	393.86	100.31
S312CM	2339	1200	312.18	7.89	56.72	172.15	21.17	30.30	9.88	7.06	376.79	407.84	414.15	115.44
S313CM	2339	1350	323.13	8.22	64.15	194.71	22.07	31.59	10.30	7.36	395.50	427.87	434.45	130.56

HINGES

Description	Unit	Purchase Price Only	Labour Cost Only New	Labour Cost Only Repairs
Light Duty				
Steel; standard				
76 x 49 mm	pr	0.6 to 0.6	2.48	2.79
102 x 73 mm	pr	1.09 to 1.09	2.58	2.90
152 x100 mm	pr	1.52 to 1.52	2.97	3.34
Stainless steel				
76 x 49 mm	pr	1.98 to 1.98	2.77	3.12
102 x 73 mm	pr	2.75 to 2.75	2.87	3.23
Heavy Duty				
Steel; standard				
102 x 70 mm	pr	2.02 to 2.02	3.17	3.56
152 x100 mm	pr	2.8 to 2.8	3.37	3.79
Parliament Hinges				
Steel; standard				
121 x 102 mm	pr	2.98 to 2.98	3.17	3.56
155 x 127 mm	pr	3.54 to 3.54	3.37	3.79
Rising Butts				
Steel; standard				
76 x 70 mm	pr	2.64 to 2.64	3.76	4.23
102 x 76 mm	pr	2.98 to 2.98	3.86	4.34
Lift-off Butts				
Steel; standard				
76 x 70 mm	pr	2.75 to 2.75	3.76	4.23
102 x 76 mm	pr	3.03 to 3.03	3.86	4.34
Back Flap				
25 x 73 mm	pr	1.44 to 1.44	3.56	4.01
51 x 108 mm	pr	2.19 to 2.19	3.66	4.12
76 x 165 mm	pr	2.86 to 2.86	3.76	4.23
Tee Hinge;				
Japanned finish				
100 mm	pr	3.2 to 3.2	3.56	4.01
150 mm	pr	4.11 to 4.11	3.66	4.12

LOCKS

Description	Unit	Purchase Price Only	Labour Cost Only New	Labour Cost Only Repairs
Latches				
Suffolk; light				
44 x 191 mm	nr	4.66 to 4.66	1.98	2.23
48 x 203 mm	nr	6 to 6	2.08	2.34
Suffolk; heavy				
51 x 228 mm	nr	8.36 to 8.36	2.48	2.79
Rim				
102 x 73 mm	nr	8.49 to 8.49	3.47	3.90
127 x 102 mm	nr	15.08 to 15.08	3.96	4.45
Dead Locks; Rim				
102 x 73 mm	nr	11.85 to 11.85	2.48	2.79
152 x 95 mm	nr	19.56 to 19.56	2.58	2.90
Cylinder Rim Lock				
Standard; 92 x 67 mm	nr	17.94 to 17.94	3.17	3.56
D/lockg; 92 x 67 mm	nr	20.68 to 20.68	3.27	3.68
Mortice				
2 lever				
Deadlock	nr	9.53 to 9.53	4.45	5.01
Latch and dead lock	nr	10.6 to 10.6	4.55	5.12
Roller latch / d/lock	nr	13.45 to 13.45	4.65	5.23
4 lever				
Deadlock	nr	11.91 to 11.91	4.65	5.23
Latch and dead lock	nr	13.24 to 13.24	4.75	5.34
Roller latch / d/lock	nr	16.82 to 16.82	4.85	5.45
5 lever; insurance approved				
Deadlock	nr	12.61 to 12.61	4.65	5.23
Latch and dead lock	nr	18.44 to 18.44	4.75	5.34
Cylinder Mortice Lock				
Standard; 92 x 67 mm	nr	20.68 to 20.68	3.17	3.56
ditto; double locking	nr	25.85 to 25.85	3.27	3.68

HANDLES AND BOLTS

Description	Unit	Purchase Price Only	Labour Cost Only New	Labour Cost Only Repairs
Handles				
Aluminium; set,				
knob	nr	5.05 to 5.05	1.49	1.67
pull; D; 191 mm	nr	5.55 to 5.55	2.18	2.45
pull; D; 228 mm	nr	8.13 to 8.13	2.48	2.79
flush pull; 165 mm	nr	6.39 to 6.39	3.17	3.56
flush pull; 228 mm	nr	9.35 to 9.35	3.76	4.23
lever handle, latch	pair	8.36 to 8.36	1.98	2.23
lever handle, locking	pair	9.67 to 9.67	1.98	2.23
Brass; set,				
knob	nr	10.85 to 10.85	1.86	2.09
pull; D; 191 mm	nr	11.93 to 11.93	2.72	3.06
pull; D; 228 mm	nr	17.48 to 17.48	3.09	3.48
flush pull; 165 mm	nr	13.72 to 13.72	3.96	4.45
flush pull; 228 mm	nr	20.1 to 20.1	4.70	5.29
lever handle, latch	pair	17.96 to 17.96	2.48	2.79
lever handle, locking	pair	20.78 to 20.78	2.48	2.79
Bolts				
Aluminium; set,				
Barrel bolt; 75 mm	nr	4.67 to 4.67	1.78	2.01
Barrel bolt; 100 mm	nr	6.11 to 6.11	2.18	2.45
Barrel bolt; 150 mm	nr	7.71 to 7.71	2.48	2.79
Flush bolt; 100 mm	nr	12.58 to 12.58	3.17	3.56
Flush bolt; 150 mm	nr	14.1 to 14.1	3.56	4.01
Indicating bolt	nr	16.31 to 16.31	2.77	3.12
Panic bolts; set				
single door set	nr	61.65 to 61.65	13.35	15.02
double door set	nr	72.86 to 72.86	24.23	27.25
Brass				
Barrel bolt; 75 mm	nr	7.7 to 7.7	2.05	2.31
Barrel bolt; 100 mm	nr	10.08 to 10.08	2.51	2.82
Barrel bolt; 150 mm	nr	12.71 to 12.71	2.85	3.20
Flush bolt; 100 mm	nr	20.75 to 20.75	3.64	4.10
Flush bolt; 150 mm	nr	23.27 to 23.27	4.10	4.61
Indicating bolt	nr	26.91 to 26.91	3.19	3.59

DOOR CLOSURES

Description	Unit	Purchase Price Only	Labour Cost Only New	Labour Cost Only Repairs
Overhead Door Closures				
Japanned iron;				
Helical springs;				
102 mm,	nr	10.65 to 10.65	3.47	3.90
150 mm,	nr	12.33 to 12.33	4.45	5.01
Springs				
229 mm	nr	5.22 to 5.22	4.45	5.01
305 mm	nr	7.01 to 7.01	5.15	5.79
356 mm	nr	8.24 to 8.24	5.74	6.46
Aluminium				
Springs; lever				
250 mm; corridor	nr	17.94 to 17.94	5.24	5.90
350 mm; external	nr	20.18 to 20.18	6.43	7.23
Perko	nr	3.87 to 3.87	3.47	3.90
Pneumatic;				
light	nr	65.01 to 65.01	6.43	7.23
heavy	nr	84.07 to 84.07	11.37	12.80
Dictator	nr	8.97 to 8.97	3.76	4.23
Door selectors	nr	14.58 to 14.58	4.45	5.01
Brass;				
Springs; lever				
250 mm; corridor	nr	29.59 to 29.59	6.03	6.78
350 mm; external	nr	33.29 to 33.29	7.40	8.32
Perko	nr	6.39 to 6.39	3.98	4.48
Pneumatic;				
light	nr	94.27 to 94.27	7.07	7.96
heavy	nr	121.9 to 121.9	12.51	14.07
Dictator	nr	14.8 to 14.8	4.14	4.65
Door selectors	nr	24.05 to 24.05	4.90	5.51
Floor Springs				
Aluminium cover plates				
Single action	nr	130.1 to 130.1	12.36	13.91
Double action	nr	151.4 to 151.4	17.31	19.47
door holders	nr	7.23 to 7.23	6.43	7.23

WINDOW FITTINGS

Description	Unit	Purchase Price Only	Labour Cost Only New	Labour Cost Only Repairs
Sash Fittings				
Aluminium				
Lift;	nr	1.52 to 1.52	1.49	1.67
Eye	nr	1.85 to 1.85	1.69	1.90
Fastener	nr	3.54 to 3.54	3.47	3.90
Stop	nr	2.75 to 2.75	1.19	1.34
Unique spiral balance	nr	7.29 to 7.29	7.42	8.35
Axle pulley	nr	2.81 to 2.81	3.37	3.79
Brass				
Lift; 57 mm	nr	2.5 to 2.5	1.71	1.92
Ring lift; 22 mm	nr	3.06 to 3.06	3.47	3.90
Eye	nr	5.83 to 5.83	1.37	1.54
Fastener	nr	4.54 to 4.54	8.53	9.60
Stop	nr	12.03 to 12.03	3.87	4.35
Casement Fittings				
Fasteners				
Japanned iron	nr	2.19 to 2.19	3.17	3.56
Aluminium	nr	2.41 to 2.41	3.47	3.90
Brass	nr	3.98 to 3.98	3.66	4.12
Stays; peg type				
Japanned iron	nr	4.99 to 4.99	2.79	3.14
Aluminium	nr	7.71 to 7.71	3.05	3.43
Brass	nr	12.71 to 12.71	3.22	3.63
Stays; sliding type				
Brass	nr	15.7 to 15.7	4.45	5.01
Espagnolette bolts; brass				
1200 mm	nr	19.62 to 19.62	8.41	9.46
1600 mm	nr	25.22 to 25.22	9.89	11.13
Budget locks				
Aluminium	nr	3.93 to 3.93	7.42	8.35
Brass	nr	6.48 to 6.48	8.11	9.13
Handles				
Aluminium	nr	8.36 to 8.36	3.47	3.90
Brass	nr	13.78 to 13.78	3.86	4.34

FANLIGHTS AND SKY LIGHTS

Description	Unit	Purchase Price Only	Labour Cost Only New	Labour Cost Only Repairs
Fanlights				
Cam openers				
Iron	nr	12.84 to 12.84	5.44	6.12
Aluminium	nr	16.82 to 16.82	6.43	7.23
Straight pinion and pulley system				
Iron	nr	25.78 to 25.78	7.42	8.35
Aluminium	nr	29.15 to 29.15	8.41	9.46
Segmental pinion and pulley system				
Iron	nr	34.75 to 34.75	11.37	12.80
Aluminium	nr	40.35 to 40.35	12.36	13.91
Stays				
Iron	nr	5.1 to 5.1	3.17	3.56
Aluminium	nr	7.57 to 7.57	3.47	3.90
Catch				
Iron	nr	3.93 to 3.93	2.97	3.34
Aluminium	nr	4.77 to 4.77	3.17	3.56
Friction pivots	nr	2.75 to 2.75	2.48	2.79
Skylight				
Pinion and pulley system				
355 mm	nr	24.66 to 24.66	10.39	11.68
406 mm	nr	29.15 to 29.15	10.68	12.02
457 mm	nr	35.87 to 35.87	10.98	12.35
Pole and hook	nr	15.7 to 15.7	0.50	0.56
Grills				
Aluminium				
Aluminium mesh				
and frame	m2	4.49 to 4.49	6.43	7.23
Zinc perforated sheet				
on metal frame	m2	6.73 to 6.73	7.42	8.35
Brass mesh on				
brass frame	m2	13.45 to 13.45	11.37	12.80

USE AND PERFORMANCE

The schedules of key unit rates enable quick comparisons to be made between the cost of various materials. Their performance and maintenance values are indicated by 'Icons' shown, where appropriate, for each schedule of rates.

BEST USE

For economic use and optimum performance the Manufacturers suggest specification of these finishes in the following areas.

Finish	Activity Area
Cement & Sand	M5; M6;
Granolithic	M5; M6; T7; T10;
Terrazzo	H5; H6; H7; H8; M7;
Asphalt	H6; H7; H8;
Quarry Tile	H4; H5; H6; H7; H8;T6; T10:
Ceramic Tile	M4; H4; H5; H6; H7; H8;T6; T10:
Cork	H1; H2; T3; T11;
Rubber	M4; M6; M7; H4; T6; T10;
Linoleum	M4; M6; M7; H4; H5; H6; T6; T10;
Vinyl	M1 - M4; M6; M7; T7; T9; T10; T12
Hardwood	H1; H2; H3; H4; T1; T3; T4; S1; S2:
Carpet	M1 - 4; M7; H2; H3; T4; T8; T11;

PREPARATION WORKS

ADD the cost of the following preparation work to the appropriate key unit rates.

	Unit	New	Refurb
Remove existing floors and finishings			
Concrete beds (not reinforced)			
100 mm thick	m2	5.78	6.51
150 mm thick	m2	9.58	10.78
Finishings			
Screeds	m2	4.77	5.37
Clay tiles	m2	7.72	8.68
Sheet or plastic	m2	6.25	7.03
Insitu cement based	m2	6.25	7.03
Wood block	m2	5.50	6.19
Boarding	m2	6.10	6.86
Carpets (loose)	m2	0.88	0.99
Preparation to receive new finishings			
Remove adhesive	m2	2.72	3.06
Grit blasting	m2	3.30	3.72
Damp proof treatment	m2	4.06	4.56

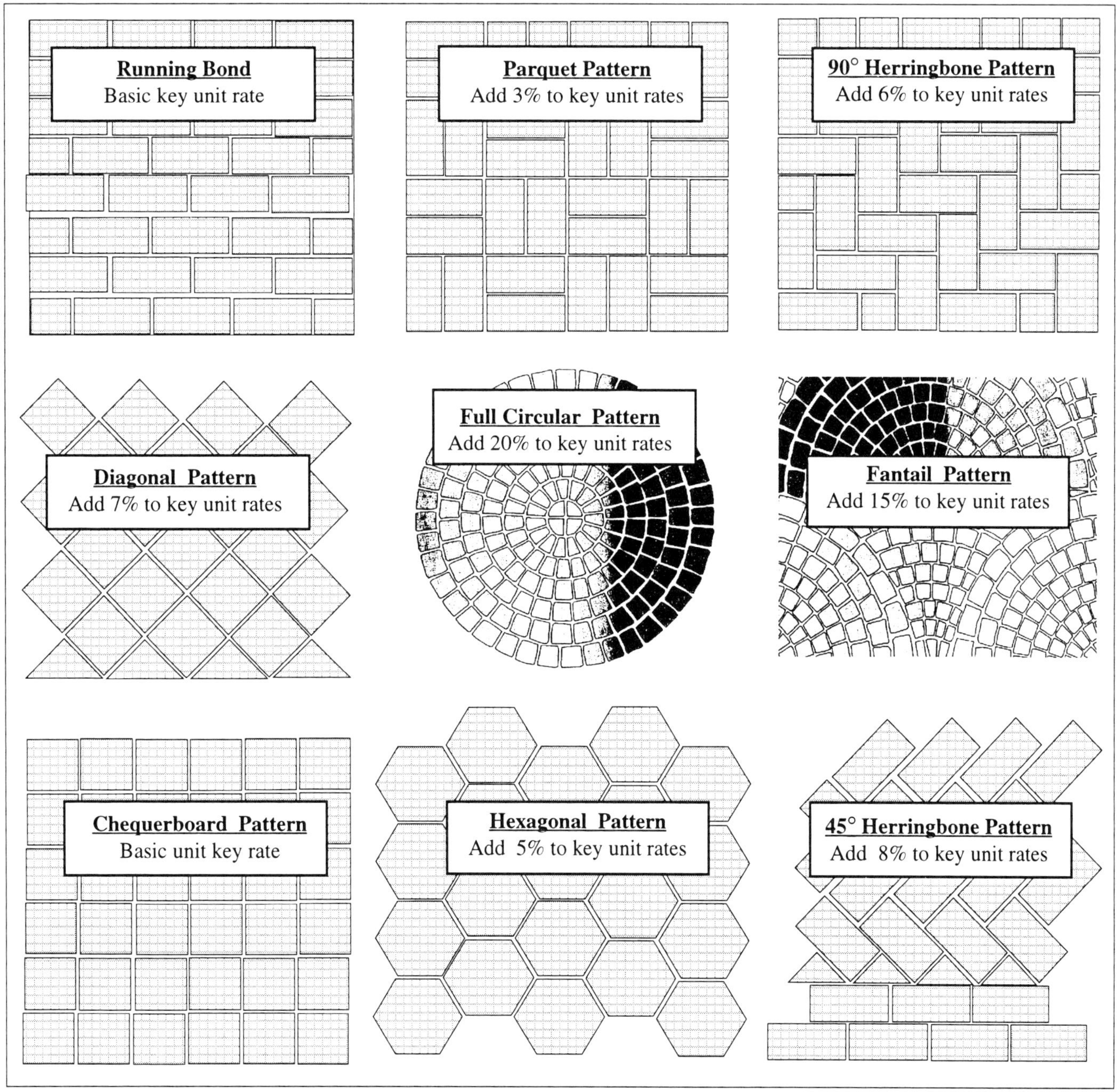

CEMENT AND SAND

- Cement and sand (1:3);
- Tamped finish.

	Unit	New	Refurb
General Floor Screeds			
Level			
25 mm thick	m2	5.49	6.18
38 mm thick	m2	6.64	7.47
50 mm thick	m2	7.84	8.82
75 mm thick	m2	10.72	12.06
Finished to falls			
32 mm (average) thick	m2	6.29	7.08
38 mm (average) thick	m2	6.99	7.86
50 mm (average) thick	m2	8.47	9.53
63 mm (average) thick	m2	9.88	11.12
Finished to falls and currents			
32 mm (average) thick	m2	6.36	7.16
38 mm (average) thick	m2	7.06	7.94
50 mm (average) thick	m2	8.54	9.61
63 mm (average) thick	m2	9.95	11.20
Landings			
38 mm thick	m2	6.45	7.26
50 mm thick	m2	7.80	8.78
Treads to steps or staircases			
25 mm thick			
225 mm wide	m	1.79	2.01
275 mm wide	m	2.19	2.46
Riser to steps or staircase			
25 mm thick			
150 mm high	m	1.36	1.53
Special finishings to floor screeds			
Add extra cost of;			
Wood float finish	m2	0.16	0.19
Steel trowel finish	m2	0.25	0.28
Power float finish	m2	0.62	0.69
Water repellent additive	m2	0.89	1.00

LEVELLING SCREEDS

- Resdev Pumaseal P2 sealer; brush or roller;
- Laybond latex Screedmaster 1; prime; trowelled;
- Repair work; Laybond Plycopatch; trowelled.

	Unit	New	Refurb
Polyurethane P1 concrete sealer; 2 mm thick			
Floors	m2	0.31	0.35
Landings	m2	0.36	0.41
Treads			
225 mm wide	m	0.10	0.12
275 mm wide	m	0.12	0.13
Latex; Screedmaster; prime; trowelled finish			
Floors			
3 mm thick	m2	3.93	4.43
5 mm thick	m2	5.64	6.34
7 mm thick	m2	6.82	7.67
Landings			
3 mm thick	m2	4.52	5.09
5 mm thick	m2	6.48	7.30
7 mm thick	m2	7.84	8.82
Treads of steps or staircases			
3 mm thick			
225 mm wide	m	1.14	1.28
275 mm wide	m	1.40	1.58
5 mm thick			
225 mm wide	m	1.64	1.84
275 mm wide	m	2.01	2.26
7 mm thick			
225 mm wide	m	1.99	2.24
275 mm wide	m	2.44	2.74
Repairing Screeds; Plycopatch; 6 mm thick			
Areas < 1 m2	m2	10.43	11.74
Areas > 1 < 10 m2	m2	8.18	9.20
Treads; average < 1 m lengths			
225 mm wide	m	3.05	3.43
275 mm wide	m	3.73	4.20

PROPRIETARY FLOOR FINISHINGS

- Tilcon Truscreed; ready mixed; trowelled;
- Ardex; Ardurapid 35; (1:4); trowelled;
- Lightweight aggregate screed; (1:4); trowelled.

	Unit	New	Refurb
Truscreed; ready mix; high strength			
Floors; level			
50 mm thick	m2	15.74	17.70
75 mm thick	m2	18.36	20.65
Floors; to falls			
35 mm (average) thick	m2	17.31	19.47
50 mm (average) thick	m2	20.19	22.72
Landings; level			
50 mm thick	m2	18.10	20.36
75 mm thick	m2	21.11	23.75
Treads; 20 mm thick			
225 mm wide	m	4.60	5.18
Ardurapid 35; Rapid setting; (1:4); trowelled			
Floors; level			
50 mm thick	m2	18.49	20.80
75 mm thick	m2	21.57	24.27
Floors; to falls			
35 mm (average) thick	m2	20.34	22.88
50 mm (average) thick	m2	23.73	26.70
Landings; level			
50 mm thick	m2	21.26	23.92
75 mm thick	m2	24.81	27.91
Treads; 20 mm thick			
225 mm wide	m	5.41	6.08
Lightweight Screeds; Premixed			
Floors; level			
50 mm thick	m2	19.01	21.39
75 mm thick	m2	20.65	23.24
Floors; to falls			
40 mm (average) thick	m2	19.47	21.91
63 mm (average) thick	m2	21.31	23.97

CEMENT AND SAND PAVING

- Mix, cement and sand (1:3);
- Trowelled surface;
- Temporary screeds.

	Unit	New	Refurb
General Flooring			
Floors; level			
38 mm thick	m2	7.04	7.92
50 mm thick	m2	8.26	9.30
Floors; finished to falls			
32 mm (average) thick	m2	6.68	7.52
38 mm (average) thick	m2	7.39	8.32
50 mm (average) thick	m2	8.92	10.03
Floors; finished to falls and currents			
32 mm (average) thick	m2	6.76	7.60
38 mm (average) thick	m2	7.47	8.40
50 mm (average) thick	m2	8.99	10.11
Landings			
38 mm thick	m2	6.85	7.70
50 mm thick	m2	8.23	9.26
Skirtings; coved next to floor; angles			
20 mm thick; 100 mm high	m2	2.37	2.67
20 mm thick; 150 mm high	m2	3.31	3.72
Treads			
25 mm thick			
225 mm wide	m2	2.11	2.37
275 mm wide	m2	2.57	2.89
Riser; 25 mm thick; 150 mm high	m2	1.39	1.56
Special Finishings to Floor Screeds			
Add extra cost of;			
Power float finish	m2	0.62	0.69
Sealed surface; dust proof	m2	0.89	1.00
Carborundum; non slip finish	m2	0.79	0.88

Properties

Management

30 Maintenance Comprehensive

GRANOLITHIC PAVING

- Mix, cement and dust free granite chippings (2:5);
- Trowelled surface;
- Temporary screeds.

	Unit	New	Refurb
General Flooring			
Floors; level			
38 mm thick	m2	18.66	20.99
50 mm thick	m2	21.44	24.12
Floors; finished to falls			
32 mm (average) thick	m2	17.19	19.34
38 mm (average) thick	m2	19.34	21.75
50 mm (average) thick	m2	21.86	24.59
Floors; finished to falls and currents			
32 mm (average) thick	m2	17.27	19.43
38 mm (average) thick	m2	19.41	21.84
50 mm (average) thick	m2	21.94	24.68
Landings			
38 mm thick	m2	21.46	24.14
50 mm thick	m2	24.66	27.74
Skirtings; coved next to floor; angles			
20 mm thick; 100 mm high	m	3.84	4.32
20 mm thick; 150 mm high	m	5.76	6.48
Treads			
25 mm thick			
225 mm wide	m	5.46	6.14
275 mm wide	m	6.27	7.05
Riser; 25 mm thick; 150 mm high	m2	5.18	5.82
Add for Special Finishings			
Add extra cost of;			
Power float finish	m2	0.52	0.59
Sealed surface; dust proof	m2	0.75	0.85
Carborundum; non slip finish	m2	0.60	0.68

Properties

Management

3 Warranty Comprehensive

100 Maintenance Comprehensive

TERRAZZO PAVING

- Marble cost; **£325.00** per ton delivered site;
- White Carrarra chippings; polished surface;
- White cement; laid by Specialist contractor;
- Including builder's attendance and profit.

	Unit	New	Refurb
General Flooring			
Floors; 16 mm thick			
level	m2	34.48	38.79
to falls	m2	40.17	45.19
to falls and cross falls	m2	43.10	48.48
to falls and currents	m2	46.54	52.36
Landings; 16 mm thick			
level	m2	49.99	56.24
Skirtings; coved next to floor; angles			
100 mm high	m	8.63	9.71
150 mm high	m	11.33	12.75
Treads of steps and staircases; 16 mm thick			
225 mm wide	m	18.62	20.95
275 mm wide	m	22.76	25.60
Riser; 150 mm high	m	9.28	10.44
String; 16 mm thick, angles and coves			
inner; 220 mm wide (extreme)	m	21.84	24.57
outer; 220 mm wide (extreme)	m	17.07	19.20
Making good around;			
small pipe	nr	1.74	1.96
large pipe	nr	2.90	3.26
Add for Special Finishings			
Paving set out in panels; average 1 m2			
plastic dividing strips	m2	4.06	4.56
brass dividing strips	m2	7.53	8.47
Selected aggregates	m2	4.06	4.56
Coloured cements	m2	2.43	2.74
Carborundum; non slip finish	m2	1.45	1.63

Properties

Management

7 Warranty Comprehensive

80 Maintenance Comprehensive

MASTIC ASPHALT

- One coat work; to concrete base;
- Rubbed surface with sanded finish;
- Felt isolating membrane under.

	Unit	New	Refurb
Mastic Asphalt; BS 6925; Black			
Floors;			
level	m2	10.36	11.66
to falls	m2	10.42	11.72
to falls and cross falls	m2	10.50	11.81
to falls and currents	m2	11.81	13.29
Landings; 16 mm thick			
level	m2	11.40	12.82
Skirtings; angle fillet next to floor; angles			
100 mm high	m	8.69	9.78
150 mm high	m	10.43	11.73
Treads of steps and staircases; 20 mm thick			
225 mm wide	m	18.11	20.37
275 mm wide	m	22.45	25.26
Riser; 150 mm high	m	13.04	14.67
String; angles and coves			
inner; 220 mm wide (extreme)	m	21.99	24.74
outer; 220 mm wide (extreme)	m	17.21	19.36
Making good around;			
small pipe	nr	1.56	1.76
large pipe	nr	4.06	4.56
floor gulley	nr	5.79	6.52
Add for Special Finishings			
Add extra cost of;			
Acid resistant asphalt	m2	38%	42%

Properties — *Management*: 10 Warranty Comprehensive; 70 Maintenance Comprehensive

COLOURED MASTIC ASPHALT

- One coat work; to concrete base;
- Rubbed surface with sanded finish;
- Felt isolating membrane under.

	Unit	New	Refurb
Mastic Asphalt; BS 6925; Red			
Floors;			
level	m2	17.06	19.19
to falls	m2	28.09	31.60
to falls and cross falls	m2	28.09	31.60
to falls and currents	m2	28.09	31.60
Landings; 16 mm thick			
level	m2	18.76	21.11
Skirtings; angle fillet next to floor; angles			
100 mm high	m	14.31	16.10
150 mm high	m	17.17	19.32
Treads of steps and staircases; 20 mm thick			
225 mm wide	m	30.75	34.59
275 mm wide	m	37.91	42.64
Riser; 150 mm high	m	21.47	24.15
String; angles and coves			
inner; 220 mm wide (extreme)	m	36.21	40.73
outer; 220 mm wide (extreme)	m	30.81	34.66
Making good around;			
small pipe	nr	1.66	1.87
large pipe	nr	4.30	4.84
floor gulley	nr	6.14	6.91

Properties — *Management*: 10 Warranty Comprehensive; 80 Maintenance Comprehensive

SEAMLESS EPOXY COVERING

- Jointless; epoxy resin; 3 mm thick;
- Leybond Products Ltd; coloured;
- Two coats; brush or roller application.

	Unit	New	Refurb
Abrasive Resistant; hard wearing			
Floors	m2	18.54	20.86
Landings	m2	20.40	22.95
Treads			
225 mm wide	m	5.42	6.10
275 mm wide	m	6.63	7.46
Skirtings			
100 mm high	m	2.41	2.71
150 mm high	m	3.62	4.07
Hygienic and chemical resistant			
Floors	m2	22.02	24.77
Landings	m2	24.22	27.25
Treads; 225 mm wide	m	6.44	7.25
Treads; 275 mm wide	m	7.87	8.86
Skirtings			
100 mm high	m	2.86	3.22
150 mm high	m	4.29	4.83
Non Slip			
Floors	m2	24.34	27.38
Landings	m2	26.77	30.12
Treads; 225 mm wide	m	7.12	8.01
Treads; 275 mm wide	m	8.70	9.79
Skirtings			
100 mm high	m	3.16	3.56
150 mm high	m	4.75	5.34

Properties — *Management*: 15 Warranty Comprehensive; 100 Maintenance Comprehensive

QUARRY TILING

- Cost **£14.50** per m2 delivered site; regular pattern;
- Dennis Ruabon Ltd; Heatherbrown;
- Bedding in Quarry Fix; 5 mm thick;
- Grout in Quarry grout; flush pointing.

	Unit	New	Refurb
Tile; 150 x 150 x 12.5 mm			
Floors			
level	m2	27.94	31.43
laid to falls	m2	28.02	31.52
laid to falls and currents	m2	28.21	31.73
Staircases			
landings	m2	29.00	32.62
treads; rounded nosings			
225 mm wide	m	9.46	10.64
275 mm wide	m	11.56	13.01
riser; 150 mm high	m	6.31	7.09
Skirtings; 138 mm high			
rounded top	m	6.32	7.11
rounded top; coved base	m	7.71	8.68
Tile; 194 x 194 x 19 mm			
Floors			
level	m2	34.24	38.53
laid to falls	m2	34.33	38.62
laid to falls and currents	m2	34.51	38.83
Staircases			
landings	m2	35.31	39.72
treads; rounded nosings			
225 mm wide	m	11.52	12.96
275 mm wide	m	14.08	15.84
riser; 150 mm high	m	7.68	8.64
Skirtings; 150 mm high			
rounded top	m	7.59	8.54
rounded top; coved base	m	8.98	10.10

Properties *Management*

5 Warranty Materials · 5 Warranty Comprehensive · 90 Maintenance Comprehensive

CERAMIC TILING

- Cost **£20.50** per m2 delivered site; regular pattern;
- Glazed finish; colour range;
- Bedding Bal cement adhesive; 3 mm thick;
- Grout in coloured grout; flush pointing.

	Unit	New	Refurb
Tile; 152 x 152 x 9.5 mm			
Floors			
level	m2	26.65	29.99
laid to falls	m2	26.74	30.08
laid to falls and currents	m2	26.92	30.29
Staircases			
landings	m2	30.65	34.48
treads; rounded nosings			
225 mm wide	m	10.00	11.25
275 mm wide	m	12.22	13.75
riser; 150 mm high	m	6.67	7.50
Skirtings; 138 mm high			
rounded top	m	6.66	7.49
rounded top; coved base	m	8.09	9.10
Tile; 200 x 200 x 8 mm			
Floors			
level	m2	31.29	35.20
laid to falls	m2	31.37	35.29
laid to falls and currents	m2	31.56	35.50
Staircases			
landings	m2	35.98	40.48
treads; rounded nosings			
225 mm wide	m	11.74	13.21
275 mm wide	m	14.35	16.14
riser; 150 mm high	m	7.83	8.80
Skirtings; 150 mm high			
rounded top	m	7.72	8.69
rounded top; coved base	m	9.11	10.25

Properties *Management*

80 Maintenance Comprehensive

TERRAZZO TILING

- Cost **£22.50** per m2 delivered site; regular pattern;
- Carrarra marble; polished:
- Bedding in cement and sand (1:3); 5 mm thick;
- Grout joints tinted mortar (1:1); flush joints.

	Unit	New	Refurb
Tile; 200 x 200 x 25 mm			
Floors			
level	m2	48.67	54.76
laid to falls	m2	48.75	54.85
laid to falls and currents	m2	48.94	55.06
Staircases			
landings	m2	55.97	62.97
treads; rounded nosings			
225 mm wide	m	18.26	20.54
275 mm wide	m	22.32	25.11
riser; 150 mm high	m	12.17	13.70
Skirtings; 138 mm high			
rounded top	m	11.72	13.19
rounded top; coved base	m	14.24	16.02
Tile; 305 x 305 x 25 mm			
Floors			
level	m2	37.08	41.72
laid to falls	m2	37.17	41.81
laid to falls and currents	m2	37.35	42.02
Staircases			
landings	m2	42.65	47.98
treads; rounded nosings			
225 mm wide	m	13.91	15.65
275 mm wide	m	17.01	19.13
riser; 150 mm high	m	9.28	10.44
Skirtings; 150 mm high			
rounded top	m	9.06	10.19
rounded top; coved base	m	11.00	12.38

Properties *Management*

10 Warranty Materials · 10 Warranty Comprehensive · 90 Maintenance Comprehensive

CORK TILING

- Cost (as stated below with the item) per m2;
- Laid to regular patterns;
- Transparent vinyl wearing surface;
- Manufacturer's approved adhesive;
- Wood surfaces; tongued and grooved.

	Unit	New	Refurb
Parquet Tile; 600 x 300 x 4 mm			
Floors			
Traditional (£8.50/ m2)	m2	14.36	16.15
Designer styles (£12.00/ m2)	m2	18.72	21.06
Extra for varnished tiles	m2	3.77	4.24
Corkmaster Tile; 300 x 300 x 3.2 mm			
Floors			
Natural (£21.25/ m2)	m2	30.73	34.57
Designer styles (£30.00/ m2)	m2	41.63	46.83
Special designs (£34.00/ m2)	m2	46.61	52.44
Cork-O-Wood; 900 x 150 x 3.2 mm			
Floors			
Natural (£29.00/ m2)	m2	40.13	45.14
Designer style (£34.00/ m2)	m2	46.36	52.15
Cork-O-Floor; 900 x 185 x 9 mm			
Floors			
Natural (£30.00/ m2)	m2	42.76	48.11
Designer style (£32.25/ m2)	m2	45.57	51.26
Special design (£42.50/ m2)	m2	58.33	65.63
Wood-O-Floor; 900 x 185 x 9 mm			
Floors			
Mahogany (£41.50/m2)	m2	59.41	66.83
Beech (£42.50/m2)	m2	60.65	68.23
Oak (£44.00/m2)	m2	62.52	70.34
Maple (£45.00/m2)	m2	63.77	71.74

Properties *Management*

10 Warranty Materials | 5 Warranty Comprehensive | 70 Maintenance Comprehensive

RUBBER TILING

- Cost as stated below with the item;
- Laid to regular pattern;
- Laying as instructed by the manufacturers;
- Full colour ranges; cutting and waste.

	Unit	New	Refurb
Tile; 610 x 610 x 2 mm			
Floors (£18.50/m2)	m2	29.04	32.67
Staircases			
landings	m2	33.40	37.57
treads; 275 mm wide	m	10.38	11.68
riser; 100 mm high	m	10.38	11.68
Tile; 610 x 610 x 3 mm			
Floors (£20.00/m2)	m2	31.41	35.33
Staircases			
landings	m2	36.12	40.63
treads; 275 mm wide	m	11.23	12.63
riser; 100 mm high	m	11.23	12.63
Tile; 610 x 610 x 4 mm; flexible			
Floors (£22.00/m2)	m2	33.61	37.81
landings	m2	36.97	41.59
Studded Tile; 500 x 500 x 2.5 mm			
Floors (£24.00/m2)	m2	35.93	40.42
Staircases			
landings	m2	39.52	44.46
treads; 275 mm wide	m	12.84	14.45
Studded Tile; 500 x 500 x 4 mm			
Floors (£26.00/m2)	m2	38.24	43.02
Staircases			
landings	m2	42.07	47.33
treads; 275 mm wide	m	13.67	15.38

Properties *Management*

15 Warranty Materials | 10 Warranty Comprehensive | 80 Maintenance Comprehensive

LINOLEUM TILING

- Cost **£22.50** per m2 delivered site;
- Laid to regular pattern;
- Colour ranges;
- F Ball Ltd; Stycobond F53 adhesive.

	Unit	New	Refurb
Tile; 300 x 300 x 2 mm			
Floors	m2	16.35	18.40
Staircases			
landings	m2	18.80	21.16
treads; 225 mm wide	m	6.13	6.90
treads; 275 mm wide	m	7.50	8.44
riser; 150 mm high	m	4.09	4.60
Skirtings;			
100 mm high	m	4.28	4.82
Tile; 300 x 300 x 2.5 mm			
Floors	m2	17.09	19.23
Staircases			
landings	m2	19.66	22.11
treads; 225 mm wide	m	6.41	7.21
treads; 275 mm wide	m	7.84	8.82
riser; 150 mm high	m	4.28	4.81
Skirtings;			
100 mm high	m	4.45	5.01
Tile; 300 x 300 x 3.2 mm			
Floors	m2	20.48	23.04
Staircases			
landings	m2	23.55	26.49
treads; 225 mm wide	m	7.68	8.64
treads; 275 mm wide	m	9.39	10.56
riser; 150 mm high	m	5.12	5.76
Skirtings;			
100 mm high	m	5.23	5.89

Properties *Management*

10 Warranty Comprehensive | 10 Warranty Materials | 70 Maintenance Comprehensive

VINYL TILES

- Cost **£4.25** per m2 delivered site; regular pattern;
- Fixing with approved flooring mastic;
- Clean and polish at completion.

	Unit	New	Refurb
Tile; 300 x 300 x 2 mm; Standard			
Floors	m2	7.53	8.47
Staircases			
landings	m2	8.66	9.74
treads; 225 mm wide	m	2.83	3.18
treads; 275 mm wide	m	3.45	3.89
riser; 150 mm high	m	1.88	2.12
Skirtings;			
100 mm high	m	2.37	2.67
Tile; 300 x 300 x 2.5 mm; Heavy Duty			
Floors	m2	8.53	9.60
Staircases			
landings	m2	9.81	11.04
treads; 225 mm wide	m	3.20	3.60
treads; 275 mm wide	m	3.91	4.40
riser; 150 mm high	m	2.13	2.40
Skirtings;			
100 mm high	m	2.46	2.77
Tile; 300 x 300 x 3.2 mm; Anti static			
Floors	m2	15.46	17.39
Staircases			
landings	m2	17.78	20.00
treads; 225 mm wide	m	5.80	6.53
treads; 275 mm wide	m	7.09	7.98
riser; 150 mm high	m	3.87	4.35
Skirtings;			
100 mm high	m	3.65	4.11

Properties *Management*

3 Warranty Materials | 1 Warranty Comprehensive | 50 Maintenance Comprehensive

VINYL SHEET

- Cost **£8.50** per m2 delivered site; regular widths;
- Fixing with approved flooring mastic;
- Sealed watertight edges;
- Clean and polish at completion.

	Unit	New	Refurb
2 mm Sheet; welded seams			
Floors	m2	13.65	15.36
Staircases			
landings	m2	15.70	17.66
treads; 225 mm wide	m	5.12	5.76
treads; 275 mm wide	m	6.26	7.04
riser; 150 mm high	m	3.42	3.84
Skirtings;			
100 mm high	m	3.65	4.11
2.5 mm Sheet; welded seams			
Floors	m2	15.23	17.13
Staircases			
landings	m2	17.51	19.70
treads; 225 mm wide	m	5.71	6.43
treads; 275 mm wide	m	6.98	7.86
riser; 150 mm high	m	3.81	4.29
Skirtings;			
100 mm high	m	4.03	4.54
2. mm Sheet; welded seams; non slip			
Floors	m2	20.35	22.90
Staircases			
landings	m2	23.41	26.33
treads; 225 mm wide	m	7.64	8.59
treads; 275 mm wide	m	9.33	10.50
riser; 150 mm high	m	5.09	5.73
Skirtings;			
100 mm high	m	4.55	5.12

Properties *Management*

5 Warranty Materials | 3 Warranty Comprehensive | 60 Maintenance Comprehensive

ALTRO NON SLIP FLOORING

- Cost **£16.50** per m2 delivered site; regular widths;
- Fixing with approved flooring mastic;
- Sealed watertight edges;
- Clean and polish at completion.

	Unit	New	Refurb
2 mm Sheet; welded seams			
Floors	m2	20.20	22.73
Staircases			
landings	m2	23.23	26.14
treads; 225 mm wide	m	7.58	8.53
treads; 275 mm wide	m	9.26	10.42
riser; 150 mm high	m	5.05	5.68
Skirtings;			
100 mm high	m	5.05	5.68
2.5 mm Sheet; welded seams			
Floors	m2	24.38	27.43
Staircases			
landings	m2	28.04	31.55
treads; 225 mm wide	m	9.15	10.29
treads; 275 mm wide	m	11.18	12.58
riser; 150 mm high	m	6.10	6.86
Skirtings;			
100 mm high	m	6.04	6.79
2. mm Sheet; welded seams; non slip			
Floors	m2	33.32	37.49
Staircases			
landings	m2	38.32	43.11
treads; 225 mm wide	m	12.50	14.06
treads; 275 mm wide	m	15.28	17.19
riser; 150 mm high	m	8.33	9.38
Skirtings;			
100 mm high	m	7.14	8.03

Properties *Management*

10 Warranty Materials | 5 Warranty Comprehensive | 80 Maintenance Comprehensive

CORK SHEET UNDERLAY

- Cost **£2.50** per m2 delivered site; regular widths;
- Fixing with approved flooring mastic;
- Insulating; sound proof.

	Unit	New	Refurb
2 mm Thick underlay			
Floors	m2	4.00	4.50
Staircases			
landings	m2	4.60	5.17
treads; 225 mm wide	m	1.50	1.69
treads; 275 mm wide	m	1.83	2.06
riser; 150 mm high	m	1.00	1.13
3.2 mm Thick underlay			
Floors	m2	4.81	5.41
Staircases			
landings	m2	5.53	6.22
treads; 225 mm wide	m	1.80	2.03
treads; 275 mm wide	m	2.21	2.48
riser; 150 mm high	m	1.20	1.35

Properties *Management*

7 Warranty Materials | 5 Warranty Comprehensive | 60 Maintenance Comprehensive

RUBBER SHEETING

- Cost **£22.00** per m2 delivered site; regular widths;
- Fixing with Manufacturer's flooring adhesive;
- Sealed watertight edges;
- Clean and polish at completion.

	Unit	New	Refurb
2 mm Sheet			
Floors	m2	31.22	35.12
Staircases			
landings	m2	35.90	40.39
treads; 275 mm wide	m	12.45	14.00
riser; 100 mm high	m	4.53	5.09
3 mm Sheet			
Floors	m2	31.89	35.88
Staircases			
landings	m2	36.68	41.26
treads; 275 mm wide	m	12.72	14.31
riser; 100 mm high	m	4.62	5.20
4 mm Flexible sheet			
Floors	m2	33.61	37.81
Landings	m2	38.65	43.48
2.5 mm Studded sheet			
Floors	m2	37.93	42.68
Staircases			
landings	m2	43.62	49.08
treads; 275 mm wide	m	15.13	17.02
4 mm Studded sheet			
Floors	m2	40.84	45.94
Staircases			
landings	m2	46.96	52.83
treads; 275 mm wide	m	16.28	18.32

Properties *Management*

15 Warranty Materials | 10 Warranty Comprehensive | 90 Maintenance Comprehensive

LINOLEUM SHEETING

- Cost **£10.50** per m2 delivered site; regular widths;
- Fixing with approved flooring mastic;
- Sealed watertight edges.
- Clean and polish at completion:

	Unit	New	Refurb
2 mm Sheet			
Floors	m2	15.37	17.29
Staircases			
landings	m2	17.68	19.89
treads; 225 mm wide	m	5.77	6.49
treads; 275 mm wide	m	7.05	7.93
riser; 150 mm high	m	3.84	4.33
Skirtings;			
100 mm high	m	5.33	6.00
2.5 mm Sheet			
Floors	m2	16.07	18.08
Staircases			
landings	m2	18.48	20.79
treads; 225 mm wide	m	6.03	6.78
treads; 275 mm wide	m	7.37	8.29
riser; 150 mm high	m	4.02	4.52
Skirtings;			
100 mm high	m	6.08	6.84
3.2 mm Sheet			
Floors	m2	19.25	21.65
Staircases			
landings	m2	22.14	24.90
treads; 225 mm wide	m	7.22	8.12
treads; 275 mm wide	m	8.83	9.93
riser; 150 mm high	m	4.81	5.42
Skirtings;			
100 mm high	m	6.66	7.50

Properties *Management*

10 Warranty Materials | 10 Warranty Comprehensive | 70 Maintenance Comprehensive

STONE FLOORING

- Cost **£30.00** per m2 delivered site; irregular pattern;
- Bedding in cement mortar (1;3); 10 mm thick;
- Flush pointing in general mortar.

	Unit	New	Refurb
50 mm Crazy paving; flush pointed			
Floors	m2	63.74	71.71
Staircases			
landings	m2	73.30	82.46
treads; 225 mm wide	m	20.79	23.39
treads; 275 mm wide	m	25.42	28.59
riser; 150 mm high	m	13.86	15.60
50 mm Regular squared paving; flush pointed			
Floors	m2	75.33	84.74
Staircases			
landings	m2	86.63	97.45
treads; 225 mm wide	m	24.58	27.65
treads; 275 mm wide	m	30.04	33.79
riser; 150 mm high	m	16.38	18.43
75 mm Regular squared paving; flush pointed			
Floors	m2	86.92	97.78
Staircases			
landings	m2	99.95	112.45
treads; 225 mm wide	m	28.36	31.90
treads; 275 mm wide	m	34.66	38.99
riser; 150 mm high	m	18.90	21.27

Properties *Management*

80 Maintenance Comprehensive

SLATE FLOORING

- Cost as stated with the item; delivered site;
- Laid to irregular pattern;
- Bedding in cement mortar (1;3); 10 mm thick;
- Flush pointing in tinted mortar.

	Unit	New	Refurb
Spanish; 10 mm thick; 600 x 300 mm; riven.			
Floors (£50.00/m2)	m2	85.76	96.48
Staircases			
landings	m2	98.62	110.95
treads; 225 mm wide	m	27.98	31.48
treads; 275 mm wide	m	34.20	38.47
riser; 150 mm high	m	18.65	20.98
Skirtings;			
100 mm high	m	20.88	23.49
Spanish; 15 mm thick; 600 x 300 mm; polished.			
Floors (£57.00/m2)	m2	95.03	106.91
Staircases			
landings	m2	109.28	122.94
treads; 225 mm wide	m	31.00	34.88
treads; 275 mm wide	m	37.89	42.63
riser; 150 mm high	m	20.67	23.25
Skirtings;			
100 mm high	m	22.04	24.80
Cumbria; 15 mm thick; 600 x 300 mm; polished			
Floors (£100.00/m2)	m2	144.86	162.97
Staircases			
landings	m2	166.59	187.41
treads; 225 mm wide	m	47.26	53.17
treads; 275 mm wide	m	57.76	64.98
riser; 150 mm high	m	31.51	35.45
Skirtings;			
100 mm high	m	32.45	36.50

Properties *Management*

15 Warranty Materials | 10 Warranty Comprehensive | 90 Maintenance Comprehensive

MARBLE FLOORING

- Cost as stated with the item; delivered site;
- Laid to irregular pattern;
- Bedding in cement mortar (1;3); 10 mm thick;
- Flush pointing in tinted mortar.

	Unit	New	Refurb
Travertine; 20 mm thick; 600 x 300 mm; polished.			
Floors (£135.00/m2)	m2	168.04	189.04
Staircases			
landings	m2	193.24	217.40
treads; 225 mm wide	m	54.82	61.67
treads; 275 mm wide	m	67.00	75.38
riser; 150 mm high	m	36.55	41.12
Skirtings;			
100 mm high	m	38.24	43.02
Carrarra; 20 mm thick; 600 x 300 mm; polished.			
Floors (£155.00/m2)	m2	191.21	215.12
Staircases			
landings	m2	219.90	247.38
treads; 225 mm wide	m	62.38	70.18
treads; 275 mm wide	m	76.25	85.78
riser; 150 mm high	m	41.59	46.79
Skirtings;			
100 mm high	m	44.04	49.54
Carrarra; 20 mm thick; designed to pattern; polished.			
Floors (£170.00/m2)	m2	202.80	228.15
Staircases			
landings	m2	233.22	262.38
treads; 225 mm wide	m	66.16	74.44
treads; 275 mm wide	m	80.87	90.98
riser; 150 mm high	m	44.11	49.62
Skirtings;			
100 mm high	m	46.36	52.15

Properties *Management*

70 Maintenance Comprehensive

SOFTWOOD BOARDING

- Boards well cramped up; butt end joints;
- Fixing to each joist; 2 floor brads punched home;
- Clean off.

	Unit	New	Refurb
18 mm Sawn boarding; butt jointed; brads.			
Floors	m2	8.11	9.13
Staircases			
landings	m2	9.33	10.50
treads; 225 mm wide	m	2.65	2.98
treads; 275 mm wide	m	3.23	3.64
riser; 150 mm high	m	1.76	1.98
Skirtings;			
100 mm high	m	2.14	2.41
18 mm Tongued and grooved boarding; brads			
Floors	m2	11.59	13.04
Staircases			
landings	m2	13.33	14.99
treads; 225 mm wide	m	3.78	4.25
treads; 275 mm wide	m	4.62	5.20
riser; 150 mm high	m	2.52	2.84
Skirtings;			
100 mm high	m	2.84	3.20
22 mm Tongued and grooved boarding; brads			
Floors	m2	13.91	15.64
Staircases			
landings	m2	15.99	17.99
treads; 225 mm wide	m	4.54	5.10
treads; 275 mm wide	m	5.55	6.24
riser; 150 mm high	m	3.02	3.40
Skirtings;			
100 mm high	m	3.30	3.72

Properties — *Management*: 70 Maintenance Comprehensive

HARDWOOD BLOCK

- 25 mm Thick; 60 x 180 mm; laid to pattern;
- Bed in bitumastic on screed (measured separately);
- Sand down; prepare for polishing.

	Unit	New	Refurb
Iroko			
Floors; running;	m2	43.46	48.89
Floors; herringbone;	m2	46.07	51.82
Floors; parquet	m2	44.76	50.36
Floors; diagonal parquet	m2	46.50	52.31
Staircases			
landings	m2	49.98	56.22
Margins			
two blocks wide; running	m	2.49	2.80
Meranti			
Floors; running;	m2	48.67	54.76
Floors; herringbone;	m2	51.59	58.04
Floors; parquet	m2	50.13	56.40
Floors; diagonal parquet	m2	52.08	58.59
Staircases			
landings	m2	55.97	62.97
Margins			
two blocks wide; running	m	3.07	3.45
Oak			
Floors; running;	m2	64.90	73.01
Floors; herringbone;	m2	68.79	77.39
Floors; parquet	m2	66.84	75.20
Floors; diagonal parquet	m2	69.44	78.12
Staircases			
landings	m2	74.63	83.96
Margins			
two blocks wide; running	m	3.77	4.24

Properties — *Management*: 80 Maintenance Comprehensive

HARDWOOD STRIP

- 25 mm Boards 18 mm thick; biscuit end joints;
- Tongued and grooved; well cramped up together;
- Secret nailed with cadmium plated nails;
- Sand down; prepare for polishing.

	Unit	New	Refurb
Oak			
Floors	m2	61.42	69.10
Staircases			
landings	m2	70.63	79.46
treads; 225 mm wide	m	20.04	22.54
treads; 275 mm wide	m	24.49	27.55
riser; 150 mm high	m	13.36	15.03
nosings; rounded edges	m	17.80	20.03
Utile			
Floors	m2	56.78	63.88
Staircases			
landings	m2	65.30	73.47
treads; 225 mm wide	m	18.53	20.84
treads; 275 mm wide	m	22.64	25.47
riser; 150 mm high	m	12.35	13.89
nosings; rounded edges	m	16.40	18.45
Utile; floating with sound deadening insulation			
Floors	m2	99.66	112.12
Staircases			
landings	m2	114.61	128.94
treads; 225 mm wide	m	32.52	36.58
treads; 275 mm wide	m	39.74	44.71
riser; 150 mm high	m	21.68	24.39
nosings; rounded edges	m	20.69	23.27

Properties — *Management*: 80 Maintenance Comprehensive

CONTRACT CARPETS

- Cost (As stated) per m2 delivered site;
- Full colour ranges;
- Broadloom; Laybond 49 adhesive;
- Carpet tiles; Stycobond 79 tackifier:

	Unit	New	Refurb
Broadloom			
Fine ribbed fibre bonded (£6.65/ m2)			
general floor areas	m2	11.77	13.24
landings	m2	12.95	14.57
staircases	m2	14.71	16.55
Heavy duty ribbed bonded (£7.45/m2)			
general floor areas	m2	13.36	15.03
landings	m2	14.69	16.53
staircases	m2	16.70	18.78
Velour pile bonded (£8.70/ m2)			
general floor areas	m2	15.49	17.43
landings	m2	17.04	19.17
staircases	m2	19.37	21.79
Tiles; 500 x 500 mm			
Fine ribbed fibre bonded (£9.76/ m2)			
general floor areas	m2	16.18	18.20
landings	m2	17.79	20.02
treads 275 mm wide	m	5.56	6.26
risers; 150 mm high	m	3.34	3.75
Velour pile spike resistant (£15.60/ m2)			
general floor areas	m2	23.91	26.90
landings	m2	26.30	29.59
treads 275 mm wide	m	8.22	9.25
Tufted; heavy duty (£20.96/ m2)			
general floor areas	m2	31.06	34.94
landings	m2	34.16	38.43
treads 275 mm wide	m	10.68	12.01
risers; 150 mm high	m	5.82	6.55

Properties *Management*

5 Warranty Materials; 3 Warranty Comprehensive; 60 Maintenance Comprehensive

GENERAL CARPETS

- Cost (As stated) per m2 delivered site;
- Full colour ranges;
- Broadloom; Laybond 49 adhesive;
- Carpet tiles; Stycobond 79 tackifier.

	Unit	New	Refurb
Broadloom			
Foam backed; nylon (£4.35/ m2)			
general floor areas	m2	9.09	10.22
landings	m2	10.00	11.25
staircases	m2	11.36	12.78
Hessian backed; nylon (£7.00/m2)			
general floor areas	m2	12.81	14.41
landings	m2	14.09	15.85
staircases	m2	16.01	18.01
Axminster (£15.00/ m2)			
general floor areas	m2	23.79	26.76
landings	m2	26.17	29.44
staircases	m2	29.73	33.45
Tiles; 500 x 500 mm			
Tufted; fibre bonded (£7.45/ m2)			
general floor areas	m2	13.61	15.32
landings	m2	14.98	16.85
treads 275 mm wide	m	4.68	5.26
risers; 150 mm high	m	2.81	3.16
Tufted; hessian backed; (£12.00/ m2)			
general floor areas	m2	19.70	22.16
landings	m2	21.67	24.38
treads 275 mm wide	m	6.77	7.62
Tufted; heavy duty (£16.50/ m2)			
general floor areas	m2	25.81	29.04
landings	m2	28.39	31.94
treads 275 mm wide	m	8.87	9.98
risers; 150 mm high	m	4.84	5.44

Properties *Management*

5 Warranty Comprehensive; 70 Maintenance Comprehensive

SPECIAL CARPETS

- Cost (As stated) per m2 delivered site;
- Full colour ranges;
- Broadloom; Laybond 49 adhesive;
- Carpet tiles; Stycobond 79 tackifier.

	Unit	New	Refurb
Barrier Matting			
Nylon matting			
Sheet (£23.30/ m2)			
general floor areas	m2	39.00	43.87
matwells < 2 m2	m2	46.02	51.77
Tiles; 500 x 500 mm (£18.64/ m2)			
general floor areas	m2	33.16	37.31
matwells < 2 m2	m2	39.13	44.02
Coir yarn matting (£8.00/ m2)			
general floor areas	m2	14.03	15.78
matwells < 2 m2	m2	16.55	18.62
Coir yarn designer matting (£16.00/ m2)			
General floor areas	m2	29.46	33.14
Computer Flooring Systems			
Antistatic tiles; 615 x 615 (£5.94/ m2)			
general floor areas	m2	13.94	15.69
landings	m2	16.04	18.04
staircases	m2	20.22	22.75
Cable management systems (£54.00/ m2)			
General floor areas	m2	86.98	97.85
Special Purpose Carpets			
Leisure carpets (indoor/outdoor)			
UB stabilised carpet (£11.20/ m2)			
general floor areas	m2	19.99	22.49
Exhibition carpets			
Heavy ribbed; bonded fireproof (£10.48/ m2)			
general floor areas	m2	19.42	21.84

Properties *Management*

5 Warranty Materials; 3 Warranty Comprehensive; 80 Maintenance Comprehensive

STAIR EDGINGS - ALUMINIUM

	Unit	New	Refurb
Stair edgings; light duty; aluminium; non slip insert; cut to length; plugging and screwing to tread.			
Square edging;			
50 mm wide; single insert	m	21.72	24.44
Extra; end trim	nr	7.27	8.17
Extra; matching end to riser	nr	7.82	8.80
96 mm wide; double insert	m	27.24	30.64
Extra; end trim	nr	8.14	9.16
Extra; matching end to riser	nr	8.78	9.88
Bevel shaped edging			
53 mm wide; single insert	m	19.43	21.86
Extra; end trim	nr	6.44	7.24
Extra; matching end to riser	nr	6.88	7.74
100 mm wide; double insert	m	26.17	29.45
Extra; end trim	nr	7.75	8.72
Extra; matching end to riser	nr	8.34	9.39
Undercut edging			
50 mm wide; single insert	m	23.04	25.92
Extra; end trim	nr	7.74	8.71
Extra; matching end to riser	nr	8.35	9.40
97 mm wide; double insert	m	40.14	45.16
Extra; end trim	nr	12.80	14.40
Extra; matching end to riser	nr	14.03	15.79
Rounded edging			
55 mm wide; single insert	m	34.89	39.25
Extra; end trim	nr	12.02	13.52
Extra; matching end to riser	nr	13.18	14.83
89 mm wide; double insert	m	41.75	46.97
Extra; end trim	nr	13.38	15.05
Extra; matching end to riser	nr	14.69	16.53
ADD for extra cost of bronze finish.		65%	75%

Properties *Management*

70 Maintenance Comprehensive

STAIR EDGINGS - ALUMINIUM

	Unit	New	Refurb
Stair edgings; heavy duty; aluminium; non slip insert; cut to length; plugging and screwing to tread.			
Square edging;			
64 mm wide; single insert	m	23.61	26.56
Extra; end trim	nr	7.97	8.97
Extra; matching end to riser	nr	8.61	9.68
96 mm wide; double insert	m	30.60	34.43
Extra; end trim	nr	9.38	10.56
Extra; matching end to riser	nr	10.17	11.44
Bevel shaped edging			
78 mm wide; single insert	m	26.11	29.37
Extra; end trim	nr	8.88	9.99
Extra; matching end to riser	nr	9.62	10.83
101 mm wide; double insert	m	33.63	37.84
Extra; end trim	nr	10.48	11.79
Extra; matching end to riser	nr	11.40	12.83
Undercut edging			
64 mm wide; single insert	m	34.66	38.99
Extra; end trim	nr	11.96	13.46
Extra; matching end to riser	nr	13.11	14.74
101 mm wide; double insert	m	46.51	52.33
Extra; end trim	nr	15.13	17.02
Extra; matching end to riser	nr	16.65	18.73
Rounded edging			
64 mm wide; single insert	m	35.89	40.37
Extra; end trim	nr	12.41	13.96
Extra; matching end to riser	nr	13.61	15.31
98 mm wide; double insert	m	42.41	47.72
Extra; end trim	nr	13.65	15.36
Extra; matching end to riser	nr	14.98	16.85
ADD for extra cost of bronze finish.		65%	75%

Properties *Management*

90 Maintenance Comprehensive

STAIR EDGINGS - uPVC

	Unit	New	Refurb
Stair edgings; light duty; PVC; non slip insert; cut to length; plugging and screwing to tread.			
Square edging;			
50 mm wide; single insert	m	9.37	10.54
Extra; end trim	nr	3.96	4.45
Extra; matching end to riser	nr	3.69	4.15
Bevel shaped edging			
53 mm wide; single insert	m	9.69	10.91
Extra; end trim	nr	4.19	4.71
Extra; matching end to riser	nr	3.90	4.39
Undercut edging			
50 mm wide; single insert	m	12.58	14.15
Extra; end trim	nr	6.23	7.00
Extra; matching end to riser	nr	5.81	6.53
Stair edgings; textured; PVC; non slip insert; cut to length; plugging and screwing to tread.			
Square edging;			
60 mm wide; single insert	m	13.77	15.49
Extra; end trim	nr	7.07	7.95
Extra; matching end to riser	nr	6.60	7.42
Bevel shaped edging			
60 mm wide; single insert	m	14.39	16.19
Extra; end trim	nr	7.51	8.45
Extra; matching end to riser	nr	7.01	7.89
Undercut edging			
55 mm wide; single insert	m	19.93	22.42
Extra; end trim	nr	11.42	12.85
Extra; matching end to riser	nr	10.66	12.00

Properties *Management*

50 Maintenance Comprehensive

PERFORMANCE APPLICATIONS

BEST USE

For economic use and optimum performance the Manufacturers suggest specification of these finishes in the following areas. See page 21 for Area References.

Finish	Activity Area
Premix Plaster	
Standard	All areas except H4-H8, S2-4:
High Impact	T7; S1-2:
One Coat	H4; H5; H6; H7; H8;T6; T10:
Rendering	
Plain	M5; M6:
Waterproof	M5; M6; H5-8; S3-4:
Suspended Ceilings	
Fibre Board	M5-6:
Plasterboard	All areas:
Mineral Fibre	M1-4; M7; H1-4; T1-13:
Metal	M2; H3; H5; T6-8; S2:

REMOVING EXISTING FINISHINGS

	Unit	New	Refurb
Remove existing wall finishings			
Finishings			
Cement and sand rendering	m2	4.23	4.76
Wall plastering	m2	2.14	2.41
Hardwood wall panelling	m2	2.55	2.87
Ceramic Wall tiling	m2	3.59	4.04
Tyrolean rendering	m2	2.49	2.80
Preparation to receive new finishings			
Remove adhesive	m2	2.72	3.06
Grit blasting	m2	3.30	3.72
Damp Proof treatment	m2	4.06	4.56

TYPES OF BACKGROUNDS

High Suction;	common brickwork
	lightweight concrete blocks
	no fines concrete
Low Suction;	insitu concrete
	heavy weight concrete blocks
	wall boarding
Expanded Metal	lining to walls and ceilings
	wrapping to beams and columns.

PREMIXED LIGHTWEIGHT PLASTER

Standard.

	Unit	New	Refurb
13 mm thick - 2 coats; high suction backgrounds 1st coat; - 11 mm thick; 2nd coat - 2 mm thick			
Walls > 300 mm wide	m2	7.00	7.87
returns < 150 mm wide	m	1.40	1.57
returns > 150 < 300 mm wide	m	2.19	2.46
Piers > 300 mm wide	m2	8.62	9.70
sides < 150 mm wide	m	1.72	1.94
sides > 150 < 300 mm wide	m	2.69	3.03
10 mm thick - 2 coats; low suction backgrounds 1st coat; - 8 mm thick; 2nd coat - 2 mm thick			
Walls > 300 mm wide	m2	7.41	8.33
returns < 150 mm wide	m	1.48	1.66
returns > 150 < 300 mm wide	m	2.31	2.60
Ceilings > 300 mm wide	m2	7.50	8.44
returns < 150 mm wide	m	1.50	1.68
returns > 150 < 300 mm wide	m	2.34	2.64
Piers; beams > 300 mm wide	m2	9.12	10.26
sides < 150 mm wide	m	1.50	1.68
sides > 150 < 300 mm wide	m	2.34	2.64
17 mm thick - 3 coats; expanded metal backgrounds Coats; 1st - 4 mm; 2nd - 11 mm; 3rd - 2 mm thick			
Walls > 300 mm wide	m2	12.48	14.04
returns < 150 mm wide	m	2.49	2.80
returns > 150 < 300 mm wide	m	3.90	4.39
Ceilings > 300 mm wide	m2	12.67	14.25
returns < 150 mm wide	m	2.49	2.80
returns > 150 < 300 mm wide	m	3.90	4.39
Piers; beams > 300 mm wide	m2	14.97	16.84
sides < 150 mm wide	m	2.49	2.80
sides > 150 < 300 mm wide	m	3.90	4.39

PREMIXED LIGHTWEIGHT PLASTER

High Impact.

	Unit	New	Refurb
13 mm thick - 2 coats; high suction backgrounds 1st coat; - 11 mm thick; 2nd coat - 2 mm thick			
Walls > 300 mm wide	m2	7.61	8.57
returns < 150 mm wide	m	1.52	1.71
returns > 150 < 300 mm wide	m	2.38	2.68
Piers > 300 mm wide	m2	8.62	9.70
sides < 150 mm wide	m	1.72	1.94
sides > 150 < 300 mm wide	m	2.69	3.03
10 mm thick - 2 coats; low suction backgrounds 1st coat; - 8 mm thick; 2nd coat - 2 mm thick			
Walls > 300 mm wide	m2	8.02	9.02
returns < 150 mm wide	m	1.60	1.80
returns > 150 < 300 mm wide	m	2.51	2.82
Ceilings > 300 mm wide	m2	7.50	8.44
returns < 150 mm wide	m	1.50	1.68
returns > 150 < 300 mm wide	m	2.34	2.64
Piers; beams > 300 mm wide	m2	9.12	10.26
sides < 150 mm wide	m	1.50	1.68
sides > 150 < 300 mm wide	m	2.34	2.64
21 mm thick - 3 coats; expanded metal backgrounds Coats; 1st - 8 mm; 2nd - 11 mm; 3rd - 2 mm thick			
Walls > 300 mm wide	m2	13.16	14.80
returns < 150 mm wide	m	2.62	2.95
returns > 150 < 300 mm wide	m	4.11	4.62
Ceilings > 300 mm wide	m2	13.34	15.01
returns < 150 mm wide	m	2.62	2.95
returns > 150 < 300 mm wide	m	4.11	4.62
Piers; beams > 300 mm wide	m2	15.65	17.60
sides < 150 mm wide	m	2.62	2.95
sides > 150 < 300 mm wide	m	4.11	4.62

PREMIXED LIGHTWEIGHT PLASTER

Board Finishing.

	Unit	New	Refurb
2 mm thick - 1 coat; base or plasterboard 1st coat - 2 mm thick			
Walls > 300 mm wide	m2	3.66	4.12
returns < 150 mm wide	m	0.73	0.82
returns > 150 < 300 mm wide	m	1.14	1.29
Ceilings > 300 mm wide	m2	3.75	4.22
returns < 150 mm wide	m	0.73	0.82
returns > 150 < 300 mm wide	m	1.14	1.29
Piers; beams > 300 mm wide	m2	5.28	5.95
sides < 150 mm wide	m	0.73	0.82
sides > 150 < 300 mm wide	m	1.14	1.29
3 mm thick - 1 coat; base or plasterboard 1st coat - 3 mm thick			
Walls > 300 mm wide	m2	3.97	4.47
returns < 150 mm wide	m	0.79	0.89
returns > 150 < 300 mm wide	m	1.24	1.40
Ceilings > 300 mm wide	m2	4.07	4.58
returns < 150 mm wide	m	0.79	0.89
returns > 150 < 300 mm wide	m	1.24	1.40
Piers; beams > 300 mm wide	m2	5.60	6.30
sides < 150 mm wide	m	0.79	0.89
sides > 150 < 300 mm wide	m	1.24	1.40
5 mm thick - 2 coats; base or plasterboard 1st coat - 3 mm thick; 2nd coat - 2 mm thick			
Walls > 300 mm wide	m2	4.60	5.18
returns < 150 mm wide	m	0.92	1.03
returns > 150 < 300 mm wide	m	1.44	1.62
Ceilings > 300 mm wide	m2	4.72	5.31
returns < 150 mm wide	m	0.92	1.03
returns > 150 < 300 mm wide	m	1.44	1.62
Piers; beams > 300 mm wide	m2	6.22	7.00
sides < 150 mm wide	m	0.92	1.03
sides > 150 < 300 mm wide	m	1.44	1.62

PREMIXED LIGHTWEIGHT PLASTER

One Coat Plasterwork.

	Unit	New	Refurb
13 mm thick - 1 coat; high suction background 1st coat; - 13 mm thick			
Walls > 300 mm wide	m2	8.12	9.14
returns < 150 mm wide	m	1.62	1.82
returns > 150 < 300 mm wide	m	2.54	2.86
Ceilings > 300 mm wide	m2	8.33	9.37
returns < 150 mm wide	m	0.32	0.36
returns > 150 < 300 mm wide	m	0.51	0.57
10 mm thick - 1 coat; low suction backgrounds 1st coat; - 10 mm thick;			
Walls > 300 mm wide	m2	8.53	9.60
returns < 150 mm wide	m	1.70	1.91
returns > 150 < 300 mm wide	m	2.67	3.00
Ceilings > 300 mm wide	m2	8.74	9.84
returns < 150 mm wide	m	0.34	0.38
returns > 150 < 300 mm wide	m	0.53	0.60
Piers; beams > 300 mm wide	m2	10.15	11.42
sides < 150 mm wide	m	1.70	1.91
sides > 150 < 300 mm wide	m	2.67	3.00
13 mm thick - 1 coat; expanded metal backgrounds Coats; 1st - 13 mm;			
Walls > 300 mm wide	m2	10.66	11.99
returns < 150 mm wide	m	2.13	2.39
returns > 150 < 300 mm wide	m	3.33	3.75
Ceilings > 300 mm wide	m2	10.92	12.29
returns < 150 mm wide	m	0.42	0.48
returns > 150 < 300 mm wide	m	0.66	0.75
Piers; beams > 300 mm wide	m2	12.28	13.81
sides < 150 mm wide	m	2.13	2.39
sides > 150 < 300 mm wide	m	3.33	3.75

PREMIXED LIGHTWEIGHT PLASTER

Renovation Work.

	Unit	New	Refurb
13 mm thick - 2 coats; high suction backgrounds 1st coat; - 11 mm thick; 2nd coat - 2 mm thick			
Walls > 300 mm wide	m2	11.22	12.62
returns < 150 mm wide	m	2.24	2.52
returns > 150 < 300 mm wide	m	3.51	3.94
Piers > 300 mm wide	m2	11.50	12.94
sides < 150 mm wide	m	0.45	0.50
sides > 150 < 300 mm wide	m	0.70	0.79
10 mm thick - 2 coats; low suction backgrounds 1st coat; - 8 mm thick; 2nd coat - 2 mm thick			
Walls > 300 mm wide	m2	11.62	13.08
returns < 150 mm wide	m	2.32	2.61
returns > 150 < 300 mm wide	m	3.63	4.09
Ceilings > 300 mm wide	m2	11.91	13.40
returns < 150 mm wide	m	0.46	0.52
returns > 150 < 300 mm wide	m	0.72	0.82
Piers; beams > 300 mm wide	m2	13.25	14.90
sides < 150 mm wide	m	2.32	2.61
sides > 150 < 300 mm wide	m	3.63	4.09
Dubbing out; single coat; high suction background Coats - 6; 9; 12 mm thick			
6 mm thick > 300 mm wide	m2	3.37	3.79
returns < 150 mm wide	m	0.67	0.76
returns > 150 < 300 mm wide	m	1.05	1.19
9 mm thick > 300 mm wide	m2	3.46	3.89
returns < 150 mm wide	m	0.13	0.15
returns > 150 < 300 mm wide	m	0.21	0.24
12 mm thick > 300 mm wide	m2	4.99	5.62
returns < 150 mm wide	m	0.67	0.76
returns > 150 < 300 mm wide	m	1.05	1.19

RENDERING

Plain; to walls.

	Unit	New	Refurb
One coat work; 10 mm thick; as backing; Cement and sand (1:3); wood float.			
Walls > 300 mm wide	m2	8.23	9.26
returns < 150 mm wide	m	1.64	1.85
returns > 150 < 300 mm wide	m	2.57	2.89
Piers > 300 mm wide	m2	10.29	11.57
sides < 150 mm wide	m	1.64	1.85
sides > 150 < 300 mm wide	m	2.57	2.89
Two coat work; 13 mm thick; as backing; Cement and sand (1:3); wood float.			
Walls > 300 mm wide	m2	9.85	11.08
returns < 150 mm wide	m	1.97	2.21
returns > 150 < 300 mm wide	m	3.08	3.46
Piers > 300 mm wide	m2	12.31	13.85
sides < 150 mm wide	m	1.97	2.21
sides > 150 < 300 mm wide	m	3.08	3.46
Two coat work; 13 mm thick; as rendering; Cement and sand (1:3); steel trowelled finish.			
Walls > 300 mm wide	m2	10.37	11.67
returns < 150 mm wide	m	2.07	2.33
returns > 150 < 300 mm wide	m	3.24	3.65
Piers > 300 mm wide	m2	12.96	14.59
sides < 150 mm wide	m	2.07	2.33
sides > 150 < 300 mm wide	m	3.24	3.65
Two coat work; 15 mm thick; as rendering; Cement and sand (1:3); steel trowelled finish.			
Walls > 300 mm wide	m2	10.89	12.26
returns < 150 mm wide	m	2.17	2.44
returns > 150 < 300 mm wide	m	3.40	3.83
Piers > 300 mm wide	m2	13.62	15.32
sides < 150 mm wide	m	2.17	2.44
sides > 150 < 300 mm wide	m	3.40	3.83

RENDERING

Waterproof; to walls.

	Unit	New	Refurb
Two coat work; 13 mm thick; as backing; Cement and sand (1:3); wood float.			
Walls > 300 mm wide	m2	11.11	12.50
returns < 150 mm wide	m	2.22	2.49
returns > 150 < 300 mm wide	m	3.47	3.91
Piers > 300 mm wide	m2	13.88	15.62
sides < 150 mm wide	m	2.22	2.49
sides > 150 < 300 mm wide	m	3.47	3.91
Two coat work; 13 mm thick; as rendering; Cement and sand (1:3); steel trowelled finish.			
Walls > 300 mm wide	m2	13.30	14.96
returns < 150 mm wide	m	2.65	2.98
returns > 150 < 300 mm wide	m	4.16	4.68
Piers > 300 mm wide	m2	16.62	18.70
sides < 150 mm wide	m	2.65	2.98
sides > 150 < 300 mm wide	m	4.16	4.68
Specialist rendering; two coats; 18 mm thick; Cement and sand (1:3); steel trowelled finish.			
Walls > 300 mm wide	m2	31.45	35.38
returns < 150 mm wide	m	6.27	7.06
returns > 150 < 300 mm wide	m	9.83	11.06
Piers > 300 mm wide	m2	39.31	44.23
sides < 150 mm wide	m	6.27	7.06
sides > 150 < 300 mm wide	m	9.83	11.06
Specialist rendering; three coats; 25mm thick; Cement and sand (1:3); steel trowelled finish.			
Walls > 300 mm wide	m2	35.52	39.96
returns < 150 mm wide	m	7.09	7.97
returns > 150 < 300 mm wide	m	11.10	12.49

PLASTERBOARD

Wall and ceiling linings.

	Unit	New	Refurb
Plasterboard; 9.5 mm thick; plain grade; fixing with nails; open joints.			
Ceilings > 300 mm wide	m2	3.70	4.16
returns < 150 mm wide	m	0.74	0.83
returns > 150 < 300 mm wide	m	1.16	1.30
Piers; beams > 300 mm wide	m2	4.46	5.02
sides < 150 mm wide	m	0.89	1.00
sides > 150 < 300 mm wide	m	1.39	1.57
Plasterboard; 9.5 mm thick; plain grade; tapered edges; fixing with nails; filled joints; taped.			
Walls > 300 mm wide	m2	6.33	7.12
returns < 150 mm wide	m	1.26	1.42
returns > 150 < 300 mm wide	m	1.98	2.22
Ceilings > 300 mm wide	m2	5.95	6.69
returns < 150 mm wide	m	1.19	1.33
returns > 150 < 300 mm wide	m	1.86	2.09
Piers; beams > 300 mm wide	m2	8.03	9.03
sides < 150 mm wide	m	1.60	1.80
sides > 150 < 300 mm wide	m	2.51	2.82
Plasterboard; 12.5 mm thick; plain grade; tapered edges; fixing with nails; filled joints; taped.			
Walls > 300 mm wide	m2	6.79	7.64
returns < 150 mm wide	m	1.35	1.52
returns > 150 < 300 mm wide	m	2.12	2.39
Ceilings > 300 mm wide	m2	6.39	7.18
returns < 150 mm wide	m	1.27	1.43
returns > 150 < 300 mm wide	m	2.00	2.24
Piers; beams > 300 mm wide	m2	8.62	9.70
sides < 150 mm wide	m	1.72	1.93
sides > 150 < 300 mm wide	m	2.69	3.03

WALL BOARDING

	Unit	New	Refurb
Fireline Board; plain grade; square edges; fixing with nails; filled joints; taped.			
12.5 mm thick > 300 mm wide	m2	7.47	8.41
returns < 150 mm wide	m	1.49	1.68
returns > 150 < 300 mm wide	m	2.34	2.63
15 mm thick > 300 mm wide	m2	7.76	8.74
returns < 150 mm wide	m	1.55	1.74
returns > 150 < 300 mm wide	m	2.43	2.73
Fireline Board; moisture resistant; square edges; fixing with nails; filled joints; taped.			
12.5 mm thick > 300 mm wide	m2	7.81	8.79
returns < 150 mm wide	m	1.56	1.75
returns > 150 < 300 mm wide	m	2.44	2.75
15 mm thick > 300 mm wide	m2	8.20	9.23
returns < 150 mm wide	m	1.64	1.84
returns > 150 < 300 mm wide	m	2.56	2.88
Thermal Board; plain grade; square edges; fixing with nails; filled joints; taped.			
25 mm thick > 300 mm wide	m2	9.97	11.21
returns < 150 mm wide	m	1.99	2.24
returns > 150 < 300 mm wide	m	3.11	3.50
32 mm thick > 300 mm wide	m2	10.49	11.80
returns < 150 mm wide	m	2.09	2.35
returns > 150 < 300 mm wide	m	3.28	3.69
Thermal Board Plus; plain grade; square edges; fixing with nails; filled joints; taped.			
25 mm thick > 300 mm wide	m2	11.70	13.17
returns < 150 mm wide	m	2.34	2.63
returns > 150 < 300 mm wide	m	3.66	4.11
30 mm thick > 300 mm wide	m2	12.40	13.95
returns < 150 mm wide	m	2.47	2.78
returns > 150 < 300 mm wide	m	3.87	4.36

	Unit	New	Refurb
Blockboard; birch faced; fixing with cadmium plated nails; adhesive.			
12 mm thick > 300 mm wide	m2	38.84	43.69
returns < 150 mm wide	m	7.75	8.72
returns > 150 < 300 mm wide	m	12.14	13.65
18 mm thick > 300 mm wide	m2	50.24	56.52
returns < 150 mm wide	m	10.02	11.27
returns > 150 < 300 mm wide	m	15.70	17.66
Plywood; external quality; birch faced; fixing with cadmium plated nails; adhesive.			
12 mm thick > 300 mm wide	m2	35.04	39.42
returns < 150 mm wide	m	6.99	7.86
returns > 150 < 300 mm wide	m	10.95	12.32
18 mm thick > 300 mm wide	m2	47.90	53.89
returns < 150 mm wide	m	9.56	10.75
returns > 150 < 300 mm wide	m	14.97	16.84
Chipboard; plain; butt joints; fixing with cadmium plated nails; adhesive.			
12 mm thick > 300 mm wide	m2	16.60	18.68
returns < 150 mm wide	m	3.31	3.73
returns > 150 < 300 mm wide	m	5.19	5.84
18 mm thick > 300 mm wide	m2	20.34	22.88
returns < 150 mm wide	m	4.06	4.56
returns > 150 < 300 mm wide	m	6.36	7.15
Chipboard; fire retardent; butt joints; fixing with cadmium plated nails; adhesive.			
12 mm thick > 300 mm wide	m2	29.84	33.57
returns < 150 mm wide	m	5.95	6.70
returns > 150 < 300 mm wide	m	9.32	10.49
18 mm thick > 300 mm wide	m2	37.37	42.05
returns < 150 mm wide	m	7.46	8.39
returns > 150 < 300 mm wide	m	11.68	13.14

	Unit	New	Refurb
Superlux; non asbestos; sand finish drilled and screwed; adhesive.			
6 mm thick > 300 mm wide	m2	12.50	14.07
returns < 150 mm wide	m	2.49	2.81
returns > 150 < 300 mm wide	m	3.91	4.40
9 mm thick > 300 mm wide	m2	17.34	19.50
returns < 150 mm wide	m	3.46	3.89
returns > 150 < 300 mm wide	m	5.42	6.09
MDF; standard; fixing with cadmium plated nails; adhesive.			
12 mm thick > 300 mm wide	m2	12.11	13.62
returns < 150 mm wide	m	2.42	2.72
returns > 150 < 300 mm wide	m	3.78	4.26
18 mm thick > 300 mm wide	m2	16.40	18.45
returns < 150 mm wide	m	3.27	3.68
returns > 150 < 300 mm wide	m	5.12	5.76
MDF; fire retardent; fixing with cadmium plated nails; adhesive.			
12 mm thick > 300 mm wide	m2	14.43	16.23
returns < 150 mm wide	m	2.88	3.24
returns > 150 < 300 mm wide	m	4.51	5.07
18 mm thick > 300 mm wide	m2	18.60	20.92
returns < 150 mm wide	m	3.71	4.17
returns > 150 < 300 mm wide	m	5.81	6.54
Vermiculite; fire protection beams; columns; fixing with adhesive.			
25 mm thick > 300 mm wide	m2	28.97	32.59
returns < 600 mm wide	m	19.46	21.89
returns > 600 < 1200 mm wide	m	33.14	37.29
50 mm thick > 300 mm wide	m2	33.61	37.81
returns < 600 mm wide	m	20.32	22.85
returns > 600 < 1200 mm wide	m	36.25	40.78

PARTITIONING

	Unit	New	Refurb
Stud framed; sawn softwood; treated; 50 mm studs at 400 mm centres; plates; solid strutting; plasterboard; taped joints.			
Studwork; 75 mm thick;			
9.5 mm Plasterboard both sides	m2	28.97	32.59
12.5 mm Plasterboard both sides	m2	29.93	33.67
Studwork; 100 mm thick;			
9.5 mm Plasterboard both sides	m2	32.36	36.40
12.5 mm Plasterboard both sides	m2	33.31	37.47
Stud framed; sawn softwood; treated; 50 mm studs at 400 mm centres; plates; solid strutting; plasterboard; 2 mm finishing plaster.			
Studwork; 75 mm thick;			
9.5 mm Plasterboard both sides	m2	36.48	41.04
12.5 mm Plasterboard both sides	m2	37.43	42.11
Studwork; 100 mm thick;			
9.5 mm Plasterboard both sides	m2	39.86	44.85
12.5 mm Plasterboard both sides	m2	40.82	45.92
Laminated; plasterboard; cellular core; 38 mm thick construction; SW framing; plasterboard; taped joints.			
57 mm thick			
9.5 mm Plasterboard both sides	m2	14.49	16.30
63 mm thick			
12.5 mm Plasterboard both sides	m2	16.22	18.24
Laminated; plasterboard; solid core; 19 mm plasterboard plank; SW framing; plasterboard; taped joints.			
50 mm thick			
9.5 mm Plasterboard both sides	m2	17.04	19.17
65 mm thick			
12.5 mm Plasterboard both sides	m2	19.08	21.46

	Unit	New	Refurb
Stud framed; metal channel studs; plate channels to floor and head; wood plates; plugged.			
48 mm Studs; single layer boarding			
12.5 mm Plasterboard both sides	m2	27.47	30.90
15 mm Wallboard both sides	m2	28.11	31.62
12.5 mm Firelineboard both sides	m2	28.77	32.37
70 mm Studs; single layer boarding			
12.5 mm Plasterboard both sides	m2	35.14	39.53
15 mm Wallboard both sides	m2	35.78	40.26
12.5 mm Firelineboard both sides	m2	36.45	41.00
Stud framed; Metal channel studs; plate channels to floor and head; wood plates; plugged.			
48 mm Studs; two layers boarding			
12.5 mm Plasterboard both sides	m2	41.38	46.56
15 mm Wallboard both sides	m2	42.67	48.00
12.5 mm Firelineboard both sides	m2	44.00	49.50
70 mm Studs; two layers boarding			
12.5 mm Plasterboard both sides	m2	49.06	55.19
15 mm Wallboard both sides	m2	50.34	56.63
12.5 mm Firelineboard both sides	m2	51.67	58.13
Stud framed; Jumbo; Metal channel studs; plate channels to floor and head; wood plates; plugged.			
146 mm Studs; two layers boarding			
12.5 mm Plasterboard both sides	m2	56.48	63.55
15 mm Wallboard both sides	m2	57.77	64.99
12.5 mm Firelineboard both sides	m2	59.10	66.48
146 mm Boxed Studs; two layers boarding			
12.5 mm Plasterboard both sides	m2	60.58	68.16
15 mm Wallboard both sides	m2	61.87	69.60
12.5 mm Firelineboard both sides	m2	63.20	71.10

INSULATION MATERIALS

	Unit	New	Refurb
Insulation; Glass Fibre Quilt.			
25 mm thick	m2	3.27	3.68
50 mm thick	m2	4.46	5.02
60 mm thick	m2	4.54	5.11
80 mm thick	m2	5.27	5.93
100 mm thick	m2	6.04	6.79
150 mm thick	m2	7.98	8.98
Insulation; Glass Fibre Quilt; Paper Faced.			
60 mm thick	m2	5.66	6.37
80 mm thick	m2	6.72	7.56
100 mm thick	m2	7.64	8.59
Insulation; Expanded Polystyrene Board.			
25 mm thick	m2	7.47	8.40
40 mm thick	m2	9.53	10.72
50 mm thick	m2	10.91	12.27
Insulation; Stryofoam Board; to Cavity Wall.			
25 mm thick	m2	8.48	9.54
50 mm thick	m2	12.97	14.59
75 mm thick	m2	17.80	20.02
Insulation; Injection Filling to Cavity.			
50 mm thick	m2	2.35	2.64
75 mm thick	m2	3.52	3.96
Fire Barriers; Rockwool Boarding; Ceiling Spaces.			
50 mm thick	m2	12.15	13.67
100 mm thick	m2	21.18	23.82

Adjustment of key rates related to

PURCHASE PRICE OF CERAMIC WALL TILES

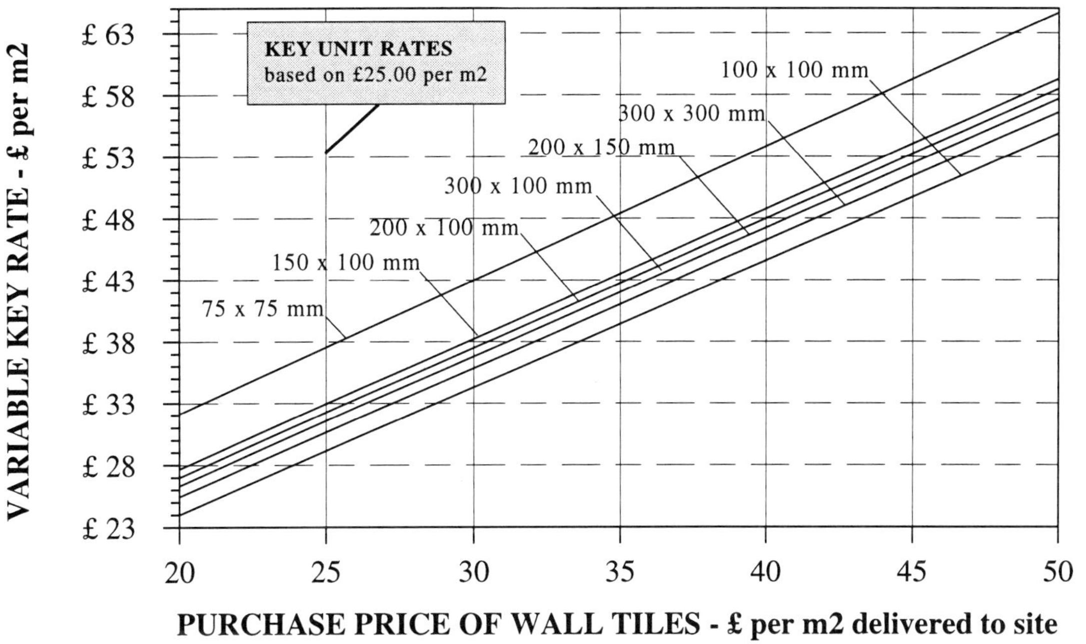

CERAMIC WALL TILING

- Walls (as stated below with the item);
- fixed to regular pattern;
- fixing with approved waterproof adhesive;
- waterproof grout; flush joint.

	Unit	New	Refurb
Ceramic tiles; 150 x 150 mm;(£25.00 per m2); approved adhesive; waterproof grout; flush pointing.			
Walls > 300 mm wide	m2	50.17	56.44
returns < 150 mm wide	m	10.01	11.26
returns > 150 < 300 mm wide	m	15.68	17.64
Piers; > 300 mm wide	m2	52.68	59.26
sides < 150 mm wide	m	10.51	11.82
sides > 150 < 300 mm wide	m	16.46	18.52
Cills; > 300 mm wide	m2	51.92	58.41
Cills < 150 mm wide	m	10.36	11.65
Cills > 150 < 300 mm wide	m	16.23	18.25
Add Extra for cost of;			
Fixing to special patterns	m	4.06	4.56
Rounded on edge tile	m	0.35	0.39
Rounded on two edge tile	no	0.75	0.85
Plastic edging; rounded.	m		
Mosaic tiles; 50 x 50 mm; (£45.00 per m2); approved adhesive; waterproof grout; flush pointing.			
Walls > 300 mm wide	m2	127.48	143.41
returns < 150 mm wide	m	25.43	28.61
returns > 150 < 300 mm wide	m	39.84	44.82
Piers; > 300 mm wide	m2	133.85	150.58
sides < 150 mm wide	m	26.70	30.04
sides > 150 < 300 mm wide	m	41.83	47.06
Cills; > 300 mm wide	m2	131.94	148.43
Cills < 150 mm wide	m	26.32	29.61
Cills > 150 < 300 mm wide	m	41.23	46.38
Add Extra for cost of;			
Fixing to special patterns	m	12.75	14.34
Plastic edging; rounded.	m	0.41	0.46

TRAVERTINE MARBLE CLADDINGS

	Unit	New	Refurb
Tiles; 200 x 100 x 10 mm; polished; regular pattern; approved adhesive; waterproof grout; flush pointing.			
Walls > 300 mm wide	m2	84.60	95.17
returns < 150 mm wide	m	16.88	18.99
returns > 150 < 300 mm wide	m	26.44	29.74
Cladding; 20 mm; polished; regular pattern; secret lug fixings; tinted compound; flush pointing.			
Walls > 300 mm wide	m2	133.27	149.93
returns < 150 mm wide	m	26.59	29.91
returns > 150 < 300 mm wide	m	41.65	46.85

SLATE WALL LININGS

	Unit	New	Refurb
Spanish; tiles; 300 x 300 X 18 mm; riven; approved adhesive; waterproof grout; flush pointing.			
Walls > 300 mm wide	m2	78.80	88.65
returns < 150 mm wide	m	15.72	17.69
returns > 150 < 300 mm wide	m	24.63	27.70
Cumbria; 20 mm; polished; regular pattern; secret lug fixings; tinted compound; flush pointing.			
Walls > 300 mm wide	m2	113.57	127.77
returns < 150 mm wide	m	22.66	25.49
returns > 150 < 300 mm wide	m	35.49	39.93

FIBRE BOARDS

- Plain insulating board; 1800 x 1200 mm panels:
- cutting; fixing; cadmium plated nails;
- taped joints; ready for decoration.

	Unit	New	Refurb
19 mm Insulation board; taped joints fixed to soffit floor joists			
level			
> 300 mm wide	m2	6.11	6.87
sloping; >10° <10°			
> 300 mm wide	m2	6.41	7.21
linings and beam casings			
> 300 mm wide	m2	7.02	7.90
< 300 mm wide	m2	7.73	8.69
19 mm Insulation board; taped joints battens; 38 x 50 mm; treated; plugged.			
level			
> 300 mm wide	m2	12.72	14.31
sloping; >10° <10°			
> 300 mm wide	m2	13.35	15.02
linings and beam casings			
> 300 mm wide	m2	14.63	16.45
< 300 mm wide	m2	17.23	19.38
12 mm Superlux non asbestos board fire resistance; Class (0) battens; 38 x 50 mm; treated; plugged.			
level			
> 300 mm wide	m2	26.93	30.29
sloping; >10° <10°			
> 300 mm wide	m2	28.27	31.81
linings and beam casings			
> 300 mm wide	m2	30.97	34.84
< 300 mm wide	m2	35.20	39.60

PLASTER BOARDS

- Gyproc plasterboard; 1200 x 1200 sheets;
- cutting; fixing; plasterboard clout nails;
- taped joints; ready for decoration.

	Unit	New	Refurb
9.5 mm Plasterboard; taped joints fixed to soffit floor joists.			
level			
> 300 mm wide	m2	6.80	7.65
sloping; >10° <30°			
> 300 mm wide	m2	7.14	8.04
linings and beam casings			
> 300 mm wide	m2	7.82	8.80
< 300 mm wide	m2	8.61	9.68
9.5 mm Plasterboard; taped joints battens; 38 x 50 mm; treated; plugged.			
level			
> 300 mm wide	m2	13.41	15.09
sloping; >10° <30°			
> 300 mm wide	m2	14.08	15.85
linings and beam casings			
> 300 mm wide	m2	15.43	17.35
< 300 mm wide	m2	18.11	20.37
9.5 mm Plasterboard; 2 mm plaster finish ceiling joists; 38 x 100 mm; treated.			
level			
> 300 mm wide	m2	19.82	22.30
sloping; >10° <30°			
> 300 mm wide	m2	20.81	23.42
linings and beam casings			
> 300 mm wide	m2	22.80	25.65
< 300 mm wide	m2	26.58	29.90
Extra cost of			
insulated boarding	m2	1.53	1.72

PLASTER BOARDS

- Gyproc plasterboard; 1200 x 1200 sheets;
- cutting; fixing; plasterboard clout nails;
- taped joints; ready for decoration.

	Unit	New	Refurb
12.5 mm Plasterboard; taped joints fixed to soffit floor joists.			
level			
> 300 mm wide	m2	7.31	8.23
sloping; >10° <30°			
> 300 mm wide	m2	7.68	8.64
linings and beam casings			
> 300 mm wide	m2	8.41	9.46
< 300 mm wide	m2	9.25	10.41
12.5 mm Plasterboard; taped joints battens; 38 x 50 mm; treated; plugged.			
level			
> 300 mm wide	m2	13.92	15.66
sloping; >10° <30°			
> 300 mm wide	m2	14.62	16.45
linings and beam casings			
> 300 mm wide	m2	16.01	18.01
< 300 mm wide	m2	18.75	21.10
12.5 mm Plasterboard; 2 mm plaster finish ceiling joists; 38 x 100 mm; treated.			
level			
> 300 mm wide	m2	20.33	22.87
sloping; >10° <30°			
> 300 mm wide	m2	21.35	24.02
linings and beam casings			
> 300 mm wide	m2	23.38	26.31
< 300 mm wide	m2	26.58	29.90
Extra cost of			
insulated boarding	m2	1.53	1.72

MINERAL FIBRE; Plain tiles

- lay in tiles; £11.50 per m2 delivered to site;
- exposed suspension; modular; aluminium trim;
- galvanised heavy duty primary channels;
- galvanised hangers; 1200 mm centres.

	Unit	New	Refurb
Tiles; 600 x 600 mm			
level			
> 300 mm wide	m2	22.02	24.77
sloping; >10° <30°			
> 300 mm wide	m2	26.09	29.35
linings and beam casings			
> 300 mm wide	m2	27.52	30.96
< 300 mm wide	m2	34.40	38.70
Tiles; 1200 x 300 mm			
level			
> 300 mm wide	m2	21.03	23.66
sloping; >10° <30°			
> 300 mm wide	m2	24.92	28.03
linings and beam casings			
> 300 mm wide	m2	26.28	29.57
< 300 mm wide	m2	32.86	36.96
Tiles; 1200 x 600 mm			
level			
> 300 mm wide	m2	20.15	22.67
sloping; >10° <30°			
> 300 mm wide	m2	23.87	26.86
linings and beam casings			
> 300 mm wide	m2	25.18	28.33
< 300 mm wide	m2	31.48	35.41

MINERAL FIBRE; Tegular tiles

- fitted tiles; £14.00 per m2 delivered to site;
- concealed suspension; modular; aluminium trim;
- galvanised heavy duty primary channels;
- galvanised hangers; 1200 mm centres.

	Unit	New	Refurb
Tiles; 600 x 600 mm			
level			
> 300 mm wide	m2	25.87	29.11
sloping; >10° <30°			
> 300 mm wide	m2	30.66	34.49
linings and beam casings			
> 300 mm wide	m2	32.34	36.38
< 300 mm wide	m2	40.42	45.48
Tiles; 1200 x 300 mm			
level			
> 300 mm wide	m2	24.71	27.80
sloping; >10° <30°			
> 300 mm wide	m2	29.28	32.94
linings and beam casings			
> 300 mm wide	m2	30.88	34.75
< 300 mm wide	m2	38.61	43.43
Tiles; 1200 x 600 mm			
level			
> 300 mm wide	m2	23.67	26.63
sloping; >10° <30°			
> 300 mm wide	m2	28.05	31.56
linings and beam casings			
> 300 mm wide	m2	29.59	33.29
< 300 mm wide	m2	36.99	41.61

METAL; Decorative tiles

- self finish tiles; £32.00 per m2 delivered to site;
- concealed suspension; modular; aluminium trim;
- galvanised heavy duty primary channels;
- galvanised hangers; 1200 mm centres.

	Unit	New	Refurb
Tiles; 600 x 600 mm			
level			
> 300 mm wide	m2	43.52	48.96
sloping; >10° <30°			
> 300 mm wide	m2	49.83	56.05
linings and beam casings			
> 300 mm wide	m2	50.04	56.30
< 300 mm wide	m2	57.55	64.74
Tiles; 1200 x 300 mm			
level			
> 300 mm wide	m2	41.56	46.75
sloping; >10° <30°			
> 300 mm wide	m2	47.58	53.53
linings and beam casings			
> 300 mm wide	m2	47.79	53.77
< 300 mm wide	m2	54.96	61.83
Tiles; 1200 x 600 mm			
level			
> 300 mm wide	m2	39.82	44.79
sloping; >10° <30°			
> 300 mm wide	m2	45.59	51.29
linings and beam casings			
> 300 mm wide	m2	45.79	51.51
< 300 mm wide	m2	52.66	59.24

GLOSSARY

acid etching	surface of polished glass is etched by hydrofluoric acid.
acoustic laminate	a laminate glass with special accoustic interlayers.
annealed glass	ordinary glass not treated.
blast resistant	special glass toughened to be able to resist explosion and pressure waves.
brilliant cutting	V notch cut into the surface of polished glass for decorative purposes.
bullet resistant	special glass toughened to be able to resist bullets and armed attack.
condensation	the formation of water droplets on cold surfaces
conduction	transmission of heat or electricity.
curtain walling	a complete glazing system in which the facade is glazed into frames fixed to the building.
data protection	electro-magnetic shielding of computers.
decibels	unit of the intensity of sound.
fanlight	a glazed panel immediately above a window or door.
laminated glass	combination of layers of glass with plastic or resin interlayers.
reflective glass	coated glass which reflects light or infra red radiation
safety film	an organic film applied to the surface of glass to provide solar protection, safety, insulation and privacy.
sand blasting	process for etching the surface of glass for decorative purposes.
silicone sealant	sealing or glazing compound.
silvering	application of a silver backing paint to form a mirror.
thermal break	gap of low thermal conductivity in a metal frame separating the inner and outer layers.
toughened glass	heat treated glass with improved surface compressive stress and strength.
translucent	letting through light but obscuring clear vision.

Product	Primary Function							Cost
	Solar Control	Thermal Insulation	Noise Control	Fire Resistance	Safety and Security	Privacy & Decoration	Mechanical Strength	per m2
Acid etched						*		£ 92.00
Antisun; 6 mm	*							£ 68.00
Armourscreen					*			£ 94.00
Audio screen			*					£ 106.00
Bevelled						*		£ 38.00
Brilliant cut						*		£ 43.00
Cast; georgian wired				*	*	*	*	£ 32.00
Cast; rough							*	£ 29.00
Datastop					*			£ 120.00
Eclipse	*							£ 39.00
Float; clear (Standard); 3 mm						*		£ 19.00
Float; thick; 6 mm			*				*	£ 33.00
Float; thick; 8 mm							*	£ 81.00
Insulating units		*	*					£ 129.00
K Glass		*						£ 50.00
Laminated; 8 mm			*		*	*	*	£ 85.00
Leaded						*		£ 99.00
Mirage						*		£ 50.00
Mirror						*		£ 50.00
Mirror; surveillance					*			£ 71.00
Mirror; venetian striped					*			£ 83.00
Oriel						*		£ 36.00
Planar							*	£ 44.00
Planarclad							*	£ 58.00
Polished; georgian wired				*	*	*	*	£ 62.00
Polycarbonate				*	*		*	£ 106.00
Polycarbonate; Margard coated	*			*	*		*	£ 132.00
Pryosheild; textured				*			*	£ 131.00
Pyrodur				*				£ 199.00
Pyroshield; clear				*			*	£ 143.00
Pyroshield; safety clear				*	*		*	£ 221.00
Pyrosheild; safety textured				*	*		*	£ 243.00
Pyrostop				*				£ 269.00
Reflectafloat	*							£ 72.00
Sand blasted						*		£ 60.00
Secure-vu					*			£ 136.00
Shopbond					*			£ 60.00
Stained						*		£ 80.00
Suncool; classic	*							£ 44.00
Suncool; high performance	*	*						£ 78.00
Suncool; low reflection	*							£ 89.00
Textured; 6 mm						*		£ 38.00
Toughened; 6 mm					*		*	£ 40.00
Toughened; blast resistant					*		*	£ 135.00
Toughened; bullet resistant					*		*	£ 160.00
UV Screen	*							£ 85.00

SINGLE GLAZING TO WOOD

New Work	Panes < 0.15 m2; glazed with;				Panes > 0.15 < 4.00 m2; glazed with;			
	linseed putty; pins	beads; pinned	beads; screwed	beads & washleather; screwed	linseed putty; pins	beads; pinned	beads; screwed	beads & washleather; screwed
CLEAR; Float								
4 mm thick	23.57	24.63	26.51	27.45	22.12	23.12	24.88	25.77
5 mm thick	32.75	34.23	36.85	38.16	31.09	32.49	34.98	36.22
6 mm thick	37.35	39.03	42.01	43.51	35.58	37.18	40.03	41.45
8 mm thick	64.91	67.83	73.02	75.62	62.50	65.31	70.31	72.81
10 mm thick	92.47	96.63	104.03	107.73	89.42	93.45	100.60	104.18
TEXTURED								
Type A								
3 mm thick	24.66	25.77	27.74	28.72	23.19	24.23	26.08	27.01
4 mm thick	26.72	27.93	30.06	31.13	25.20	26.34	28.36	29.36
6 mm thick	43.95	45.93	49.44	51.20	42.03	43.92	47.28	48.97
TINTED; Float								
4 mm thick	30.46	31.83	34.26	35.48	28.85	30.15	32.46	33.61
6 mm thick	50.27	52.53	56.55	58.56	48.20	50.37	54.22	56.15
10 mm thick	71.80	75.03	80.77	83.65	69.23	72.35	77.89	80.66
CAST; Rough								
6 mm thick	30.46	31.83	34.26	35.48	28.85	30.15	32.46	33.61
10 mm thick	77.54	81.03	87.23	90.34	74.84	78.21	84.20	87.19
CAST; Georgian wired								
7 mm thick	24.71	25.83	27.80	28.79	23.24	24.29	26.15	27.08
10 mm thick	51.13	53.43	57.52	59.56	49.04	51.25	55.17	57.13
POLISHED PLATE; Georgian wired								
6 mm thick	70.65	73.83	79.48	82.31	68.11	71.18	76.62	79.35

SINGLE GLAZING TO METAL

New Work	Panes < 0.15 m2; glazed with;			Panes < 0.15 m2; glazed with;		
	linseed putty; clips	metal beads & compound; screwed	metal beads & rubber gaskets; clip on	linseed putty; clips	metal beads & compound; screwed	metal beads & rubber gaskets; clip on
CLEAR; Float						
4 mm thick	23.57	26.98	27.22	22.12	23.12	24.88
5 mm thick	32.75	37.50	37.83	31.09	32.49	34.98
6 mm thick	37.35	42.76	43.13	35.58	37.18	40.03
8 mm thick	64.91	74.32	74.97	62.50	65.31	70.31
10 mm thick	92.47	105.88	106.80	89.42	93.45	100.60
TEXTURED						
Type A						
3 mm thick	24.66	28.23	28.48	23.19	24.23	26.08
4 mm thick	26.72	30.60	30.87	25.20	26.34	28.36
6 mm thick	43.95	50.32	50.76	42.03	43.92	47.28
TINTED; Float						
4 mm thick	30.46	34.87	35.18	28.85	30.15	32.46
6 mm thick	50.27	57.55	58.06	48.20	50.37	54.22
10 mm thick	71.80	82.21	82.93	69.23	72.35	77.89
CAST; Rough						
6 mm thick	30.46	34.87	35.18	28.85	30.15	32.46
10 mm thick	77.54	88.78	89.56	74.84	78.21	84.20
CAST; Georgian wired						
7 mm thick	24.71	28.30	28.54	23.24	24.29	26.15
10 mm thick	51.13	58.54	59.05	49.04	51.25	55.17
POLISHED PLATE; Georgian wired						
6 mm thick	70.65	80.89	81.60	68.11	71.18	76.62

SAFETY GLAZING TO WOOD

New Work	Panes < 0.15 m2; glazed with; linseed putty; pins	beads; pinned	beads; screwed	beads & washleather; screwed	Panes > 0.15 < 4.00 m2; glazed with; linseed putty; pins	beads; pinned	beads; screwed	beads & washleather; screwed
TOUGHENED; Clear; Georgian wired								
4 mm thick	35.05	36.63	39.43	40.83	33.34	34.84	37.50	38.84
6 mm thick	43.66	45.63	49.12	50.87	41.75	43.63	46.97	48.64
8 mm thick	69.50	72.63	78.19	80.97	66.99	70.00	75.36	78.04
10 mm thick	83.28	87.03	93.69	97.03	80.45	84.07	90.51	93.72
TOUGHENED; Textured								
4 mm thick	39.24	41.01	44.15	45.72	37.43	39.12	42.11	43.61
6 mm thick	47.85	50.01	53.84	55.75	45.84	47.91	51.57	53.41
TOUGHENED; Tinted								
4 mm thick	44.70	46.71	50.28	52.07	42.76	44.68	48.10	49.81
6 mm thick	55.89	58.41	62.88	65.12	53.70	56.11	60.41	62.56
LAMINATED; Clear								
4.4 mm thick	59.04	61.70	66.42	68.78	56.77	59.32	63.86	66.14
5.4 mm thick	61.14	63.90	68.79	71.23	58.83	61.47	66.18	68.53
6.4 mm thick	63.25	66.10	71.16	73.69	60.88	63.62	68.49	70.93
8.8 mm thick	96.94	101.30	109.05	112.93	93.79	98.01	105.51	109.26
LAMINATED; Textured								
5.4 mm thick	66.90	69.91	75.26	77.94	64.45	67.35	72.50	75.08
6.4 mm thick	69.00	72.11	77.63	80.39	66.50	69.50	74.82	77.48
8.8 mm thick	71.11	74.31	80.00	82.84	68.56	71.64	77.13	79.87
LAMINATED; Clear; Security								
7.5 mm thick	94.77	99.03	106.61	110.40	91.67	95.79	103.12	106.79
9.5 mm thick	100.51	105.03	113.07	117.09	97.28	101.65	109.43	113.33
11.5 mm thick	106.25	111.03	119.53	123.78	102.88	107.51	115.74	119.86

SAFETY GLAZING TO METAL

New Work	Panes < 0.15 m2; glazed with; linseed putty; clips	metal beads & compound; screwed	metal beads & rubber gaskets; clip on	Panes < 0.15 m2; glazed with; linseed putty; clips	metal beads & compound; screwed	metal beads & rubber gaskets; clip on
TOUGHENED; Clear; Georgian wired						
4 mm thick	35.05	40.13	40.48	33.34	34.84	37.50
6 mm thick	43.66	49.99	50.43	41.75	43.63	46.97
8 mm thick	69.50	79.58	80.27	66.99	70.00	75.36
10 mm thick	83.28	95.36	96.19	80.45	84.07	90.51
TOUGHENED; Textured						
4 mm thick	39.24	44.93	45.32	37.43	39.12	42.11
6 mm thick	47.85	54.79	55.27	45.84	47.91	51.57
TOUGHENED; Tinted						
4 mm thick	44.70	51.18	51.62	42.76	44.68	48.10
6 mm thick	55.89	64.00	64.56	53.70	56.11	60.41
LAMINATED; Clear						
4.4 mm thick	59.04	67.60	68.19	56.77	59.32	63.86
5.4 mm thick	61.14	70.01	70.62	58.83	61.47	66.18
6.4 mm thick	63.25	72.42	73.05	60.88	63.62	68.49
8.8 mm thick	96.94	110.99	111.96	93.79	98.01	105.51
LAMINATED; Textured						
5.4 mm thick	66.90	76.60	77.27	64.45	67.35	72.50
6.4 mm thick	69.00	79.01	79.70	66.50	69.50	74.82
8.8 mm thick	71.11	81.42	82.13	68.56	71.64	77.13
LAMINATED; Clear; Security						
7.5 mm thick	94.77	108.51	109.46	91.67	95.79	103.12
9.5 mm thick	100.51	115.08	116.09	97.28	101.65	109.43
11.5 mm thick	106.25	121.66	122.72	102.88	107.51	115.74

SPECIAL GLAZING TO WOOD

New Work	Panes < 0.15 m2; glazed with:				Panes > 0.15 < 4.00 m2; glazed with:			
	linseed putty; pins	beads; pinned	beads; screwed	beads & washleather; screwed	linseed putty; pins	beads; pinned	beads; screwed	beads & washleather; screwed
SOLAR CONTROL; Anti-sun								
4 mm thick	60.31	63.03	67.85	70.27	58.01	60.63	65.27	67.59
6 mm thick	77.54	81.03	87.23	90.34	74.84	78.21	84.20	87.19
SOLAR CONTROL; Suncool Classic								
4 mm thick	86.15	90.03	96.92	100.37	83.25	87.00	93.66	96.99
6 mm thick	96.20	100.53	108.23	112.08	93.07	97.26	104.70	108.42
SOLAR CONTROL; Eclipse								
6 mm thick	100.51	105.03	113.07	117.09	97.28	101.65	109.43	113.33
FIRE RESISTANT; Pyroshield								
Clear								
6 mm thick	157.93	165.04	177.67	183.99	153.36	160.26	172.53	178.67
Textured								
7 mm thick	165.97	173.44	186.72	193.36	161.21	168.47	181.36	187.81
SAFETY; Pyroshield								
Clear								
6 mm thick	232.58	243.04	261.65	270.95	226.27	236.46	254.56	263.61
Textured								
7 mm thick	241.77	252.65	271.99	281.66	235.25	245.83	264.65	274.06
FIRE RESISTANT; Pyrodur								
10 mm thick	276.22	288.65	310.75	321.80	268.90	281.00	302.51	313.27
13 mm thick	280.81	293.45	315.91	327.15	273.39	285.69	307.56	318.49
FIRE RESISTANT; Pyrostop								
15 mm thick	341.68	357.06	384.39	398.06	332.84	347.81	374.44	387.75
18 mm thick	364.65	381.06	410.23	424.81	355.27	371.26	399.68	413.89
21 mm thick	387.62	405.06	436.07	451.57	377.71	394.70	424.92	440.03
27 mm thick	473.75	495.07	532.97	551.92	461.83	482.62	519.56	538.04

SPECIAL GLAZING TO METAL

New Work	Panes < 0.15 m2; glazed with:			Panes < 0.15 m2; glazed with:		
	linseed putty; clips	metal beads & compound; screwed	metal beads & rubber gaskets; clip on	linseed putty; clips	metal beads & compound; screwed	metal beads & rubber gaskets; clip on
SOLAR CONTROL; Anti-sun						
4 mm thick	69.06	69.06	69.66	58.01	60.63	65.27
6 mm thick	88.78	88.78	89.56	74.84	78.21	84.20
SOLAR CONTROL; Suncool Classic						
4 mm thick	98.65	98.65	99.51	83.25	87.00	93.66
6 mm thick	110.15	110.15	111.11	93.07	97.26	104.70
SOLAR CONTROL; Eclipse						
6 mm thick	115.08	115.08	116.09	97.28	101.65	109.43
FIRE RESISTANT; Pyroshield						
Clear						
6 mm thick	180.83	180.83	182.41	153.36	160.26	172.53
Textured						
7 mm thick	190.04	190.04	191.70	161.21	168.47	181.36
SAFETY; Pyroshield						
Clear						
6 mm thick	266.30	266.30	268.63	226.27	236.46	254.56
Textured						
7 mm thick	276.82	276.82	279.24	235.25	245.83	264.65
FIRE RESISTANT; Pyrodur						
10 mm thick	316.27	316.27	319.03	268.90	281.00	302.51
13 mm thick	321.53	321.53	324.34	273.39	285.69	307.56
FIRE RESISTANT; Pyrostop						
15 mm thick	391.22	391.22	394.64	332.84	347.81	374.44
18 mm thick	417.52	417.52	421.17	355.27	371.26	399.68
21 mm thick	443.82	443.82	447.70	377.71	394.70	424.92
27 mm thick	542.44	542.44	547.18	461.83	482.62	519.56

SPECIAL GLAZING TO WOOD

New Work	Panes < 0.15 m2; glazed with;				Panes > 0.15 < 4.00 m2; glazed with;			
	linseed putty; pins	beads; pinned	beads; screwed	beads & washleather; screwed	linseed putty; pins	beads; pinned	beads; screwed	beads & washleather; screwed
DATA PROTECTION; Clear								
6 mm thick	133.81	139.84	150.54	155.89	129.80	135.65	146.03	151.22
10 mm thick	168.27	175.84	189.30	196.03	163.46	170.81	183.89	190.43
MIRRORS								
Plain								
6 mm thick	55.72	58.23	62.69	64.91	53.53	55.94	60.22	62.36
10 mm thick	93.62	97.83	105.32	109.07	90.54	94.62	101.86	105.48
Bevelled								
6 mm thick	71.57	74.79	80.52	83.38	69.01	72.11	77.63	80.39
10 mm thick	120.84	126.27	135.94	140.77	117.13	122.40	131.77	136.46
Surveillance								
6 mm thick	79.84	83.43	89.82	93.01	77.08	80.55	86.72	89.80
10 mm thick	117.74	123.03	132.45	137.16	114.10	119.24	128.36	132.93
Venetian striped								
6 mm thick	91.38	95.49	102.80	106.46	88.36	92.33	99.40	102.94
10 mm thick	134.96	141.04	151.83	157.23	130.93	136.82	147.29	152.53
Secur-Vu								
6 mm thick	136.11	142.24	153.12	158.57	132.05	137.99	148.55	153.84
10 mm thick	170.56	178.24	191.88	198.71	165.70	173.16	186.41	193.04
SURFACE TREATMENTS; Extra								
Acid etching	105.10	109.83	118.24	122.45	101.76	106.34	114.48	118.55
Stained glass	89.03	93.03	100.15	103.71	86.06	89.93	96.82	100.26
POLYCARBONATE								
4 mm thick	109.70	114.63	123.41	127.80	106.25	111.03	119.53	123.78
6 mm thick	118.88	124.23	133.75	138.50	115.22	120.41	129.63	134.23
10 mm thick	136.11	142.24	153.12	158.57	132.05	137.99	148.55	153.84
POLYCARBONATE; Surface coated								
4 mm thick	136.40	142.54	153.45	158.90	132.33	138.28	148.87	154.16
6 mm thick	147.88	154.54	166.37	172.28	143.55	150.01	161.49	167.23
10 mm thick	169.42	177.04	190.59	197.37	164.58	171.98	185.15	191.73

SPECIAL GLAZING TO METAL

New Work	Panes < 0.15 m2; glazed with;			Panes < 0.15 m2; glazed with;		
	linseed putty; clips	metal beads & compound; screwed	metal beads & rubber gaskets; clip on	linseed putty; clips	metal beads & compound; screwed	metal beads & rubber gaskets; clip on
DATA PROTECTION; Clear						
6 mm thick	133.81	153.22	154.56	129.80	135.65	146.03
10 mm thick	168.27	192.67	194.35	163.46	170.81	183.89
MIRRORS						
Plain						
6 mm thick	55.72	63.80	64.36	53.53	55.94	60.22
10 mm thick	93.62	107.19	108.13	90.54	94.62	101.86
Bevelled						
6 mm thick	71.57	81.95	82.66	69.01	72.11	77.63
10 mm thick	120.84	138.36	139.57	117.13	122.40	131.77
Surveillance						
6 mm thick	79.84	91.41	92.21	77.08	80.55	86.72
10 mm thick	117.74	134.81	135.99	114.10	119.24	128.36
Venetian striped						
6 mm thick	91.38	104.63	105.54	88.36	92.33	99.40
10 mm thick	134.96	154.53	155.88	130.93	136.82	147.29
Secur-Vu						
6 mm thick	136.11	155.85	157.21	132.05	137.99	148.55
10 mm thick	170.56	195.30	197.00	165.70	173.16	186.41
SURFACE TREATMENTS; Extra						
Acid etching	105.10	120.34	121.39	101.76	106.34	114.48
Stained glass	89.03	101.93	102.82	86.06	89.93	96.82
POLYCARBONATE						
4 mm thick	109.70	125.60	126.70	106.25	111.03	119.53
6 mm thick	118.88	136.12	137.31	115.22	120.41	129.63
10 mm thick	136.11	155.85	157.21	132.05	137.99	148.55
POLYCARBONATE; Surface coated						
4 mm thick	136.40	156.18	157.54	132.33	138.28	148.87
6 mm thick	147.88	169.33	170.80	143.55	150.01	161.49
10 mm thick	169.42	193.98	195.67	164.58	171.98	185.15

DOUBLE GLAZING TO WOOD

NOTE;- All double glazing units are hermetically sealed.

New Work	Panes < 0.15 m2; glazed with;				Panes > 0.15 < 4.00 m2; glazed with;			
	linseed putty; pins	beads; pinned	beads; screwed	beads & washleather; screwed	linseed putty; pins	beads; pinned	beads; screwed	beads & washleather; screwed
SEALED UNITS; Clear Glass; 5 - 10 mm space.								
3 mm Inner; 3 mm Outer	64.13	67.01	72.14	74.71	61.74	64.52	69.46	71.93
4 mm Inner; 4 mm Outer	71.96	75.20	80.95	83.83	69.39	72.51	78.06	80.84
5 mm Inner; 4 mm Outer	96.88	101.24	108.99	112.87	93.73	97.95	105.45	109.20
6 mm Inner; 5 mm Outer	114.68	119.84	129.02	133.60	111.12	116.12	125.01	129.45
SEALED UNITS; Clear Glass; 15 - 20 mm space.								
3 mm Inner; 3 mm Outer	79.44	83.01	89.37	92.54	76.69	80.14	86.28	89.35
4 mm Inner; 4 mm Outer	89.23	93.24	100.38	103.95	86.25	90.14	97.04	100.49
5 mm Inner; 4 mm Outer	110.98	115.97	124.85	129.29	107.50	112.34	120.94	125.24
6 mm Inner; 6 mm Outer	131.45	137.36	147.88	153.14	127.50	133.23	143.43	148.53
SEALED UNITS; Solar Glass; 5 - 10 mm space.								
3 mm Inner; 4 mm Solar Outer	122.51	128.03	137.83	142.73	118.77	124.11	133.61	138.36
4 mm Inner; 4 mm Solar Outer	123.94	129.51	139.43	144.39	120.16	125.57	135.18	139.98
5 mm Inner; 6 mm Solar Outer	143.98	150.46	161.97	167.73	139.73	146.02	157.20	162.79
6 mm Inner; 6 mm Solar Outer	150.18	156.94	168.95	174.96	145.79	152.35	164.01	169.84
SEALED UNITS; Solar Glass; 15 - 20 mm space.								
3 mm Inner; 4 mm Solar Outer	140.46	146.78	158.01	163.63	136.29	142.43	153.33	158.78
4 mm Inner; 4 mm Solar Outer	142.09	148.49	159.86	165.54	137.89	144.10	155.13	160.64
5 mm Inner; 6 mm Solar Outer	154.56	161.51	173.88	180.06	150.07	156.82	168.83	174.83
6 mm Inner; 6 mm Solar Outer	172.27	180.02	193.81	200.70	167.37	174.90	188.29	194.98
SEALED UNITS; Safety Glass; 5 - 10 mm space.								
4 mm Inner; 6 mm Safety Outer	105.25	109.98	118.40	122.61	101.90	106.49	114.64	118.72
5 mm Inner; 6 mm Safety Outer	98.13	102.54	110.39	114.32	94.95	99.22	106.82	110.61
6 mm Inner; 6 mm Safety Outer	98.24	102.66	110.52	114.45	95.06	99.34	106.94	110.74
6 mm Inner; 8 mm Safety Outer	139.33	145.60	156.74	162.32	135.19	141.27	152.09	157.50
SEALED UNITS; Safety Glass; 15 - 20 mm space.								
4 mm Inner; 6 mm Safety Outer	120.60	126.03	135.67	140.50	116.90	122.16	131.51	136.19
5 mm Inner; 6 mm Safety Outer	112.41	117.47	126.46	130.96	108.90	113.80	122.51	126.87
6 mm Inner; 6 mm Safety Outer	105.39	110.14	118.57	122.78	102.04	106.64	114.80	118.88
6 mm Inner; 8 mm Safety Outer	159.79	166.98	179.77	186.16	155.18	162.16	174.58	180.78

DOUBLE GLAZING TO METAL

New Work	Panes < 0.15 m2; glazed with;			Panes < 0.15 m2; glazed with;		
	linseed putty; clips	metal beads & compound; screwed	metal beads & rubber gaskets; clip on	linseed putty; clips	metal beads & compound; screwed	metal beads & rubber gaskets; clip on
SEALED UNITS; Clear Glass; 5 - 10 mm space.						
3 mm Inner; 3 mm Outer	64.13	73.43	74.07	61.74	64.52	69.46
4 mm Inner; 4 mm Outer	71.96	82.39	83.11	69.39	72.51	78.06
5 mm Inner; 4 mm Outer	96.88	110.93	111.90	93.73	97.95	105.45
6 mm Inner; 5 mm Outer	114.68	131.31	132.46	111.12	116.12	125.01
SEALED UNITS; Clear Glass; 15 - 20 mm space.						
3 mm Inner; 3 mm Outer	79.44	90.95	91.75	76.69	80.14	86.28
4 mm Inner; 4 mm Outer	89.23	102.16	103.06	86.25	90.14	97.04
5 mm Inner; 4 mm Outer	110.98	127.07	128.18	107.50	112.34	120.94
6 mm Inner; 6 mm Outer	131.45	150.51	151.82	127.50	133.23	143.43
SEALED UNITS; Solar Glass; 5 - 10 mm space.						
3 mm Inner; 4 mm Solar Outer	122.51	140.28	141.50	118.77	124.11	133.61
4 mm Inner; 4 mm Solar Outer	123.94	141.91	143.15	120.16	125.57	135.18
5 mm Inner; 6 mm Solar Outer	143.98	164.85	166.29	139.73	146.02	157.20
6 mm Inner; 6 mm Solar Outer	150.18	171.96	173.46	145.79	152.35	164.01
SEALED UNITS; Solar Glass; 15 - 20 mm space.						
3 mm Inner; 4 mm Solar Outer	140.46	160.82	162.23	136.29	142.43	153.33
4 mm Inner; 4 mm Solar Outer	142.09	162.70	164.12	137.89	144.10	155.13
5 mm Inner; 6 mm Solar Outer	154.56	176.97	178.52	150.07	156.82	168.83
6 mm Inner; 6 mm Solar Outer	172.27	197.25	198.97	167.37	174.90	188.29
SEALED UNITS; Safety Glass; 5 - 10 mm space.						
4 mm Inner; 6 mm Safety Outer	105.25	120.51	121.56	101.90	106.49	114.64
5 mm Inner; 6 mm Safety Outer	98.13	112.36	113.34	94.95	99.22	106.82
6 mm Inner; 6 mm Safety Outer	98.24	112.49	113.47	95.06	99.34	106.94
6 mm Inner; 8 mm Safety Outer	139.33	159.53	160.92	135.19	141.27	152.09
SEALED UNITS; Safety Glass; 15 - 20 mm space.						
4 mm Inner; 6 mm Safety Outer	120.60	138.09	139.29	116.90	122.16	131.51
5 mm Inner; 6 mm Safety Outer	112.41	128.71	129.84	108.90	113.80	122.51
6 mm Inner; 6 mm Safety Outer	105.39	120.67	121.73	102.04	106.64	114.80
6 mm Inner; 8 mm Safety Outer	159.79	182.96	184.56	155.18	162.16	174.58

PERFORMANCE APPLICATIONS

BEST USE

For economic use and optimum performance the Manufacturers suggest specification of these finishes in the following areas.
See page 21 for Key Area References.

Finish	Activity Area
Emulsion	All areas except H5:
Eggshell	All areas:
Oil	Walls; H5-8; T6,9,10: All wood:
Linseed Oil	All hardwood:
Varnish	All internal hardwood:
Wood Preservative	All external woodwork:
Wax Polish	M1-3; H1-4; S1:
French Polish	M1-2; H3:
Cement	M5-6:
Masonary	External brickwork or rendering:
Artex	M1-4; H6-8; T6-13:
Plastic	H5; T12-13; S2:
Intumescent	Fire protection where required:
Wall Paper	M1-3; H1-4:

WORKING WITH OLD DECORATIONS

Preparing old painted surfaces for re-decoration.

	Unit	Walls	Ceilings
Prepare old painted surfaces for re-decoration Wash down; remove loose paint/ paper; fill cracks.			
Whitewash; two coats sealant	m2	2.40	2.70
Emulsion paint	m2	0.81	0.91
Oil bound paint	m2	0.74	0.83
Wall paper	m2	0.60	0.68
Remove old paintwork Remove all old paint; clean down; fill cracks.			
Whitewash	m2	3.62	4.07
Emulsion paint	m2	1.80	2.02
Oil bound paint	m2	4.42	4.97
Oil bound; burning off	m2	6.08	6.84
Remove old wall paper Remove all paper and linings; clean down; fill cracks.			
Standard			
one layer	m2	1.45	1.63
two layers	m2	1.67	1.88
Embossed			
one layer	m2	2.14	2.41
two layers	m2	2.36	2.66
Lincrusta dado lining	m2	2.97	3.34

EMULSION - Matt or Silk Vinyl

- WALLS;
- Unpainted; smooth background;
- Clean down; roller applications.

	Unit	New	Refurb
1st Application; one mist coat.			
General areas			
> 300 mm wide	m2	0.38	0.43
< 150 mm wide	m	0.07	0.08
> 150 < 300 mm wide	m	0.12	0.14
Isolated areas			
< 1 m2	m2	0.44	0.50
Rails or Pipes			
< 50 mm diameter	m	0.22	0.25
2nd Application; one full coat.			
General areas			
> 300 mm wide	m2	1.18	1.33
< 150 mm wide	m	0.23	0.26
> 150 < 300 mm wide	m	0.37	0.41
Isolated areas			
< 1 m2	m2	1.36	1.53
Rails or Pipes			
< 50 mm diameter	m	0.69	0.77
3rd Application; one full coat.			
General areas			
> 300 mm wide	m2	0.97	1.10
< 150 mm wide	m	0.19	0.21
> 150 < 300 mm wide	m	0.30	0.34
Isolated areas			
< 1 m2	m2	1.12	1.26
Rails or Pipes			
< 50 mm diameter	m	0.57	0.64

Properties *Management*

60 Maintenance Comprehensive

EMULSION - Matt or Silk Vinyl

- CEILINGS;
- Unpainted; smooth background;
- Clean down; roller applications.

	Unit	New	Refurb
1st Application; one mist coat.			
General areas			
> 300 mm wide	m2	0.41	0.46
< 150 mm wide	m	0.08	0.09
> 150 < 300 mm wide	m	0.13	0.15
Isolated areas			
< 1 m2	m2	0.48	0.53
Rails or Pipes			
< 50 mm diameter	m	0.24	0.27
2nd Application; one full coat.			
General areas			
> 300 mm wide	m2	1.27	1.43
< 150 mm wide	m	0.25	0.28
> 150 < 300 mm wide	m	0.40	0.45
Isolated areas			
< 1 m2	m2	1.46	1.64
Rails or Pipes			
< 50 mm diameter	m	0.74	0.83
3rd Application; one full coat.			
General areas			
> 300 mm wide	m2	1.05	1.18
< 150 mm wide	m	0.20	0.23
> 150 < 300 mm wide	m	0.33	0.37
Isolated areas			
< 1 m2	m2	1.20	1.36
Rails or Pipes			
< 50 mm diameter	m	0.61	0.68

Properties *Management*

70 Maintenance Comprehensive

RATE ADJUSTMENTS

Additional cost of carrying out work in varied locations.

Internally	Key Rates
Externally	+7.50%
Staircase Areas	+5.00%
Heights >3.5 m	+6.00%

Additional cost of applications on varied surfaces.

	1st App'n	2nd App'n	3rd App'n
Plaster; plasterboard; wood	Key Rates (no adjustment)		
Smooth concrete	+6.50%	+2.00%	+0.50%
Coarse rendering	+8.00%	+2.50%	+0.75%
Tyrolean rendering	+5.50%	+1.50%	+0.40%
Coarse concrete	+7.50%	+2.25%	+0.55%
Brick or blockwork	+9.50%	+3.25%	+0.75%
Rough cast rendering	+12.00%	+4.50%	+0.15%

Additional cost of colour variations.

1 colour application	Key Rates
2 colour applications	+ 7.50%
> 2 colour applications	+ 5.00%

TYPICAL EXAMPLE

a) Select Key Rate	Page	Unit	£
Emulsion; Matt; > 300 mm wide; Ceilings.			
1st application	182	m2	0.46
2nd application	182	m2	1.43
3rd application	182	m2	1.18
b) Prepare old paint surface			
Old emulsion painted	182	m2	0.91
Total Key Rate			3.98
Additional allowances:			
c) Location; staircase	183		
d) Back ground; render	183	+ 8%	
e) Multi-colours; >2 colours	183		
f) Material prices	107		
g) Historic restoration	107		
h) Size of Contract	108	+ 2%	
i) Regional Variation	12	+ 2%	
Total Variation %		+ 12%	0.48
ADJUSTED UNIT RATE			**4.46**

NOTE:- To adjust your unit rates to take account of these special conditions refer to the appropriate pages listed above.

EGGSHELL - Matt or Silk Vinyl

- WALLS; High quality finish;
- Water or oil based paint;
- Unpainted; smooth background;
- Clean down; roller applications.

	Unit	New	Refurb
1st Application; one mist coat.			
General areas			
> 300 mm wide	m2	0.40	0.45
< 150 mm wide	m	0.08	0.09
> 150 < 300 mm wide	m	0.13	0.14
Isolated areas			
< 1 m2	m2	0.46	0.52
Rails or Pipes			
< 50 mm diameter	m	0.23	0.26
2nd Application; one full coat.			
General areas			
> 300 mm wide	m2	1.23	1.38
< 150 mm wide	m	0.24	0.27
> 150 < 300 mm wide	m	0.38	0.43
Isolated areas			
< 1 m2	m2	1.42	1.59
Rails or Pipes			
< 50 mm diameter	m	0.72	0.80
3rd Application; one full coat.			
General areas			
> 300 mm wide	m2	1.02	1.14
< 150 mm wide	m	0.20	0.22
> 150 < 300 mm wide	m	0.32	0.36
Isolated areas			
< 1 m2	m2	1.17	1.31
Rails or Pipes			
< 50 mm diameter	m	0.59	0.66

Properties *Management*

70 Maintenance Comprehensive

EGGSHELL - Matt or Silk Vinyl

- CEILINGS; High quality finish;
- Water or oil based paint;
- Unpainted; smooth background;
- Clean down; roller applications.

	Unit	New	Refurb
1st Application; one mist coat.			
General areas			
> 300 mm wide	m2	0.42	0.47
< 150 mm wide	m	0.08	0.09
> 150 < 300 mm wide	m	0.13	0.15
Isolated areas			
< 1 m2	m2	0.48	0.55
Rails or Pipes			
< 50 mm diameter	m	0.24	0.28
2nd Application; one full coat.			
General areas			
> 300 mm wide	m2	1.29	1.46
< 150 mm wide	m	0.25	0.28
> 150 < 300 mm wide	m	0.40	0.45
Isolated areas			
< 1 m2	m2	1.49	1.67
Rails or Pipes			
< 50 mm diameter	m	0.75	0.85
3rd Application; one full coat.			
General areas			
> 300 mm wide	m2	1.07	1.20
< 150 mm wide	m	0.21	0.23
> 150 < 300 mm wide	m	0.33	0.38
Isolated areas			
< 1 m2	m2	1.23	1.38
Rails or Pipes			
< 50 mm diameter	m	0.62	0.70

Properties *Management*

80 Maintenance Comprehensive

OIL PAINT

- **WALLS**;
- Unpainted; smooth background;
- Clean down; brush applications.

	Unit	New	Refurb
1st Application; One coat primer.			
General areas			
> 300 mm wide	m2	1.77	1.99
< 150 mm wide	m	0.35	0.39
> 150 < 300 mm wide	m	0.55	0.62
Isolated areas			
< 1 m2	m2	2.04	2.29
Cornices; moulded rails			
< 300 mm girth	m	0.69	0.77
2nd Application; One coat undercoat.			
General areas			
> 300 mm wide	m2	1.39	1.56
< 150 mm wide	m	0.27	0.30
> 150 < 300 mm wide	m	0.43	0.49
Isolated areas			
< 1 m2	m2	1.59	1.79
Cornices; moulded rails			
< 300 mm girth	m	0.54	0.60
3rd Application; One coat gloss paint.			
General areas			
> 300 mm wide	m2	1.25	1.40
< 150 mm wide	m	0.24	0.27
> 150 < 300 mm wide	m	0.39	0.44
Isolated areas			
< 1 m2	m2	1.44	1.62
Cornices; moulded rails			
< 300 mm girth	m	0.48	0.54

Properties *Management* 80 Maintenance Comprehensive

OIL PAINT

- **CEILINGS;**
- Unpainted; smooth background;
- Clean down; brush applications.

	Unit	New	Refurb
1st Application; One coat primer.			
General areas			
> 300 mm wide	m2	1.90	2.14
< 150 mm wide	m	0.37	0.42
> 150 < 300 mm wide	m	0.59	0.67
Isolated areas			
< 1 m2	m2	2.19	2.46
Cornices; moulded rails			
< 300 mm girth	m	0.74	0.83
2nd Application; One coat undercoat.			
General areas			
> 300 mm wide	m2	1.49	1.68
< 150 mm wide	m	0.29	0.33
> 150 < 300 mm wide	m	0.47	0.52
Isolated areas			
< 1 m2	m2	1.71	1.93
Cornices; moulded rails			
< 300 mm girth	m	0.58	0.65
3rd Application; One coat gloss paint.			
General areas			
> 300 mm wide	m2	1.34	1.51
< 150 mm wide	m	0.26	0.29
> 150 < 300 mm wide	m	0.42	0.47
Isolated areas			
< 1 m2	m2	1.54	1.74
Cornices; moulded rails			
< 300 mm girth	m	0.52	0.59

Properties *Management* 80 Maintenance Comprehensive

OIL PAINT

- **WOODWORK;**
- Unpainted; smooth background;
- Clean down; roller applications.

	Unit	New	Refurb
1st Application; Treat knots; fill; one coat primer.			
General areas			
> 300 mm wide	m2	1.76	1.98
< 150 mm wide	m	0.34	0.39
> 150 < 300 mm wide	m	0.55	0.62
Glazed windows and doors			
panes < 0.5 m2	m2	2.04	2.29
panes > 0.5 < 1.0 m2	m2	1.89	2.13
panes >1.0 m2	m2	1.58	1.78
2nd Application; One undercoat.			
General areas			
> 300 mm wide	m2	1.32	1.48
< 150 mm wide	m	0.26	0.29
> 150 < 300 mm wide	m	0.41	0.46
Glazed windows and doors			
panes < 0.5 m2	m2	1.65	1.86
panes > 0.5 < 1.0 m2	m2	1.37	1.54
panes >1.0 m2	m2	1.19	1.34
3rd Application; One coat gloss paint.			
General areas			
> 300 mm wide	m2	1.23	1.39
< 150 mm wide	m	0.24	0.27
> 150 < 300 mm wide	m	0.39	0.43
Glazed windows and doors			
panes < 0.5 m2	m2	1.53	1.72
panes > 0.5 < 1.0 m2	m2	1.45	1.63
panes >1.0 m2	m2	1.05	1.18

Properties *Management* 90 Maintenance Comprehensive

RATE ADJUSTMENTS

Additional cost of carrying out work in varied locations.	
Internally	Key Rates (no adjustment)
Externally	+7.50%
Staircase Areas	+5.00%
Heights >3.5 m	+6.00%
Historic Restoration	+15.00%

Additional cost of applications on varied surfaces.

	1st App'n	2nd App'n	3rd App'n
Plaster; plasterboard; wood;	Key Rates (no adjustment)		
Smooth concrete	+6.50%	+2.00%	+0.50%
Coarse rendering	+8.00%	+2.50%	+0.75%
Tyrolean rendering	+5.50%	+1.50%	+0.40%
Coarse concrete	+7.50%	+2.25%	+0.55%
Brick or blockwork	+9.50%	+3.25%	+0.75%
Rough cast rendering	+12.00%	+4.50%	+0.15%

Additional cost of colour variations.

1 colour application	Key Rates (no adjustment)
2 colour applications	+ 7.50%
> 2 colour applications	+ 5.00%

TYPICAL EXAMPLE

a) Select Key Rate	Page	Unit	£
Oil paint; Gloss; > 300 mm wide; Woodwork.			
1st application	184	m2	1.98
2nd application	184	m2	1.48
3rd application	184	m2	1.39
b) Prepare old paint surface			
Old oil paint	182	m2	4.97
Total Key Rate			9.82
Additional allowances:			
c) Location; staircase	185		
d) Back ground; render	185	+ 8%	
e) Multi-colours; >2 colours	185		
f) Material prices	107		
g) Historic restoration	107		
h) Size of Contract	108	+ 2%	
i) Regional Variation	12	+ 2%	
Total Variation %		+ 12%	1.18
ADJUSTED UNIT RATE			**11.00**

NOTE:- To adjust your unit rates to take account of these special conditions refer to the appropriate pages listed above.

OIL PAINT

- **METALWORK;**
- Unpainted; smooth background;
- Clean down; roller applications.

	Unit	New	Refurb
1st Application; One coat zinc phosphate primer.			
General areas			
> 300 mm wide	m2	1.70	1.91
< 150 mm girth	m	0.33	0.37
> 150 < 300 mm girth	m	0.53	0.60
Glazed windows and doors			
panes < 0.5 m2	m2	1.97	2.22
panes > 0.5 < 1.0 m2	m2	1.83	2.06
panes >1.0 m2	m2	1.53	1.72
2nd Application; One undercoat.			
General areas			
> 300 mm wide	m2	1.27	1.43
< 150 mm girth	m	0.25	0.28
> 150 < 300 mm girth	m	0.40	0.45
Glazed windows and doors			
panes < 0.5 m2	m2	1.60	1.79
panes > 0.5 < 1.0 m2	m2	1.32	1.49
panes >1.0 m2	m2	1.15	1.30
3rd Application; One coat gloss paint.			
General areas			
> 300 mm wide	m2	1.19	1.34
< 150 mm girth	m	0.23	0.26
> 150 < 300 mm girth	m	0.37	0.42
Glazed windows and doors			
panes < 0.5 m2	m2	1.48	1.66
panes > 0.5 < 1.0 m2	m2	1.40	1.58
panes >1.0 m2	m2	1.01	1.14

Properties *Management*

90 Maintenance Comprehensive

OIL PAINT

- **RADIATORS AND PIPEWORK;**
- Unpainted; smooth background;
- Clean down; roller applications.

	Unit	New	Refurb
1st Application; One coat chromate primer.			
General surfaces radiators			
> 300 mm wide	m2	1.70	1.91
< 150 mm girth	m	0.33	0.37
> 150 < 300 mm girth	m	0.53	0.60
Isolated areas			
< 1 m2	m2	1.95	2.20
Rails or Pipes			
< 50 mm diameter	m	0.99	1.11
2nd Application; One undercoat.			
General surfaces radiators			
> 300 mm wide	m2	1.27	1.43
< 150 mm girth	m	0.25	0.28
> 150 < 300 mm girth	m	0.40	0.45
Isolated areas			
< 1 m2	m2	1.46	1.64
Rails or Pipes			
< 50 mm diameter	m	0.74	0.83
3rd Application; One coat gloss paint.			
General surfaces radiators			
> 300 mm wide	m2	1.19	1.34
< 150 mm girth	m	0.23	0.26
> 150 < 300 mm girth	m	0.37	0.42
Isolated areas			
< 1 m2	m2	1.37	1.54
Rails or Pipes			
< 50 mm diameter	m	0.69	0.78

Properties *Management*

90 Maintenance Comprehensive

LINSEED OIL

- Linseed oil;
- Unpainted smooth background;
- Rub down with wire wool; brush applications.

	Unit	New	Refurb
1st Application; Rub down with wire wool.			
General areas			
> 300 mm wide	m2	0.59	0.67
< 150 mm wide	m	0.59	0.67
> 150 < 300 mm wide	m	0.19	0.21
Isolated areas			
< 1 m2	m2	0.68	0.77
Cornices; moulded rails			
< 300 mm girth	m	0.23	0.26
2nd Application; One coat linseed oil; rub down.			
General areas			
> 300 mm wide	m2	0.55	0.62
< 150 mm wide	m	0.11	0.12
> 150 < 300 mm wide	m	0.17	0.19
Isolated areas			
< 1 m2	m2	0.63	0.71
Cornices; moulded rails			
< 300 mm girth	m	0.21	0.24
3rd Application; One coat linseed oil; rub down.			
General areas			
> 300 mm wide	m2	0.54	0.60
< 150 mm wide	m	0.10	0.12
> 150 < 300 mm wide	m	0.17	0.19
Isolated areas			
< 1 m2	m2	0.62	0.69
Cornices; moulded rails			
< 300 mm girth	m	0.21	0.23

Properties *Management*

60 Maintenance Comprehensive

VARNISH

- Polyurethane paint or varnish; satin or matt;
- Unpainted smooth background;
- Rub down with wire wool; brush applications.

	Unit	New	Refurb
1st Application; Rub down; one coat varnish.			
General areas			
> 300 mm wide	m2	1.19	1.34
< 150 mm wide	m	0.23	0.26
> 150 < 300 mm wide	m	0.37	0.42
Isolated areas			
< 1 m2	m2	1.37	1.54
Cornices; moulded rails			
< 300 mm girth	m	0.46	0.52
2nd Application; One coat varnish; rub down.			
General areas			
> 300 mm wide	m2	1.10	1.24
< 150 mm wide	m	0.21	0.24
> 150 < 300 mm wide	m	0.34	0.39
Isolated areas			
< 1 m2	m2	1.27	1.43
Cornices; moulded rails			
< 300 mm girth	m	0.43	0.48
3rd Application; One coat varnish; rub down.			
General areas			
> 300 mm wide	m2	1.07	1.21
< 150 mm wide	m	0.21	0.24
> 150 < 300 mm wide	m	0.34	0.38
Isolated areas			
< 1 m2	m2	1.24	1.39
Cornices; moulded rails			
< 300 mm girth	m	0.42	0.47

Properties *Management*

70 Maintenance Comprehensive

WOOD PRESERVATIVE

- Preservative; coloured;
- Unpainted smooth background;
- Clean down; brush applications.

	Unit	New	Refurb
1st Application; One coat wood preservative.			
General areas			
> 300 mm wide	m2	2.69	3.02
< 150 mm wide	m	0.52	0.59
> 150 < 300 mm wide	m	0.84	0.95
Glazed windows and doors			
panes < 0.5 m2	m2	2.55	2.87
panes > 0.5 < 1.0 m2	m2	2.43	2.73
panes >1.0 m2	m2	2.17	2.44
2nd Application; One coat wood preservative.			
General areas			
> 300 mm wide	m2	2.42	2.72
< 150 mm wide	m	0.47	0.53
> 150 < 300 mm wide	m	0.76	0.85
Glazed windows and doors			
panes < 0.5 m2	m2	2.30	2.59
panes > 0.5 < 1.0 m2	m2	2.18	2.46
panes >1.0 m2	m2	1.95	2.20
Fire Retardent treatment; Application in two coats.			
General areas			
> 300 mm wide	m2	5.46	6.14
< 150 mm wide	m	1.06	1.20
> 150 < 300 mm wide	m	1.71	1.92
Glazed windows and doors			
panes < 0.5 m2	m2	5.19	5.83
panes > 0.5 < 1.0 m2	m2	4.93	5.54
panes >1.0 m2	m2	4.41	4.96

Properties *Management*

100 Maintenance Comprehensive

RATE ADJUSTMENTS

Additional cost of carrying out work in varied locations.	
Internally	Key Rates (no adjustment)
Externally	+7.50%
Staircase Areas	+5.00%
Heights >3.5 m	+6.00%
Historic restoration	+15.00%

Additional cost of applications on varied surfaces.

	1st App'n	2nd App'n	3rd App'n
Plaster; plasterboard; wood	Key Rates (no adjustment)		
Smooth concrete	+6.50%	+2.00%	+0.50%
Coarse rendering	+8.00%	+2.50%	+0.75%
Tyrolean rendering	+5.50%	+1.50%	+0.40%
Coarse concrete	+7.50%	+2.25%	+0.55%
Brick or blockwork	+9.50%	+3.25%	+0.75%
Rough cast rendering	+12.00%	+4.50%	+0.15%

Additional cost of colour variations.

1 colour application	Key Rates (no adjustment)
2 colour applications	+ 7.50%
> 2 colour applications	+ 5.00%

TYPICAL EXAMPLE

a) Select Key Rate	Page	Unit	£
Varnish; Gloss; > 300 mm wide; Woodwork.			
1st application	186	m2	1.34
2nd application	186	m2	1.24
3rd application	186	not required	
b) Remove old paint surface			
Oil bound; burning off.	182	m2	6.84
Total Key Rate			9.42
Additional allowances:			
c) Location; staircase	187		
d) Back ground; render	187	+ 8%	
e) Multi-colours; >2 colours	187		
f) Material prices	107		
g) Historic restoration	107		
h) Size of Contract	108	+ 2%	
i) Regional Variation	12	+ 2%	
Total Variation %		+ 12%	1.13
ADJUSTED UNIT RATE			**10.55**

NOTE:- To adjust your unit rates to take account of these special conditions refer to the appropriate pages listed above.

WAX POLISH

- Rub down with wire wool;
- Unpainted; smooth background;
- Hand application.

	Unit	New	Refurb
1st Application; Stain and rub down.			
General areas			
> 300 mm wide	m2	2.49	2.80
< 150 mm girth	m	0.49	0.55
> 150 < 300 mm girth	m	0.78	0.88
Glazed windows and doors			
panes < 0.5 m2	m2	2.37	2.66
panes > 0.5 < 1.0 m2	m2	2.25	2.53
panes >1.0 m2	m2	2.01	2.26
2nd Application; Fill & apply one coat wax; rub down.			
General areas			
> 300 mm wide	m2	4.46	5.02
< 150 mm girth	m	0.87	0.98
> 150 < 300 mm girth	m	1.39	1.57
Glazed windows and doors			
panes < 0.5 m2	m2	4.24	4.77
panes > 0.5 < 1.0 m2	m2	4.03	4.53
panes >1.0 m2	m2	3.60	4.05
3rd Application; Apply one coat wax; rub down.			
General areas			
> 300 mm wide	m2	4.24	4.77
< 150 mm girth	m	0.83	0.93
> 150 < 300 mm girth	m	1.32	1.49
Glazed windows and doors			
panes < 0.5 m2	m2	4.03	4.53
panes > 0.5 < 1.0 m2	m2	3.83	4.30
panes >1.0 m2	m2	3.42	3.85

Properties *Management* 80 Maintenance Comprehensive

FRENCH POLISH

- Rub down with wire wool;
- Unpainted; smooth background;
- Multiple applications by hand.

	Unit	New	Refurb
1st Application; Stain and rub down.			
General surfaces radiators			
> 300 mm wide	m2	4.12	4.63
< 150 mm girth	m	0.80	0.90
> 150 < 300 mm girth	m	1.29	1.45
Isolated areas			
< 1 m2	m2	3.91	4.40
Rails or Pipes			
< 50 mm diameter	m	3.33	3.74
2nd Application; Fill and French polish; rub down.			
General surfaces radiators			
> 300 mm wide	m2	9.13	10.27
< 150 mm girth	m	1.78	2.00
> 150 < 300 mm girth	m	2.85	3.21
Isolated areas			
< 1 m2	m2	8.67	9.76
Rails or Pipes			
< 50 mm diameter	m	7.37	8.29
3rd Application; Fill and French polish; rub down.			
General surfaces radiators			
> 300 mm wide	m2	8.67	9.76
< 150 mm girth	m	1.69	1.90
> 150 < 300 mm girth	m	2.71	3.05
Isolated areas			
< 1 m2	m2	8.24	9.27
Rails or Pipes			
< 50 mm diameter	m	7.00	7.88

Properties *Management* 90 Maintenance Comprehensive

CEMENT PAINT

- Cement paint; wall surfaces;
- Unpainted; smooth background;
- Clean down; brush applications.

	Unit	New	Refurb
1st Application; Clean down; one coat Sandtex.			
General areas			
> 300 mm wide	m2	1.81	2.04
< 150 mm wide	m	1.81	2.04
> 150 < 300 mm wide	m	0.57	0.64
Isolated areas			
< 1 m2	m2	2.08	2.34
Cornices; moulded rails			
< 300 mm wide	m	0.70	0.79
2nd Application; One coat Sandtex.			
General areas			
> 300 mm wide	m2	1.49	1.68
< 150 mm wide	m	0.29	0.33
> 150 < 300 mm wide	m	0.47	0.52
Isolated areas			
< 1 m2	m2	1.72	1.93
Cornices; moulded rails			
< 300 mm wide	m	0.58	0.65
3rd Application; One coat Sandtex.			
General areas			
> 300 mm wide	m2	1.27	1.43
< 150 mm wide	m	0.25	0.28
> 150 < 300 mm wide	m	0.40	0.45
Isolated areas			
< 1 m2	m2	1.46	1.65
Cornices; moulded rails			
< 300 mm wide	m	0.49	0.56

Properties *Management*

50 Maintenance Comprehensive

MASONRY PAINT

- Masonary paint; wall surfaces;
- Unpainted; smooth background;
- Clean down; brush applications.

	Unit	New	Refurb
1st Application; Clean down; One coat sealer.			
General areas			
> 300 mm wide	m2	2.62	2.95
< 150 mm wide	m	0.51	0.58
> 150 < 300 mm wide	m	0.82	0.92
Isolated areas			
< 1 m2	m2	3.02	3.39
Cornices; moulded rails			
< 300 mm wide	m	1.02	1.14
2nd Application; One coat Weathershield paint.			
General areas			
> 300 mm wide	m2	2.87	3.23
< 150 mm wide	m	0.56	0.63
> 150 < 300 mm wide	m	0.90	1.01
Isolated areas			
< 1 m2	m2	3.30	3.71
Cornices; moulded rails			
< 300 mm wide	m	1.11	1.25
3rd Application; One coat Weathershield paint.			
General areas			
> 300 mm wide	m2	2.68	3.02
< 150 mm wide	m	0.52	0.59
> 150 < 300 mm wide	m	0.84	0.94
Isolated areas			
< 1 m2	m2	3.09	3.47
Cornices; moulded rails			
< 300 mm wide	m	1.04	1.17

Properties *Management*

80 Maintenance Comprehensive

ARTEX

- Clean down surfaces ready for decoration;
- One coat Artex sealer;
- Artex compound; standard stipple finish.

	Unit	New	Refurb
1st Application; Clean down.			
General areas			
> 300 mm wide	m2	0.38	0.43
< 150 mm wide	m	0.07	0.08
> 150 < 300 mm wide	m	0.12	0.14
Isolated areas			
< 1 m2	m2	0.44	0.50
Cornices; moulded rails			
< 300 mm wide	m	0.15	0.17
2nd Application; One coat Artex; standard pattern.			
General areas			
> 300 mm wide	m2	2.36	2.66
< 150 mm wide	m	0.46	0.52
> 150 < 300 mm wide	m	0.74	0.83
Isolated areas			
< 1 m2	m2	2.71	3.05
Cornices; moulded rails			
< 300 mm wide	m	0.91	1.03
2nd Application; One coat Artex; designer pattern.			
General areas			
> 300 mm wide	m2	2.71	3.05
< 150 mm wide	m	0.53	0.60
> 150 < 300 mm wide	m	0.85	0.95
Isolated areas			
< 1 m2	m2	3.12	3.51
Cornices; moulded rails			
< 300 mm wide	m	0.91	1.03

Properties *Management*

80 Maintenance Comprehensive

RATE ADJUSTMENTS

Additional cost of carrying out work in varied locations.	
Internally	Key Rates (no adjustment)
Externally	+7.50%
Staircase Areas	+5.00%
Heights >3.5 m	+6.00%
Historic Restoration	+15.00%

Additional cost of applications on varied surfaces.	1st App'n	2nd App'n	3rd App'n
Plaster; plasterboard; wood;	Key Rates (no adjustment)		
Smooth concrete	+6.50%	+2.00%	+0.50%
Coarse rendering	+8.00%	+2.50%	+0.75%
Tyrolean rendering	+5.50%	+1.50%	+0.40%
Coarse concrete	+7.50%	+2.25%	+0.55%
Brick or blockwork	+9.50%	+3.25%	+0.75%
Rough cast rendering	+12.00%	+4.50%	+0.15%

Additional cost of colour variations.	
1 colour application	Key Rates (no adjustment)
2 colour applications	+ 7.50%
> 2 colour applications	+ 5.00%

TYPICAL EXAMPLE

	Page	Unit	£
a) Select Key Rate			
Intumescent paint; metalwork.			
1st application	189	m2	2.09
2nd application	189	m2	14.60
3rd application	189	m2	10.79
b) Remove old paint surface (Not applicable)	182	m2	
Total Key Rate			27.48
Additional allowances:			
c) Location; staircase	189		
d) Back ground; render	189		
e) Multi-colours; >2 colours	189		
f) Material prices	107		
g) Historic restoration	107		
h) Size of Contract	108	+ 1%	
i) Regional Variation	12	+ 4%	
Total Variation %		+ 5%	1.37
ADJUSTED UNIT RATE			**28.85**

NOTE;- To adjust your unit rates to take account of these special conditions refer to the appropriate pages listed above.

PLASTIC PAINT

- Metalwork;
- Unpainted; smooth background;
- Clean down; brush applications.

	Unit	New	Refurb
1st Application; Clean down; one coat sealer.			
General areas			
> 300 mm wide	m2	2.49	2.80
< 150 mm wide	m	0.49	0.55
> 150 < 300 mm wide	m	0.78	0.88
Glazed windows and doors			
panes < 0.5 m2	m2	2.37	2.66
panes > 0.5 < 1.0 m2	m2	2.25	2.53
panes >1.0 m2	m2	2.01	2.26
2nd Application; One coat plastic paint.			
General areas			
> 300 mm wide	m2	4.46	5.02
< 150 mm wide	m	0.87	0.98
> 150 < 300 mm wide	m	1.39	1.57
Glazed windows and doors			
panes < 0.5 m2	m2	4.24	4.77
panes > 0.5 < 1.0 m2	m2	4.03	4.53
panes >1.0 m2	m2	3.60	4.05
3rd Application; One coat plastic paint.			
General areas			
> 300 mm wide	m2	4.24	4.77
< 150 mm wide	m	0.83	0.93
> 150 < 300 mm wide	m	1.32	1.49
Glazed windows and doors			
panes < 0.5 m2	m2	4.03	4.53
panes > 0.5 < 1.0 m2	m2	3.83	4.30
panes >1.0 m2	m2	3.42	3.85

Properties — *Management*: 80 Maintenance Comprehensive

INTUMESCENT PAINT

- Structural Steelwork; Fire Protection;
- Unpainted; smooth background;
- Clean down; spray applications.

	Unit	New	Refurb
1st Application; Rub down; prime.			
General surfaces steelwork			
> 300 mm wide	m2	1.86	2.09
< 150 mm girth	m	0.36	0.41
> 150 < 300 mm girth	m	0.58	0.65
Isolated areas			
< 1 m2	m2	2.14	2.41
Rails or Pipes			
< 50 mm diameter	m	0.39	0.44
2nd Application; One coat intumescent paint.			
General surfaces steelwork			
> 300 mm wide	m2	12.98	14.60
< 150 mm girth	m	2.53	2.85
> 150 < 300 mm girth	m	4.06	4.56
Isolated areas			
< 1 m2	m2	14.92	16.79
Rails or Pipes			
< 50 mm diameter	m	2.71	3.05
3rd Application; One coat intumescent paint.			
General surfaces steelwork			
> 300 mm wide	m2	9.59	10.79
< 150 mm girth	m	1.87	2.10
> 150 < 300 mm girth	m	3.00	3.37
Isolated areas			
< 1 m2	m2	11.03	12.41
Rails or Pipes			
< 50 mm diameter	m	2.00	2.25

Properties — *Management*: 10 Warranty Materials; 10 Warranty Comprehensive; 100 Maintenance Comprehensive

PERFORMANCE APPLICATIONS

BEST USE

For economic use and optimum performance the Manufacturers suggest specification of these finishes in the following areas.

See page 21 for Key Area References.

Finish	Activity Area
Lining	Better finish on poor surfaces
Cross Lining	Bad surfaces.
Woodchip	Poor surfaces; cheap finish.
Polystyrene	Insulation; avoid condensatio
Standard	M1-4; T11:
Embossed	M1-4; H1-4; T11:
Vinyl	M1-4; H1-4; T11:
Washable	M4; M8; H4; T4:

WORKING WITH OLD DECORATIONS

Preparing old painted surfaces for re-decoration.

	Unit	Walls	Ceilings
Prepare old painted surfaces for re-decoration. Wash down; remove loose paint/paper; fill cracks.			
Whitewash; two coats sealant	m2	2.40	2.70
Emulsion paint	m2	0.81	0.91
Oil bound paint	m2	0.74	0.83
Wall paper	m2	0.60	0.68
Remove old paintwork Remove all old paint; clean down; fill cracks.			
Whitewash	m2	3.62	4.07
Emulsion paint	m2	1.80	2.02
Oil bound paint	m2	4.42	4.97
Oil bound; burning off	m2	6.08	6.84
Remove old wall paper Remove all paper and linings; clean down; fill cracks.			
Standard			
one layer	m2	1.45	1.63
two layers	m2	1.67	1.88
Embossed			
one layer	m2	2.14	2.41
two layers	m2	2.36	2.66
Lincrusta dado lining	m2	2.97	3.34

WALL PAPER

(Supply and hang wallpaper - Unit rate per m2)

Price	Lining Papers				Wall Paper (5.2 square metres per roll)							
					Standard paper		Embossed paper		Vinyl paper		Washable paper	
Cost per roll	Lining	Cross lining	Wood-chip	Poly-styrene	Small pattern	Large pattern	Small pattern	Large pattern	Small pattern	Large pattern	Small pattern	Large pattern
£ 2.00	2.29	4.45	2.71	-	-	-	-	-	-	-	-	-
£ 3.00	2.54	4.95	2.97	-	3.46	4.24	-	-	-	-	-	-
£ 4.00	2.77	5.41	3.20	5.55	3.70	4.50	-	-	-	-	-	-
£ 5.00	2.99	5.85	3.43	5.79	3.93	4.74	4.44	-	4.32	5.24	4.37	5.29
£ 6.00	3.22	6.30	3.66	6.06	4.17	4.99	4.69	5.65	4.57	5.50	4.62	5.56
£ 7.00	-	-	3.88	6.29	4.39	5.21	4.91	5.87	4.79	5.72	4.84	5.78
£ 8.00	-	-	4.10	6.53	4.62	5.45	5.14	6.12	5.02	5.96	5.07	6.02
£ 9.00	-	-	4.31	6.76	4.83	5.67	5.36	6.34	5.24	6.19	5.29	6.24
£ 10.00	-	-	4.53	6.98	5.05	5.89	5.58	6.56	5.46	6.41	5.50	6.47
£ 11.00	-	-	-	7.20	5.26	6.10	5.79	6.78	5.67	6.62	5.72	6.68
£ 12.00	-	-	-	7.40	5.46	6.30	6.00	6.98	5.87	6.83	5.92	6.88
£ 13.00	-	-	-	-	5.67	6.50	6.20	7.18	6.08	7.03	6.12	7.09
£ 14.00	-	-	-	-	6.00	6.87	6.55	7.58	6.42	7.41	6.47	7.48
£ 15.00	-	-	-	-	6.20	7.07	6.75	7.78	6.62	7.62	6.67	7.68
£ 16.00	-	-	-	-	6.40	7.27	6.95	7.98	6.83	7.82	6.87	7.88
£ 17.00	-	-	-	-	6.67	7.56	7.24	8.29	7.11	8.12	7.16	8.18
£ 18.00	-	-	-	-	6.90	7.80	7.47	8.53	7.34	8.36	7.39	8.42
£ 19.00	-	-	-	-	7.11	8.02	7.69	8.75	7.56	8.58	7.61	8.65
£ 20.00	-	-	-	-	7.33	8.24	7.91	8.97	7.78	8.81	7.82	8.87

RATE ADJUSTMENTS

Additional cost of carrying out work in varied locations.	
Internally	Key Rates
Staircase Areas	+5%
Heights >3.5 m	+6%
Additional cost of applications on varied surfaces.	
Plaster	Key Rates
Plasterboard	Key Rates
Wood boarding	+ 5%
Smooth concrete	+ 5%
Additional cost of colour variations.	
Extended patterns	+ 10%
> 2 finishing wall coverings	+ 4.5%
Patterned borders	1.5% to 2.25%
Restoration work	+ 15%

EXAMPLE

	Page	Unit	£
a) Select Key Rate			
Lining paper; one layer.	191	m2	2.54
Vinyl wall paper	190	m2	5.72
b) Remove old wall paper	190	m2	1.67
Total Key Rate			9.93
Additional allowances:			
c) Location; staircase	190		Nil
d) Back ground; render	190		Nil
e) Multi-colours; >2 colours	190	+ 6%	0.60
f) Material prices	107		Nil
g) Historic restoration	107		Nil
h) Size of Contract	108	+ 1%	0.09
Sub Total			10.62
i) Regional Variation	12	0.85	1.59
ADJUSTED UNIT RATE			**9.03**

LINING PAPERS

- Cutting and hanging; size; adhesive;
- Unpainted; smooth background;
- Clean down; hand applications.

	Unit	New	Refurb
Lining paper; 80 g; one layer; £3.00 per roll delivered to site.			
Walls; ceilings			
> 300 mm wide	m2	2.54	2.86
< 150 mm wide	m	0.52	0.59
> 150 < 300 mm wide	m	0.84	0.94
Isolated areas			
< 1 m2	m2	3.21	3.61
Lining paper; 80 g; two layers; £3.00 per roll delivered to site.			
Walls; ceilings			
> 300 mm wide	m2	4.95	5.57
< 150 mm wide	m	1.02	1.15
> 150 < 300 mm wide	m	1.63	1.83
Isolated areas			
< 1 m2	m2	6.25	7.03
Woodchip paper; standard; one layer; £4.00 per roll delivered to site.			
Walls; ceilings			
> 300 mm wide	m2	3.20	3.60
< 150 mm wide	m	0.66	0.74
> 150 < 300 mm wide	m	1.05	1.19
Isolated areas			
< 1 m2	m2	4.04	4.55
Polystyrene insulation; 2 mm thick; £4.60 per roll delivered to site.			
Walls; ceilings			
> 300 mm wide	m2	5.76	6.48
< 150 mm wide	m	1.19	1.33
> 150 < 300 mm wide	m	1.90	2.13
Isolated areas			
< 1 m2	m2	7.27	8.18

DECORATIVE WALL PAPERS

- Cutting and hanging; size; adhesive;
- Unpainted; smooth background; filling;
- Clean down; hand application.

	Unit	New	Refurb
Normal pulp paper; standard pattern; £6.00 per roll delivered to site.			
Walls; ceilings			
> 300 mm wide	m2	4.17	4.69
< 150 mm wide	m	0.86	0.97
> 150 < 300 mm wide	m	1.37	1.55
Isolated areas			
< 1 m2	m2	5.26	5.92
Embossed paper; standard pattern; £5.00 per roll delivered to site.			
Walls; ceilings			
> 300 mm wide	m2	4.44	5.00
< 150 mm wide	m	0.91	1.03
> 150 < 300 mm wide	m	1.46	1.65
Isolated areas			
< 1 m2	m2	5.61	6.31
Vinyl paper; standard pattern; £7.00 per roll delivered to site.			
Walls; ceilings			
> 300 mm wide	m2	4.79	5.39
< 150 mm wide	m	0.99	1.11
> 150 < 300 mm wide	m	1.58	1.77
Isolated areas			
< 1 m2	m2	6.05	6.80
Vinyl paper; standard pattern; £12.00 per roll delivered to site.			
Walls; ceilings			
> 300 mm wide	m2	5.87	6.61
< 150 mm wide	m	1.21	1.36
> 150 < 300 mm wide	m	1.93	2.18
Isolated areas			
< 1 m2	m2	7.41	8.34

DECORATIVE WALL COVERINGS

- Cutting and hanging; adhesive;
- Unpainted; smooth background; filling;
- Hand application; protection.

	Unit	New	Refurb
Vinyl sheet wall covering; paper backed; £4.50 per m2 delivered to site.			
Walls; ceilings			
> 300 mm wide	m2	9.31	10.47
< 150 mm wide	m	1.99	2.24
> 150 < 300 mm wide	m	3.19	3.59
Isolated areas			
< 1 m2	m2	12.77	14.37
Vinyl sheet wall covering; fabric backed; £5.35 per m2 delivered to site.			
Walls; ceilings			
> 300 mm wide	m2	10.41	11.71
< 150 mm wide	m	2.23	2.51
> 150 < 300 mm wide	m	3.57	4.02
Isolated areas			
< 1 m2	m2	14.29	16.07
Hessian decorative covering; £7.50 per m2 delivered to site.			
Walls; ceilings			
> 300 mm wide	m2	13.43	15.11
< 150 mm wide	m	2.87	3.23
> 150 < 300 mm wide	m	4.61	5.18
Isolated areas			
< 1 m2	m2	18.42	20.73
Textile decorative covering; £15.00 per m2 delivered to site.			
Walls; ceilings			
> 300 mm wide	m2	22.91	25.77
< 150 mm wide	m	4.90	5.52
> 150 < 300 mm wide	m	7.86	8.84
Isolated areas			
< 1 m2	m2	31.43	35.36

TESTING PLUMBING AND DRAINS

Testing — To test the soundness of drainage pipework and sanitary fittings.

i) Water Test
Seal all outlets and charge the pipework and fittings with water, noting the amount the water subsides. This is a very severe test.

ii) Air Test
Seal all outlets and pump air air into the pipework. Measure the reduction in pressure over a period of time. This is the most severe test which can be applied.

iii) Smoke Test
Fill the pipework with smoke using smoke rockets or machines. The defects being are exposed by the escape of the smoke. Very useful for testing existing pipework installations.

iv) Coloured liquids
Carried out by pouringcoloured or volatile liquid into the pipework system. The appearance of stains or fumes help trace the source of leaks and defects.

PIPEWORK

Access pipes — Pipes with screwed or sealed removable fittings to allow access for the removal of blockages and cleaning out the system.

Traps — A 'P' or 'S' shaped bend in pipework or a specially formed fitting filled with water to prevent foul air from the drainage system entering the building.

Bottle trap — A trap in the shape of a bottle comprising a removable container which acts as a 'P' trap and can be removed for cleaning wastes to the sinks, basins and the urinals.

Drain plug —
i) an inflatable rubber bag;
ii) a metal fitting with two discs and an india rubber ring which can be tightened by a wing nut.

Cess Pools — A container of brick, concrete or metal suitable for storage of sewage until emptied. The design requirements are strictly controlled by the Building Regulations.

Anti syphonage pipework — A secondary system of pipes conecting the traps to the ventilation outlets to prevent the water seal in the traps from being broken by pressures which can develop when branches discharge waste water.

FITTINGS

Discharge Stacks — The main drainage pipework in a building for discharging waste from sanitary appliances on the upper floors of a building.

Branches — The pipework connecting the main drainage stack to appliances or ranges of appliances;
The length and size of branches are controlled by the Building Regulations.

Watertightness — Pipework is required by the Building Regulations to withstand an air or smoke test of positive pressure of > 38 mm water gauge for at least 3 minutes.

Pipe Gradients — The fall necessary to carry the flow of waste through the drainage system and which are generally required to meet the following minimum standards;

75 mm pipe	1 in 40
100 mm diameter pipe	1 in 40 to 1 in 80
150 mm diameter pipe	1 in 150

W.C. SUITES

- vitreous china pan; fixing; jointing to drains;
- black plastic cistern; high pressure ball valve;
- flush pipe; connections; overflow;
- plastic seat; cover; chromium plated fittings.

		Unit	New	Refurb
W.C. Suite; high level cistern.				
Key stage 1	(£155.00 per set)	nr	217	245
Key stage 2	(£152.00 per set)	nr	214	241
Key stages 3/4	(£148.00 per set)	nr	208	234
Adult	(£140.00 per set)	nr	196	221
W.C. Suite; low level cistern.				
Key stage 1	(£115.00 per set)	nr	166	187
Key stage 2	(£113.00 per set)	nr	164	185
Key stages 3/4	(£108.00 per set)	nr	157	177
Adult	(£100.00 per set)	nr	145	164
W.C. Suite; concealed cistern.				
Key stage 1	(£105.00 per set)	nr	151	170
Key stage 2	(£102.00 per set)	nr	148	167
Key stages 3/4	(£ 98.00 per set)	nr	142	160
Adult	(£ 92.00 per set	nr	130	147
W.C. Suite; electronic flushing system; (No cistern, flush pipes or overflows).				
Key stage 1	(£80.00 per set)	nr	568	639
Key stage 2	(£80.00 per set)	nr	568	639
Key stages 3/4	(£73.00 per set)	nr	559	629
Adult	(£65.00 per set)	nr	547	616
Seats; extra cost.				
Polished hardwood with cover			55	62

URINALS AND SINKS

- slab urinal; glazed fireclay; floor channel; grating;
- urinal bowl; vitreous enamel; locking strainer;
- cistern; chromium plated flush and sparge pipes;
- sinks; glazed fireclay; chromium plated taps; wastes.

	Unit	New	Refurb
Urinal stalls; slab type; range.			
Key stage 1/2 (£132.00 per set)			
single	nr	195	220
range of 5	nr	1,060	1,195
range of 10	nr	2,115	2,380
Key stage 3/4/Adult (£150.00 per set)			
single	nr	215	245
range of 5	nr	1,060	1,195
range of 10	nr	2,115	2,380
Extra; electronic flushing	nr	140	160
Urinal stalls; bowl type; range.			
Key stage 1 & 2 (£180.00 per set)			
single	nr	250	285
range of 5	nr	1,060	1,195
range of 10	nr	2,115	2,380
Key stage 3, 4 and Adult (£210.00 per set)			
single	nr	285	325
range of 5	nr	1,060	1,195
range of 10	nr	2,115	2,380
Extra; electronic flushing	nr	140	160
Sinks.			
610 x 455 x 255 mm	nr	160	180
760 x 455 x 255 mm	nr	210	240
Bucket sink with grating			
760 x 455 x 255 mm	nr	400	450
Sinks; stainless steel			
Single drainer			
500 x 1050 mm	nr	140	160
Double drainer			
500 x 1550 mm	nr	160	180

WASHBASINS

- basins; vitreous enamelled; chromium plated waste;
- plug and chain; taps; water economy; iron brackets;
- shower trays; slip resistant; waste fitting; bedding;
- thermostatic mixer valve; spray fitting; wall fixing.

	Unit	New	Refurb
Lavatory Basin; single.			
560 x 405 mm	nr	80	90
635 x 455 mm	nr	105	120
Lavatory Basin; range.			
Ranges; basins; 560 x 405 mm			
range of 5	nr	410	465
range of 10	nr	820	925
Ranges; basins; 635 x 455 mm			
range of 5	nr	550	620
range of 10	nr	1,095	1,235
Shower Tray.			
Plastic; coloured			
750 x 750	nr	280	315
800 x 800	nr	305	345
Glazed fireclay; white			
750 x 750	nr	340	385
800 x 800	nr	375	425
Drinking fountain.			
Vitreous china; white			
Key stage 1 & 2	nr	195	220
Key stage 3 & 4	nr	200	225
Adult	nr	215	245
Stainless steel			
Wall fixing	nr	155	175
Projecting wall fixing	nr	170	195
Standard; 855 mm heigh	nr	245	280
Fittings			
Toilet roll holder	nr	10	15
soap dispenser	nr	20	25
paper towel dispenser	nr	30	35

NEW WORK

Size of Pipe	Pipework			Bends				Couplings and Connectors					Tee junctions		Sundries		
diameter	clipped to walls	laid in duct	laid in floor or roof space	made bend	bend coupling	backplate elbow	overflow bend	straight	tank	uPVC to iron	union	expansion	equal	swept	Access door	Cement joint to stoneware	Roof connector - asphalt
	m	m	m	nr	nr	nr	nr	nr	nr	nr	nr	nr	nr	nr	nr	nr	nr
muPVC; solvent welded joints. (Waste Pipes)																	
32 mm	4.39	4.74	5.34	3.60	3.60	-	-	2.60		4.80	5.26	3.74	4.84	4.84	-	-	-
40 mm	5.05	5.46	6.14	4.23	4.23	-	-	3.06		4.97	6.58	4.46	5.92	5.92	-	-	-
50 mm	7.03	7.60	8.55	5.45	5.45	-	-	3.59		6.17	9.03	5.44	8.32	8.32	-	-	-
Polypropylene; push fit joints. (Waste Pipes)																	
32 mm	2.97	3.20	3.60	2.60	2.60	-	-	2.38	-	3.60	3.95	-	3.40	3.40	-	-	-
40 mm	3.07	3.32	3.73	3.22	3.22	-	-	2.80	-	3.73	4.94	-	4.39	4.39	-	-	-
50 mm	4.96	5.36	6.03	4.22	4.22	-	-	3.29	-	4.62	6.77	-	5.41	5.41	-	-	-
uPVC; solvent welded joints. (Overflow pipes)																	
19 mm	3.19	3.44	3.87	2.36	-	-	-	2.18	2.99	2.92	-	-	2.99	2.99	-	-	-
uPVC; solvent welded joints. (Soil and Waste Pipes)																	
82 mm	9.62	10.39	11.69	-	9.58	11.02	-	6.12	-	6.78	-	-	13.12	13.12	13.25	-	-
110 mm	8.92	9.64	10.84	-	10.99	12.63	-	7.47	-	7.24	-	-	15.97	15.97	14.43	-	-
160 mm	5.99	6.47	7.28	-	21.45	24.67	-	8.53	-	8.05	-	-	18.83	18.83	18.83	-	-
Cast iron; socketed joints; caulked; fixing with brackets. (Soil and Waste Pipes)																	
50 mm	25.02	27.02	30.40	18.68	9.58	32.81	-	-	-	6.78	-	-	30.43		30.83	3.77	35.67
75 mm	25.58	27.62	31.08	19.76	10.99	33.40	-	-	-	7.24	-	-	32.47	50.93	31.65	4.81	36.42
100 mm	34.23	36.97	41.59	25.60	21.45	40.51	-	-	-	8.05	-	-	42.75	58.52	34.35	5.85	35.00
150 mm	63.54	68.62	77.20	43.16	21.45	59.43	-	-	-	8.05	-	-	77.46	116.41	53.42	7.24	68.74

RE-MODELLED WORK

Size of Pipe	Pipework			Bends				Couplings and Connectors					Tee junctions		Sundries		
diameter	clipped to walls	laid in duct	laid in floor or roof space	made bend	bend coupling	backplate elbow	overflow bend	straight	tank	uPVC to iron	union	expansion	equal	swept	Access door	Cement joint to stoneware	Roof connector - asphalt
	m	m	m	nr	nr	nr	nr	nr	nr	nr	nr	nr	nr	nr	nr	nr	nr
muPVC; solvent welded joints. (Waste Pipes)																	
32 mm	4.94	5.34	6.00	4.05	4.05	-	-	2.92	-	5.40	5.92	4.21	5.45	5.45	-	-	-
40 mm	5.68	6.14	6.91	4.76	4.76	-	-	3.44	-	5.59	7.41	5.02	6.66	6.66	-	-	-
50 mm	7.91	8.55	9.62	6.13	6.13	-	-	4.04	-	6.94	10.16	6.11	9.36	9.36	-	-	-
Polypropylene; push fit joints. (Waste Pipes)																	
32 mm	3.34	3.60	4.06	2.92	2.92	-	-	2.67	-	4.05	4.44	-	3.82	3.82	-	-	-
40 mm	3.45	3.73	4.20	3.62	3.62	-	-	3.15	-	4.19	5.55	-	4.94	4.94	-	-	-
50 mm	5.58	6.03	6.78	4.75	4.75	-	-	3.70	-	5.20	7.62	-	6.09	6.09	-	-	-
uPVC; solvent welded joints. (Overflow pipes)																	
19 mm	3.59	3.87	4.36	2.66	-	-	-	2.45	3.36	3.29	-	-	3.36	3.36	-	-	-
uPVC; solvent welded joints. (Soil and Waste Pipes)																	
82 mm	10.82	11.69	13.15	-	10.78	12.40	-	6.88	-	7.63	-	-	14.76	14.76	14.90	-	-
110 mm	10.04	10.84	12.20	-	12.36	14.21	-	8.41	-	8.15	-	-	17.97	17.97	16.23	-	-
160 mm	6.74	7.28	8.19	-	24.13	27.75	-	9.60	-	9.06	-	-	21.19	21.19	21.19	-	-
Cast iron; socketed joints; caulked; fixing with brackets. (Soil and Waste Pipes)																	
50 mm	28.15	30.40	34.20	21.02	10.78	36.91	-	-	-	7.63	-	-	34.24	-	34.68	4.24	40.13
75 mm	28.77	31.08	34.96	22.23	12.36	37.57	-	-	-	8.15	-	-	36.53	57.30	35.60	5.41	40.98
100 mm	38.51	41.59	46.79	28.80	24.13	45.58	-	-	-	9.06	-	-	48.09	65.84	38.64	6.58	39.37
150 mm	71.48	77.20	86.85	48.55	24.13	66.86	-	-	-	9.06	-	-	87.14	130.96	60.10	8.15	77.34

NEW WORK

Size of Pipe diameter	Pipework: clipped to walls (m)	Pipework: laid in duct (m)	Pipework: laid in floor or roof (m)	Bends: made bend (nr)	Bends: bend coupling (nr)	Bends: backplate elbow (nr)	Couplings and Connectors: straight (nr)	Couplings and Connectors: reducing (nr)	Couplings and Connectors: copper to lead (nr)	Couplings and Connectors: tap (nr)	Couplings and Connectors: tank (nr)	Tee junctions: equal (nr)	Tee junctions: reducing (nr)	Stopcocks: stop cocks (nr)	Stopcocks: gate valve (nr)	Stopcocks: fullway valve (nr)
COPPER; capillary fittings (Hot and Cold water)																
15 mm	3.88	4.19	4.72	1.49	2.00	6.78	1.85	2.83	3.49	6.49	6.13	3.02	5.63	5.05	10.09	11.57
22 mm	5.31	5.73	6.45	1.99	2.89	9.94	1.85	2.83	3.49	6.49	-	4.62	4.57	8.41	12.26	13.77
28 mm	6.48	7.00	7.87	2.49	4.30	-	3.73	4.80	6.52	-	-	7.37	7.35	19.68	19.40	18.81
35 mm	11.55	12.48	14.04	2.99	8.23	-	5.79	8.37	9.57	-	14.42	13.15	13.81	-	-	38.50
42 mm	13.86	14.97	16.84	3.99	12.79	-	7.52	11.03	11.08	-	17.28	18.17	17.28	-	-	47.73
54 mm	17.48	18.87	21.23	5.47	22.69	-	11.88	14.89	13.35	-	24.16	30.58	30.00	-	-	67.71
COPPER; compression fittings (Hot and Cold water)																
15 mm	3.78	4.08	4.59	1.49	3.13	6.83	2.85	1.91	-	-	5.98	4.48	8.62	5.56	11.10	12.72
22 mm	5.21	5.63	6.34	1.99	4.59	-	4.21	2.07	-	-	7.02	6.79	8.24	9.25	13.49	15.14
28 mm	6.39	6.90	7.76	2.49	8.12	-	6.84	7.24	-	-	10.64	12.63	12.33	21.65	21.34	20.69
35 mm	11.46	12.38	13.93	2.99	9.57	-	12.03	13.30	-	-	12.45	29.61	28.64	-	-	42.35
42 mm	13.77	14.87	16.73	3.99	10.95	-	15.37	16.47	-	-	13.73	29.61	28.64	-	-	52.51
54 mm	17.37	18.76	21.11	5.47	16.87	-	21.75	22.33	-	-	20.22	45.24	45.24	-	-	74.48
GALVANISED STEEL; screwed and socketed joints (Cold water)																
15 mm	7.12	7.69	8.65	-	6.04	-	-	-	-	-	-	7.51		7.88	9.51	-
20 mm	8.39	9.06	10.19	-	7.56	-	-	-	-	-	-	9.43	9.89	10.54	11.47	-
25 mm	10.94	11.81	13.29	-	9.53	-	-	-	-	-	-	12.00	12.67	19.76	15.00	-
32 mm	13.55	14.63	16.46	-	12.50	-	-	-	-	-	-	15.91	16.90	28.79	20.29	-
40 mm	15.80	17.07	19.20	-	15.91	-	-	-	-	-	-	20.37	21.89	35.80	28.05	-
50 mm	19.59	21.15	23.80	-	20.92	-	-	-	-	-	-	26.46	29.01	57.01	37.94	-
POLYTHENE; brass compression fittings (Cold water)																
10 mm	5.32	5.74	6.46	-	6.65	-	9.82	-	-	-	-	6.33	6.04	10.83	-	-
12 mm	6.09	6.57	7.39	-	7.45	-	11.01	-	-	-	-	7.10	6.90	11.67	-	-
18 mm	7.43	8.03	9.03	-	10.14	-	14.99	-	-	-	-	10.17	10.01	16.51	-	-
25 mm	10.31	11.13	12.53	-	13.80	-	20.40	-	-	-	-	12.83	12.81	22.59	-	-
32 mm	14.06	15.19	17.09	-	19.65	-	29.05	-	-	-	-	3.77	3.77	10.83	-	-
37 mm	15.82	17.08	19.22	-	25.45	-	37.63	-	-	-	-	3.77	3.77	10.83	-	-

RE-MODELLED WORK

Size of Pipe	Pipework			Bends			Couplings and Connectors					Tee junctions		Stopcocks		
diameter	clipped to walls	laid in duct	laid in floor or roof	made bend	bend coupling	backplate elbow	straight	reducing	copper to lead	tap	tank	equal	reducing	stop cocks	gate valve	fullway valve
	m	m	m	nr	nr	nr	nr	nr	nr	nr	nr	nr	nr	nr	nr	nr
COPPER; capillary fittings (Hot and Cold water)																
15 mm	4.37	4.72	5.31	1.68	2.26	7.63	2.09	3.18	3.92	7.30	6.90	3.40	6.34	5.68	11.36	13.01
22 mm	5.97	6.45	7.25	2.24	3.25	11.19	2.09	3.18	3.92	7.30	-	5.20	5.14	9.47	13.79	15.49
28 mm	7.29	7.87	8.85	2.80	4.84	-	4.20	5.40	7.34	-	-	8.29	8.27	22.14	21.82	21.16
35 mm	13.00	14.04	15.79	3.36	9.26	-	6.52	9.41	10.77	-	16.22	14.80	15.54	-	-	43.31
42 mm	15.59	16.84	18.95	4.48	14.39	-	8.46	12.41	12.46	-	19.44	20.44	19.44	-	-	53.70
54 mm	19.66	21.23	23.89	6.15	25.53	-	13.36	16.75	15.02	-	27.18	34.41	33.75	-	-	76.18
COPPER; compression fittings (Hot and Cold water)																
15 mm	4.25	4.59	5.16	1.68	3.52	7.68	3.21	2.15	-	-	6.73	5.05	9.70	6.25	12.49	14.31
22 mm	5.87	6.34	7.13	2.24	5.16	-	4.73	2.33	-	-	7.90	7.64	9.27	10.41	15.17	17.04
28 mm	7.18	7.76	8.73	2.80	9.14	-	7.69	8.15	-	-	11.97	14.21	13.87	24.35	24.01	23.28
35 mm	12.89	13.93	15.67	3.36	10.77	-	13.53	14.97	-	-	14.00	33.31	32.22	-	-	47.64
42 mm	15.49	16.73	18.82	4.48	12.32	-	17.29	18.53	-	-	15.45	33.31	32.22	-	-	59.07
54 mm	19.54	21.11	23.74	6.15	18.98	-	24.47	25.12	-	-	22.75	50.90	50.90	-	-	83.79
GALVANISED STEEL; screwed and socketed joints (Cold water)																
15 mm	8.01	8.65	9.73	-	6.79	-	-	-	-	-	-	8.45	-	8.87	10.70	-
20 mm	9.43	10.19	11.46	-	8.50	-	-	-	-	-	-	10.61	11.12	11.86	12.90	-
25 mm	12.30	13.29	14.95	-	10.72	-	-	-	-	-	-	13.50	14.26	22.23	16.88	-
32 mm	15.24	16.46	18.52	-	14.06	-	-	-	-	-	-	17.90	19.01	32.38	22.83	-
40 mm	17.78	19.20	21.60	-	17.90	-	-	-	-	-	-	22.92	24.62	40.27	31.56	-
50 mm	22.04	23.80	26.77	-	23.53	-	-	-	-	-	-	29.76	32.63	64.14	42.68	-
POLYTHENE; brass compression fittings (Cold water)																
10 mm	5.98	6.46	7.27	-	7.48	-	11.05	-	-	-	-	7.12	6.79	12.18	-	-
12 mm	6.85	7.39	8.32	-	8.38	-	12.39	-	-	-	-	7.99	7.76	13.13	-	-
18 mm	8.36	9.03	10.16	-	11.41	-	16.86	-	-	-	-	11.45	11.26	18.57	-	-
25 mm	11.60	12.53	14.09	-	15.52	-	22.95	-	-	-	-	14.43	14.41	25.42	-	-
32 mm	15.82	17.09	19.22	-	22.11	-	32.68	-	-	-	-	4.24	4.24	12.18	-	-
37 mm	17.79	19.22	21.62	-	28.64	-	42.33	-	-	-	-	4.24	4.24	12.18	-	-

BOILERS

- complete installation; burners; control systems;
- connections; electricity; water; heating system;
- enamelled casings; insulation; commissioning;
- fixings; bases; housings; builder's work.

	Unit	New	Refurb
Boilers; Gas Fired			
Rating			
147 kW	nr	5,300	7,500
220 kW	nr	5,700	8,200
293 kW	nr	6,400	9,000
440 kW	nr	9,400	13,300
586 kW	nr	11,800	16,700
Boilers; Oil Fired			
Rating			
147 kW	nr	4,300	6,200
220 kW	nr	4,900	7,000
293 kW	nr	5,500	7,800
440 kW	nr	8,300	11,800
586 kW	nr	10,200	14,400
Boilers; Solid Fuel			
Rating			
147 kW	nr	7,600	10,800
220 kW	nr	9,200	13,000
293 kW	nr	10,900	15,400
440 kW	nr	12,800	18,000
586 kW	nr	15,200	21,400

RADIATORS

- steel; epoxy paint finish; connectors;
- fixings; wall brackets; plugged and screwed;
- cast iron; natural finish; require painting;
- fixings; wall brackets; plugged and screwed.

	Unit	New	Refurb
Radiators; Steel			
Single			
305 x 800 mm	nr	35	40
305 x 1440 mm	nr	45	55
305 x 2080 mm	nr	75	85
432 x 800 mm	nr	30	35
432 x 1280 mm	nr	50	60
432 x 2080 mm	nr	80	90
Double			
305 x 800 mm	nr	50	60
305 x 1440 mm	nr	75	85
305 x 2080 mm	nr	115	130
432 x 800 mm	nr	45	55
432 x 1280 mm	nr	80	90
432 x 2080 mm	nr	130	150
Radiators; Cast Iron			
Single			
305 x 800 mm	nr	45	55
305 x 1440 mm	nr	60	70
305 x 2080 mm	nr	95	110
432 x 800 mm	nr	40	45
432 x 1280 mm	nr	65	75
432 x 2080 mm	nr	110	125
Double			
305 x 800 mm	nr	70	80
305 x 1440 mm	nr	95	110
305 x 2080 mm	nr	150	170
432 x 800 mm	nr	60	70
432 x 1280 mm	nr	105	120
432 x 2080 mm	nr	175	200

HEATERS

- electric; portable or fixed; connectors;
- night storage; floor mounted; connections;
- water heaters; casing; self finish; insulation;
- wall or floor mounted; fixings; connections.

	Unit	New	Refurb
Radiators; Electric			
Single			
305 x 800 mm	nr	100	115
305 x 1440 mm	nr	140	160
432 x 800 mm	nr	40	45
432 x 1280 mm	nr	50	60
Radiators; Electric; Night Storage (excluding cost economy wiring installation)			
Single			
432 x 800 mm	nr	380	430
432 x 1280 mm	nr	570	645
Water Heaters; Electric			
Direct supply			
3 kW - 7 litres	nr	140	160
3 kW - 10 litres	nr	150	170
Storage			
1.5 kW - 13.6 litres	nr	100	115
3.0 kW - 23 litres	nr	125	145
3.0 kW - 55 litres	nr	125	145
Unvented; system heater			
70 - 300 litres	nr	660	745
450 - 3000 litres	nr	2,260	2,545
Multi point			
3 kW - 15 litres	nr	125	145
Driers; Electric			
Hand; wall fixing			
1.40 kW	nr	110	125
Hair; wall fixing			
1.00 kW	nr	100	115

W.C. CUBICLES

- panels; solid; laminate finish or resin bonded;
- doors; matching; hinges; indicator sliding bolts;
- aluminium or steel frame; assembling;
- fixings; rawl bolts to floors and walls.

	Unit	New	Refurb
W.C. Cubicles; laminated; aluminium frame; self finish			
Key Stage 1 - 2			
Corner cubicle (1 door; 1 pnl)	set	395	445
Cubicle between walls (1 door)	set	265	300
Island cubicle (1 door; 2 pnls)	set	495	560
Key Stage 3 - 4			
Corner cubicle (1 door; 1 pnl)	set	430	485
Cubicle between walls (1 door)	set	240	270
Island cubicle (1 door; 2 pnls)	set	535	605
Adult			
Corner cubicle (1 door; 1 pnl)	set	510	575
Cubicle between walls (1 door)	set	240	270
Island cubicle (1 door; 2 pnls)	set	760	855
W.C. Cubicles; heavy duty resin bonded; steel frame; epoxy paint finish.			
Key Stage 1 - 2			
Corner cubicle (1 door; 1 pnl)	set	535	605
Cubicle between walls (1 door)	set	360	405
Island cubicle (1 door; 2 pnls)	set	660	745
Key Stage 3 - 4			
Corner cubicle (1 door; 1 pnl)	set	565	640
Cubicle between walls (1 door)	set	355	400
Island cubicle (1 door; 2 pnls)	set	700	790
Adult			
Corner cubicle (1 door; 1 pnl)	set	705	795
Cubicle between walls (1 door)	set	400	450
Island cubicle (1 door; 2 pnls)	set	1,035	1,165

FITTINGS

- taps; chromium plated; spray nozzle; economy;
- stop valves; standard fittings;
- disabled fittings; standard; assembling;
- secured fixings; rawl plugs.

	Unit	New	Refurb
Taps chromium plated; jointing to service pipes;			
Bib taps; wash basin			
13 mm	nr	22.00	25.00
19 mm	nr	30.00	35.00
Pillar taps; sink tops.			
13 mm	nr	30.00	34.00
19 mm	nr	40.00	45.00
Stop Valves gunmetal and brass; jointing to pipe.			
Stop valves			
25 mm diameter	nr	21.00	24.00
32 mm	nr	35.00	40.00
40 mm	nr	40.00	45.00
50 mm	nr	65.00	75.00
Wheel valves			
25 mm diameter	nr	16.00	18.00
32 mm	nr	25.00	30.00
40 mm	nr	30.00	35.00
50 mm	nr	40.00	45.00
Disabled Equipment chromium plated; fixed to walls and floor.			
Arm supports; upright.			
Support arm	nr	116.00	131.00
Support arm with leg	nr	135.00	155.00
Toilet support rail			
32 mm PVC construction.			
150 mm rail	nr	43.00	49.00
300 mm rail	nr	50.00	60.00
450 mm rail	nr	55.00	62.00

FIRE EQUIPMENT

- standard equipment; Fire Officer's requirement;
- free standing or wall fixings;
- commissioning; servicing.

	Unit	New	Refurb
Hose Reels; automatic enamelled fittings; reinforced rubber hoses; 25 metres; fixing to walls; connecting to water main.			
fixed drum	nr	201	227
swing drum	nr	270	305
swinging recess drum	nr	510	575
swinging; recessed cupboard	nr	555	625
Fire Extinguishers complete; fixing brackets plugged to walls.			
Water; 9 litres	nr	92	104
CO2; 2.3 Kg	nr	125	145
CO2; 4.4 Kg	nr	205	235
Foam; 9 litres	nr	120	135
BCF; 1 Kg	nr	125	145
BCF; 2.5 Kg	nr	125	145
Dry powder; 1 Kg	nr	55	65
Dry powder; 6 Kg	nr	130	150
Dry powder; 12 Kg	nr	165	190
Multi purpose; 5.5 litre	nr	105	120
Fire Blankets storage cyclinder; enamelled.			
1200 x 1200 mm	nr	30	35
1800 x 1800 mm	nr	55	65
Fire Curtains bonded fibreglass quilt; hanging battens; sealing.			
50 mm thick	nr	25	30
75 mm thick	nr	30	35

Description	Unit	Composite Rates	Symbol
SWITCHES			
1 Way	nr	£10 to £15	
2 Way	nr	£15 to £20	
3 Way	nr	£20 to £25	
Dimmer	nr	£20 to £30	
Immersion heater	nr	£10 to £15	
Pull	nr	£10 to £15	
Shower	nr	£15 to £20	
Timer	nr	£30 to £40	
POWER SOCKETS			
Single; switch; low level	nr	£10 to £15	
Double; switch; low level	nr	£15 to £20	
Shaver	nr	£55 to £75	
Single; switch; low level	nr	£10 to £15	
Double; switch; low level	nr	£20 to £25	
Lockable - landlord's	nr	£10 to £15	
FUSED SPUR OUTLETS			
Unswitched	nr	£10 to £10	
Switched	nr	£10 to £15	

Description	Unit	Composite Rates	Symbol
LIGHT FITTINGS			
Pendant	nr	£10 to £15	
Fluorescent	nr	£10 to £15	
Shaver socket/light	nr	£15 to £20	
External bulkhead	nr	£20 to £25	
External bulkhead - solar	nr	£25 to £30	
ACCESSORIES			
Door bell	nr	£35 to £45	
Door bell push	nr	£20 to £25	
Door entry caller	nr	£35 to £45	
Door entry control	nr	£140 to £190	
Cooker control	nr	£75 to £100	C
Heater programmer	nr	£55 to £75	P
Thermostat	nr	£45 to £60	T
Immersion heater switch	nr	£45 to £60	IH
Immersion heater	nr	£165 to £220	
Storage heater	nr	£260 to £350	SH
Fan heater	nr	£115 to £155	FH
Radiant heater	nr	£65 to £90	RH

Description	Unit	Composite Rates	Symbol
SWITCHGEAR			
Meter unit	nr	£65 to £85	E
Consumer unit	nr	£60 to £80	
TELEVISION			
TV point	nr	£75 to £95	
Aerial	nr	£10 to £15	
FIRE ALARM SYSTEM			
Control point	nr	£220 to £300	FP
Call point	nr	£30 to £40	
Bell	nr	£35 to £45	
Emergency exit point	nr	£55 to £70	EE
Emergency light fitting	nr	£110 to £145	EL
Smoke detector	nr	£65 to £90	SD
Heat detector	nr	£45 to £60	HD
Magnetic door closure	nr	£65 to £90	MC
Extinguisher - water	nr	£80 to £110	W
Extinguisher - dry powder	nr	£115 to £150	P
Extinguisher - CO2	nr	£110 to £145	C
Fire blanket	nr	£35 to £50	B

DISTRIBUTION BOARDS

	Unit	New	Refurb
Switchgear; breakers; boxes; conduits; covers; wiring to mains and sub-mains; complete.			
Distribution panel; isolators; square D.			
6-way; TP & N 1 line	nr	1,685	1,975
4-way; TP & N MCB	nr	435	515
12-way; TP & N MCB	nr	505	595
18-way; TP & N MCB	nr	730	855
36-way; TP & N line	nr	1,105	1,300
Control apparatus; wiring to mains and equipment; complete.			
Automatic time switched.	nr	1,455	1,710
Distribution Trunking; surface fixing.			
Galvanised; 300 x 150 mm.	m	34.77	40.85
Earth Bonding; conductors; clamps; brackets; earthing rods; wiring; testing; complete.			
Metal roofing	nr	170.00	200.00
Steel framing	nr	280.00	330.00
Taps and pipework	nr	35.00	40.00
Tanks	nr	40.00	45.00
Radiators	nr	35.00	45.00
Fixed apparatus	nr	35.00	45.00
Mains fuse panel	nr	160.00	185.00
Mains switch gear	nr	340.00	395.00
Maintenance of Light Fittings			
Take down, clean light fitting and replace fluorescent lamps.	nr	7.24	8.51

DUCTWORK

	Unit	New	Refurb
Conduit; steel; galvanised; fixed			
20 mm diameter	m	3.77	4.43
25 mm diameter	m	4.64	5.45
Extra for ;			
junction box	nr	3.44	4.04
Conduit; PVC ; fixed			
20 mm diameter	m	1.80	2.12
25 mm diameter	m	2.13	2.50
Extra for ;			
junction box	nr	2.02	2.37
Trunking; steel; galvanised; single compartment.			
size 50 x 50 mm	m	5.15	6.06
Extra; bend;	nr	3.79	4.45
Extra; T junction;	nr	4.07	4.79
size 100 x 75 mm	m	7.04	8.27
Extra; bend;	nr	4.74	5.56
Extra; T junction;	nr	5.09	5.98
Trunking; steel; galvanised; twin compartment.			
size 50 x 50 mm	m	6.96	8.17
Extra; bend;	nr	5.11	6.01
Extra; T junction;	nr	5.50	6.46
size 100 x 75 mm	m	9.51	11.17
Extra; bend;	nr	6.39	7.51
Extra; T junction;	nr	6.87	8.08
Trunking; PVC; single compartment; fixed			
size 50 x 50 mm	m	3.87	4.54
Extra; bend;	nr	2.84	3.34
Extra; T junction;	nr	3.05	3.59
size 100 x 75 mm	m	5.28	6.21
Extra; bend;	nr	3.55	4.17
Extra; T junction;	nr	3.82	4.49
Trunking; PVC; twin compartment; fixed			
size 50 x 50 mm	m	5.22	6.13
Extra; bend;	nr	3.84	4.51
Extra; T junction;	nr	4.12	4.85
size 100 x 75 mm	m	7.13	8.38
Extra; bend;	nr	4.80	5.63
Extra; T junction;	nr	5.15	6.06

ACCESSORIES

	Unit	New	Refurb
Accessories			
Door bell	nr	36.71	43.13
Door bell push	nr	19.42	22.81
Door entry caller	nr	38.75	45.54
Door entry control	nr	163.05	191.58
Cooker control	nr	84.25	98.99
Thermostat	nr	51.79	60.86
Heater programmer	nr	70.46	82.79
Immersion heater	nr	188.40	221.37
Storage heater	nr	300.44	353.02
Fan heater	nr	132.72	155.95
Radiant heater	nr	74.66	87.72
Dado or Skirting Trunking; cover.			
Three compartment			
70 x 30 mm	m	24.96	29.33
Extra for end	nr	6.12	7.19
Extra for bend	nr	9.25	10.87
Mini Trunking; uPVC			
25 x 16 mm	m	13.17	15.47
Extra for end	nr	2.13	2.50
Extra for bend	nr	2.85	3.35
Switch Gear			
Meter unit	nr	70.87	83.27
Consumer unit	nr	65.69	77.18
RCD; mounted with enclosure	nr	132.72	155.95
Television Installations			
TV Point	nr	26.56	31.21
Aerial	nr	82.51	96.95
Fire Alarm Systems			
Control Panel	nr	257.67	302.76
Call Point	nr	33.18	38.99
Bell	nr	39.01	45.84
Smoke detector	nr	59.10	69.45
Heat detector	nr	50.42	59.24
Magnetic door closure	nr	76.34	89.70
Electronic locking	nr	915	1,075

CABLE MANAGEMENT

	Unit	New	Refurb
Solid cored coductor; PVC Insulated; PVC sheathed; Non armoured; 300 to 500V.			
Single cored			
2.5 mm 2	m	1.70	2.00
10 mm 2	m	2.18	2.57
25 mm2	m	2.36	2.77
35 mm2	m	2.88	3.38
Two cored			
2.5 mm 2	m	2.64	3.10
10 mm2	m	3.03	3.57
16 mm2	m	3.31	3.89
Three cored			
1 mm 2	m	2.18	2.57
1.5 mm 2	m	2.20	2.59
Stranded cored cable; PVC Insulated; Non sheathed; Non armoured; 450 to 750V.			
single-cored			
1.5 mm 2	m	0.45	0.53
2.5 mm 2	m	0.49	0.57
10 mm 2	m	0.82	0.96
16 mm2	m	0.97	1.14
25 mm2	m	1.09	1.28
35 mm2	m	1.70	2.00
Solid cored cable; Mineral Insulated; Copper sheathed; Non armoured; 750V.			
Single core			
6 mm 2	m	4.87	5.73
10 mm 2	m	5.05	5.93
25 mm2	m	5.83	6.85
35 mm2	m	6.02	7.07
two-core			
6 mm 2	m	5.72	6.72
10 mm 2	m	6.05	7.11
25 mm2	m	6.68	7.84
three-core			
6 mm 2	m	5.90	6.93
10 mm 2	m	6.09	7.15
16 mm2	m	6.33	7.44
25 mm2	m	7.56	8.88

	Unit	New	Refurb
Solid cored conductor; Cross linked Polythene Insulat PVC Oversheathed; Armoured; 600 to 1000V.			
Two cored			
35 mm2	m	4.51	5.30
50 mm2	m	5.56	6.54
70 mm2	m	7.20	8.46
95 mm2	m	9.24	10.86
Three cored			
50 mm2	m	12.86	15.12
70 mm2	m	7.20	8.46
95 mm2	m	9.24	10.86
Four cored			
70 mm2	m	16.12	18.94
120 mm2	m	20.41	23.99
Solid cored cable; PVC Insulated; PVC Oversheathed; Armoured; 600 to 1000V.			
single-cored			
50 mm2	m	4.61	5.42
70 mm2	m	5.37	6.31
95 mm2	m	6.35	7.46
120 mm2	m	7.37	8.66
Stranded conductors; PVC Oversheathed; Armoured; 600/1000V.			
Three-cored			
10 mm2	m	3.67	4.31
16 mm2	m	4.03	4.74
25 mm2	m	4.75	5.58
35 mm2	m	5.75	6.76
50 mm2	m	7.45	8.75
70 mm2	m	10.03	11.79
Four-cored			
10 mm2	m	5.80	6.81
16 mm2	m	7.08	8.32
25 mm2	m	7.52	8.84
35 mm2	m	8.21	9.64
50 mm2	m	9.13	10.72
70 mm2	m	12.60	14.81

	Unit	New	Refurb
Armoured cables; XLPE/SWA/LSF; laid in trench (by others).			
Two cored			
5 mm 2	m	4.80	5.64
16 mm2	m	5.05	5.93
Three cored			
10 mm 2	m	5.00	5.87
Four cored			
25 mm2	m	6.09	7.15
35 mm2	m	7.41	8.71
95 mm2	m	14.10	16.57
Armoured cables; XLPE/SWA/LSF; cleated to structural surfaces.			
Two cored			
4 mm 2	m	4.80	5.64
16 mm2	m	7.12	8.37
Three cored			
10 mm 2	m	6.71	7.88
Four cored			
2 5 mm 2	m	8.16	9.59
35 mm 2	m	9.49	11.15
95 mm2	m	16.18	19.01
Glands; termination for Armoured cables;			
Two cored			
4 mm 2	nr	7.12	8.37
16 mm2	nr	22.76	26.74
Three cored			
10 mm 2	nr	10.22	12.01
Four cored			
25 mm2	nr	25.07	29.46
35 mm2	nr	28.96	34.03
95 mm2	nr	37.42	43.97

LIGHTING

	Unit	New	Refurb
Tubular Fluorescent; single luminaire; metal cover plates; enclosed control gear.			
Batten only			
600 mm long	nr	22.96	26.97
1200 mm long	nr	27.66	32.50
1800 mm long	nr	32.67	38.39
Batten with deflector			
600 mm long	nr	25.38	29.82
1200 mm long	nr	31.29	36.77
1800 mm long	nr	37.51	44.08
Batten with prismatic diffuser			
600 mm long	nr	26.60	31.25
1200 mm long	nr	33.25	39.07
1800 mm long	nr	40.68	47.80
Batten with acrylic prismatic diffuser			
600 mm long	nr	35.91	42.19
1200 mm long	nr	44.89	52.75
1800 mm long	nr	54.92	64.54
58 W; reflector; wire covers.			
1200 mm long	nr	73.90	86.83
Tubular Fluorescent; twin luminaires; metal cover plates; enclosed control gear.			
Batten only			
600 mm long	nr	37.76	44.36
1200 mm long	nr	50.92	59.83
1800 mm long	nr	63.01	74.04
Batten with deflector			
600 mm long	nr	41.74	49.04
1200 mm long	nr	57.60	67.69
1800 mm long	nr	72.35	85.01
Batten with prismatic diffuser			
600 mm long	nr	43.75	51.40
1200 mm long	nr	61.21	71.93
1800 mm long	nr	78.46	92.19
Batten with acrylic prismatic diffuser			
600 mm long	nr	59.06	69.39
1200 mm long	nr	82.64	97.10
1800 mm long	nr	105.92	124.46

	Unit	New	Refurb
Modular ceiling luminaires; metal boxes; switch start.			
Recessed fittings;			
3 x 18 watt	nr	50.17	58.95
4 x 18 watt	nr	53.71	63.11
3 x 36 watt	nr	60.30	70.85
4 x 36 watt	nr	65.43	76.89
Recessed fittings; louvres; Category 2.			
3 x 18 watt	nr	57.63	67.71
4 x 18 watt	nr	61.16	71.87
3 x 36 watt	nr	67.75	79.61
4 x 36 watt	nr	72.89	85.65
Recessed fittings; prismatic controllers			
3 x 18 watt	nr	61.95	72.79
4 x 18 watt	nr	65.75	77.26
3 x 36 watt	nr	72.84	85.58
4 x 36 watt	nr	78.36	92.07
Extra for 3 hour integral emergency maintained packs;			
3 x 18 watt	nr	124.74	146.57
3 x 36 watt	nr	149.15	175.25
Compact fluorescent luminaires.			
Down lighter; recessed; low energy			
18 watt x 2	nr	52.88	62.13
28 watt	nr	62.37	73.29
Down lighter; recessed; enery masters			
21 watt	nr	52.88	62.13
28 watt	nr	62.37	73.29
Ceiling fitting; perspex covers.			
recessed	nr	59.79	70.26
surface fixed	nr	62.79	73.78
Bulkhead Luminaires			
28 watt	nr	54.76	64.34
28 watt; maintained	nr	80.48	94.57
250 watt; loading bay; guard.	nr	111.19	130.65
Security Floodlights; external			
Photocell	nr	77.64	91.23

SWITCH AND SOCKETS

	Unit	New	Refurb
Switches			
1 way	nr	10.28	12.07
2 way	nr	14.13	16.60
3 way	nr	17.98	21.13
dimmer	nr	23.12	27.17
pull	nr	10.28	12.07
timer	nr	32.11	37.73
Weatherproof; 1 way	nr	231.65	272.19
Power Sockets			
Single - switched - low level	nr	10.28	12.07
Double - switched - low level	nr	14.13	16.60
Shaver	nr	61.66	72.45
Single - switched - high level	nr	11.56	13.58
Double - switched - high level	nr	20.55	24.15
Lockable - landlord's	nr	10.28	12.07
Fused spur outlets			
Unswitched	nr	7.24	8.50
Switched	nr	7.85	9.22
Lighting roses			
Pendant	nr	11.56	13.58
Fluorescent	nr	10.28	12.07
Shaver socket/ light	nr	15.41	18.11
External bulkhead	nr	19.27	22.64
External bulkhead - solar	nr	25.69	30.19

GLOSSARY

Definition of 'falls'...

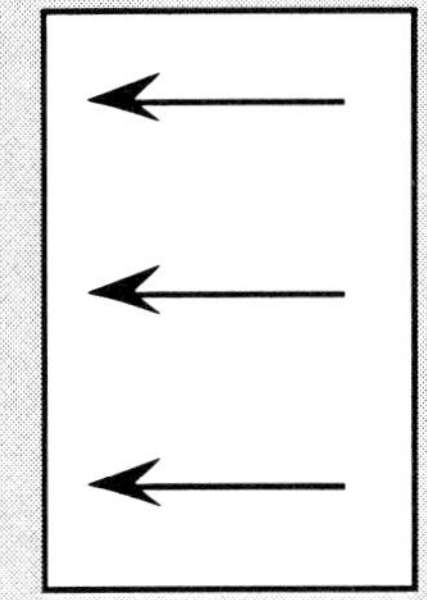

Paved areas laid to falls

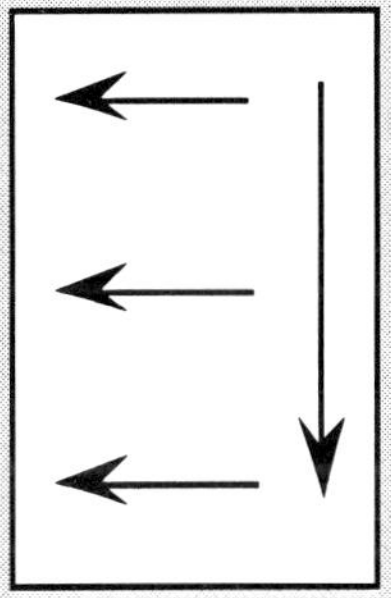

Paved areas laid to falls and cross falls

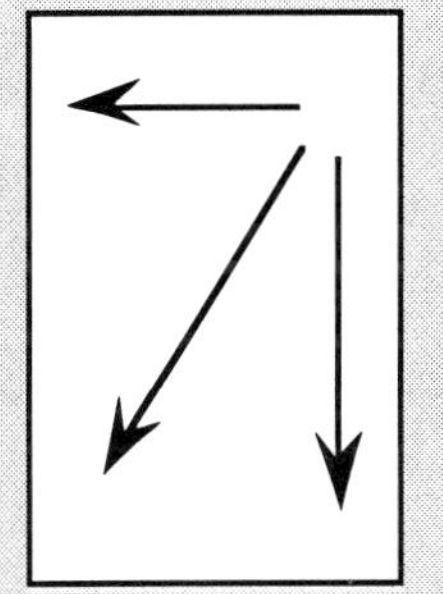

Paved areas laid to falls and currents

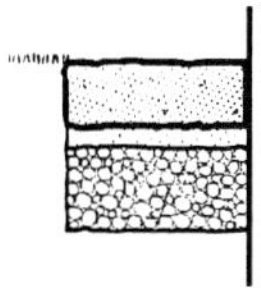

Mowing Margin
In situ Concrete
£12.00 per metre

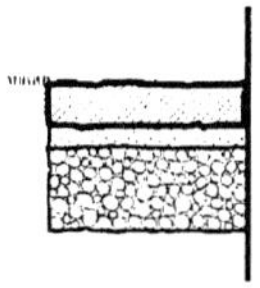

Mowing Margin
Paving slab
£11.00 per metre

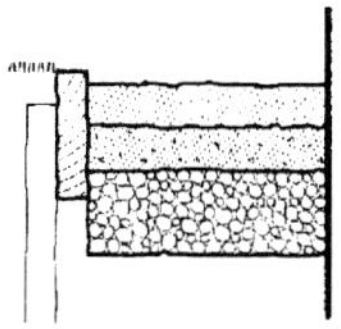

Mowing Margin
Hoggin
£9.00 per metre

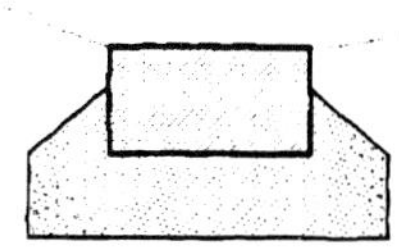

Flush kerb
Precast Concrete
£14.00 per metre

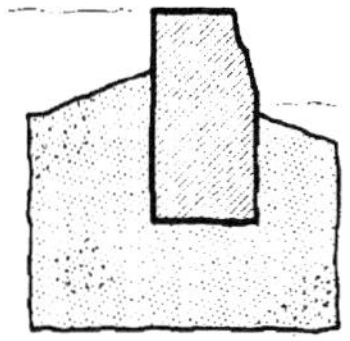

Kerb
Precast Concrete
£16.00 per metre

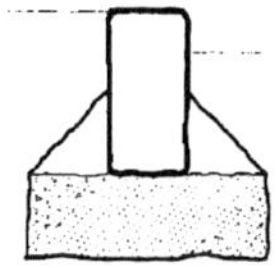

Edging
Precast Concrete
£10.00 per metre

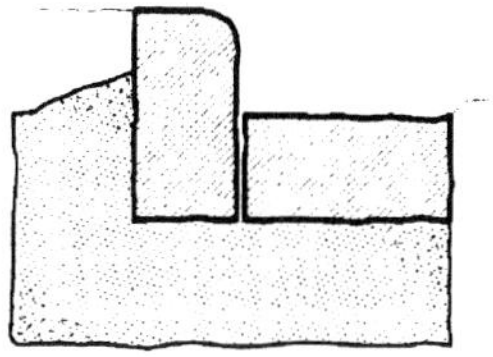

Kerb
Precast Concrete
£19.00 per metre

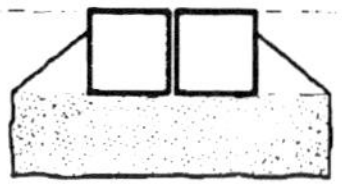

Trim
Granite sett
£12.00 per metre

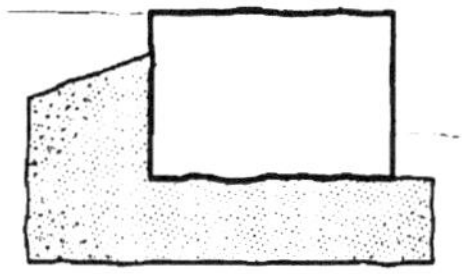

Kerb
Granite
£.00 per metre

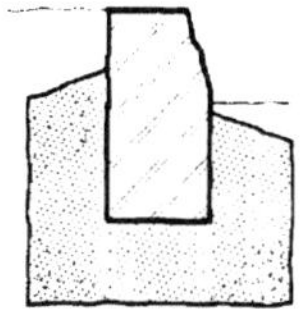

Kerb
Brick on End
£14.00 per metre

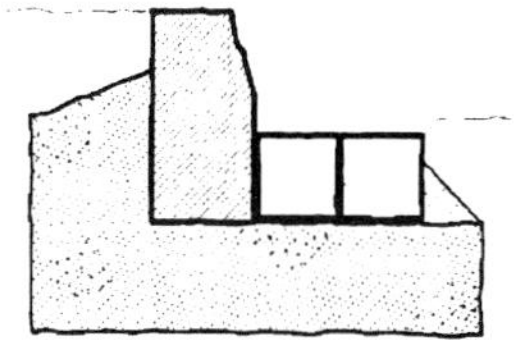

Kerb
Precast Concrete
Granite sett gutter
£27.00 per metre

Trim
Facing Brick
£10.00 per metre

SERVICE ROADS

- Macadam surface; rolled and consolidated;
- Laid to falls and cross falls;
- Hardcore base; granular; consolidated; 5 ton roller;
- Excavation; as necessary; clear away surplus.

	Unit	New	Refurb
Light duty roads; Macadam; 1st base course; 50 mm thick; Macadam; 2nd wearing course; 30 mm thick; Hardcore filling; 200 mm thick.			
Limestone aggregate	m2	21.86	23.44
Extra for cost of;			
White spar chippings	m2	1.92	2.05
Heavy duty roads; Macadam; 1st base course; 70 mm thick; Macadam; 2nd wearing course; 30 mm thick; Hardcore filling; 250 mm thick.			
Limestone aggregate	m2	26.34	28.24
Resurfacing Work; Break up wearing course macadam paving; scarify surface of base course; make good where required; New macadam wearing course.			
30 mm; Limestone aggregate	m2	9.54	10.23
50 mm; Limestone aggregate	m2	12.12	13.00
Extra for cost of;			
White spar chippings	m2	2.20	2.36
Repairs; Break up defective areas of existing macadam paving; scarify surface of base; patch with new macadam wearing course; roll; clear away debris; areas < 2 m2.			
30 mm; Limestone aggregate	m2	15.74	16.88
50 mm; Limestone aggregate	m2	20.00	21.44
Extra for cost of;			
White spar chippings	m2	2.75	2.95

PARKING AREAS

- Macadam surface; rolled and consolidated;
- Laid to falls and cross falls;
- Hardcore base; granular; consolidated; 5 ton roller;
- Excavation; as necessary; clear away surplus.

	Unit	New	Refurb
Light duty; Macadam; One wearing course; 50 mm thick; Hardcore filling; 150 mm thick.			
Limestone aggregate	m2	17.99	19.29
Extra for cost of;			
White spar chippings	m2	1.92	2.05
Heavy duty; Macadam; 1st base course; 45 mm thick; Macadam; 2nd wearing course; 30 mm thick; Hardcore filling; 150 mm thick.			
Limestone aggregate	m2	20.17	21.62
Grassgrid Concrete Block system; 366 x 274 x 100 mm thick; interlocking grid; sand bed; 40 mm thick; gravel base; 100 mm thick.			
560 x 280 mm (heavy use)	m2	17.19	18.43
Extra for cost of;			
60 mm top soil; grass seed	m2	2.05	2.20
Resurfacing Work; Break up wearing course macadam paving; scarify surface; make good; New macadam wearing course.			
30 mm; Limestone aggregate	m2	8.43	9.04
50 mm; Limestone aggregate	m2	11.79	12.64
Repairs; Break up defective areas of macadam; scarify surface; patch with new macadam; roll; clear away; areas < 2 m2.			
30 mm; Limestone aggregate	m2	15.59	16.72
50 mm; Limestone aggregate	m2	21.81	23.39

FOOTPATHS

- Macadam surface; rolled and consolidated;
- Mastic asphalt; rubbed with fine sand;
- Precast concrete; spot bedded in cement and sand;
- Excavation; clear away hardcore base; consolidated.

	Unit	New	Refurb
Light duty; Macadam; One wearing course; 30 mm thick; laid to falls; Hardcore bed; 100 mm thick.			
Limestone aggregate	m2	12.43	13.33
Heavy duty; Macadam; base course; 40 mm thick; wearing course; 20 mm thick; laid to falls; Hardcore bed; 150 mm thick.			
Limestone aggregate	m2	18.24	19.56
Light Duty; Mastic asphalt; One wearing course; laid to falls; isolating membrane; 75 mm in situ concrete base; hardcore bed; 100 mm thick.			
20 mm thick; one coat;			
> 300 mm wide	m2	23.47	25.16
> 150 < 300 mm	m2	7.80	8.37
< 150 mm wide	m2	3.90	4.18
Skirting; internal angle fillet			
< 200 mm (av) high	m	12.75	13.67
Precast Concrete Paving Slabs; Laid running bond; to falls; spot bedded and jointed in cement mortar (1:3); recessed joint; 50 mm sand base.			
Plain finish			
600 x 600 x 50 mm	m2	13.00	13.94
750 x 600 x 50 mm	m2	10.64	11.40
900 x 600 x 50 mm	m2	9.95	10.67
Coloured			
600 x 600 x 50 mm	m2	14.04	15.05
750 x 600 x 50 mm	m2	11.67	12.52
900 x 600 x 50 mm	m2	10.99	11.78

BRICK PAVIORS

- Cost **£350.00** per 1000 delivered to site;
- Laid running bond; to falls.

	Unit	New	Refurb
Brick Paviors; bedding and jointing in cement mortar (1:3); 75 mm insitu concrete bed; 100 mm hardcore bed.			
215 x 104 x 65 mm thick			
laid flat	m2	42.54	45.62
extra cost of laying to;			
herringbone patterns	m2	1.50	1.61
diagonal patterns	m2	2.10	2.25
extra cost of forming;			
two brick border; straight	m	5.18	5.56
two brick border; curved	m	7.87	8.44
laid on edge	m2	56.29	60.36
extra cost of laying;			
herringbone patterns	m2	45.88	49.20
diagonal patterns	m2	46.75	50.13
extra cost of forming;			
two brick border; straight	m	5.96	6.39
two brick border; curved	m	9.05	9.70
Brick Paviors; 50 mm bed of sand; sand brushed into joints.			
215 x 104 x 65 mm thick			
laid flat	m2	28.22	30.26
extra cost of laying to;			
herringbone patterns	m2	1.31	1.41
diagonal patterns	m2	1.84	1.97
laid on edge	m2	45.28	48.56
extra cost of laying;			
herringbone patterns	m2	1.91	2.05
diagonal patterns	m2	2.68	2.87

CONCRETE BLOCK PAVIORS (1)

- Pedestrian use only;
- Cost as stated per m2 delivered to site;
- Laid running bond; base as described.

	Unit	New	Refurb
Concrete Block Paviors. 50 mm bed of sand; sand brushed into joints.			
Cost £6.60 per m2 d/d site;			
Plain; natural aggregate;			
200 x 100 x 60 mm thick;	m2	20.57	22.05
extra cost of laying;			
herringbone patterns	m2	0.93	1.00
diagonal patterns	m2	1.30	1.39
Cost £9.45 per m2 d/d site;			
Textured; granite aggregate;			
200 x 100 x 65 mm thick;	m2	24.83	26.63
extra cost of laying;			
herringbone patterns	m2	1.14	1.23
diagonal patterns	m2	1.60	1.72
Paving Setts. 50 mm bed of sand; sand brushed into joints.			
Cost £15 per m2 d/d site;			
Plain; mottled effect; laid			
herringbone pattern;			
100 x 134 x 60 mm	m2	30.04	32.22
134 x 134 x 60 mm	m2	28.21	30.25
201 x 134 x 60 mm	m2	26.38	28.29
Cost £17.5 per m2 d/d site;			
Textured; granite aggregate;			
Mixed sizes x 60 mm thick	m2	32.48	34.83
Borders and Patterns. average extra cost			
Border; two block; straight	m	4.21	4.51
Border; two block; curved	m	6.40	6.87
Circles and fans	nr	21.96	23.55

CONCRETE BLOCK PAVIORS (2)

- Pedestrian and light traffic;
- Cost as stated per m2 delivered to site;
- Laid running bond; base as described.

	Unit	New	Refurb
Concrete Block Paviors. bedding and jointing in cement mortar (1:3); 100 mm in situ concrete bed; 100 mm hardcore bed;			
Cost £8.10 per m2 d/d site;			
Plain; natural aggregate;			
200 x 100 x 80 mm thick;			
heavy traffic use;	m2	35.46	38.03
extra cost of laying;			
herringbone patterns	m2	1.02	1.09
diagonal patterns	m2	1.43	1.53
Paving Setts. bedding and jointing in cement mortar (1:3); 75 mm in situ concrete bed; 100 mm hardcore bed;			
Cost £22 per m2 d/d site;			
Plain; mottled effect; laid			
herringbone pattern;			
100 x 134 x 80 mm	m2	49.16	52.71
134 x 134 x 80 mm	m2	47.33	50.75
201 x 134 x 80 mm	m2	45.50	48.79
Borders and Patterns. average extra cost			
Border; two block; straight	m	6.10	6.54
Border; two block; curved	m	9.29	9.96
Circles and fans	nr	31.84	34.14

STONE PAVING

- Natural stone; Cost £55.00 per m2 d/d site;
- Honed finish; 5 mm joints; recessed pointing;
- Reconstructed stone;
- Cost £22.00 per m2 delivered to site;
- Plain finish; 5 mm joints; flush pointing.

	Unit	New	Refurb
Natural Stone; 50 mm thick; running bond; to falls; bedded and jointed in cement and sand (1:3); pointed; tinted mortar; 75 mm in situ concrete base; hardcore.			
600 x 600 mm	m2	90.67	97.22
900 x 600 mm	m2	88.50	94.89
Extra cost of;			
laid to falls and cross falls	m2	1.71	1.83
Tread; 275 mm wide; fair edge	m	28.60	30.67
Tread; 450 mm wide; fair edge	m	45.04	48.30
Riser; 150 mm high;	m	14.05	15.07
Reconstructed Stone; 50 mm thick; running bond; to falls; bedded and jointed in cement and sand (1:3); pointed; tinted mortar; 75 mm in situ concrete; hardcore.			
600 x 600 mm	m2	39.90	42.78
900 x 600 mm	m2	39.14	41.97
Extra cost of;			
laid to falls and cross falls	m2	0.72	0.77
Tread; 275 mm wide; fair edge	m	13.24	14.20
Tread; 450 mm wide; fair edge	m	21.05	22.57
Riser; 150 mm high;	m	5.68	6.09
Crazy Paving; stone; random pattern; to falls; bedded and jointed in cement and sand (1:3); pointed; tinted mortar; 50 mm bed sand; 75 mm bed hardcore.			
Natural; 38 mm thick	m2	56.36	60.44
Natural; 50 mm thick	m2	65.15	69.86
Reconstructed; 38 mm thick	m2	23.99	25.73
Reconstructed; 50 mm thick	m2	26.50	28.42

SLATE PAVING

- Cost as stated per m2 delivered to site;
- Running bond; bedding in cement mortar (1:3);
- Pointing; recessed joints; tinted mortar;
- 75 mm Concrete base; hardcore bed; 100 mm thick.

	Unit	New	Refurb
Cumbria Slate.			
Standard slates; riven.			
300 x 300 x 12 mm (£42 m2)	m2	76.88	82.44
300 x 300 x 20 mm (£54 m2)	m2	92.25	98.92
600 x 300 x 20 mm (£60 m2)	m2	99.93	107.16
Regular pattern slabs; riven.			
12 mm thick (£48 per m2)	m2	84.56	90.68
20 mm thick (£65 per m2)	m2	106.34	114.03
Welsh Slate.			
Standard slates; riven.			
250 x 250 x 12 mm (£28 m2)	m2	58.95	63.21
600 x 300 x 15 mm (£32 m2)	m2	64.07	68.70
Regular pattern slabs; riven.			
12 mm thick (£34 per m2)	m2	66.63	71.45
15 mm thick (£40 per m2)	m2	74.32	79.69
Spanish Slates.			
Standard slates; riven.			
600 x 300 x 10 mm (£34 m2)	m2	66.63	71.45
600 x 300 x 15 mm (£38 m2)	m2	71.75	76.94
Regular pattern slabs; riven.			
12 mm thick (£40 per m2)	m2	74.32	79.69
15 mm thick (£48 per m2)	m2	84.56	90.68
Italian Slates.			
Standard slates; riven.			
600 x 300 x 10 mm (£37 m2)	m2	70.47	75.57
600 x 300 x 15 mm (£39 m2)	m2	73.03	78.32
Regular pattern slabs; riven.			
12 mm thick (£43 per m2)	m2	78.16	83.81
15 mm thick (£52 per m2)	m2	89.69	96.17

KERBS AND CHANNELS

- Facing bricks; feature areas; pointing all round;
- Precast concrete; standard sections;
- Granite aggregate; bedding in cement mortar (1:3);
- Including angles; end; excavation; concrete base.

	Unit	New	Refurb
Facing Brick; Cost £350.00 per 1000 delivered to site; bedding and jointing in cement mortar (1:3); flush pointing; in situ concrete base; flaunching up one side; excavation and clear away.			
Brick on edge; straight; ends.	m	9.72	10.42
Brick on edge; curved; ends.	m	11.25	12.06
Brick on end; straight; ends.	m	13.15	14.10
Brick on end; curved; ends.	m	15.54	16.66
Precast Concrete; granite aggregate; bedding, jointing and pointing in cement mortar (1:3); in situ concrete base; flaunching one side; excavation and clear away.			
Edging; bullnosed; Play areas			
51 x 203 mm; ends.	m	9.41	10.09
Ditto; curved.	m	11.77	12.62
51 x 254 mm; ends.	m	10.50	11.26
Ditto; curved.	m	13.26	14.22
Kerb; splayed; Service roads;			
125 x 255 mm; ends; quadrants	m	15.78	16.92
Ditto; curved.	m	19.89	21.33
152 x 305 mm; ends; quadrants	m	18.10	19.41
Ditto; curved.	m	22.96	24.62
Channel; iron grille; Play Areas.			
125 x 152 mm; ends; quadrants	m	11.77	12.62
Ditto; curved.	m	14.54	15.60
Channel; iron grille; Service Roads			
125 x 250 mm; ends; quadrants	m	15.41	16.52
Ditto; curved.	m	19.66	21.08
150 x 305 mm; ends; quadrants	m	17.27	18.51
Ditto; curved.	m	22.26	23.87

SPORTS FIELDS

- Excavation by machine; cut and fill;
- Preparation ready for planting, seeding or turf;
- Seeding; low maintenance mixture; first cut;
- Turf; low maintenance; rolling; first cut.

	Unit	New	Refurb
Excavation Work; Bulk excavation work; by machine; cut and fill; levelling and compacting; spreading vegetable soil.			
Excavate to reduce site levels			
< 1 metre deep	m3	0.84	0.90
> 1 metre < 2 metres	m3	0.77	0.82
Filling; wheel, spread & level excavated material; well consolidate; layers 200 mm			
< 50 metre distance	m3	0.70	0.75
> 50 < 100 metres distance	m3	1.17	1.25
Load and transport off site			
< 1 km distance	m3	11.20	12.01
> 1 > 5 km distance	m3	13.53	14.51
Import vegetable soil, spread and level; average 200 mm	m2	11.87	12.73
Harrow top soil to a fine tilth; remove stones and old roots.	m2	0.08	0.09
Seeding and Turfing.			
Grass Seed; lightly roll; 1st cut;			
playing fields	m2	0.04	0.05
low maintenance playing fields	m2	0.05	0.05
low maintenance lawns	m2	0.08	0.09
cricket square	m2	0.10	0.10
bowls	m2	0.12	0.13
Turf; 19 mm thick; lay and roll;			
playing fields	m2	2.49	2.67
low maintenance playing fields	m2	2.75	2.95
low maintenance lawns	m2	2.96	3.17
cricket square	m2	3.94	4.23
bowls	m2	4.25	4.56

SPORTING ACTIVITY AREAS (1)

- Playgrounds;
- 5 A Side Football; Netball; Basketball;
- Tennis; Grass Courts; Hard Courts.

	Unit	New	Refurb
General Macadam Areas; Base course 40 mm thick; wearing course 20 mm thick; laid to falls; hardcore bed 150 mm thick.			
Limestone aggregate	m2	18.24	19.56
> 150 < 300 mm	m	5.99	6.43
< 150 mm wide	m	3.30	3.54
Tennis Courts; Grass; Low maintenance grass seeding; harrow top soil and level and roll.			
General areas > 300 mm wide	m2	0.38	0.41
Extra cost of;			
Site drainage if required	Sum	1,025	1,500
Line markings; paint	Sum	100.00	150.00
Line markings; plastic	Sum	175.00	275.00
Tennis Courts; Macadam; Porous open graded macadam; to falls 1/120; 40 mm wearing course; 20 mm blinding; 150 mm base course; single size stones.			
General areas > 300 mm wide	m2	26.99	28.94
Extra cost of;			
Acrylic protection	m2	4.21	4.51
Coloured paving	m2	4.88	5.23
Site drainage if required	Sum	1,025	1,500
Line markings; paint	Sum	100.00	150.00
Line markings; plastic	Sum	225.00	350.00

SPORTING ACTIVITY AREAS (2)

- Athletic tracks; hard surface;
- Safety play areas.

	Unit	New	Refurb
Athletic Track. 20 mm Shredded rubber bonded wearing course; laid to falls 1/120; 40 mm porous macadam course; 150 mm single size stone base;			
General areas > 300 mm wide	m2	30.28	32.47
Extra cost of;			
Site drainage if required	Sum	1,525	2,225
Line markings; 8 lane; paint	Sum	1,675	2,425
Safety Play Areas. 75 mm Bark Chippings; 100 mm hardcore base			
General areas > 300 mm wide	m2	20.07	21.52
Wood edge trim; 38 x 100 mm	m	4.00	4.29
Safety Play Areas. Continuous shredded rubber bonded wearing course; to falls 1/120; 75 mm in situ no fines concrete; 75 mm hardcore base.			
General areas > 300 mm wide			
40 mm thick; 1.2 m fall	m2	28.80	30.88
60 mm thick; 1.5 m fall	m2	33.68	36.12
110 mm thick; 2.5 m fall	m2	45.88	49.20
Safety Play Areas. Safety tiles; 1 x 1 metre; bonded with adhesive; laid to falls 1/120; 75 mm in situ no fines concrete; 75 mm hardcore base.			
General areas > 300 mm wide			
43 mm thick; 1.2 m fall	m2	42.22	45.27
60 mm thick; 1.5 m fall	m2	45.88	49.20
100 mm thick; 2.5 m fall	m2	53.20	57.04
Extra cost of coloured tiles	m2	4.88	5.23

Sports Area	Grade	Age Group	Total Cost £
Existing grassed areas; drained; rolled and levelled; 3 cuttings; markings and goal posts.			
Association Football	Small	8 - 13	1,740.00
	Medium	13 - 15	1,910.00
	Large	15 - 16+	2,380.00
	Medium	18 and over	2,920.00
	Adult	18 and over	3,350.00
	5-a-side	8 - 14+	470.00
Rugby League	Small	11 - 13	1,720.00
	Medium	13 - 15	2,150.00
	Large	15 and over	3,310.00
	mini league	under 9	1,480.00
	mod league	under 11	2,030.00
Rugby Union	Small	11 - 13	1,810.00
	Medium	13 - 15	2,130.00
	Large	15 - 16+	3,090.00
	Adult	16 and over	3,680.00
	mini rugby	under 7 & 8	810.00
	mini rugby	under 9 & 10	1,320.00
	mini rugby	uner 11	1,490.00
	mini rugby	uner 12	1,580.00
Hockey	Small	11 - 13	1,640.00
	Medium	13 - 15	1,950.00
	Large	15 - 16	2,130.00
	Club	16 and over	2,330.00
	6-a-side	8 - 14+	640.00
	7-a-side	8 - 14+	880.00

Sports Area	Grade	Age Group	Total Cost £
NEW grassed areas; surface harrowed; new top soil; drained; rolled and levelled; seeding; 3 cuttings; markings and goal posts.			
Association Football	Small	8 - 13	2,970.00
	Medium	13 - 15	3,300.00
	Large	15 - 16+	4,140.00
	Medium	18 and over	5,170.00
	Adult	18 and over	6,000.00
	5-a-side	8 - 14+	680.00
Rugby League	Small	11 - 13	2,770.00
	Medium	13 - 15	3,500.00
	Large	15 and over	5,180.00
	mini league	under 9	2,340.00
	mod league	under 11	3,130.00
Rugby Union	Small	11 - 13	2,980.00
	Medium	13 - 15	3,490.00
	Large	15 - 16+	4,810.00
	Adult	16 and over	5,840.00
	mini rugby	under 7 & 8	1,090.00
	mini rugby	under 9 & 10	2,070.00
	mini rugby	uner 11	2,290.00
	mini rugby	uner 12	2,470.00
Hockey	Small	11 - 13	2,740.00
	Medium	13 - 15	3,280.00
	Large	15 - 16	3,600.00
	Club	16 and over	3,940.00
	6-a-side	8 - 14+	830.00
	7-a-side	8 - 14+	2,020.00

Sports Area	Grade	Age Group	Total Cost £
NEW grassed areas; surface harrowed; new top soil; drained; rolled and levelled; seeding; 3 cuttings; nets; 2750 mm fencing.			
Tennis	short	5 - 11	1,480.00
	padder	5 - 11	1,480.00
	lawn	11 - 16+	3,530.00
NEW hard court areas; Porous open graded macadam; hardcore base; plastic line markings; nets; 2750 mm high fencing.			
Tennis	short	5 - 11	6,110.00
	padder	5 - 11	5,460.00
	lawn	11 - 16+	29,370.00
NEW hard court areas; Porous open graded macadam; hardcore base; plastic line markings; posts; 2750 mm high fencing.			
Basketball		12 and over	12,100.00
Netball	first step	8 - 11	5,770.00
	netball	12 and over	13,840.00
Existing grassed areas; drained; rolled and levelled; 3 cuttings; markings and net.			
Volleyball	volleyball	11 and over	350.00
	mini	9 - 13	250.00
Existing grassed areas; drained; rolled and levelled; 3 cuttings; markings.			
Cricket		under 7	430.00
		under 9 - 11	440.00
		under 12 - 13	580.00
		under 14 & 15	720.00

GLOSSARY

Arm barriers;	A post with a boom which can be raised or swung aside to allow access for traffic; either manually or mechanically/automatically operated.
Barriers;	Free standing steel fixed/moveable units used to restrict the movement of crowds;
Bollard; angle,	A post formed of wood, steel or precast concrete and set in the ground to limit vehicular access.
Fencing Post;	Plain post formed of wood, steel angle, or precast concrete.
- End straining post;	A post with a single strut and used at the end of a fencing run.
- Two way straining post;	A post with two struts used to strengthen long runs of fencing.
- Corner straining post;	A post with two struts used at corners at the intersection of two fencing runs.
- Cranked post;	A post with a cranked head with perforations for barbed wire security wiring.
Gravel board;	A wood board set vertically in the ground at the edge of gravel paving
Height restrictor;	A horizontal bar or swinging boom set to limit the height of vehicles using an access route, supported by steel posts.
Telescopic security post;	A lockable retractable steel post set in the paving to allow intermittent access to parking or service roads.

Fencing Contractors Association
St. John's House,
23 St. John's Road,
Watford, WD1 1PY
Telephone; 01923 48895

TYPES OF FENCING

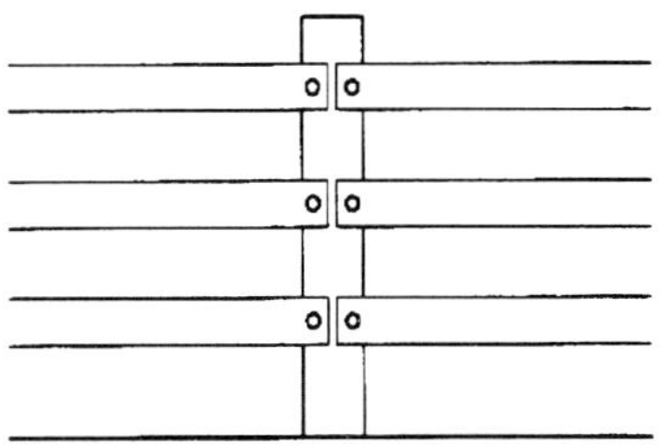

Timber Post and Rail Fence
£23.00 per metre

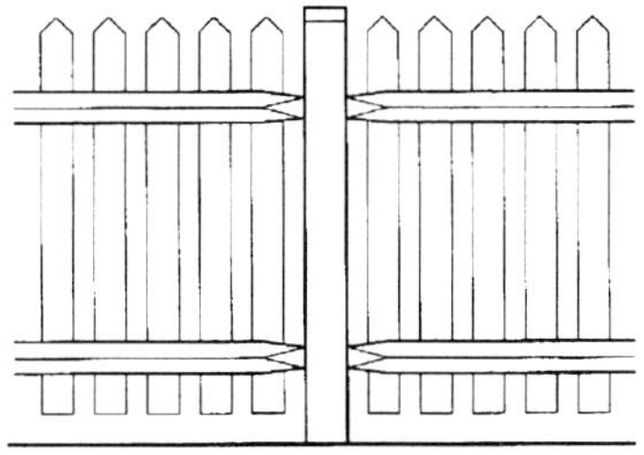

Timber Palisade Fence
£26.00 per metre

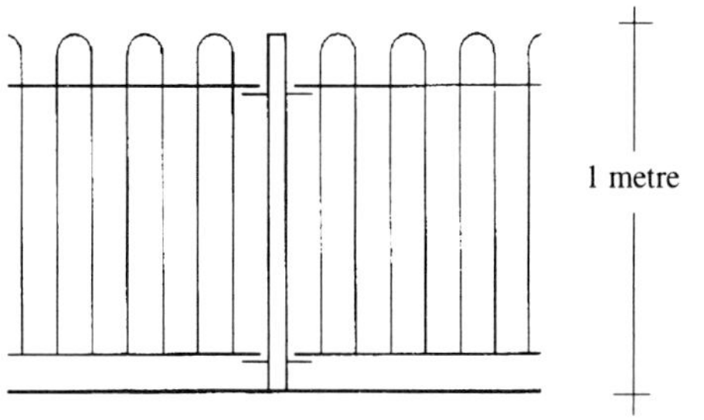

Bow Topped Steel Fence
£35.00 per metre

Close Boarded Fence
£37.00 per metre

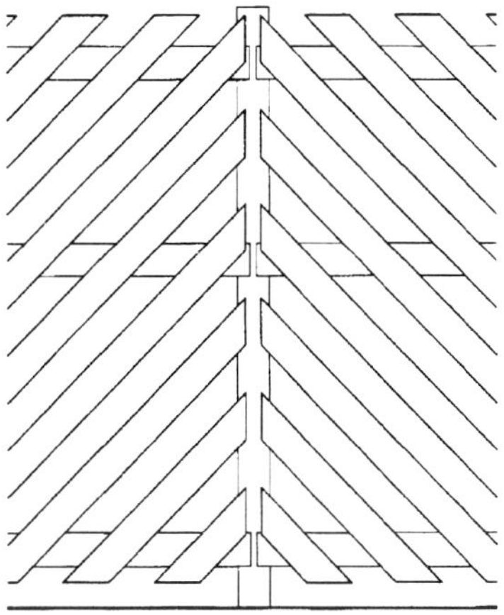

Timber Diagonal Boarded Fence
£42.00 per metre

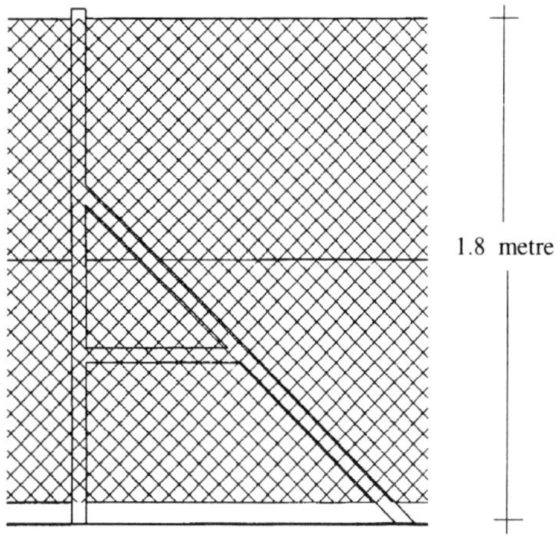

Chain Link Fence
£30.00 per metre

POST AND WIRE FENCING

	Unit	New	Refurb
Strained wire; timber stakes; 1.8 metre centres.			
Fence; 1000 mm high	m	6.73	7.03
Extra for;-			
End straining post; strut	nr	7.88	8.24
Angle straining post; struts	nr	13.12	13.71
Fence; 1200 mm high	m	7.69	8.04
Extra for;-			
End straining post; strut	nr	8.81	9.20
Angle straining post; struts	nr	14.53	15.18
Strained wire; timber posts; 1.8 metre centres.			
Fence; 1000 mm high	m	8.60	8.99
Extra for;-			
End straining post; strut	nr	12.03	12.57
Angle straining post; struts	nr	20.76	21.69
Fence; 1200 mm high	m	9.78	10.22
Extra for;-			
End straining post; strut	nr	13.43	14.04
Angle straining post; struts	nr	23.01	24.04
Strained wire; concrete posts; 1.8 metre centres.			
Fence; 1000 mm high	m	11.67	12.20
Extra for;-			
End straining post; strut	nr	9.32	9.74
Angle straining post; struts	nr	14.84	15.50
Fence; 1200 mm high	m	13.48	14.09
Extra for;-			
End straining post; strut	nr	10.78	11.27
Angle straining post; struts	nr	17.13	17.90
Concrete filling to post hole	nr	1.71	1.79

POST AND RAIL FENCING

	Unit	New	Refurb
Riven timber; rails; posts; 1.8 m centres; nailed.			
Fence			
1000 mm high; two rails	m	5.86	6.13
1000 mm high; three rails	m	7.14	7.46
Fence			
1200 mm high; two rails	m	7.04	7.35
1200 mm high; three rails	m	8.57	8.95
Riven timber; rails; posts; 1.8 m centres; morticed.			
Fence			
1000 mm high; two rails	m	8.12	8.49
1000 mm high; three rails	m	10.42	10.89
Fence			
1200 mm high; two rails	m	9.75	10.19
1200 mm high; three rails	m	12.50	13.06
Sawn timber; rails; posts; 1.8 m centres; nailed.			
Fence			
1000 mm high; two rails	m	6.64	6.94
1000 mm high; three rails	m	8.14	8.50
Fence			
1200 mm high; two rails	m	7.20	7.53
1200 mm high; three rails	m	8.83	9.22
Sawn timber; rails; posts; 1.8 m centres; morticed.			
Fence			
1000 mm high; two rails	m	10.24	10.71
1000 mm high; three rails	m	12.96	13.54
Fence			
1200 mm high; two rails	m	12.29	12.85
1200 mm high; three rails	m	15.55	16.25

TIMBER BOARDED FENCING

	Unit	New	Refurb
Cleft chestnut pale fencing.			
Fence; 900 mm high	m	6.96	7.28
Extra for;-			
End straining post; strut	nr	12.70	13.27
Angle straining post; struts	nr	16.94	17.71
Fence; 1000 mm high	m	7.65	7.99
Extra for;-			
End straining post; strut	nr	14.00	14.63
Angle straining post; struts	nr	18.65	19.49
Close boarded fencing; two rails; hardwood posts.			
1000 mm high	m	21.03	21.98
1200 mm high	m	24.05	25.13
Close boarded fencing; two rails; concrete posts.			
1000 mm high	m	29.09	30.40
1200 mm high	m	30.06	31.41
Close boarded fencing; three rails; hardwood posts.			
1400 mm high	m	27.05	28.27
1600 mm high	m	30.74	32.13
1800 mm high	m	34.29	35.83
Close boarded fencing; three rails; concrete posts.			
1400 mm high	m	39.08	40.84
1600 mm high	m	40.57	42.40
1800 mm high	m	42.14	44.03
Diagonal close boarded fencing; 2/3 rails; hardwood posts.			
1200 mm high	m	36.79	38.44
1400 mm high	m	41.81	43.69
1800 mm high	m	46.64	48.73
Diagonal close boarded fencing; 2/3 rails; concrete posts.			
1200 mm high	m	53.15	55.54
1400 mm high	m	55.18	57.66
1800 mm high	m	57.31	59.88

CHAIN LINK FENCING

	Unit	New	Refurb
Galvanised mesh; steel angle posts; concrete bases.			
Fence; 900 mm high	m	9.99	10.44
Extra for;-			
End straining post; strut	nr	35.66	37.27
Angle straining post; struts	nr	48.05	50.21
Fence; 1200 mm high	m	12.03	12.57
Extra for;-			
End straining post; strut	nr	40.32	42.13
Angle straining post; struts	nr	52.71	55.08
Fence; 1800 mm high	m	15.19	15.87
Extra for;-			
End straining post; strut	nr	57.36	59.94
Angle straining post; struts	nr	79.07	82.63
Fence; 2745 mm high	m	25.05	26.18
Extra for;-			
End straining post; strut	nr	82.75	86.47
Angle straining post; struts	nr	112.91	117.99
Fence; 3660 mm high	m	33.82	35.34
Extra for;-			
End straining post; strut	nr	95.16	99.44
Angle straining post; struts	nr	129.85	135.69
Gate; matching mesh and posts; hinges; lock.			
Single gate			
740 x 900 mm high	nr	263.39	275.24
740 x 1200 mm high	nr	276.06	288.48
740 x 1800 mm high	nr	299.28	312.75
740 x 2000 mm high	nr	344.79	360.30
Pair gate			
2000 x 1200 mm high	nr	669.68	699.82
2000 x 1800 mm high	nr	760.62	794.85
2000 x 2000 mm high	nr	792.49	828.16

	Unit	New	Refurb
Galvanised mesh; concrete posts; concrete bases.			
Fence; 900 mm high	m	10.55	11.02
Extra for;-			
End straining post; strut	nr	40.29	42.10
Angle straining post; struts	nr	54.29	56.73
Fence; 1200 mm high	m	13.63	14.24
Extra for;-			
End straining post; strut	nr	45.55	47.60
Angle straining post; struts	nr	59.54	62.22
Fence; 1800 mm high	m	17.28	18.06
Extra for;-			
End straining post; strut	nr	64.80	67.72
Angle straining post; struts	nr	89.33	93.35
Fence; 2745 mm high	m	27.01	28.23
Extra for;-			
End straining post; strut	nr	93.48	97.69
Angle straining post; struts	nr	127.56	133.30
Fence; 3660 mm high	m	Not used	
Extra for;-			
End straining post; strut	nr	Not used	
Angle straining post; struts	nr	Not used	
Gate; matching mesh and posts; hinges; lock.			
Single gate			
740 x 900 mm high	nr	316.86	331.12
740 x 1200 mm high	nr	329.87	344.72
740 x 1800 mm high	nr	365.05	381.47
740 x 2000 mm high	nr	376.63	393.58
Pair gate			
2000 x 1200 mm high	nr	745.72	779.28
2000 x 1800 mm high	nr	815.94	852.66
2000 x 2000 mm high	nr	836.71	874.36

	Unit	New	Refurb
Plastic coated mesh & steel angle posts; concrete bases.			
Fence; 900 mm high	m	10.67	11.15
Extra for;-			
End straining post; strut	nr	48.84	51.03
Angle straining post; struts	nr	65.80	68.76
Fence; 1200 mm high	m	12.85	13.43
Extra for;-			
End straining post; strut	nr	55.21	57.69
Angle straining post; struts	nr	72.17	75.42
Fence; 1800 mm high	m	18.30	19.12
Extra for;-			
End straining post; strut	nr	78.55	82.08
Angle straining post; struts	nr	108.27	113.15
Fence; 2745 mm high	m	27.19	28.41
Extra for;-			
End straining post; strut	nr	113.31	118.41
Angle straining post; struts	nr	154.62	161.57
Fence; 3660 mm high	m	36.70	38.35
Extra for;-			
End straining post; strut	nr	152.98	159.86
Angle straining post; struts	nr	208.73	218.13
Gate; plastic coated steel mesh and posts; hinges; lock.			
Single gate			
740 x 900 mm high	nr	154.62	161.57
740 x 1200 mm high	nr	167.63	175.17
740 x 1800 mm high	nr	191.48	200.10
740 x 2000 mm high	nr	238.22	248.94
Pair gate			
2000 x 1200 mm high	nr	700.38	731.89
2000 x 1800 mm high	nr	797.32	833.20
2000 x 2000 mm high	nr	813.91	850.54

CHAIN LINK FENCING

	Unit	New	Refurb
Plastic coated mesh; concrete posts; concrete bases			
Fence; 900 mm high	m	10.96	11.46
Extra for;-			
End straining post; strut	nr	44.58	46.59
Angle straining post; struts	nr	61.55	64.32
Fence; 1200 mm high	m	16.11	16.83
Extra for;-			
End straining post; strut	nr	59.43	62.10
Angle straining post; struts	nr	78.55	82.08
Fence; 1800 mm high	m	18.84	19.69
Extra for;-			
End straining post; strut	nr	72.17	75.42
Angle straining post; struts	nr	99.80	104.29
Fence; 2745 mm high	m	28.82	30.12
Extra for;-			
End straining post; strut	nr	113.31	118.41
Angle straining post; struts	nr	154.62	161.57
Fence; 3660 mm high	m	Not used	
Extra for;-			
End straining post; strut	nr	Not used	
Angle straining post; struts	nr	Not used	
Gate; plastic coated mesh; concrete posts; hinges; lock			
Single gate			
740 x 900 mm high	nr	316.86	331.12
740 x 1200 mm high	nr	329.87	344.72
740 x 1800 mm high	nr	353.46	369.36
740 x 2000 mm high	nr	388.22	405.69
Pair gate			
2000 x 1200 mm high	nr	826.85	864.05
2000 x 1800 mm high	nr	873.88	913.21
2000 x 2000 mm high	nr	883.06	922.80

SECURITY FENCING

	Unit	New	Refurb
Galvanised mesh; concrete posts; concrete bases; splayed barbed wire top security barrier 600 mm wide.			
Fence; 1800 mm high	m	20.11	21.01
Extra for;-			
End straining post; strut	nr	64.67	67.58
Angle straining post; struts	nr	89.62	93.65
Electrification	m	5.21	5.45
Fence; 2745 mm high	m	30.37	31.74
Extra for;-			
End straining post; strut	nr	96.74	101.10
Angle straining post; struts	nr	129.28	135.10
Electrification	m	5.91	6.18
Steel palisade security; steel posts and rails; pointed and turned tops; concrete bases.			
Fence; 1800 mm high	m	33.03	34.51
Extra for;-			
Splayed security top barrier	nr	6.20	6.48
Electrification	nr	7.53	7.87
Fence; 2745 mm high	m	44.59	46.59
Extra for;-			
Splayed security top barrier	nr	6.66	6.96
Electrification	m	6.08	6.36
Gate; hinges; security locks.			
Single gate; chain link			
740 x 1800 mm high	nr	365.05	381.47
740 x 2000 mm high	nr	400.97	419.01
Pair gate; chain link			
2000 x 1800 mm high	nr	830.16	867.52
2000 x 2000 mm high	nr	844.30	882.29
Single gate; palisade			
740 x 1800 mm high	nr	399.81	417.80
740 x 2000 mm high	nr	422.99	442.02
Pair gate; palisade			
2000 x 1800 mm high	nr	1,095.1	1,144.4
2000 x 2000 mm high	nr	1,303.7	1,362.4

SUNDRY FENCING

	Unit	New	Refurb
Rebound panels; weldmesh ball catch; steel posts.			
Fence; galvanised			
3600 mm high.	m	81.82	85.50
Single gate; galvanised			
1200 x 2000 mm high	nr	375.02	391.90
3000 x 2000 mm high	nr	477.13	498.60
Fence; plastic coated			
3600 mm high.	m	115.42	120.62
Single gate; plastic coated			
1200 x 2000 mm high	nr	488.84	510.83
3000 x 2000 mm high	nr	610.73	638.21
Bow topped steel fence; painted.			
Fence; 900 mm high	m	20.86	21.80
Extra for;-			
single gate; latch	nr	28.39	29.67
Fence; 1000 mm high	m	24.92	26.04
Extra for;-			
single gate; latch	nr	30.71	32.09
Fence; 1800 mm high	m	44.04	46.02
Extra for;-			
single gate; latch	nr	63.74	66.61
double gate; latch	nr	99.66	104.15
Ornamental steel railings.			
Fence; 900 mm high	m	37.08	38.75
Extra for;-			
single gate; latch	nr	75.33	78.72
Fence; 1000 mm high	m	41.72	43.60
Extra for;-			
single gate; latch	nr	52.15	54.50
Fence; 1800 mm high	m	83.44	87.19
Extra for;-			
single gate; latch	nr	98.50	102.94
double gate; latch	nr	136.75	142.90

BOLLARDS

	Unit	New	Refurb
Bollards and Control Posts			
Wood; plain; 900 mm high			
round; 100 mm diameter	nr	12.05	12.59
round; 150 mm diameter	nr	17.42	18.21
Wood; plain;1200 mm high			
round; 100 mm diameter	nr	13.61	14.22
round; 150 mm diameter	nr	20.40	21.31
Wood; plain; 1800 mm high			
round; 100 mm diameter	nr	15.73	16.43
round; 150 mm diameter	nr	25.40	26.55
Wood; grooved; 900 mm high			
round; 100 mm diameter	nr	14.69	15.35
round; 150 mm diameter	nr	17.89	18.69
Wood; grooved;1200 mm high			
round; 100 mm diameter	nr	18.55	19.39
round; 150 mm diameter	nr	23.19	24.23
Wood; grooved; 1800 mm high			
round; 100 mm diameter	nr	20.59	21.52
round; 150 mm diameter	nr	25.87	27.03
Metal; flat top; 1100 mm high			
round; 168 mm diameter	nr	72.77	76.04
round; 150 mm diameter	nr	90.76	94.84
Metal; pyramid top; 1200 mm high			
square; 100 x 100 mm	nr	55.39	57.89
square; 150 x 150 mm	nr	63.92	66.79
Marker Rails			
Silver rail; 42 mm diameter steel tube; softwood posts.	m	51.11	53.41
Softwood diamond rail; posts; treated; iron straps.	m	13.27	13.87

GATES AND SECURITY FITTINGS

	Unit	New	Refurb
Entrance Gates; posts; ironmongery; stained.			
Pedestrian gates;			
900 mm wide	nr		
1200 mm wide	nr	152.97	159.86
1500 mm wide	nr	166.88	174.39
Vehicle gates;			
1800 mm wide	nr	169.72	177.35
2400 mm wide	nr	194.23	202.97
3000 mm wide	nr	218.97	228.82
3600 mm wide	nr	241.51	252.38
Temporary Fencing; galvanised steel.			
Unit width; 3500 mm			
2000 mm high	m	13.18	13.77
1000 mm high	m	16.26	16.99
Extra for gate	nr	46.34	48.43
Security toppings for steel fencing			
Multi spike type	m	33.03	34.51
Spiked capping to brickwork	m	15.82	16.53
RT Barriers			
Bollards; cast into concrete base	nr	32.45	33.91
Sliding barriers; standards cast into concrete bases.	nr	347.66	363.31
Hinged barriers; standards cast into concrete bases.	nr	805.42	841.66
Sound Preventative Fencing			
Wood; plain;1200 mm high Close boarded fencing; three rails; concrete posts			
2000 mm high	m	49.25	51.47
2500 mm high	m	58.82	61.47
3000 mm high	m	65.82	68.79

SUMMER HOUSES

	Unit	New	Refurb
Garden Sheds; softwood; stained.			
Single door; window			
1800 x 1200 mm	nr	300.21	313.72
2100 x 1500 mm	nr	290.88	303.97
2400 x 1800 mm	nr	362.73	379.05
3000 x 1800 mm	nr	362.73	379.05
3600 x 1800 mm	nr	636.22	664.85
3900 x 1800 mm	nr	701.12	732.67
4200 x 1800 mm	nr	757.90	792.01
extra for floor and bearers	m2	15.64	16.35
Summer Houses; glazed doors; windows; fitted base with bearers;			
Small summer house			
2700 x 1800 mm	nr	981	1,025
3000 x 1800 mm	nr	897	937
3000 x 2400 mm	nr	990	1,034
Octagonal Summer House; glazed doors; windows; fitted base and bearers			
2400 x 1800 mm	nr	1,077	1,126
Wendy House			
1500 x 1500 mm	nr	415	434
Play cabins			
Den; 1800 x 1200 mm	nr	363	380
Cabin; 2400 x 1500 mm	nr	626	654

Trenches Depth	NEW WORK								REFURBISHMENT WORKS							
	Pipe diameter								Pipe diameter							
	100 mm	150 mm	225 mm	300 mm	375 mm	450 mm	600 mm	1200 mm	100 mm	150 mm	225 mm	300 mm	375 mm	450 mm	600 mm	1200 mm
DIGGING; HAND	Trenches; hand excavation; grade bottom to falls; back filling; earth supports; cart away surplus															
0.75 m deep	10.26	10.52	10.78	12.28	13.68	15.43	NA	NA	11.03	11.31	11.59	13.20	14.70	16.58	NA	NA
1.00 m deep	16.12	16.53	16.95	18.99	21.10	23.42	25.66	NA	17.33	17.77	18.22	20.42	22.68	25.17	27.58	NA
1.50 m deep	30.87	31.66	32.45	35.53	39.24	42.89	48.84	59.72	33.18	34.03	34.88	38.19	42.18	46.11	52.51	64.20
2.00 m deep	46.20	47.38	48.57	47.38	52.29	57.21	64.57	79.30	49.66	50.94	52.21	50.94	56.21	61.50	69.41	85.24
2.50 m deep	69.43	71.21	72.99	71.21	78.92	86.30	91.55	119.96	74.64	76.55	78.46	76.55	84.84	92.77	98.42	128.95
3.00 m deep	83.24	85.37	87.50	85.37	94.49	103.60	116.93	143.59	89.48	91.77	94.07	91.77	101.58	111.37	125.70	154.36
3.50 m deep	97.04	99.53	102.02	99.53	110.75	120.57	136.35	167.56	104.32	106.99	109.67	106.99	119.05	129.61	146.57	180.12
DIGGING; MACHINE	Trenches; machine excavation; grade bottom to falls; back filling; earth supports; cart away surplus															
0.75 m deep	4.05	4.16	4.26	4.95	5.66	5.88	NA	NA	4.36	4.47	4.58	5.32	6.09	6.32	NA	NA
1.00 m deep	8.01	8.22	8.42	8.92	9.80	10.48	11.17	NA	8.61	8.84	9.06	9.59	10.53	11.27	12.01	NA
1.50 m deep	12.95	13.28	13.61	14.16	15.40	16.75	18.14	21.27	13.92	14.28	14.63	15.22	16.56	18.01	19.50	22.87
2.00 m deep	17.71	18.16	18.61	18.68	19.56	21.11	23.20	27.02	19.03	19.52	20.01	20.08	21.03	22.70	24.94	29.04
2.50 m deep	25.86	26.52	27.19	26.69	28.78	30.69	33.30	38.85	27.80	28.51	29.23	28.69	30.94	33.00	35.79	41.77
3.00 m deep	31.42	32.22	33.03	32.39	34.14	36.41	40.56	46.64	33.78	34.64	35.51	34.82	36.70	39.14	43.61	50.14
3.50 m deep	35.98	36.90	37.82	37.24	39.67	42.81	46.44	54.27	38.68	39.67	40.66	40.03	42.65	46.02	49.93	58.34
GRAVEL; BEDS	Granular Filling; bed															
100 mm thick	2.37	2.66	3.09	3.49	3.92	4.34	4.76	8.81	2.55	2.86	3.32	3.76	4.21	4.66	5.12	9.47
150 mm thick	3.49	3.92	4.34	4.76	5.88	6.30	7.42	13.73	3.76	4.21	4.66	5.12	6.32	6.78	7.98	14.76
GRAVEL; SURROUND	Granular Covering; bed and surround															
100 mm thick	6.72	8.19	11.20	13.45	16.17	19.19	24.64	45.59	7.23	8.81	12.04	14.46	17.39	20.62	26.49	49.01
150 mm thick	9.94	11.20	14.43	17.15	20.17	24.16	29.13	53.89	10.69	12.04	15.51	18.43	21.69	25.97	31.31	57.93
CONCRETE; BED	Concrete; in situ; 20 N/mm2; bed															
100 mm thick	4.53	5.00	5.46	6.33	7.25	8.12	9.05	14.52	4.87	5.38	5.87	6.80	7.80	8.73	9.73	15.61
150 mm thick	6.33	7.25	8.12	9.05	10.85	11.79	13.58	20.84	6.80	7.80	8.73	9.73	11.66	12.67	14.60	22.40
CONCRETE; HAUNCH	Concrete; in situ; 20 N/mm2; bed and haunching both sides															
100 mm thick	8.42	9.42	11.30	13.17	16.92	19.80	25.43	61.13	9.06	10.13	12.15	14.16	18.19	21.28	27.33	65.72
150 mm thick	9.42	10.38	13.17	16.92	19.80	23.55	30.09	67.69	10.13	11.15	14.16	18.19	21.28	25.31	32.35	72.77
CONCRETE SURROUND	Concrete; insitu; 20 N/mm2; beds and surround															
100 mm thick	11.68	13.63	19.47	23.37	28.27	33.10	42.84	89.58	12.55	14.65	20.93	25.12	30.39	35.59	46.05	96.29
150 mm thick	17.53	19.47	25.32	30.21	35.04	41.90	50.64	116.83	18.85	20.93	27.22	32.48	37.67	45.04	54.44	125.60

PIPEWORK (Nominal diameters)	NEW BUILDING WORK — Pipe diameter - laid straight lines								REFURBISHMENT WORKS — Pipe diameter; short branches or refurbishment work							
	100 mm	150 mm	225 mm	300 mm	375 mm	450 mm	600 mm	1200 mm	100 mm	150 mm	225 mm	300 mm	375 mm	450 mm	600 mm	1200 mm
									Note; NA means 'Not Available"							
LAND DRAINS	Vitrified clay pipes; socketed; with push fit flexible joints															
Pipework; perforated	6.76	11.41	20.05	NA	NA	NA	NA	NA	7.27	12.27	21.56	NA	NA	NA	NA	NA
Pipework; butt joints	5.24	8.92	15.00	NA	NA	NA	NA	NA	5.64	9.59	16.13	NA	NA	NA	NA	NA
CLAY PIPES	Vitrified clay pipes; socketed; with push fit flexible joints															
Pipework	10.22	13.06	23.95	36.74	72.78	94.31	NA	NA	10.99	14.04	25.74	39.50	78.23	101.38	NA	NA
Extra for fittings;-																
bend	11.10	18.43	36.79	73.85	238.68	314.39	NA	NA	11.93	19.81	39.55	79.39	256.58	337.97	NA	NA
access pipe	24.37	38.25	97.26	133.10	NA	NA	NA	NA	26.19	41.12	104.56	143.09	NA	NA	NA	NA
single junction	15.30	23.93	55.94	117.47	216.79	256.82	NA	NA	16.45	25.73	60.14	126.28	233.04	276.08	NA	NA
double junction	42.42	51.22	113.29	243.17	NA	NA	NA	NA	45.60	55.06	121.79	261.40	NA	NA	NA	NA
CLAY PIPES	Vitrified clay pipes; socketed; cement and sand (1:2) joints; tarred yarn															
Pipework	8.04	13.06	23.95	36.74	72.78	94.31	NA	NA	8.64	14.04	25.74	39.50	78.23	101.38	NA	NA
Extra for fittings;-																
bend	5.93	18.43	36.79	73.85	238.68	314.39	NA	NA	6.37	19.81	39.55	79.39	256.58	337.97	NA	NA
access pipe	31.87	38.25	97.26	133.10	NA	NA	NA	NA	34.27	41.12	104.56	143.09	NA	NA	NA	NA
single junction	9.55	19.86	46.42	97.48	179.89	213.11	NA	NA	10.27	21.35	49.90	104.79	193.38	229.09	NA	NA
double junction	17.30	42.50	94.01	201.78	NA	NA	NA	NA	18.60	45.69	101.06	216.91	NA	NA	NA	NA
CONCRETE	Precast concrete drains															
Pipework	NA	NA	NA	18.49	NA	37.08	46.70	NA	NA	NA	NA	19.87	NA	39.86	50.20	NA
Extra for fittings;-																
bend	NA	NA	NA	56.28	NA	118.55	145.24	NA	NA	NA	NA	60.50	NA	127.44	156.13	NA
access pipe	NA	NA	NA	NA	NA	NA	NA	NA								
single junction	NA	NA	NA	29.05	NA	33.09	34.76	NA	NA	NA	NA	31.23	NA	35.57	37.36	NA
double junction	NA	NA	NA	39.11	NA	44.44	46.07	NA	NA	NA	NA	42.04	NA	47.78	49.53	NA
CAST IRON	Coated cast iron; mechanical coupling joints															
Pipework	28.21	47.91	NA	NA	NA	NA	NA	NA	30.33	51.51	NA	NA	NA	NA	NA	NA
Extra for fittings;-																
bend	37.70	62.71	NA	NA	NA	NA	NA	NA	40.52	67.42	NA	NA	NA	NA	NA	NA
access pipe	28.42	40.31	NA	NA	NA	NA	NA	NA	30.55	43.33	NA	NA	NA	NA	NA	NA
W.C connector	34.38								36.96					Not available		
single junction	57.39	75.50	NA	NA	NA	NA	NA	NA	61.69	81.16	NA	NA	NA	NA	NA	NA
single access junction	99.53	165.59	NA	NA	NA	NA	NA	NA	106.99	178.00	NA	NA	NA	NA	NA	NA
double junction	85.96	143.01	NA	NA	NA	NA	NA	NA	92.41	153.74	NA	NA	NA	NA	NA	NA
double access junction	126.45	210.37	NA	NA	NA	NA	NA	NA	135.93	226.15	NA	NA	NA	NA	NA	NA

Type	Internal Size		Excavation Work		Concrete work			Brickwork		Precast Concrete			Polypropylene		Channels		Covers		Total
Depth to invert of drainage channel	Width mm	Length mm	Digging and disposal £	Filling £	Base £	Cover £	Bench for channels £	Half Brick £	One Brick £	Base £	Rings £	Cover £	£	£	Straight £	Curved £	Light £	Light recessed £	£
									SOAKAWAYS										
Hardcore Filling			Excavate to form pit and cart away; fill with broken brick hardcore (18 to 50 mm stones; no fine materials); cover with hessian; replace vegetable soil > 300 mm thick or pavings.																
1,000 mm	900	900	15.59	14.24	nil	1.25	nil	nil	nil	nil	nil	nil	nil	nil	nil	nil	nil	nil	31.09
2,000 mm	1,200	1,200	44.32	49.46	nil	2.22	nil	nil	nil	nil	nil	nil	nil	nil	nil	nil	nil	nil	96.01
Precast Concrete			Excavate to form pit and cart away; assemble precast concrete units with cover; set in pit; replace vegetable soil > 300 mm thick or pavings.																
1,000 mm	675		12.38	0.82	3.23	nil	nil	nil	nil	45.85	nil	32.96	nil	nil	nil	nil	nil	nil	95.24
1,500 mm	1,050		31.73	2.52	7.80	nil	nil	nil	nil	86.40	nil	39.14	nil	nil	nil	nil	nil	nil	167.59
									MANHOLES										
Brickwork			Excavate for manhole; cart away surplus; backfill around new sides; build walls in semi engineering brickwork in cement mortar (1:3); fair face and pointing internally; benching; channels and covers.																
1,000 mm	450	600	16.13	1.19	6.26	17.80	3.80	47.90	nil	nil	nil	nil	nil	nil	2.75	4.64	24.72	nil	125.19
1,000 mm	450	750	17.79	1.37	7.19	20.92	4.75	53.61	nil	nil	nil	nil	nil	nil	3.44	9.27	24.72	nil	143.05
1,500 mm	450	600	31.38	2.73	9.58	22.65	3.80	nil	144.18	nil	nil	nil	nil	nil	2.75	4.64	24.72	nil	246.42
1,500 mm	600	900	40.00	3.87	13.57	34.89	7.59	nil	195.68	nil	nil	nil	nil	nil	4.13	13.91	24.72	nil	338.35
Brickwork			Excavate for manhole; cart away surplus; backfill around new sides; build walls in semi engineering brickwork in cement mortar (1:3); fair face and pointing internally; benching; channels and covers.																
1,000 mm	450	600	21.49	1.82	9.58	22.65	3.80	nil	102.04	nil	nil	nil	nil	nil	2.75	4.64	55.62	nil	224.38
1,500 mm	600	900	40.00	3.87	13.57	34.89	7.59	nil	225.03	nil	nil	nil	nil	nil	4.13	9.27	55.62	nil	393.97
2,000 mm	900	1,200	69.51	7.67	20.19	56.96	15.18	nil	420.90	nil	nil	nil	nil	nil	5.50	13.91	55.62	nil	665.45
Precast Concrete Inspection Chambers			Excavate for inspection chamber; cart away surplus; backfill around new sides; assembly precast concrete units and set in cement mortar (1:3); benching; channels and covers.																
450 mm	460	610	6.94	0.43	4.99	nil	3.95	nil	nil	84.39	nil	nil	nil	nil	2.80	4.64	24.72	nil	132.84
600 mm	460	610	8.96	0.57	4.99	nil	3.95	nil	nil	115.29	nil	nil	nil	nil	2.80	9.27	24.72	nil	170.53
450 mm	690	1,030	10.82	0.81	9.49	nil	9.99	nil	nil	109.70	nil	nil	nil	nil	4.72	4.64	24.72	nil	174.89
600 mm	690	1,030	13.86	1.08	9.49	nil	9.99	nil	nil	149.87	nil	nil	nil	nil	4.72	9.27	24.72	nil	223.01
Precast Concrete Manholes			Excavate for manhole; cart away surplus; backfill around new sides; assembly precast concrete units and set in cement mortar (1:3); benching; channels and covers.																
1,000 mm	1,050		26.36	1.57	8.77	nil	12.46	nil	nil	57.68	139.85	57.68	nil	nil	nil	4.64	55.62	nil	364.63
1,500 mm	1,050		37.51	2.29	8.77	nil	12.46	nil	nil	57.68	266.99	57.68	nil	nil	nil	9.27	55.62	nil	508.26
2,000 mm	1,500		75.03	5.34	15.60	nil	25.42	nil	nil	74.98	489.17	74.98	nil	nil	nil	4.64	55.62	nil	820.78
									INSPECTION CHAMBERS										
uPVC Inspection Chambers			Excavate for inspection chamber; cart away surplus; backfill around new sides; assembly uPVC units and set in cement mortar (1:3); benching; channels and covers.																
600 mm	250 mm diameter		3.10	0.08	0.39	nil	nil	nil	nil	nil	nil	nil	68.45	nil	nil	nil	18.54	nil	90.56
500 mm	460 mm diameter		4.55	0.16	0.90	nil	nil	nil	nil	nil	nil	nil	103.89	nil	nil	nil	18.54	nil	128.03
730 mm	460 mm diameter		6.30	0.23	0.90	nil	nil	nil	nil	nil	nil	nil	119.07	nil	nil	nil	18.54	nil	145.03
960 mm	460 mm diameter		8.04	0.29	0.90	nil	nil	nil	nil	nil	nil	nil	134.37	nil	nil	nil	18.54	nil	162.15

MANHOLE COVERS

	Unit	New	Refurb
Light Duty; Grade C; solid cover; coated metal; rectangular; single seal. for use in pedestrian and decorative paved areas.			
450 x 450 mm	nr	44.07	47.38
600 x 450 mm	nr	44.37	47.70
600 x 600 mm	nr	83.99	90.29
Light Duty; Grade C; solid cover; coated metal; rectangular; double seal. for use in pedestrian and decorative paved areas.			
450 x 450 mm	nr	62.89	67.61
600 x 450 mm	nr	79.63	85.60
600 x 600 mm	nr	108.59	116.74
Light Duty; Grade C; recessed cover; coated metal; rectangular; single seal. for use in pedestrian and decorative paved areas.			
600 x 450 mm	nr	80.88	86.94
600 x 600 mm	nr	108.25	116.37
Light Duty; Grade C; recessed cover; coated metal; rectangular; double seal. for use in pedestrian and decorative paved areas.			
450 x 450 mm	nr	98.41	105.79
600 x 450 mm	nr	115.61	124.28
600 x 600 mm	nr	128.95	138.62
Medium Duty; Grade B; solid cover; for use in light vehicular traffic areas.			
600 x 450 mm	nr	73.90	79.45
600 x 600 mm	nr	91.39	98.24
300 mm diameter	nr	115.20	123.84
600 mm diameter	nr	80.81	86.87

	Unit	New	Refurb
Medium Duty; Grade B; solid cover; coated iron; rectangular; double seal. for use in light vehicular traffic areas.			
600 x 450 mm	nr	82.45	88.63
600 x 600 mm	nr	101.95	109.60
300 mm diameter	nr	128.51	138.15
600 mm diameter	nr	90.15	96.92
Medium Duty; Grade B; recessed cover; coated iron; rectangular; single seal. for use in light vehicular traffic areas.			
600 x 450 mm	nr	117.62	126.44
600 x 600 mm	nr	132.26	142.18
300 mm diameter	nr	169.64	182.36
600 mm diameter	nr	119.00	127.93
Medium Duty; Grade B; recessed cover; coated iron; rectangular; double seal. for use in light vehicular traffic areas.			
600 x 450 mm	nr	135.26	145.40
600 x 600 mm	nr	152.10	163.51
300 mm diameter	nr	195.08	209.72
600 mm diameter	nr	136.85	147.12
Heavy Duty; Grade A; solid cover; coated iron; double seal. for use in heavy vehicular traffic areas.			
single triangular type;			
550 x 495 mm	nr	135.93	146.12
double triangular type;			
550 x 495 mm	nr	126.67	136.17
550 x 550 mm	nr	142.25	152.91

GULLY GRATINGS

	Unit	New	Refurb
Light Duty; coated iron; open grating for use in pedestrian and decorative paved areas.			
Cast Iron:			
150 x 150 mm	nr	1.83	1.96
225 x 225 mm	nr	5.31	5.71
250 x 250 mm	nr	6.02	6.47
150 mm diameter	nr	1.45	1.56
200 mm diameter	nr	3.00	3.22
282 mm diameter	nr	5.61	6.03
Galvanised:			
150 x 150 mm	nr	2.47	2.65
225 x 225 mm	nr	7.17	7.71
250 x 250 mm	nr	8.13	8.74
150 mm diameter	nr	1.96	2.11
200 mm diameter	nr	4.05	4.35
282 mm diameter	nr	7.57	8.14
Light Duty; sealed grating with frame; locking; for use in pedestrian and decorative paved areas.			
Cast Iron:			
150 x 150 mm	nr	9.30	10.00
225 x 225 mm	nr	16.50	17.74
250 x 250 mm	nr	18.38	19.76
150 mm diameter	nr	8.94	9.61
200 mm diameter	nr	12.42	13.36
282 mm diameter	nr	24.16	25.97
Galvanised:			
150 x 150 mm	nr	12.55	13.50
225 x 225 mm	nr	22.28	23.95
250 x 250 mm	nr	24.81	26.67
150 mm diameter	nr	12.07	12.97
200 mm diameter	nr	16.77	18.03
282 mm diameter	nr	32.61	35.06

Chapter 5

THE ADMINISTRATION of BUILDING PROJECTS

THE COST BASE

The Prices, Cost Analyses and Unit Rates for Measured Building Work are all based upon the cost of wages and prices of materials ruling during the first quarter of 1999 and within the Home Counties.

GENERAL

This chapter covers the administration necessary for the following types of building projects:

i) New building works;
(new extensions, small repair work and re-modelling existing premises)

ii) Cyclical maintenance work.

Governors and head teachers should be wary of undertaking the direct management of all but the smallest building project. They should always consider the appointment of the appropriate building consultant who should be qualified and used to dealing with educational premises.

But, irrespective of the type of project, it is important that records are kept of all correspondence and financial transactions which may be needed for reference or inspection at a later stage. Wherever possible all notes, transactions, minutes of meetings and instructions should be made in writing. Contemporaneous records are invaluable if disputes arise. Records must be kept for not less than seven years.

Records are necessary for audit, but will also prove invaluable should problems or latent defects arise in the work at a later stage, or simply as a record of the materials used which may need replacement.

For larger building and maintenance contracts an independent filing system should be set up so that correspondence with each consultant, contractor and supplier and the instructions, variations, financial reports and the like can be filed systematically enabling them to be found quickly and referred to without delay.

THE PROJECT BRIEF

A clear brief should be defined by the 'Employer' at the outset of each project. This is often the very essence of the success of a project. Even minor repair works should form a part of an overall clearly defined maintenance programme.

The brief should set out the following:

i) the person appointed to oversee the project and special conditions;
ii) the scope and extent of the works;
iii) the budget cost;
iv) the date work is to start and the completion date.

The brief may be drawn up by the head teacher, bursar, principal or governors, or by a committee of staff from the members, which may also include consultants acting in an 'advisory' capacity depending upon the scope of the proposed works.

Time should be taken to discuss each option, the programme and costs involved. Notes of these discussions and meetings should be retained since they can be very helpful at a later stage.

For minor work the brief may take the form of a hand written note or letter. For larger projects settling the brief may take more time. Initially it may take the form of an 'agenda' or programme setting out the development stages, the appointments to be made and the critical dates.

A typical brief for a large project is set out in Appendix 'U'. It can be useful to circulate the brief to all those involved including each consultant and, if relevant, the private investors. If the brief is attached to their contract of appointment, they become bound by its content and the intention of all parties is clear.

APPOINTMENT OF CONSULTANTS

Figure 5.3 on page 229 shows the preferred involvement of the various resource controllers at each of the planning, design and construction stages of a project. Ideally these arrangements should also include new cyclical maintenance and facilities management contracts.

If it is decided to appoint consultants, competitive proposals should be invited from the potential consultants in each discipline. Their proposals should provide the following information:

i) The nature of their business and portfolio;
ii) Address, trading name and VAT No;
iii) If incorporated, the Company Registration No;
iv) Professional Indemnity Certificate;
v) Number of employees;
vi) Qualifications and experience;
vii) An indication of the services they would provide;
viii) A fee proposal;
ix) Three references.

After interview and selection, a formal agreement should be drawn up, either by letter or upon standard agreements which can usually be obtained from the relevant professional associations.

However, the harsh reality of our economy and restricted funding often prohibits the appointment of professional advisers, but clients should be clear about the value of an independent consultant. They usually help to avoid pitfalls and enhance the performance and design of the work involved.

A guide to the agreement of professional fees to be paid to Consultants is set out on pages 234 - 236.

ARCHITECT

The Architect is usually the key consultant on large projects and his decisions are legally binding.

the Architect's duties include:

- ♦ Design, space planning and surveys;
- ♦ Planning and Building Regulation approvals;
- ♦ Contract drawings;
- ♦ Supervising the works;
- ♦ Instructions and variation orders;
- ♦ Payment certificates;
- ♦ Completion and/or delay certificates.

Details of members from:
Royal Institute of British Architects
66, Portland Place
London, W1N 4AD
Tel; 0171 580 5533

QUANTITY SURVEYOR

The Quantity Surveyor is employed to protect the financial interests of the Employer.

the Quantity Surveyor's duties include:

- ♦ Benchmark estimates and cost plans;
- ♦ Financial reports;
- ♦ Bills of Quantities and Schedules of Works;
- ♦ Tender procedures and quotations;
- ♦ Payment recommendations;
- ♦ Final accounts;
- ♦ Investment portfolios.

Details of members from:
Royal Institution of Chartered Surveyors
12, Great George Street
London, SW1P 3AD
Tel; 0171 222 7000

STRUCTURAL ENGINEER

The Structural Engineer is normally appointed, if required, on the recommendation of the Architect.

the Structural Engineer's duties include:

- ♦ Structural report on existing premises;
- ♦ Soil reports;
- ♦ Structural design work;
- ♦ Structural calculations for regulation approval;
- ♦ Supervising the structural works;
- ♦ Structural certificate.

Details of members from:
Institution of Structural Engineers
12, Upper Belgrave Street
London, SW1X 8BH
Tel; 0171 235 4535

PLANNING SUPERVISOR

The Construction (Design and Management) Regulations 1994 require the appointment of a Planning Supervisor if the duration of the works exceeds more than one month or the labour employed exceeds 30 man days.

the Planning Supervisor's duties include:

- ♦ Risk assessment including site and design; hazards (Unstable buildings, asbestos etc);
- ♦ Health and Safety Plan for the works;
- ♦ Questionnaires to prospective contractors;
- ♦ Health and Safety file;
- ♦ Final report and records and drawings.

Details of members from:
Royal Institution of Chartered Surveyors
12, Great George Street
London, SW1P 3AD
Tel; 0171 222 7000

SERVICES ENGINEER

Very large projects, or complicated remodelling work may require a consultant to deal with specialist electrical and mechanical installations and is usually recommended by the Architect.

the Services Engineer's duties include:

- ♦ Reporting the condition of existing installations;
- ♦ Designing new installations;
- ♦ Specification for specialist work;
- ♦ Supervising the specialist work and advising the Architect on standards and progress;
- ♦ Instructions and variation orders.

Details of members from:
Chartered Institute of Building Service Engineers
Delta House
222 Balham High Road
London SW12 9BS
Tele; 0181 675 5211

CLERK OF WORKS

A Clerk of Works is appointed to enable the works to be supervised on a day to day basis on behalf of the Architect and is an additional increment on their fees.

the Clerk of Work's duties include:

- ♦ Ensuring that the works comply with the contract specification and the drawings;
- ♦ Ensuring the Contractor complies with all special site restrictions;
- ♦ Maintaining a site diary of progress and events;
- ♦ Maintaining records of site labour;
- ♦ Issuing verbal instructions on behalf of the Architect.

Details of members from:
The Institute of Clerks of Works
41, The Mall, Ealing
London, W5 3TJ
Tele; 0181 579 2917

TENDER INVITATIONS

Small Works.
Invitations to submit quotations for small building and repair works may be called for and accepted verbally. But it is safer and much more preferable to make both the invitation and the order to carry out the works in writing so that a record of your instructions is available in the future.

Quotations for small works should state clearly the extent of the work required to be carried out and any special conditions that may apply.

Tenders should always be called for on a 'fixed sum' basis and so that quotations are not subject to variations in the work requiring re-measurement, delays due to inclement weather, inflation or changes in the price of the materials.

In every case the contractor must be instructed to visit the site to inspect the premises and the nature of the proposed work for himself. The scope of the work and the conditions under which he must work must not be varied without also informing the other contractors.

The order for the work must refer to the quotation and its date, the agreed amount of the lump sum, the terms of payment and the start and completion dates for the works. Any other special conditions regarding carrying out the work, access, protection, temporary works and any other safety precautions should be made clear.

There must be written confirmation of the situation regarding the deduction of tax. Value Added Tax should only be paid if the contractor is registered for VAT.

Main Building Works.
Tenders for building work should be invited under the rules contained under the Code of Procedure for Single Stage Selective Tendering issued by NJCC which sets out the procedures for inviting and assessing competitive bids.

Figure 5.1
TENDER DOCUMENTS REQUIRED FOR A STANDARD TENDER INVITATION FOR BUILDING WORKS.

i) Tender invitation setting out the terms of the tender submission;
ii) Specification for the works including details of the Form of Agreement to be used in the Contract;
iii) Detailed Conditions to be incorporated into the Contract;
iv) Schedule of the Works or Measured Bill of Quantities;
v) A numbered set of drawings showing the extent and basic construction of the work;
vi) Details of 'Named' specialists, the specification of their work and their tender invitations;
vii) Health and Safety Plan for the works;
viii) Form of Tender and Tender envelope.

When the Contractor's involvement is considered to be helpful at an earlier stage, the Code for Two Stage Selective Tendering allows for inviting and assessing alternative contractors' preliminary bids prior to the preparation of the full documentation.

Tender Documents.
The invitations, in whatever form, should include the documents listed in Figure 5.1 where available and state clearly:

i) the extent of the works involved;
ii) start and completion dates;
iii) method of payment;
iv) any special conditions which may affect carrying out the work.

Inviting tenders for larger contracts under the selective tendering procedure is more involved since the risks and the costs of preparing tender submissions for both parties are much greater and the invitations should provide:

i) a detailed specification including a description of the Conditions of Contract;
ii) set of drawings (numbered);
iii) Health and Safety Plan for the works;
iv) a specification for specialist installations.

Tenders should be submitted on a pre-selected date and time to the head of the school or university who may be the Head Teacher or the Chairman of the Board of Governors. All the tenders should be opened at the same time in front of witnesses so that the procedure can be seen to be entirely fair and above board.

Bids which are late should be rejected, although this may be difficult if the offending contractor's bid is particularly attractive. Technically, if it is wished to consider such a tender, the other contractors should also be offered the opportunity of reviewing their tenders, but this is a dangerous procedure and not one that can be recommended.

TENDER SUBMISSIONS

The Form of Tender

Tenders for large contracts are usually required to be submitted on a formal tender document, a copy of which is set out in Appendix T. But even small quotations and tenders are better submitted in a standard form so that alternative offers and quotations can be easily reconciled, one with the other and with the school's requirements.

Main Contractors are normally asked to submit their tenders in the following parts:

i) Form of Tender clearly stating the amount of their tender;

ii) Detailed breakdown of their lump sum;

iii) Method Statement showing how they propose to carry out the work;

iv) Health and Safety Plan;

v) Insurance and Tax Certificates.

The 'lump sum' tender submission, if accepted, binds the Contractor to the Conditions of Contract set out in the tender documents and forming the basis of the proposed agreement.

The detailed price build-up may be based upon a formal Bill of Quantities prepared by the Quantity Surveyor or the Contractor's own analysis. Either should be in a form that provides:

i) a means of ensuring that the tender is arithmetically correct;

ii) so that the basis of the lump sum tender can be seen to be fair and reasonable;

iii) a check on the content of the bid;

iv) a basis for pricing variations which may be ordered at a later stage.

Approval of Tenders and Quotations

The approval of tenders and quotations should be considered both on the basis of the competitive price submitted and the details of the submission.

Quotations for Minor Works

Quotations for minor works should be listed out in ascending order of value, but selection should not necessarily be limited to the most competitive bid.

The job should be discussed with the Contractor to ensure that he can do the job properly for the money. He may have misunderstood the scope of the work or specification requirements, or his price may simply be too low. Placing an order for the work under these conditions is fraught with difficulties for the future.

Providing it can be seen to be transparently fair, it is preferable to accept the tender bid that fairly represents proper value for money from the Contractor who provides some security that the works will be completed on time and in accordance with the specification and who also has experience in working in an environment which needs to provide protection for the pupils and staff.

Tenders for Larger Contracts

Tenders for larger projects should not be judged only on the basis of the lowest lump sum tender. It is more often considered appropriate to call for detailed price breakdowns from the three bidding Contractors which will provide a better understanding of the offers of the most competitive bids.

Arithmetical Mistakes

Strict rules must be applied in the case of arithmetical mistakes and alterations and omissions to the specification of the works.

Arithmetical mistakes should be corrected and the revised total should be increased or decreased accordingly by a percentage adjustment to equal the original tender amount submitted.

The correction should be explained to the Contractor without indicating whether this would effect the acceptability of his tender. He should then be asked if he wishes to stand by his original tender amount. If not his tender may be discarded.

In general, any significant deviation by a Contractor from the tender invitation should cause that tender to be rejected.

In practice and for many reasons there can be a wide divergence in tender values, even when the bids are based upon the most detailed specification. The acceptance of a tender and the approval of the Contractor has to be considered very carefully.

The Contractor's attitude must be fully understood. He may or may not want to obtain the order for the work. The contract may be very convenient and/or lucrative in the right location and the programme for the work may coincide with his arrangements for other projects.

Alternatively a particular Contractor may be desperately in need of work. Where an exceptionally low bid has been submitted, the selection of the Contractor should be even more carefully considered.

SELECTION OF THE MAIN CONTRACTOR

Compiling a List of Suitable Contractors.
The selection of a suitable contractor to carry out the work is the single most important ingredient in ensuring the successful completion of a project, big or small, to the standard required and without unnecessary delay.

Care must therefore be taken by the employer together with his consultants, if appointed, to compile a list of those considered suitable who have the resources to carry out the work.

The Interviews
It is generally considered good practice, whatever the nature of the work, to interview each of the selected contractors, to outline the proposed scheme together with all the advantages and difficulties entailed in the project and any particular restrictions which may be necessary. The contractors should also be taken to see the site.

This gives the employer and his consultants an opportunity to get to know the contractor's team, to ask how they would propose to carry out the work and whether they are interested in submitting a serious tender.

Most contractors are very happy to hand over a copy of their current brochure and provide the information listed in Figure 5.2 which will be required, but clearly these procedures may be too laborious in the case of minor repair works. The same principles generally should be applied and certainly must be followed when placing orders for long term cyclical maintenance work or other management agreements.

Selection of Small Jobbing Contractors.
Only bona fide contractors should be invited to submit quotations, preferably only those who belong to a recognised Trade Association. This may not eliminate all the risks, but it is certainly a step in the right direction.

Figure 5.2
INFORMATION TO BE PROVIDED BY BUILDERS AND SUB-CONTRACTORS

i) The nature of their business and portfolio;
ii) Address, trading name and VAT No;
iii) If incorporated, the Company Registration No;
iv) Tax exemption certificates;
v) Insurance and Bond certificates;
vi) Quality, Health and Safety documentation;
vii) Number of employees and if directly employed;
viii) Name of proposed foreman;
ix) Method of working and availability;
x) Three trade references;
xi) Three references from previous Clients.

Very small builders and sub-contractors are notoriously risky and often trade illegally without even a proper trading address. While many are genuine, they are all too often short of the resources and skill necessary to carry out the work properly. They frequently don't have a tax exemption certificate (see later) and usually ask for payments on account in order to buy materials before they can start or complete the work. But their low overheads mean they are cheap to employ. There is a balance to be drawn between the risk of accepting a cheap tender submitted by a small builder or specialist and the price, normally higher, submitted by a more established company.

Selection of a Main Contractor.
The problem with larger Contractors is that while they display a high profile based upon their wide experience and usually have the resources and skill to carry out the work, in practice their performance is very much dependent on the site staff. And even then the agent or foreman are often subjected to pressure from their own directors or accounts department who may operate a "pay when paid' policy which may also penalise the sub-contractors' performances.

The Construction Industry's Tax Deduction Scheme
In order to reduce widespread tax evasion by self-employed building operatives working either on 'the lump', or for 'cash in hand', the Inland Revenue set up a self regulating 'tax deduction scheme' operated with the use of what are known as 714 Certificates issued by the Tax Office.

The contractor must provide you with sight of their tax deduction certificate. Care should be taken to ensure that it is a current certificate held in the name of the person or company you propose to employ.

If the contractor cannot, for any reason, provide you with his 714 tax exemption certificate you are required by law to deduct 25% in respect of tax from any payments which may fall due to them and for which you become liable.

DESIGN AND BUILD PROJECTS

Development of Design and Build Contracts.
The roots of 'design and build contracts' in the modern construction industry lie in the commercial development of houses and offices.

Developers and builders began by assembling packages incorporating all the financial, design and building resources necessary to complete a project, which would then be offered for sale or lease in the open market.

Houses are sold on after completion to private clients with the aid of mortgages which may have provided assistance with the building finance. Office and retail developments are either sold on privately or to a major investment fund which then leases out the property on a commercially viable basis.

Loans and Profit
Developers look for profits of 10 to 15% and often seek to borrow between 35% and 50% of the land or development cost at interest rates not exceeding 0.75% to 2% above the standard bank rate.

Sale and Lease of Development Property
The builder/developer's profit accrues by reducing the risk and finance costs and achieving a sale with the minimum of delay, generally no later than 12 months after completion of the work.

Commercial projects are often 'sold on' even before a brick is laid on site. 'Pre let' situations or sales made prior to the commencement of work, such as work on school buildings, clearly reduce the risk and should enable developers to reduce their costs and tender bids.

The Design Brief
Commercial developers clearly respond to market demand by providing desirable premises at the most attractive prices. Specifications have to meet these criteria and cost and profit are their benchmarks. Designs are usually attractive, but specifications for materials and services often do not meet the user's full requirements.

The Professional Consultants
Design and build means, as the name implies, that the consultants are appointed and paid by the developer/ builder. This is an important aspect because, although the professional consultants will be bound by their codes of conduct, they are in effect employed by the builder and the extent and remuneration they receive may be limited. Clients considering 'design and build' bids are best advised to seek professional assistance when examining the proposals and selecting the best scheme.

The Architect.
It is always sensible to meet the architect and other consultants to be appointed by the builder offering a design and build contract to understand the terms of the consultant's appointment. Generally it is important to know the extent to which they are being used, their resources and experience.

The Quantity Surveyor
The developer/builder will wish to use 'in house' estimators to cost the works, both because that is their trade practice, but to avoid the additional cost of an independent consultant.

The Client needs to look closely at any costs prepared in this way.

Tender Invitations
A tender invitation takes the form of a 'performance specification', telling the contractor what the building has to provide by way of gross floor area, numbers of teaching and non-teaching spaces, the special requirements of various areas such as laboratories and multi material workshops and the facilities that these areas will provide for school activities.

The tender invitation should also set out the general contractual conditions, start and completion dates, terms of payment and possibly the ongoing service arrangements for the buildings, the legal implications of which may be quite different.

Tenders are normally invited in two stages because of the high cost of preparing the submissions and because these proposals include the interpretation and approval of the planning criteria.

The preliminary submission would provide outline indications of the proposals together with a programme for the completion of the design and then the construction work, and a target benchmark for the cost of the works.

Following the selection of the most attractive of the preliminary proposals, stage two would involve the detailed design and planning of the works and may allow for an initial payment to the contractor for the work he has carried out up to that time.

The principles for tendering and for the examination of the proposals must be rigorously followed. Design and build and management contracts are generally not recommended or appropriate for schools projects.

PUBLIC AND PRIVATE INVESTMENT PARTNERSHIP

The Private Finance Initiative
The Government's initiative to attract private finance offers new opportunities for the operation and development of educational estates.

Relaxing regulations for raising finance and the use of equity by the Government, the Funding Councils and the Local Authorities can help with:

- running and operational costs;
- upkeep and maintenance;
- re-modelling existing premises;
- building new premises.

Finance for Re-Modelling or Building NewPremises.
Many financial institutions are favourably inclined towards lending money directly to educational institutions for such projects, but are not too inclined to provide finance for builders for long periods if at all.

Commercial bankers are usually prepared to offer finance based upon the accounting records provided by schools and universities, depending upon performance and their development plans. They are generally not keen on 'collateral' since valuations of educational premises are usually not satisfactory and their alternative use (if possession in the event of the school's failure is allowed) is often difficult to envisage.

Should a developer or builder be asked to offer finance, he will have to provide his bank with collateral and the interest charges will be part of the consideration provided for in his agreement.

A 'contract agreement' with the school may be sufficient for the developer to fund building costs.

But if long term finance is to be provided without an interest in the revenue to be derived from the use of the premises, then it will be necessary to consider transferring to the developer or builder an interest in the fixed assets such as land and/or buildings.

This method might provide the collateral to develop and/or expand the project.

Partnership implies a business agreement to share the equity and risks involved and requires clarity and understanding of the interests and intentions of both partners. The essentials of a successful partnership demand trust and the sharing of:

- assets,
- risk,
- equity,
- security.

Tender Conditions.
These would normally follow the lines of Design and Build Contracts, but with additional focus on proposals for providing private finance and sale or lease back arrangements. It is possible that responsibility for the ongoing serviceing and the operation, management and maintenance of the buildings would be taken up by a third party.

To test normal competitive tendering and building procedures by such complex arrangements needs careful consideration. Both legal and professional advice must be obtained which may be costly, but if at all possible this opportunity can provide education establishments, state, grant maintained, or private, with facilities not hitherto imagined possible.

Conditions Required for a PPP Agreement
Assuming these difficulties can be surmounted it is important that the arrangements, however complex, should fulfil the following criteria:

i) appointment of the Professional Consultants to be approved by both parties;

ii) selection of a builder or agreement by methods that will guarantee a transparently competitive bid;

iii) a sound financial investor with a satisfactory investment portfolio with inbuilt guarantees;

iv) agreement of competitive money lending rates and/or other arrangements;

v) approval of designs which fulfil the requirements of the curriculum and the area design guidelines recommended by the D*f*EE;

vi) satisfactory contractual arrangements for the construction or re-modelling project and ongoing operational, management and maintenance work;

vii) agreement on the use and security of the premises particularly when used outside normal school hours;

viii) approved arrangements for payment of all parties involved in the ongoing activities;

ix) guarantees in case of failure by any party;

x) arrangements for the return of the assets and provision of warranties for the design, materials and workmanship at the end of the arrangements.

Figure 5.3

THE SELECTION OF CONSULTANTS AND BUILDING CONTRACTS

Note; See Additional Amendments for the Construction and Regeneration Act 1966 where applicable.

√ To be appointed by Employer
(√) May be required by Employer
C To be appointed by Contractor

REF	TYPE OF AGREEMENT – Appointment of Consultants	Date last published	NEW BUILD WORK: Contract Value	NEW BUILD WORK: Architect or Building Surveyor	NEW BUILD WORK: Quantity Surveyor	NEW BUILD WORK: Structural Engineer	NEW BUILD WORK: Services Engineer	REMODELLING EXISTING BUILDINGS: Contract Value	REMODELLING: Architect or Building Surveyor	REMODELLING: Quantity Surveyor	REMODELLING: Structural Engineer	REMODELLING: Services Engineer	REPAIR WORK: Contract Value	REPAIR WORK: Architect or Building Surveyor	REPAIR WORK: Quantity Surveyor	REPAIR WORK: Structural Engineer	REPAIR WORK: Services Engineer	CYCLICAL MAINTENANCE CONTRACTS: Contract Value	CYCLICAL: Architect or Building Surveyor	CYCLICAL: Quantity Surveyor	CYCLICAL: Structural Engineer	CYCLICAL: Services Engineer
1.00	**Formal Letter of Engagement**																					
1.01	Standard letter		< £5,000					< £5,000	(√)				< £5,000					< £5,000				
1.02	Standard letter		< £10,000	√				< £10,000	√				< £10,000	(√)				< £10,000				
2.00	**Jobbing Agreement (1990)**																					
2.01	Standard Agreement	4.9	< £20,000	√				< £20,000	√				< £20,000	(√)								
3.00	**Minor Building Works Contract (JCT 1998)**																					
3.01	Standard Agreement		< £50,000	√	√	(√)		< £50,000	√	√	(√)		< £50,000	√	√							
4.00	**Intermediate Contract (IFC 1998)**																					
4.01	Standard Agreement		< £500,000	√	√	√	(√)	< £500,000	√	√	√	(√)	< £50,000/	√	√	(√)	(√)					
4.02	Sectional Completion Supplement	9.95		√	√				√	√			< £500,000									
5.00	**Standard Contract (JCT 1980)**																					
	Private Edition																					
5.01	With Quantities	10.95	> £500,000	√	√	√	(√)	> £500,000	√	√	√	(√)										
5.02	With Approximate Quantities	10.95	> £500,000	√	√	√	(√)	> £500,000	√	√	√	(√)										
5.03	Without Quantities	10.95	> £500,000	√	√	√	(√)	> £500,000	√	√	√	(√)										
5.04	Fluctuations Supplement	5.88																				
5.05	Contractor's Design Supplement	2.96	> £500,000	√	√	√		> £500,000	√	√	√											
	Local Authority Edition																					
5.06	With Quantities	10.95	> £500,000	√	√	√	(√)	> £500,000	√	√	√	(√)										
5.07	With Approximate Quantities	10.95	> £500,000	√	√	√	(√)	> £500,000	√	√	√	(√)										
5.08	Without Quantities	10.95	> £500,000	√	√	√	(√)	> £500,000	√	√	√	(√)										
5.09	Fluctuations Supplement	9.87																				
5.10	Contractor's Design Supplement	2.96	> £500,000	√	√	√		> £500,000	√	√	√											
6.00	**Standard Contract (JCT 1980) (Scotland)**																					
	Private Edition (Without Articles)																					
6.01	With Quantities	9.91	> £500,000	√	√	√	(√)	> £500,000	√	√	√	(√)										
6.02	Without Quantities	2.93	> £500,000	√	√	√	(√)	> £500,000	√	√	√	(√)										
6.03	Fluctuations Supplement	5.88																				
6.04	Contractor's Design Supplement	2.96	> £500,000	√	√	√		> £500,000	√	√	√											
	Local Authority Edition (Without Articles)																					
6.05	With Quantities	2.92	> £500,000	√	√	√	(√)	> £500,000	√	√	√	(√)										
6.06	Without Quantities	10.87	> £500,000	√	√	√	(√)	> £500,000	√	√	√	(√)										
6.07	Fluctuations Supplement	5.88																				
6.08	Contractor's Design Supplement	2.96	> £500,000	√	√	√		> £500,000	√	√	√											
7.00	**Management and Design and Build Contracts**																					
7.01	Contractor's Design and Build (81)	10.95	< £300,000	C	C √			< £300,000	C	C √			< £100,000	C	C √							
7.02	ditto	10.95	> £300,000	C (√)	C √			> £300,000	C (√)	C √			> £100,000	C (√)	C √		(√)					
7.03	Management (1987)	12.87	> £1M	√	√			> £1M	√	√												
7.04	Works Contract (1987)	12.87	> £1M					> £1M														
7.05	Prime Cost Contract (1992)	2.94	< £300,000	√	√	(√)	(√)	< £300,000	√	√	(√)	(√)										
8.00	**Cyclical Maintenance Contracts**																					
8.01	Measurement Term Contract (1989)	2.97																< £30,000				
8.02	ditto	2.97																> £30,000	(√)	(√)		(√)
8.03	Commercially formed agreements																	> £25,000				
8.04	ditto																	> £25,000	(√)	√	(√)	(√)

THE BUILDING CONTRACT

Building contracts in the United Kingdom may be either verbal or writen, and the rule is that they must always consist of three elements:

i) an offer;
ii) acceptance;
iii) consideration that is a payment or recompense.

The contracts must also be legal to be recognised in a Court of Law. Disputes about the terms and performance of an agreement are also decided on the basis of other legal precedents. (Decisions made in other disputes involving similar legal principles.) The offer in a building contract consists of the tender submission or quotation by the building contractor. Acceptance is provided by the employer or his agent and the consideration is the amount of the accepted offer.

The construction industry has devised a number of standard forms of contract to be used on building or civil engineering projects and these and their recommended use are set out in Fig 5.3 on page 229 but these also have to comply with the Construction Act 1998 which became compulsory in May 1998.

The use and interpretation of these standard contracts are indeed very complex and outside the general scope of this Cost Guide, but it is important and not at all difficult to understand the principles and salient conditions which generally apply to all agreements for building work, including minor projects and small repair work.

The two forms of contract most frequently used in the industry are the Intermediate JCT Standard Building Contract (IFC98) and the Standard Form for Minor Building Works (MW98). The scope of these standard contracts is set out below with appropriate references. The latter is often used beyond the scope and size of projects recommended in Figure 5.3, but in these cases it may not provide the protection expected by the parties.

THE AGREEMENT

The agreement is completed by filling out the standard forms where required.

Enter the name of the Employer (usually the name of the school or college) and their registered address in the first part and the name and registered address of the Contractor in the second part. They are then deemed to be the Employer and Contractor within the terms of the contract. It is preferable that the parties should meet to sign the agreement and agree the content and importance of each contract document.

It should be noted that every entry or amendment made to the standard form of contract should be initialled by both parties, including each of the contract drawings and the summary sheet contained in the Bill of Quantities or Specification for the Works so that there can be no misunderstanding or confusion about the intention of both parties.

RECITALS AND ARTICLES

A brief but accurate description of the works should be entered here to describe succinctly the scope of the works, with a complete list of the Contract Drawings.

The name and address of the Architect or other consultant appointed to direct the works must be entered here and the wording of the contract amended accordingly. If a consultant has not been appointed, it is normal to enter the name and address of the Estates or Building Manager or whoever will be responsible for giving instructions to the Contractor.

The references to the Quantity Surveyor, if appointed, and the description of Specification, Schedule of Works and Bills of Quantities must be amended or deleted.

THE CONTRACT AMOUNT

The cost of the works to be paid by the Employer to the Contractor, must be entered into the contract agreement. This amount is called the 'Contract Sum'.

The amount may be a 'fixed lump sum', or an agreement showing how the total price to be paid is to be calculated. In either case, this 'consideration' is an essential part of the legally binding contract.

The Contract Amount can only be varied by mutual agreement, or by reason of instructions issued by the Architect in accordance with the conditions of the contract.

- ♦ IFC98 - Lump Sum - Article 2, page 6.
  Variations - Clause - 3.6; 3.7; 4.1.
- ♦ MW98 - Article 2, page 4.
  Variations - Clause 3.6.

FLUCTUATIONS

Building Contracts may or may not be subject to adjustment to allow for changes in the cost of labour, materials or plant to be paid by the builder and used in the execution of the work. These contracts are known as either 'Fluctuation Contracts' or 'Fixed Price Contracts'.

Building Contracts make provision for agreement on either basis and must be amended accordingly, deleting the parts not required.

- ♦ IFC98 - Clause 4.9; 4.10.
  Supplemental Condition C and D.
- ♦ MW98 - Clause 4.6
  Supplementary Conditions Part A1 - 5.

CONTRACT PERIOD

The dates upon which it is agreed that work should be started and completed, or, in the case of the maintenance or management contract, the term in months or years, must be entered into the contract as follows. This is known as the contract period;

Date of Possession (or Commencement)

Date for Completion.

During this period the Contractor is entirely responsible for the safety, security and protection of his work and the premises he occupies, except that insurance for the existing buildings and associated risks are normally taken up by the Employer.

- ♦ IFC98 - Clause 2; 2.6; 2.9.
  Appendix; Page 46.
- ♦ MW98 - Clause 2.1 - 5.

DELAYS IN THE COMPLETION OF THE WORK

The completion of the works may be delayed because:

i) the employer delays the start on site;

ii) delays occur beyond the contractor's control;

iii) the contractor fails to proceed diligently with the work.

If the delays are beyond the control of the contractor, the Architect will issue a certificate extending the contract period, otherwise the contractor will be liable to pay damages.

- ♦ IFC98 - Extension of time - Clause 2.3.
  Reasons for delay - Clause 2.4.1 - 18.
- ♦ MW98 - Clause 2.2; 2,4.

DAMAGES FOR DELAY

Standard Building Contracts usually require the amount of damages to be agreed and entered into the contract as a reasonably fair estimate of the damages which may be incurred by the employer if completion of the works is delayed by the contractor

If this clause in the Appendix is left unpriced, it is difficult for the Employer to impose damages at a later date since they do not form a part of the contract.

In practice the estimate of liquidated damages should be transparently fair and show actual loss, such as rent for alternative accommodation or the extra cost of temporary services, expenses and supervision. The damages should be set out in the tender invitation.

Very high damages will only deter the contractors from submitting a reasonable tender bid. Most commonly the contractors add to their tender one or two weeks of damages at the rate quoted in the tender invitation to mitigate the consequential effect of late completion.

The events and occurrences which provide relief for the contractor from payment of liquidated damges if he is unable for reasons beyond his control to complete the works in time, are clearly set out in the contract.

Damages may be deducted by the Employer if the Architect has issued a 'certificate of delay', but the Contractor can apply to the Courts to have the damages set aside if he believes they have been applied unfairly or that the Employer has not in fact suffered any loss. If the Architect cancels his certificate of delay for any reason, damages would have to be repaid to the contractor.

- ♦ IFC98 - Clause 2.7; 2.8.
  Appendix - page 46.
- ♦ MW98 - Clause - 2.3.

PAYMENTS

Payment to the contractor may be made in several ways, but since 11 June 1998 must comply with The Late Payment of Commercial Debts Act 1998;

i) Small works:
  a) lump sum upon satisfactory competion;
  b) weekly stage payments.
  Note; Advance payments are not recommended.

ii) Main Projects:
  a) monthly intervals based upon measurement of completed work;
  b) stage payments when specified elements of the work are completed (foundations; roofs etc);
  c) deferred payment (which will attract interest).

Payments on main contracts are usually based upon the Quantity Surveyor's measured assessment of the value of work completed at set intervals (normally monthly). Difficulties arise if the Contractor is overpaid and subsequently ceases trading before completing the work. It is usually safer to pay the Contractor based upon a measured valuation.

The Contractor usually meets the Surveyor to agree his payment application and provides all necessary invoices from the specialist sub-contractors. The Surveyor makes his recommendation to the Architect who then issues his Payment Certificate. Payment is normally due 14 days from the date of the Architect's Certificate.

Retention is deducted by the Surveyor (normally 5%) as provided for in the contract and this money is released in two stages when the Architect issues his Certificate of Practical Completion (2.5%) and subsequently when he issues his Final Certificate (2.5%).

- ♦ IFC98 - Clause 4.2.1 - 3; 4.3; 4.4 - 6; 4.9.
  Appendix - page 46.
- ♦ MW98 - Clause - 4.1 - 4.8.

VARIATIONS TO THE WORKS

Wherever possible for small projects, repair work or maintenance work, quotations and tenders should be called for (in writing) on the basis that they should allow for 'everything' necessary for the entire completion of the work and provide all the appropriate guarantees. This will help to avoid misunderstandings about the scope and specification of the work.

Standard Forms of Contract make provision for instructions by the architect to vary or extend the works and the method to be used in calculating the cost of these variations.

Variations may be paid for at rates similar (pro-rata) to those in the detailed price breakdown of the tender amount, or on the basis of the Contractor's actual costs. The latter is known as 'daywork'.

The contractor may provide quotations from his sub-contractors for specialist works, or the variations may be measured by the surveyor and costed to form a 'variation account' at the completion of the contract work.

Dayworks are a very expensive way of authorising and paying for building works and the time expended must be carefully controlled and recorded.

Daywork sheets must be submitted and approved and signed by the Supervising Officer or Clerk of Works at the end of each week.

- ♦ IFC98 - Clause 3.6; 3.7; Options A or B; 3.8.
- ♦ MW98 - Clause - 3.5; 3.6; 4.2.

DEFECTS AND RETENTION

It should be clearly stated to a contractor carrying out minor works that he will be required to make good at his own expense all defects. It is better to hold back some money to ensure he does so.

Standard Contracts set out the contractor's liability for defects which arise in his work during the contract period and during the maintenance period after achieving practical completion and which must be made good at his own expense.

The period for final inspection of the work and for making good defects after practical completion should be agreed. This is normally 6 months.

The Contractor also has an extended liability for 7 years in common law in the case of 'latent' defects inherent in the materials used or because of bad workmanship, which cannot be seen at practical completion or at final inspection.

Standard contracts also make provision for the employer to withhold retention monies, usually 5% of the value of interim payments which are held in trust by the employer and released, if the work is properly completed, in two stages - 2.5% at practical completion and 2.5% when the final completion certificate is issued by the architect.

These monies should not be released if the works are not completed properly, or if there is a long snagging list.

- ♦ IFC98 - Defects - Clause 2.10.
  Retention monies - 4.2.1; 4.2.2; 4.6.
- ♦ MW98 - Defects - 2.5.
  Retention monies - 4.2; 4.4.

INSURANCES

Standard Building Contracts set out the Contractor's liability in respect of injury and damage to persons and property and the indemnity to be provided to protect the employer from these risks.

All insurance policies should be examined before the contractor's tender or quotation is accepted. They must be both current and in the name of the contracting party. Main contractors normally carry composite policies with £5,000,000 cover for any one claim or series of claims. Such policies cover most risks encountered on a building project including damage to property due to fire, theft, flood and acts of God, its subsequent reinstatement and injury to persons.

Care should be taken to ensure the appointment of a planning supervisor is in place (if required) and that all Health and Safety Regulations are carefully observed. Failure to do so may prejudice the agreement and payment of future claims.

The insurance of existing buildings to be refurbished is normally taken up by an extension of the employer's existing policies and the Governors must notify their Insurance Company of the extent and programme for the proposed work.

Working next to adjoining premises and public highways will not be covered by the normal policy and must be extended to cover these additional risks, including the design work entailed. This can be done by means of a separate policy and the premium for this cover usually forms a part of the Contract Amount.

- ♦ IFC98 - Clause 6;
  Appendix page 48:
- ♦ MW98 - Clause 6.

BONDS AND GUARANTEES

These are a form of insurance in case for any reason the contractors fail to complete the work. They may be provided by the contractor if required to do so.

The premium will vary according to the financial standing of the contractor and the risks involved, but such warranties/bonds should be read with care, preferably by the Quantity Surveyor or Solicitor. A smaller company may not be able to provide such cover.

Bonds are available from banks, insurance companies and brokers and from the Federation of Master Builders. Costs vary between 1% to 5% of the contract amount.

The bond offered by the Federation extends to members only and provides financial cover for:

i) loss due to bankruptcy;
ii) a two year guarantee of materials and workmanship;
iii) a five year warranty on structural defects caused by faulty materials;
iv) a free conciliation and arbitration service in case of disputes.

Annex 1 to IFC98 on page 50 makes provision for agreement to the terms of bonds to cover advance payment and for materials purchased by the Contractor but not brought onto the site.

Guarantees and Warranties.
These should be obtained from the contractor for all building materials and workmanship whenever possible, particularly for flooring, roofing, double glazing, special paints, electrical, heating and ventilation equipment. They should be assigned to the Employer at the completion of the works.

VALUE ADDED TAX

Value added tax is normally applicable on all building projects, except where special conditions apply, but legally it is deemed to apply when the contractor states that his output tax on services and goods supplied to the Employer will be positively or zero rated.

Normally therefore it is expected that value added tax would be payable and the words 'does not apply' must be deleted where stated.

- ♦ IFC98 - Clause 5.5.
  Supplemental Conditions; A; page 56.
- ♦ MW98 - Clause 5.2;
  Supplemental Memorandum, Part B.

STATUTORY TAX DEDUCTION SCHEME

The Finance Act 1980 is applicable to all work carried out by builders and contractors. They are therefore required to produce their tax deduction certificate, which permits the contractor to pay his own tax to the Tax Office.

Standard Building Contracts set out the Employer's responsibilities and liability. If the Contractor does not have a tax exemption certificate then the Employer is obligated to deduct the tax and pay it over to the Tax Office. The Contract also deals with the expiry of certificates or their cancellation.

- ♦ IFC98 - Clause 5.6.
  Supplemental Conditions; B; page 58.
- ♦ MW98 - Clause 5.3;
  Supplemental Memorandum, Part C.

DISPUTES AND ARBITRATION

Generally it is better to avoid building disputes wherever possible by taking the time to talk through the issues and by trying to reach a reasonable agreement. Arbitration and legal proceedings are both very arduous and expensive.

The Housing Grants, Construction and Regeneration Act 1996 became compulsory for all contracts after May 18th 1998 and now makes the settlement of disputes which arise during the course of the works mandatory by Adjudication and subsequently, if either party is not satisfied with this award, by Arbitration or in the Courts. Both the IFC98 and MW98 Standard Forms of Contract take this new law into account. A suitable person to act as Adjudicator or Arbitrator can either be named in the contract or recommended by a professional institution.

Generally adjudication proceedings are swift and require a decision by the adjudicator within 28 days of the submission by both parties. Arbitration proceedings start <u>after</u> the contractor has completed the work. The arbitrator will call for evidence from both parties and the dispute will be debated at the arbitration hearings. The rules applicable in the courts relating to the submissions and evidence will apply and this is usually a very difficult and lengthy stage in the proceedings. If necessary either or both parties may be legally represented. Submissions should be set out clearly and the evidence, documents and agreements should be carefully prepared and maintained in chronological order. The award and fee account are notified to the parties and can be obtained when the fee is paid.

- ♦ IFC98 - Ajudication; Clause 9A.1 - 8.
  Arbitration; Clause 9B.1 - 6.
- ♦ MW98 - Ajudication; Clause 4.8; 8.1.
  Arbitration; Clause 8.2.

THE ARCHITECT - SERVICES & CHARGES

Standard fees for Architectural Services are set out in the Scales of Professional Charges issued by the Royal Institute of British Architects. These scales are not mandatory and fee proposals called for from the Consultants would normally form the basis of the agreement with the Employer.

SFA/92 provides a schedule of conditions and a basis for the agreement of an inclusive fee to be paid to the Architect who would then, according to the terms of this agreement, be contracted to provide the following:

- Stage A & B - preliminary services;
- Stage C & D - outline proposals and planning applications;
- Stage E - development of scheme;
  - co-ordination of specialists;
  - cost checks & programme;
  - building regulations;
- Stage F & G - drawings and schedules for tendering purposes;
- Stage H & I - tender invitations and recommendations;
- Stage J & L - administration of the contract works.

SFA/92 provides two scales of fees to be charged for new and refurbishment work, according to the classification of the building type involved. Alternative schedules are provided for work on buildings of an historic nature and Design and Build Contracts.

Classes 3 and 4 generally cover schools and university buildings.

GUIDE TO ARCHITECT'S PROFESSIONAL FEES (£)

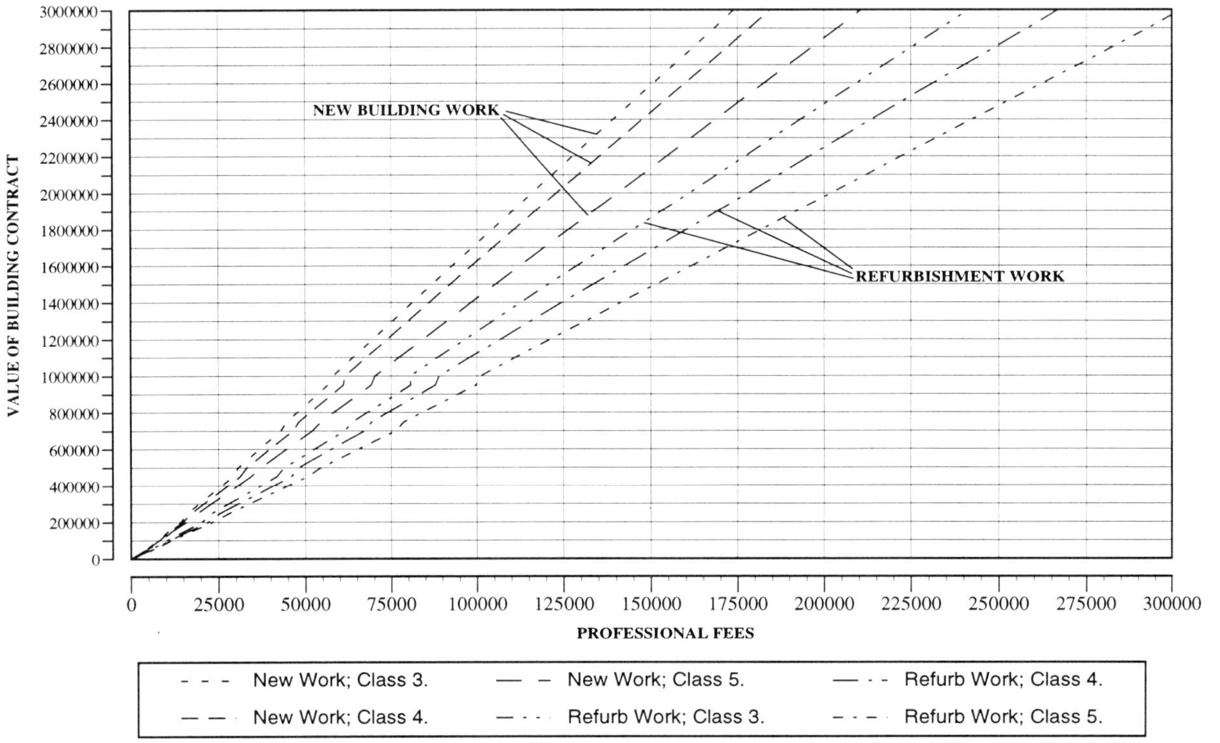

SFA/92; INCLUSIVE CHARGES

New Building Work

SFA/92 Scale of charges for new building works

Value of Work	Class 3	Class 4
> 50,000 < 75,000	8.20 - 9.15	8.60 - 9.30
> 75,000 < 100,000	7.45 - 8.10	7.60 - 8.15
> 100,000 < 250,000	6.70 - 7.45	7.10 - 7.60
> 250,000 < 500,000	6.25 - 6.70	6.75 - 7.10
> 500,000 < 750,000	6.00 - 6.25	6.50 - 6.75
> 750,000 <1,000,000	5.90 - 6.00	6.35 - 6.50
>1,000,000 <2,500,000	5.70 - 5.90	5.95 - 6.35
>2,500,000 <5,000,000	5.50 - 5.70	5.70 - 5.95

Refurbishment Work

SFA/92 Scale of charges for refurbishment works

Value of Work	Class 3	Class 4
> 50,000 < 75,000	11.70 - 11.75	12.75 - 13.25
> 75,000 < 100,000	10.65 - 11.15	11.65 - 12.45
> 100,000 < 250,000	9.60 - 10.65	10.60 - 11.40
> 250,000 < 500,000	8.95 - 9.60	9.80 - 10.60
> 500,000 < 750,000	8.60 - 8.95	9.40 - 9.80
> 750,000 <1,000,000	8.40 - 8.60	9.10 - 9.40
>1,000,000 <2,500,000	7.70 - 8.40	8.70 - 9.10
>2,500,000 <5,000,000	7.25 - 7.70	8.40 - 8.70

QUANTITY SURVEYING - SERVICES & CHARGES

Standard fees for Quantity Surveying Services are set out in the Scales of Professional Charges issued by the Royal Institution of Chartered Surveyors. These scales are not mandatory and fee proposals called for from the Consultants would normally form the basis of the agreement with the Employer.

Scale 36 provides a basis for the agreement of an inclusive fee to be paid to the Surveyor who would then be contracted to provide the following:

- budget estimates;
- cost planning and analysis;
- advice on builders and tendering;
- preparation of tender documents;
- tender reports and the contract;
- payment recommendations;
- financial reports;
- agreement of final account.

Scale 37 provides a basis for agreement of itemised fees to be paid to the Surveyor if they are only to be employed to provide a partial service as follows;-

Pre-Contract Services; para. 2.0:

- budgets, cost planning and analysis;
- advice on builders and tendering;
- preparation of tender documents;
- tender reports and the contract;
- negotiating tenders; para 3.0;
- consultative services; para 4.0;

Post-contract Services; para. 5.0:

- overall post contract service; para 5.0;
- payments; para 6.2;
- agreement of final account; para 6.3;
- monitoring costs; para 6.4.

GUIDE TO QUANTITY SURVEYOR'S PROFESSIONAL FEES (£)

SCALE 36; Inclusive Services.

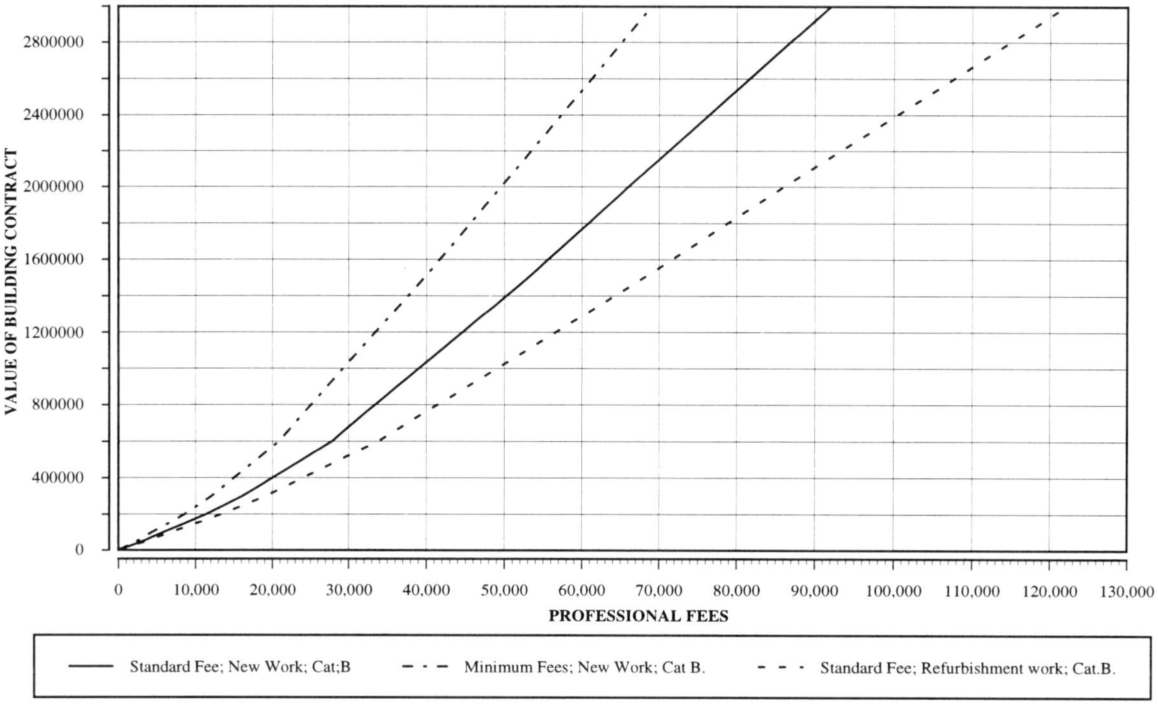

SCALE 36; INCLUSIVE CHARGES; CATEGORY B

Paragraph 2.1 Category B covers new building work (schools) and the charges to be based upon the final cost of all the work are as follows:

Value of Work		Basic fee	Balance over		
	< 150,000	£ 360	+ 5.8%	over	150,000
> 150,000	< 300,000	£ 9,060	+ 4.7%	over	150,000
> 300,000	< 600,000	£16,110	+ 3.9%	over	300,000
> 600,000	< 1,500,000	£27,810	+ 2.8%	over	600,000
> 1,500,000	< 3,000,000	£53,010	+ 2.6%	over	1,500,000
> 3,000,000	< 6,000,000	£92,010	+ 2.4%	over	3,000,000
> 6,000,000		£164,010	+ 2.0%	over	6,000,000

Additional charges;
These are provided for in paragraphs 2.2, 2.3, and 2.4 to be charged when dealing with:

- specialist services (1%),
- refurbishment work (1%), and
- redecoration work (1.5%),

each percentage only to be applied to the value of work involved. These additional charges are incorporated in the above graph.

CONSULTANT ENGINEER - SERVICES & CHARGES

Standard fees for Structural Engineer's Services are set out in the Scales of Professional Charges issued by the Association of Consulting Engineers. There are two variants as follows:

Variant 1; Civil and Structural Engineering;

Variant 2; Electrical and Mechanical Services.

These scales also provide for the following alternative appointments to act as;

A the Lead Consultant engaged by the Employer;
B a Consultant engaged by the Employer;
C a Consultant engaged by the Contractor;
D to provide reports and advice;
E a Project Manager;
F a Planning Supervisor.

Scale B is generally the most appropriate for educational projects unless they are of a substantial nature, but the scales are not mandatory and fee proposals called for from the Consultants would normally form the basis of the agreement with the Employer.

The Appendix defines the structural design services to be provided and which would normally comprise the following:

- ♦ Demolition work;
- ♦ Earthworks, ground and surface treatment;
- ♦ Foundations including piling;
- ♦ Drainage, treatment works and dewatering;
- ♦ Reinforced concrete or steelwork structures;
- ♦ Structures in masonry or blockwork;
- ♦ Structures in timber or reinforced plastics.

GUIDE TO STRUCTURAL ENGINEER'S PROFESSIONAL FEES (£)

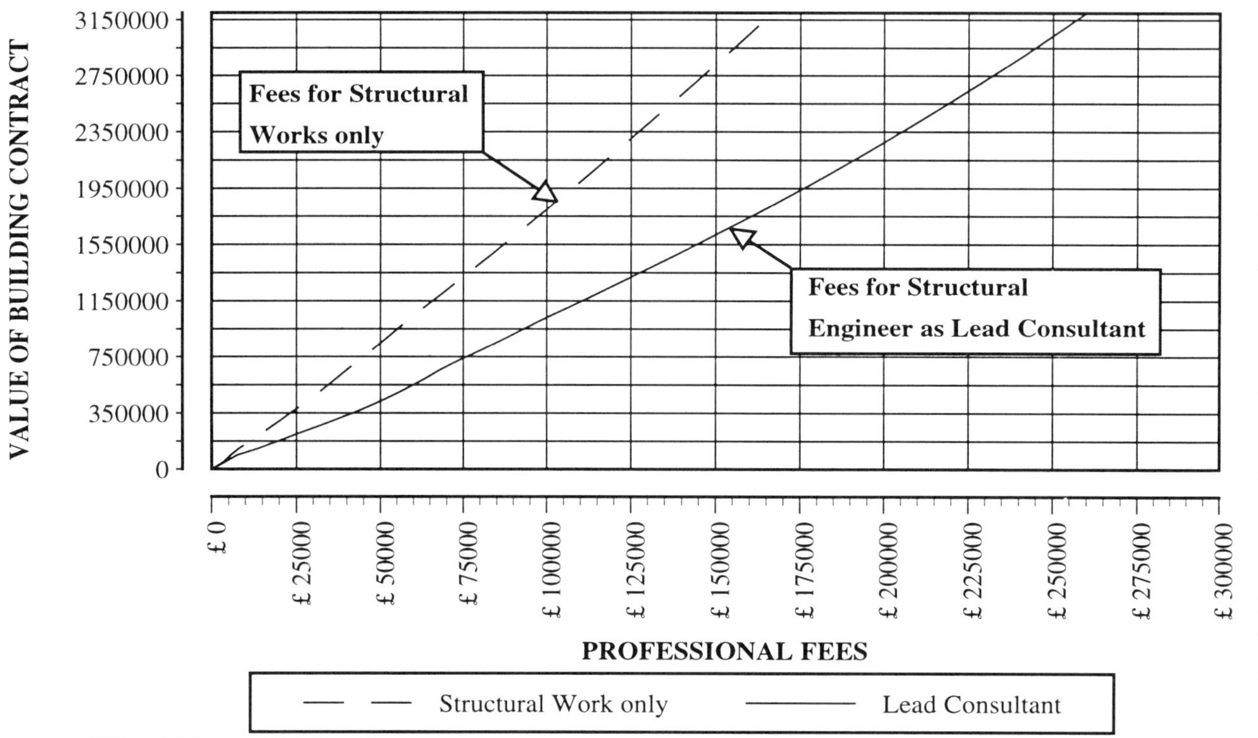

SCALE OF CHARGES

The Conditions of Engagement provide for the agreement of fees including:

- ♦ time based charges and agreed hourly rates;
- ♦ percentage based fees calculated from works costs;
- ♦ lump sum based fee agreements;
- ♦ methods of payment;
- ♦ disbursements and expenses;
- ♦ the services to be provided.

The Agreement provides graphs showing the relationship between the recommended percentage fees with the works cost divided by the Output Price Index.

Stage payments for services are usually made as following;-

I	Outline proposals	15%
II	Scheme design	35%
III	Further design	60%
IV	Production of Information	85%
V	Construction	100%

It should be noted that the above graph has been re-produced to provide a guide to Engineers' fees directly related to the building costs and does not require adjustment.

(The Index figure used was 338)

TOWN AND COUNTRY PLANNING

The Town and Country Planning Regulations 1993 require applications to be submitted to the Local Authority for consideration for the proposed erection of new buildings and for any changes in the use of land.

Permissions must be obtained before any work is started on site or changes made in the use of land or property.

Either outline or detailed applications are considered by the Planning Officers and approval is granted or rejected at monthly planning meetings, dates of which can be obtained from the Authority.

The Regulations are complicated and advice should always be obtained from the Architect or Building Surveyor.

Part 11 Scale of Fees.

The Regulations also set out the scale of fees and charges and the following is a brief guide to the likely charges;-

a) The erection of buildings; para 2;
applications for outline permission:
£180 for each 0.1 hectare of site area but < £4,500.

other applications:

no new floor space		-	£90;
< 40 m2	new floor space	-	£90;
> 40 < 75 m2	new floor space	-	£180;
> 75 m2	new floor space	-	£180;
and then for each 75 m2 increment		-	£180;
up to 3750 m2	- maximum fee		£9,000.

b) The use of land; para 11;
for one or more additional dwellings:
- £180 per dwelling with a maximum fee of £9,000;
  for disposal of rubbish:
- £90 per 0.1 hectare with a maximum fee of £13,500;
- other material changes in land or buildings £180.

THE BUILDING REGULATIONS

The Building Regulations 1994 refer to the mandatory approval and inspection of new building work including drainage work and the approval of structural calculations.

Owners are required to obtain these approvals and to give notice when work commences.

The Building Control Officer has extensive powers and is required to give consent for the following which are covered by the regulations:

- ground conditions and contaminants;
- moisture movement and resistance to weather;
- design of the building structure;
- suitability of materials;
- safety factors;
- insulation and ventilation;
- sanitation and drainage;
- fire spread and means of escape;
- health and safety.

Part 11: Fees Charged by Local Authority

The charges levied by the Authority cover the approvals which must be obtained and the regulations provide for building notice and inspection fees.

A guide to Schedule 3 charges are as follows;

Estimate cost				Fee		
<	£2,000			£60		
>	£2,000	<	£5,000	£150		
>	£5,000	<	£20,000	£150	plus £10	per £1000 over £5,000;
>	£20,000	<	£100,000	£300	plus £8	per £1000 over £20,000;
>	£100,000	<	£1M	£940	plus £5	per £1000 over £100,000;
>	£1M	<	£10M	£5,440	plus £3.50	per £1000 over £1M.

PLANNING SUPERVISOR

The CDM Regulations emanate from the European Union Directive (1992) to promote good Health and Safety Management on construction sites.

The CDM Regulations will apply to all demolition work, construction projects which exceed 30 days in duration and projects where 4 or more persons are to be involved in the work unless the Health and Safety Regulations are to be enforced by the Local Authority.

The CDM Regulations require the appointment of a 'Planning Supervisor' to co-ordinate Health and Safety aspects of the project and his duties include:

- preparation of a Health and Safety Plan;
- ensuring designers comply to avoid risks;
- ensuring completion of the Construction Phase Health and Safety Plan;
- ensuring completion of risk assessments;
- maintaining a Health and Safety File;
- advising employer on competence & resources;
- ensuring proper allocation of finance for safety;
- notices to Health & Safety Executive;
- issuing Health and Safety file at completion.

The Health and Safety Plan should contain:

- nature of project;
- details of existing environment and drawings;
- the design;
- construction methods and materials;
- site elements;
- interruption of services;
- site rules and regulations;
- continuing liaison.

LIST OF SUPPLIERS
for
EDUCATION EQUIPMENT AND SUPPLIES

The prices of furniture and fittings set out in the schedules in Chapter 2 The School Village headed 'Furniture and Fittings' are all based upon prices ruling during the first quarter of 1999 in Outer London, LEA Codes 301 - 320.

The prices have been averaged to provide a reasonable guide to the range of prices according to the standard and specification of the available equipment assuming that a 5% discount will be allowed when reasonable quantities are ordered and the prices do not relate to those of any individual supplier. Large orders may attract higher discounts and small orders may be at a premium and discounts may vary from supplier to supplier.

Other equal suppliers, not listed here, may be able to provide products of equal standard and specification and no preference is intended for the suppliers set out as follows.

	Supplier	Art Equipment	Benches	Blockwork	Book Storage	Brickwork	Carpets	Ceiling Systems	Chairs	Cladding	Coat & Bag Storage	Crowd Control	Desks Classroom	Desks Office	Display Cabinets	Display Screens	Drainage	External Paving	Flooring Materials	Food Technology	Glass	Gymnasium Equipment	Information Technology	Ironmongery	Kitchen Equipment	Laminates	Lighting	Litter Bins	Lockers	Metalwork Equipment	Music Equipment	Pinboards	Plants	Playground Equipment	Roofing	Science Equipment	Security Equipment	Shutters	Sports Equipment	Staging	Stationery	Storage	Tables	Teaching Aids	Technology Equipment	Toilet Furnishings	Tools	Wall Cladding	Woodwork Equipment
1	**ACO Technology Plc** Hitchin Road, Shefford, Bedfordshire. SG17 5TE Tel 01462 816666 Fax 01462 815895																√																																
2	**Alarm Link** Northbrook House, 17 Northbrook Road, Solihull, West Midlands. B90 3NR Tel 0121 244 3322 Fax 0121 682 1929																																				√												
3	**Allgood Plc** 297 Euston Road, London. NW1 3AQ Tel 0171 387 9951 Fax 0171 380 1232																							√																									
4	**Bishop Sports & Leisure** Bishop House, East Burnham Park, Crown Lane, Farnham Royal, Slough. SL2 3SF Tel 01753 648666 Fax 01753 648989																					√												√					√										
5	**Blockleys Brick Ltd** Sommerfeld Road, Trench Lock, Telford, Shropshire. TF1 4RY Tel 01952 251933 Fax 01952 641900					√												√																															
6	**BM Stainless Steel Drains Ltd** Station Road Industrial Estate, Tadcaster, North Yorkshire. LS24 9SG Tel 01937 834898 Fax 01937 832454																√																																
7	**Boxford Ltd** Wheatley, Halifax, West Yorkshire. HX3 5AF Tel 01422 358311 Fax 01422 355924																													√																	√		√

	Supplier	Art Equipment	Benches	Blockwork	Book Storage	Brickwork	Carpets	Ceiling Systems	Chairs	Cladding	Coat & Bag Storage	Crowd Control	Desks Classroom	Desks Office	Display Cabinets	Display Screens	Drainage
8	**British Thornton ESF** Greenholme Mills, Burley-in-Wharfdale, West Yorkshire. LS29 7DB Tel 01943 862504 Fax 01943 862706								√								
9	**Burmatex Ltd** Victoria Mills, Ossett, West Yorkshire. WF5 0AN Tel 01924 276333 Fax 01924 280033						√										
10	**Buy Direct** PO Box 106, Arkwright Road, Bicester, Oxfordshire. OX6 7ST Tel 01869 324488 Fax 01869 324012				√				√								
11	**Camas Building Materials** Hulland Ward, Asbourne, Derbyshire. DE6 3ET Tel 01335 372244 Fax 01335 370074																
12	**Commercial Lighting Systems Ltd** Units 16 & 17, Chandlers Way, Park Gate Business Centre, Swanwick, Southampton, Hampshire. SO31 1FQ Tel 01489 581002 Fax 01489 576262																
13	**Continental Sports Ltd** Paddock, Huddersfield, West Yorkshire. HD1 4SD Tel 01484 542051 Fax 01484 539148																
14	**Counties Furniture Group** The Shirehall, Abbey Foregate, Shrewsbury. SY2 6ND Tel 01743 253371 Fax 01743 253374		√		√				√					√			

	Supplier	External Paving	Flooring Materials	Food Technology	Glass	Gymnasium Equipment	Information Technology	Ironmongery	Kitchen Equipment	Laminates	Lighting	Litter Bins	Lockers	Metalwork Equipment	Music Equipment	Pinboards	Plants
8	British Thornton ESF			√			√							√			
9	Burmatex Ltd																
10	Buy Direct																
11	Camas Building Materials	√															
12	Commercial Lighting Systems Ltd										√						
13	Continental Sports Ltd					√											
14	Counties Furniture Group			√			√						√				

	Supplier	Playground Equipment	Roofing	Science Equipment	Security Equipment	Shutters	Sports Equipment	Staging	Stationery	Storage	Tables	Teaching Aids	Technology Equipment	Toilet Furnishings	Tools	Wall Cladding	Woodwork Equipment
8	British Thornton ESF			√						√	√		√				√
9	Burmatex Ltd																
10	Buy Direct							√		√	√						
11	Camas Building Materials																
12	Commercial Lighting Systems Ltd																
13	Continental Sports Ltd						√										
14	Counties Furniture Group			√						√	√		√				

	Supplier	Art Equipment	Benches	Blockwork	Book Storage	Brickwork	Carpets	Ceiling Systems	Chairs	Cladding	Coat & Bag Storage	Crowd Control	Desks Classroom	Desks Office	Display Cabinets	Display Screens	Drainage	External Paving	Flooring Materials	Food Technology	Glass	Gymnasium Equipment	Information Technology	Ironmongery	Kitchen Equipment	Laminates	Lighting	Litter Bins	Lockers	Metalwork Equipment	Music Equipment	Pinboards	Plants	Playground Equipment	Roofing	Science Equipment	Security Equipment	Shutters	Sports Equipment	Staging	Stationery	Storage	Tables	Teaching Aids	Technology Equipment	Toilet Furnishings	Tools	Wall Cladding	Woodwork Equipment
15	**Cuspo 2000 Ltd** 16-27 Pallion West Industrial Estate, Sunderland, Tyne & Wear. SR4 6SJ Tel 0191 567 9999 Fax 0191 514 1990																																				√	√											
16	**Don Gresswell Ltd** Grange House, 2 Geddings Road, Hoddesdon, Hertfordshire. EN11 ONT Tel 01992 465935 Fax 01992 465938				√				√				√	√	√	√							√					√	√			√					√				√	√	√	√					
17	**Ecophon Ltd** Old Brick Kiln, Tadley, Nr Basingstoke, Hampshire. RG26 5PP Tel 01256 850977 Fax 01256 850600							√																																									
18	**Electrolux Foodservice Equipment (UK)** Bordesley Green Road, Birmingham. B9 4UA Tel 0121 766 6655 Fax 0121 623 1801																								√																								
19	**EME (Scotland)** Blackaddie Road, Sanquhar, Dumfriesshire. DG4 6DE Tel 01659 50404 Fax 01659 50107		√		√				√		√																															√	√						
20	**Emmerich (Berlon) Limited** Wotton Road, Ashford, Kent. TN23 6JY Tel 01233 622684 Fax 01233 645801																													√												√			√		√		√
21	**ETB Furniture Ltd** Cruckfield, Shoot Hill, Ford, Shrewsbury. SY5 9NR Tel 01743 850151 Fax 01743 850367				√				√				√	√																													√						

	Supplier	Art Equipment	Benches	Blockwork	Book Storage	Brickwork	Carpets	Ceiling Systems	Chairs	Cladding	Coat & Bag Storage	Crowd Control	Desks Classroom	Desks Office	Display Cabinets	Display Screens	Drainage	External Paving	Flooring Materials	Food Technology	Glass	Gymnasium Equipment	Information Technology	Ironmongery	Kitchen Equipment
22	**Eternit UK Ltd** Meldreth, Whaddon Road, Nr. Royston, Hertfordshire. SG8 5RL Tel 01763 260421 Fax 01763 262338									√															
23	**Evans of Longton Ltd** Mercury House, Sutherland Road, Longton, Stoke-on-Trent. ST3 1JD Tel 01782 317371 Fax 01782 599281																					√			
24	**Felix Design** Unit 15, Tiverton Way, Tiverton Business Park, Tiverton, Devon. EX16 6SR Tel 01884 255420 Fax 01884 242613																								
25	**Formica Limited** Coast Road, North Shields, Tyne & Wear. NE29 8RE Tel 0191 2593000 Fax 0191 2582719																								
26	**Freshfield Lane Brickworks** Danehill, Haywards Heath, Sussex. RH17 7HH. Tel 01825 790350 Fax 01825 790779					√												√							
27	**G N Burgess & C0** Hanworth Trading Estate, Hampton Road West, Feltham, Middlesex. TW13 6EH Tel 0181 894 9231 Fax 0181 894 2943								√																
28	**Gopak Ltd** Range Road, Hythe, Kent. CT21 6HG Tel 01303 265751 Fax 01303 268282		√						√																

	Supplier	Laminates	Lighting	Litter Bins	Lockers	Metalwork Equipment	Music Equipment	Pinboards	Plants	Playground Equipment	Roofing	Science Equipment	Security Equipment	Shutters	Sports Equipment	Staging	Stationery	Storage	Tables	Teaching Aids	Technology Equipment	Toilet Furnishings	Tools	Wall Cladding	Woodwork Equipment
22	Eternit UK Ltd										√														
23	Evans of Longton Ltd									√					√										
24	Felix Design															√									
25	Formica Limited	√																				√			
26	Freshfield Lane Brickworks																								
27	G N Burgess & C0															√			√						
28	Gopak Ltd																		√						

	Supplier	Art Equipment	Benches	Blockwork	Book Storage	Brickwork	Carpets	Ceiling Systems	Chairs	Cladding	Coat & Bag Storage	Crowd Control	Desks Classroom	Desks Office	Display Cabinets	Display Screens	Drainage	External Paving	Flooring Materials	Food Technology	Glass	Gymnasium Equipment	Information Technology	Ironmongery	Kitchen Equipment	Laminates	Lighting	Litter Bins	Lockers	Metalwork Equipment	Music Equipment	Pinboards	Plants	Playground Equipment	Roofing	Science Equipment	Security Equipment	Shutters	Sports Equipment	Staging	Stationery	Storage	Tables	Teaching Aids	Technology Equipment	Toilet Furnishings	Tools	Wall Cladding	Woodwork Equipment
29	**Granit Union Ltd** PO Box 5550, Inverness. IV1 1WD Tel 01463 242090 Fax 01463 242303					√				√								√																															
30	**Hauraton Kaskade Ltd** Forge Works, The Swillett, Chorleywood, Hertfordshire. WD3 5BN Tel 01923 285601 Fax 01923 283386																√																																
31	**Ibstock Building Products Ltd** Ibstock, Leicestershire. LE67 6HS Tel 01530 261999 Fax 01530 261888					√																																											
32	**Interform Contract Furniture** 8 West Hampstead Mews, London. NW6 3BB Tel 0171 328 2340 Fax 0171 624 1777				√				√				√	√															√													√	√						
33	**International Food Service Equipment Ltd** 10 Beddington Farm Road, Croydon, Surrey. CR0 4XB Tel 0181 667 1167 Fax 0181 667 1030																								√																								
34	**Irwin-Desman Ltd** Eurocrown House, 23 Grafton Road, Croydon, Surrey. CR9 3AZ Tel 0181 680 2058 Fax 0181 681 8429																																			√									√				
35	**J T Ellis & Co Ltd** Silver Street, Hudersfield. HD5 9BA Tel 01484 514212 Fax 01484 456433				√				√				√																													√	√						

	Supplier	Art Equipment	Benches	Blockwork	Book Storage	Brickwork	Carpets	Ceiling Systems	Chairs	Cladding	Coat & Bag Storage	Crowd Control	Desks Classroom	Desks Office	Display Cabinets	Display Screens	Drainage	External Paving	Flooring Materials	Food Technology	Glass	Gymnasium Equipment	Information Technology	Ironmongery	Kitchen Equipment	Laminates	Lighting	Litter Bins	Lockers	Metalwork Equipment	Music Equipment	Pinboards	Plants	Playground Equipment	Roofing	Science Equipment	Security Equipment	Shutters	Sports Equipment	Staging	Stationery	Storage	Tables	Teaching Aids	Technology Equipment	Toilet Furnishings	Tools	Wall Cladding	Woodwork Equipment
36	**Klick Technology Ltd** Claverton Road, Wythenshawe, Manchester. M23 9FT Tel 0161 998 9726 Fax 0161 946 0419		√						√						√					√			√													√	√					√	√		√				
37	**Labofa Antocks Lairn** Lancaster Road, Cressex, High Wycombe, Buckinghamshire. HP12 3HZ Tel 01494 465454 Fax 01494 465901								√																																		√						
38	**Labspace Ltd** Piperell Way, Haverhill, Suffolk. CB9 8PH Tel 01440 702581 Fax 01440 763467																																			√									√				
39	**Laybond Products Ltd** Riverside, Saltney, Chester. CH4 8RS Tel 01244 674774 Fax 01244 680215																		√																														
40	**LEP Leisure Engineering Projects** Nene House, Barr's Street, Whittlesey, Peterborough. PE7 1DA Tel 01733 351011 Fax 01733 351027		√								√															√			√										√							√			
41	**Lune Interiors** White Lund, Morecambe, Lancashire. LA3 3DB Tel 01524 68861 Fax 01524 841067											√																√																					
42	**Magiboards Ltd** Stafford Park 2, Telford, Shropshire. TF3 3AY Tel 01952 292111 Fax 01952 292280														√																	√									√			√					

	Supplier	Art Equipment	Benches	Blockwork	Book Storage	Brickwork	Carpets	Ceiling Systems	Chairs	Cladding	Coat & Bag Storage	Crowd Control	Desks Classroom	Desks Office	Display Cabinets	Display Screens	Drainage	External Paving	Flooring Materials	Food Technology	Glass	Gymnasium Equipment	Information Technology	Ironmongery	Kitchen Equipment
43	**Maltbury Ltd** Arch 651, Tower Bridge Business Complex, Clements Road, London. SE16 4DG Tel 0171 252 3700 Fax 0171 252 3710																								
44	**MICE Kaymar Ltd** Brookhill Road, Brookhill Industrial Estate, Pinxton, Nottinghamshire. NG16 6NS Tel 01773 810107 Fax 01773 580286																			√			√		
45	**Morley's of Bicester Ltd** Arkwright Road, Bicester, Oxfordshire. OX6 7UU Tel 01869 320320 Fax 01869 320312		√		√				√		√		√	√	√	√							√		
46	**Music Education Supplies Ltd** 101 Banstead Road South, Sutton, Surrey. SM2 5LH Tel 0181 770 3866 Fax 0181 770 3554																								
47	**Naylor Clayware Ltd** Clough Green, Cawthorne, Barnsley, Yorkshire. S75 4AD Tel 01226 790591 Fax 01226 790531																√								
48	**Niels Larsen Ltd** Unit 3 Sandbeds Trading Estate, Dewsbury Road, Ossett, West Yorkshire. WF5 9ND Tel 01924 283000 Fax 01924 281872		√		√																	√			
49	**Nortek Educational Furniture** Vale Works, Priesty Fields Industrial Estate, Congleton, Cheshire. CW12 4AQ Tel 01260 298321 Fax 01260 298169		√		√				√		√				√					√			√		

	Supplier	Laminates	Lighting	Litter Bins	Lockers	Metalwork Equipment	Music Equipment	Pinboards	Plants	Playground Equipment	Roofing	Science Equipment	Security Equipment	Shutters	Sports Equipment	Staging	Stationery	Storage	Tables	Teaching Aids	Technology Equipment	Toilet Furnishings	Tools	Wall Cladding	Woodwork Equipment
43	**Maltbury Ltd**															√									
44	**MICE Kaymar Ltd**											√						√	√		√				
45	**Morley's of Bicester Ltd**			√	√	√		√	√							√		√	√	√	√				√
46	**Music Education Supplies Ltd**						√																		
47	**Naylor Clayware Ltd**																								
48	**Niels Larsen Ltd**														√	√									
49	**Nortek Educational Furniture**				√	√		√				√						√	√	√	√		√		√

No.	Supplier	Art Equipment	Benches	Blockwork	Book Storage	Brickwork	Carpets	Ceiling Systems	Chairs	Cladding	Coat & Bag Storage	Crowd Control	Desks Classroom	Desks Office	Display Cabinets	Display Screens	Drainage	External Paving	Flooring Materials	Food Technology	Glass	Gymnasium Equipment	Information Technology	Ironmongery	Kitchen Equipment	Laminates	Lighting	Litter Bins	Lockers	Metalwork Equipment	Music Equipment	Pinboards	Plants	Playground Equipment	Roofing	Science Equipment	Security Equipment	Shutters	Sports Equipment	Staging	Stationery	Storage	Tables	Teaching Aids	Technology Equipment	Toilet Furnishings	Tools	Wall Cladding	Woodwork Equipment
50	**Papworth Furniture Ltd** Ermine Street North, Papworth Everard, Cambridgeshire. CB3 8RJ Tel 01480 830095 Fax 01480 830516				√				√											√			√													√						√	√	√	√				
51	**Petrarch Claddings Ltd** Wainwright Close, Churchfields, Hastings, East Sussex. TN38 9PP Tel 01424 852641 Fax 01424 852797									√																																							
52	**PFP Fabrications Ltd** Quarry Mills, Quarry Road, Godstone, Surrey. RH9 8DQ Tel 01883 742797 Fax 01883 742943									√																√																				√			
53	**Pilkington United Kingdom Ltd** Prescott Road, St Helens, Mereyside, WA10 3TT Tel 01744 692000 Fax 01744 613049																				√																												
54	**Plannja Ltd** 69 High Street, Maidenhead, Berkshire. SL6 1JX Tel 01628 637313 Fax 01628 674940																																		√													√	
55	**Principal Furniture Ltd** Arkwright Road, Bicester, Oxfordshire. OX6 7UU. Tel 01869 324488 Fax 01869 324012																																										√						
56	**Purleystar Glazing** 118 Brighton Road, Purley, Surrey. CR8 4DB Tel 0181 668 7152 Fax 0181 668 7152																				√																												

	Supplier	Art Equipment	Benches	Blockwork	Book Storage	Brickwork	Carpets	Ceiling Systems	Chairs	Cladding	Coat & Bag Storage	Crowd Control	Desks Classroom	Desks Office	Display Cabinets	Display Screens	Drainage	External Paving	Flooring Materials	Food Technology	Glass	Gymnasium Equipment	Information Technology	Ironmongery	Kitchen Equipment	Laminates	Lighting	Litter Bins	Lockers	Metalwork Equipment	Music Equipment	Pinboards	Plants	Playground Equipment	Roofing	Science Equipment	Security Equipment	Shutters	Sports Equipment	Staging	Stationery	Storage	Tables	Teaching Aids	Technology Equipment	Toilet Furnishings	Tools	Wall Cladding	Woodwork Equipment
57	**Redland Roofing Systems Ltd** Regent House, Station Approach, Dorking, Surrey. RH4 1TG Tel 0990 601000 Fax 0990 642742																																		√														
58	**Resdev** Ainleys Industrial Estate, Elland, West Yorkshire. HX5 9JP Tel 01422 379131 Fax 01422 370943																		√																														
59	**Ring-Gard (UK) Ltd** Unit 18, Pedmore Road Industrial Estate, Brierley Hill, West Midlands. DY5 1TJ Tel 01384 74849 Fax 01384 74348																																				√	√											
60	**RJH Engineering Ltd** Artillery Street, Heckmondwike, West Yorkshire. WF16 0NR Tel 01924 402490 Fax 01924 404647																													√																	√		√
61	**RMC Concrete Products Ltd** Dale Road, Dove Holes, Buxton, Derbyshire. SK17 8BH Tel 01298 22244 Fax 01298 815221			√		√																																											
62	**Russell Roof Tiles Ltd** Wellington Road, Burton-on-Trent. DE14 2AW Tel 01283 517070 Fax 01283 516290																																		√														
63	**Sico Europe Ltd** Henwood Industrial Estate, Ashford, Kent. TN24 8DH Tel 01233 643311 Fax 01233 645143								√										√																					√			√						

	Supplier	Art Equipment	Benches	Blockwork	Book Storage	Brickwork	Carpets	Ceiling Systems	Chairs	Cladding	Coat & Bag Storage	Crowd Control	Desks Classroom	Desks Office	Display Cabinets	Display Screens	Drainage	External Paving	Flooring Materials	Food Technology	Glass	Gymnasium Equipment	Information Technology	Ironmongery	Kitchen Equipment	Laminates	Lighting	Litter Bins	Lockers	Metalwork Equipment	Music Equipment	Pinboards	Plants	Playground Equipment	Roofing	Science Equipment	Security Equipment	Shutters	Sports Equipment	Staging	Stationery	Storage	Tables	Teaching Aids	Technology Equipment	Toilet Furnishings	Tools	Wall Cladding	Woodwork Equipment
64	**SIS Jensen plc** Studio 11, Summerhouse Lane, Harefield, Middlesex. UB9 6HU Tel 01895 822458 Fax 01895 824837																						√																										
65	**Spacewise** Cavendish House, 74 Sydney Road, Watford, Hertfordshire. WD1 7PR Tel 01923 254492 Fax 01923 818637				√						√			√	√	√							√						√			√	√				√					√	√						
66	**Specialist Crafts Ltd** PO Box 247, Leicester, Leicestershire. LE1 9QS. Tel 0116 2510405 Fax 0116 2515015	√																																							√								
67	**Stadia Sports International Ltd** 19/20 Lancaster Way Business Park, Ely, Cambridgeshire. CB6 3NW Tel 01353 668686 Fax 01353 669444																					√												√					√										
68	**Stage Systems** Stage House, Prince William Road, Loughborough, Leicestershire. LE11 5GU Tel 01509 611021 Fax 01509 233146								√																															√									
69	**Steltube Ltd** Leigh Industrial Estate, The Causeway, Maldon, Essex. CM9 4LJ. Tel 01621 857273 Fax 01621 851829				√				√														√													√						√	√	√	√				
70	**Stormguard Rainwater Systems** Augustus Mill, Buckley Street, Macclesfield, Cheshire. SK11 6UH Tel 01625 665096 Fax 01625 665097																√																																

	Supplier	Art Equipment	Benches	Blockwork	Book Storage	Brickwork	Carpets	Ceiling Systems	Chairs	Cladding	Coat & Bag Storage	Crowd Control	Desks Classroom	Desks Office	Display Cabinets	Display Screens	Drainage	External Paving	Flooring Materials	Food Technology	Glass	Gymnasium Equipment	Information Technology	Ironmongery	Kitchen Equipment	Laminates	Lighting	Litter Bins	Lockers	Metalwork Equipment	Music Equipment	Pinboards	Plants	Playground Equipment	Roofing	Science Equipment	Security Equipment	Shutters	Sports Equipment	Staging	Stationery	Storage	Tables	Teaching Aids	Technology Equipment	Toilet Furnishings	Tools	Wall Cladding	Woodwork Equipment
71	**Teacherboards (1985) Ltd** Airdale Business Centre, Skipton, North Yorkshire. BD23 2TZ Tel 01756 700501 Fax 01756 700502														√	√																√									√			√					
72	**Tilcon** 40 Holborn Viaduct, London. EC1N 2PQ Tel 0171 404 20 60 Fax 0171 430 85 00			√		√												√																															
73	**Tilgear** Bridge House, 69 Station Road, Cuffley, Hertfordshire. EN6 4TG Tel 01707 873434 Fax																													√															√		√		√
74	**Ultralon TMI Ltd** PO Box 291, Watford, Hertfordshire. WD1 8QL Tel 01923 226690 Fax 01923 237596															√																√												√					
75	**Unique (Wilson & Garden Ltd)** 17 - 21 Newtown Street, Kilsyth, Glasgow. G65 0JX Tel 01236 823291 Fax 01236 825683				√										√	√																√									√			√					
76	**Wade International (UK) Ltd** 20 Broton Drive, Halstead, Essex. CO9 1HE Tel 01787 475151 Fax 01787 475579																√																																
77	**Wicksteed Leisure Ltd** Digby Street, Kettering, Northamptonshire. NN16 8YJ Tel 01536 517028 Fax 01536 410633																																	√															

LIST OF APPENDICES

THE AREA GUIDE LINES

The Area Guide Lines are based upon the graphs and formulae published by the Department for Employment and Education and contained in Bulletin 82 which can be obtained from The Stationery Office. Reference to 'Gross Areas' in these Appendices refer to the definition set out in the Glossary to Bulletin 82 on page 76.

PRIMARY SCHOOL - SITE AREAS (m2)			
School Roll	School Grounds		Playing Fields
	Min	Max	Max
60	4,890	5,800	2,500
75	5,363	6,325	2,500
90	5,835	6,850	2,500
105	6,308	7,375	2,500
120	6,780	7,900	2,500
135	7,253	8,425	2,500
150	7,725	8,950	2,500
165	8,198	9,475	2,500
180	8,670	10,000	2,500
195	9,143	10,525	2,500
210	9,615	11,050	2,500
225	10,088	11,575	2,500
240	10,560	12,100	5,000
255	11,033	12,625	5,000
270	11,505	13,150	5,000
285	11,978	13,675	5,000
300	12,450	14,200	5,000
315	12,923	14,725	5,000
330	13,395	15,250	5,000
345	13,868	15,775	5,000
360	14,340	16,300	5,000
375	14,813	16,825	5,000
390	15,285	17,350	5,000
405	15,758	17,875	5,000
420	16,230	18,400	5,000
435	16,703	18,925	5,000
450	17,175	19,450	5,000
465	17,648	19,975	5,000
480	18,120	20,500	10,000
495	18,593	21,025	10,000
510	19,065	21,550	10,000
525	19,538	22,075	10,000
540	20,010	22,600	10,000
555	20,483	23,125	10,000
570	20,955	23,650	10,000
585	21,428	24,175	10,000
600	21,900	24,700	10,000

SECONDARY SCHOOL - SITE AREAS (m2)			
School Roll	School Grounds		Playing Fields
	Min	Max	Max
450	39,200	43,000	25,000
500	42,000	46,000	25,000
550	44,800	49,000	30,000
600	47,600	52,000	30,000
650	50,400	55,000	35,000
700	53,200	58,000	35,000
750	56,000	61,000	35,000
800	58,800	64,000	40,000
850	61,600	67,000	40,000
900	64,400	70,000	40,000
950	67,200	73,000	45,000
1000	70,000	76,000	45,000
1050	72,800	79,000	45,000
1100	75,600	82,000	50,000
1150	78,400	85,000	50,000
1200	81,200	88,000	50,000
1250	84,000	91,000	55,000
1300	86,800	94,000	55,000
1350	89,600	97,000	55,000
1400	92,400	100,000	60,000
1450	95,200	103,000	60,000
1500	98,000	106,000	60,000
1550	100,800	109,000	65,000
1600	103,600	112,000	65,000
1650	106,400	115,000	65,000
1700	109,200	118,000	70,000
1750	112,000	121,000	70,000
1800	114,800	124,000	70,000

HARD SURFACED GAMES AREAS (m2)		
School Roll	Games Areas	
	Min	Max
100	480	790
150	570	885
200	660	980
250	750	1,075
300	840	1,170
350	930	1,265
400	1,020	1,360
450	1,110	1,455
500	1,200	1,550
550	1,290	1,645
600	1,380	1,740
650	1,470	1,835
700	1,560	1,930
750	1,650	2,025
800	1,740	2,120
850	1,830	2,215
900	1,920	2,310
950	2,010	2,405
1000	2,100	2,500
1050	2,190	2,595
1100	2,280	2,690
1150	2,370	2,785
1200	2,460	2,880
1250	2,550	2,975
1300	2,640	3,070
1350	2,730	3,165
1400	2,820	3,260
1450	2,910	3,355
1500	3,000	3,450
1550	3,090	3,545
1600	3,180	3,640
1650	3,270	3,735
1700	3,360	3,830
1750	3,450	3,925
1800	3,540	4,020

PRIMARY SCHOOLS - SOCIAL AREAS (m2)		
School Roll	Informal and Social Areas	
	Min	Max
65	482	658
80	558	740
95	635	823
110	711	905
125	788	988
140	864	1,070
155	941	1,153
170	1,017	1,235
185	1,094	1,318
200	1,170	1,400
215	1,247	1,483
230	1,323	1,565
245	1,400	1,648
260	1,476	1,730
275	1,553	1,813
290	1,629	1,895
305	1,706	1,978
320	1,782	2,060
335	1,859	2,143
350	1,935	2,225
365	2,012	2,308
380	2,088	2,390
395	2,165	2,473
410	2,241	2,555
425	2,318	2,638
440	2,394	2,720
455	2,471	2,803
470	2,547	2,885
485	2,624	2,968
500	2,700	3,050
515	2,777	3,133
530	2,853	3,215
545	2,930	3,298
560	3,006	3,380
575	3,083	3,463

SECONDARY SCHOOL - SOCIAL AREAS (m2)		
School Roll	Informal and Social Areas	
	Min	Max
600	3,480	3,800
650	3,620	3,950
700	3,760	4,100
750	3,900	4,250
800	4,040	4,400
850	4,180	4,550
900	4,320	4,700
950	4,460	4,850
1000	4,600	5,000
1050	4,740	5,150
1100	4,880	5,300
1150	5,020	5,450
1200	5,160	5,600
1250	5,300	5,750
1300	5,440	5,900
1350	5,580	6,050
1400	5,720	6,200
1450	5,860	6,350
1500	6,000	6,500
1550	6,140	6,650
1600	6,280	6,800
1650	6,420	6,950
1700	6,560	7,100
1750	6,700	7,250
1800	6,840	7,400

INFANTS age 5 - 7 Key Stages 1:3. 2:1

School	School Building Areas (m2)						Site Areas (m2)			
	Total		Per Pupil		Teaching Area		Total		Per Pupil	
Roll	Min	Max	Min	Max	Min	Max	Min	Max	Min	Max
90	476	542	5.29	6.02	286	325	4375	5390	48.61	59.89
120	578	656	4.82	5.47	347	394	4900	6020	40.83	50.17
150	680	770	4.53	5.13	408	462	5425	6650	36.17	44.33
180	782	884	4.34	4.91	469	530	5950	7280	33.06	40.44
210	884	998	4.21	4.75	530	599	6475	7910	30.83	37.67
240	986	1112	4.11	4.63	592	667	7000	8540	29.17	35.58
270	1088	1226	4.03	4.54	653	736	7525	9170	27.87	33.96
300	1190	1340	3.97	4.47	714	804	8050	9800	26.83	32.67
330	1292	1454	3.92	4.41	775	872	8575	10430	25.98	31.61
360	1394	1568	3.87	4.36	836	941	9100	11060	25.28	30.72
390	1496	1682	3.84	4.31	898	1009	9625	11690	24.68	29.97
420	1598	1796	3.80	4.28	959	1078	10150	12320	24.17	29.33
450	1700	1910	3.78	4.24	1020	1146	10675	12950	23.72	28.78
480	1802	2024	3.75	4.22	1081	1214	11200	13580	23.33	28.29
510	1904	2138	3.73	4.19	1142	1283	11725	14210	22.99	27.86
540	2006	2252	3.71	4.17	1204	1351	12250	14840	22.69	27.48
570	2108	2366	3.70	4.15	1265	1420	12775	15470	22.41	27.14

FIRST age 5 - 8 Key Stages 1:3. 2:1

School	School Building Areas (m2)						Site Areas (m2)			
	Total		Per Pupil		Teaching Area		Total		Per Pupil	
Roll	Min	Max	Min	Max	Min	Max	Min	Max	Min	Max
90	476	542	5.29	6.02	286	325	5011	6027	55.68	66.97
120	578	656	4.82	5.47	347	394	5719	6840	47.66	57.00
150	680	770	4.53	5.13	408	462	6427	7653	42.85	51.02
180	782	884	4.34	4.91	469	530	7135	8466	39.64	47.03
210	884	998	4.21	4.75	530	599	7843	9279	37.35	44.19
240	986	1112	4.11	4.63	592	667	8551	10092	35.63	42.05
270	1088	1226	4.03	4.54	653	736	9259	10905	34.29	40.39
300	1190	1340	3.97	4.47	714	804	9967	11718	33.22	39.06
330	1292	1454	3.92	4.41	775	872	10675	12531	32.35	37.97
360	1394	1568	3.87	4.36	836	941	11383	13344	31.62	37.07
390	1496	1682	3.84	4.31	898	1009	12091	14157	31.00	36.30
420	1598	1796	3.80	4.28	959	1078	12799	14970	30.47	35.64
450	1700	1910	3.78	4.24	1020	1146	13507	15783	30.02	35.07
480	1802	2024	3.75	4.22	1081	1214	14215	16596	29.61	34.58
510	1904	2138	3.73	4.19	1142	1283	14923	17409	29.26	34.14
540	2006	2252	3.71	4.17	1204	1351	15631	18222	28.95	33.74
570	2108	2366	3.70	4.15	1265	1420	16339	19035	28.66	33.39

FIRST age 5 - 9 Key Stages 1:3. 2:2

School	School Building Areas (m2)						Site Areas (m2)			
	Total		Per Pupil		Teaching Area		Total		Per Pupil	
Roll	Min	Max	Min	Max	Min	Max	Min	Max	Min	Max
90	476	542	5.29	6.02	286	325	5397	6412	59.97	71.24
120	578	656	4.82	5.47	347	394	6216	7336	51.80	61.13
150	680	770	4.53	5.13	408	462	7035	8260	46.90	55.07
180	782	884	4.34	4.91	469	530	7854	9184	43.63	51.02
210	884	998	4.21	4.75	530	599	8673	10108	41.30	48.13
240	986	1112	4.11	4.63	592	667	9492	11032	39.55	45.97
270	1088	1226	4.03	4.54	653	736	10311	11956	38.19	44.28
300	1190	1340	3.97	4.47	714	804	11130	12880	37.10	42.93
330	1292	1454	3.92	4.41	775	872	11949	13804	36.21	41.83
360	1394	1568	3.87	4.36	836	941	12768	14728	35.47	40.91
390	1496	1682	3.84	4.31	898	1009	13587	15652	34.84	40.13
420	1598	1796	3.80	4.28	959	1078	14406	16576	34.30	39.47
450	1700	1910	3.78	4.24	1020	1146	15225	17500	33.83	38.89
480	1802	2024	3.75	4.22	1081	1214	16044	18424	33.43	38.38
510	1904	2138	3.73	4.19	1142	1283	16863	19348	33.06	37.94
540	2006	2252	3.71	4.17	1204	1351	17682	20272	32.74	37.54
570	2108	2366	3.70	4.15	1265	1420	18501	21196	32.46	37.19

FIRST age 5 - 11 Key Stages 1:3. 2:4

School	School Building Areas (m2)						Site Areas (m2)			
	Total		Per Pupil		Teaching Area		Total		Per Pupil	
Roll	Min	Max	Min	Max	Min	Max	Min	Max	Min	Max
90	476	542	5.29	6.02	286	325	5835	6850	64.83	76.11
120	578	656	4.82	5.47	347	394	6780	7900	56.50	65.83
150	680	770	4.53	5.13	408	462	7725	8950	51.50	59.67
180	782	884	4.34	4.91	469	530	8670	10000	48.17	55.56
210	884	998	4.21	4.75	530	599	9615	11050	45.79	52.62
240	986	1112	4.11	4.63	592	667	10560	12100	44.00	50.42
270	1088	1226	4.03	4.54	653	736	11505	13150	42.61	48.70
300	1190	1340	3.97	4.47	714	804	12450	14200	41.50	47.33
330	1292	1454	3.92	4.41	775	872	13395	15250	40.59	46.21
360	1394	1568	3.87	4.36	836	941	14340	16300	39.83	45.28
390	1496	1682	3.84	4.31	898	1009	15285	17350	39.19	44.49
420	1598	1796	3.80	4.28	959	1078	16230	18400	38.64	43.81
450	1700	1910	3.78	4.24	1020	1146	17175	19450	38.17	43.22
480	1802	2024	3.75	4.22	1081	1214	18120	20500	37.75	42.71
510	1904	2138	3.73	4.19	1142	1283	19065	21550	37.38	42.25
540	2006	2252	3.71	4.17	1204	1351	20010	22600	37.06	41.85
570	2108	2366	3.70	4.15	1265	1420	20955	23650	36.76	41.49

FIRST and MIDDLE age 5 - 12 Key Stages 1:3. 2:4, 3:1

School Roll	School Building Areas (m2)						Site Areas (m2)			
	Total		Per Pupil		Teaching Area		Total		Per Pupil	
	Min	Max	Min	Max	Min	Max	Min	Max	Min	Max
90	476	542	5.29	6.02	286	325	5835	6850	64.83	76.11
120	578	656	4.82	5.47	347	394	6780	7900	56.50	65.83
150	680	770	4.53	5.13	408	462	7725	8950	51.50	59.67
180	782	884	4.34	4.91	469	530	8670	10000	48.17	55.56
210	884	998	4.21	4.75	530	599	9615	11050	45.79	52.62
240	986	1112	4.11	4.63	592	667	10560	12100	44.00	50.42
270	1088	1226	4.03	4.54	653	736	11505	13150	42.61	48.70
300	1190	1340	3.97	4.47	714	804	12450	14200	41.50	47.33
330	1292	1454	3.92	4.41	775	872	13395	15250	40.59	46.21
360	1394	1568	3.87	4.36	836	941	14340	16300	39.83	45.28
390	1496	1682	3.84	4.31	898	1009	15285	17350	39.19	44.49
420	1598	1796	3.80	4.28	959	1078	16230	18400	38.64	43.81
450	1700	1910	3.78	4.24	1020	1146	17175	19450	38.17	43.22
480	1802	2024	3.75	4.22	1081	1214	18120	20500	37.75	42.71
510	1904	2138	3.73	4.19	1142	1283	19065	21550	37.38	42.25
540	2006	2252	3.71	4.17	1204	1351	20010	22600	37.06	41.85
570	2108	2366	3.70	4.15	1265	1420	20955	23650	36.76	41.49

JUNIOR age 7 - 11 Key Stages 2:4

School Roll	School Building Areas (m2)						Site Areas (m2)			
	Total		Per Pupil		Teaching Area		Total		Per Pupil	
	Min	Max	Min	Max	Min	Max	Min	Max	Min	Max
90	476	542	5.29	6.02	286	325	5835	6850	64.83	76.11
120	578	656	4.82	5.47	347	394	6780	7900	56.50	65.83
150	680	770	4.53	5.13	408	462	7725	8950	51.50	59.67
180	782	884	4.34	4.91	469	530	8670	10000	48.17	55.56
210	884	998	4.21	4.75	530	599	9615	11050	45.79	52.62
240	986	1112	4.11	4.63	592	667	10560	12100	44.00	50.42
270	1088	1226	4.03	4.54	653	736	11505	13150	42.61	48.70
300	1190	1340	3.97	4.47	714	804	12450	14200	41.50	47.33
330	1292	1454	3.92	4.41	775	872	13395	15250	40.59	46.21
360	1394	1568	3.87	4.36	836	941	14340	16300	39.83	45.28
390	1496	1682	3.84	4.31	898	1009	15285	17350	39.19	44.49
420	1598	1796	3.80	4.28	959	1078	16230	18400	38.64	43.81
450	1700	1910	3.78	4.24	1020	1146	17175	19450	38.17	43.22
480	1802	2024	3.75	4.22	1081	1214	18120	20500	37.75	42.71
510	1904	2138	3.73	4.19	1142	1283	19065	21550	37.38	42.25
540	2006	2252	3.71	4.17	1204	1351	20010	22600	37.06	41.85
570	2108	2366	3.70	4.15	1265	1420	20955	23650	36.76	41.49

MIDDLE deemed primary age 8 - 12 Key Stages 2:3, 3:1

School	School Building Areas (m2)						Site Areas (m2)			
	Total		Per Pupil		Teaching Area		Total		Per Pupil	
Roll	Min	Max	Min	Max	Min	Max	Min	Max	Min	Max
600	2587	2893	4.31	4.82	1552	1736	32662	35848	54.44	59.75
630	2703	3021	4.29	4.80	1622	1813	34102	37396	54.13	59.36
660	2818	3150	4.27	4.77	1691	1890	35542	38944	53.85	59.01
690	2934	3278	4.25	4.75	1760	1967	36982	40492	53.60	58.68
720	3049	3407	4.23	4.73	1829	2044	38422	42040	53.36	58.39
750	3165	3535	4.22	4.71	1899	2121	39862	43588	53.15	58.12
780	3280	3663	4.21	4.70	1968	2198	41302	45136	52.95	57.87
810	3396	3792	4.19	4.68	2037	2275	42742	46684	52.77	57.63
840	3511	3920	4.18	4.67	2107	2352	44182	48232	52.60	57.42
870	3627	4049	4.17	4.65	2176	2429	45622	49780	52.44	57.22
900	3742	4177	4.16	4.64	2245	2506	47062	51328	52.29	57.03
930	3858	4305	4.15	4.63	2315	2583	48502	52876	52.15	56.86
960	3973	4434	4.14	4.62	2384	2660	49942	54424	52.02	56.69
990	4089	4562	4.13	4.61	2453	2737	51382	55972	51.90	56.54
1020	4204	4691	4.12	4.60	2522	2814	52822	57520	51.79	56.39
1050	4320	4819	4.11	4.59	2592	2891	54262	59068	51.68	56.26
1080	4435	4947	4.11	4.58	2661	2968	55702	60616	51.58	56.13
1110	4551	5076	4.10	4.57	2730	3045	57142	62164	51.48	56.00
1140	4666	5204	4.09	4.57	2800	3123	58582	63712	51.39	55.89
1170	4782	5333	4.09	4.56	2869	3200	60022	65260	51.30	55.78
1200	4897	5461	4.08	4.55	2938	3277	61462	66808	51.22	55.67
1230	5013	5589	4.08	4.54	3008	3354	62902	68356	51.14	55.57
1260	5128	5718	4.07	4.54	3077	3431	64342	69904	51.07	55.48
1290	5244	5846	4.06	4.53	3146	3508	65782	71452	50.99	55.39
1320	5359	5975	4.06	4.53	3215	3585	67222	73000	50.93	55.30
1350	5475	6103	4.06	4.52	3285	3662	68662	74548	50.86	55.22
1380	5590	6231	4.05	4.52	3354	3739	70102	76096	50.80	55.14
1410	5706	6360	4.05	4.51	3423	3816	71542	77644	50.74	55.07
1440	5821	6488	4.04	4.51	3493	3893	72982	79192	50.68	54.99
1470	5937	6617	4.04	4.50	3562	3970	74422	80740	50.63	54.93
1500	6052	6745	4.03	4.50	3631	4047	75862	82288	50.57	54.86
1530	6168	6873	4.03	4.49	3701	4124	77302	83836	50.52	54.79
1560	6283	7002	4.03	4.49	3770	4201	78742	85384	50.48	54.73
1590	6399	7130	4.02	4.48	3839	4278	80182	86932	50.43	54.67
1620	6514	7259	4.02	4.48	3908	4355	81622	88480	50.38	54.62
1650	6630	7387	4.02	4.48	3978	4432	83062	90028	50.34	54.56
1680	6745	7515	4.01	4.47	4047	4509	84502	91576	50.30	54.51
1740	6976	7772	4.01	4.47	4186	4663	87382	94672	50.22	54.41
1800	7207	8029	4.00	4.46	4324	4817	90262	97768	50.15	54.32

MIDDLE deemed secondary age 9 - 13 Key Stages 2:2, 3:2

School	School Building Areas (m2)						Site Areas (m2)			
	Total		Per Pupil		Teaching Area		Total		Per Pupil	
Roll	Min	Max	Min	Max	Min	Max	Min	Max	Min	Max
600	2965	3300	4.94	5.50	1779	1980	36975	40605	61.63	67.68
630	3094	3443	4.91	5.46	1856	2066	38595	42339	61.26	67.20
660	3223	3585	4.88	5.43	1934	2151	40215	44073	60.93	66.78
690	3352	3728	4.86	5.40	2011	2237	41835	45807	60.63	66.39
720	3481	3870	4.83	5.38	2089	2322	43455	47541	60.35	66.03
750	3610	4013	4.81	5.35	2166	2408	45075	49275	60.10	65.70
780	3739	4155	4.79	5.33	2243	2493	46695	51009	59.87	65.40
810	3868	4298	4.78	5.31	2321	2579	48315	52743	59.65	65.11
840	3997	4440	4.76	5.29	2398	2664	49935	54477	59.45	64.85
870	4126	4583	4.74	5.27	2476	2750	51555	56211	59.26	64.61
900	4255	4725	4.73	5.25	2553	2835	53175	57945	59.08	64.38
930	4384	4868	4.71	5.23	2630	2921	54795	59679	58.92	64.17
960	4513	5010	4.70	5.22	2708	3006	56415	61413	58.77	63.97
990	4642	5153	4.69	5.20	2785	3092	58035	63147	58.62	63.78
1020	4771	5295	4.68	5.19	2863	3177	59655	64881	58.49	63.61
1050	4900	5438	4.67	5.18	2940	3263	61275	66615	58.36	63.44
1080	5029	5580	4.66	5.17	3017	3348	62895	68349	58.24	63.29
1110	5158	5723	4.65	5.16	3095	3434	64515	70083	58.12	63.14
1140	5287	5865	4.64	5.14	3172	3519	66135	71817	58.01	63.00
1170	5416	6008	4.63	5.13	3250	3605	67755	73551	57.91	62.86
1200	5545	6150	4.62	5.13	3327	3690	69375	75285	57.81	62.74
1230	5674	6293	4.61	5.12	3404	3776	70995	77019	57.72	62.62
1260	5803	6435	4.61	5.11	3482	3861	72615	78753	57.63	62.50
1290	5932	6578	4.60	5.10	3559	3947	74235	80487	57.55	62.39
1320	6061	6720	4.59	5.09	3637	4032	75855	82221	57.47	62.29
1350	6190	6863	4.59	5.08	3714	4118	77475	83955	57.39	62.19
1380	6319	7005	4.58	5.08	3791	4203	79095	85689	57.32	62.09
1410	6448	7148	4.57	5.07	3869	4289	80715	87423	57.24	62.00
1440	6577	7290	4.57	5.06	3946	4374	82335	89157	57.18	61.91
1470	6706	7433	4.56	5.06	4024	4460	83955	90891	57.11	61.83
1500	6835	7575	4.56	5.05	4101	4545	85575	92625	57.05	61.75
1530	6964	7718	4.55	5.04	4178	4631	87195	94359	56.99	61.67
1560	7093	7860	4.55	5.04	4256	4716	88815	96093	56.93	61.60
1590	7222	8003	4.54	5.03	4333	4802	90435	97827	56.88	61.53
1620	7351	8145	4.54	5.03	4411	4887	92055	99561	56.82	61.46
1650	7480	8288	4.53	5.02	4488	4973	93675	101295	56.77	61.39
1680	7609	8430	4.53	5.02	4565	5058	95295	103029	56.72	61.33
1740	7867	8715	4.52	5.01	4720	5229	98535	106497	56.63	61.21
1800	8125	9000	4.51	5.00	4875	5400	101775	109965	56.54	61.09

SECONDARY (lower) age 11 - 14 Key Stages 3:3

School Roll	School Building Areas (m2)						Site Areas (m2)			
	Total		Per Pupil		Teaching Area		Total		Per Pupil	
	Min	Max	Min	Max	Min	Max	Min	Max	Min	Max
600	3720	4120	6.20	6.87	2232	2472	47600	52000	79.33	86.67
630	3876	4291	6.15	6.81	2326	2575	49280	53800	78.22	85.40
660	4032	4462	6.11	6.76	2419	2677	50960	55600	77.21	84.24
690	4188	4633	6.07	6.71	2513	2780	52640	57400	76.29	83.19
720	4344	4804	6.03	6.67	2606	2882	54320	59200	75.44	82.22
750	4500	4975	6.00	6.63	2700	2985	56000	61000	74.67	81.33
780	4656	5146	5.97	6.60	2794	3088	57680	62800	73.95	80.51
810	4812	5317	5.94	6.56	2887	3190	59360	64600	73.28	79.75
840	4968	5488	5.91	6.53	2981	3293	61040	66400	72.67	79.05
870	5124	5659	5.89	6.50	3074	3395	62720	68200	72.09	78.39
900	5280	5830	5.87	6.48	3168	3498	64400	70000	71.56	77.78
930	5436	6001	5.85	6.45	3262	3601	66080	71800	71.05	77.20
960	5592	6172	5.83	6.43	3355	3703	67760	73600	70.58	76.67
990	5748	6343	5.81	6.41	3449	3806	69440	75400	70.14	76.16
1020	5904	6514	5.79	6.39	3542	3908	71120	77200	69.73	75.69
1050	6060	6685	5.77	6.37	3636	4011	72800	79000	69.33	75.24
1080	6216	6856	5.76	6.35	3730	4114	74480	80800	68.96	74.81
1110	6372	7027	5.74	6.33	3823	4216	76160	82600	68.61	74.41
1140	6528	7198	5.73	6.31	3917	4319	77840	84400	68.28	74.04
1170	6684	7369	5.71	6.30	4010	4421	79520	86200	67.97	73.68
1200	6840	7540	5.70	6.28	4104	4524	81200	88000	67.67	73.33
1230	6996	7711	5.69	6.27	4198	4627	82880	89800	67.38	73.01
1260	7152	7882	5.68	6.26	4291	4729	84560	91600	67.11	72.70
1290	7308	8053	5.67	6.24	4385	4832	86240	93400	66.85	72.40
1320	7464	8224	5.65	6.23	4478	4934	87920	95200	66.61	72.12
1350	7620	8395	5.64	6.22	4572	5037	89600	97000	66.37	71.85
1380	7776	8566	5.63	6.21	4666	5140	91280	98800	66.14	71.59
1410	7932	8737	5.63	6.20	4759	5242	92960	100600	65.93	71.35
1440	8088	8908	5.62	6.19	4853	5345	94640	102400	65.72	71.11
1470	8244	9079	5.61	6.18	4946	5447	96320	104200	65.52	70.88
1500	8400	9250	5.60	6.17	5040	5550	98000	106000	65.33	70.67
1530	8556	9421	5.59	6.16	5134	5653	99680	107800	65.15	70.46
1560	8712	9592	5.58	6.15	5227	5755	101360	109600	64.97	70.26
1590	8868	9763	5.58	6.14	5321	5858	103040	111400	64.81	70.06
1620	9024	9934	5.57	6.13	5414	5960	104720	113200	64.64	69.88
1650	9180	10105	5.56	6.12	5508	6063	106400	115000	64.48	69.70
1680	9336	10276	5.56	6.12	5602	6166	108080	116800	64.33	69.52
1740	9648	10618	5.54	6.10	5789	6371	111440	120400	64.05	69.20
1800	9960	10960	5.53	6.09	5976	6576	114800	124000	63.78	68.89

SECONDARY age 11 - 16 Key Stages 3:3, 4:2

School Roll	School Building Areas (m2)						Site Areas (m2)			
	Total		Per Pupil		Teaching Area		Total		Per Pupil	
	Min	Max	Min	Max	Min	Max	Min	Max	Min	Max
600	4500	4900	7.50	8.17	2700	2940	47600	52000	79.33	86.67
630	4665	5080	7.40	8.06	2799	3048	49280	53800	78.22	85.40
660	4830	5260	7.32	7.97	2898	3156	50960	55600	77.21	84.24
690	4995	5440	7.24	7.88	2997	3264	52640	57400	76.29	83.19
720	5160	5620	7.17	7.81	3096	3372	54320	59200	75.44	82.22
750	5325	5800	7.10	7.73	3195	3480	56000	61000	74.67	81.33
780	5490	5980	7.04	7.67	3294	3588	57680	62800	73.95	80.51
810	5655	6160	6.98	7.60	3393	3696	59360	64600	73.28	79.75
840	5820	6340	6.93	7.55	3492	3804	61040	66400	72.67	79.05
870	5985	6520	6.88	7.49	3591	3912	62720	68200	72.09	78.39
900	6150	6700	6.83	7.44	3690	4020	64400	70000	71.56	77.78
930	6315	6880	6.79	7.40	3789	4128	66080	71800	71.05	77.20
960	6480	7060	6.75	7.35	3888	4236	67760	73600	70.58	76.67
990	6645	7240	6.71	7.31	3987	4344	69440	75400	70.14	76.16
1020	6810	7420	6.68	7.27	4086	4452	71120	77200	69.73	75.69
1050	6975	7600	6.64	7.24	4185	4560	72800	79000	69.33	75.24
1080	7140	7780	6.61	7.20	4284	4668	74480	80800	68.96	74.81
1110	7305	7960	6.58	7.17	4383	4776	76160	82600	68.61	74.41
1140	7470	8140	6.55	7.14	4482	4884	77840	84400	68.28	74.04
1170	7635	8320	6.53	7.11	4581	4992	79520	86200	67.97	73.68
1200	7800	8500	6.50	7.08	4680	5100	81200	88000	67.67	73.33
1230	7965	8680	6.48	7.06	4779	5208	82880	89800	67.38	73.01
1260	8130	8860	6.45	7.03	4878	5316	84560	91600	67.11	72.70
1290	8295	9040	6.43	7.01	4977	5424	86240	93400	66.85	72.40
1320	8460	9220	6.41	6.98	5076	5532	87920	95200	66.61	72.12
1350	8625	9400	6.39	6.96	5175	5640	89600	97000	66.37	71.85
1380	8790	9580	6.37	6.94	5274	5748	91280	98800	66.14	71.59
1410	8955	9760	6.35	6.92	5373	5856	92960	100600	65.93	71.35
1440	9120	9940	6.33	6.90	5472	5964	94640	102400	65.72	71.11
1470	9285	10120	6.32	6.88	5571	6072	96320	104200	65.52	70.88
1500	9450	10300	6.30	6.87	5670	6180	98000	106000	65.33	70.67
1530	9615	10480	6.28	6.85	5769	6288	99680	107800	65.15	70.46
1560	9780	10660	6.27	6.83	5868	6396	101360	109600	64.97	70.26
1590	9945	10840	6.25	6.82	5967	6504	103040	111400	64.81	70.06
1620	10110	11020	6.24	6.80	6066	6612	104720	113200	64.64	69.88
1650	10275	11200	6.23	6.79	6165	6720	106400	115000	64.48	69.70
1680	10440	11380	6.21	6.77	6264	6828	108080	116800	64.33	69.52
1740	10770	11740	6.19	6.75	6462	7044	111440	120400	64.05	69.20
1800	11100	12100	6.17	6.72	6660	7260	114800	124000	63.78	68.89

SECONDARY age 11 - 18 Key Stages 3:3, 4:2, 16+:2

	School Building Areas (m2)						Site Areas (m2)			
School	Total		Per Pupil		Teaching Area		Total		Per Pupil	
Roll	Min	Max	Min	Max	Min	Max	Min	Max	Min	Max
600	4588	5005	7.65	8.34	2753	3003	47600	52000	79.33	86.67
630	4753	5185	7.54	8.23	2852	3111	49280	53800	78.22	85.40
660	4918	5365	7.45	8.13	2951	3219	50960	55600	77.21	84.24
690	5083	5545	7.37	8.04	3050	3327	52640	57400	76.29	83.19
720	5248	5725	7.29	7.95	3149	3435	54320	59200	75.44	82.22
750	5413	5905	7.22	7.87	3248	3543	56000	61000	74.67	81.33
780	5578	6085	7.15	7.80	3347	3651	57680	62800	73.95	80.51
810	5743	6265	7.09	7.73	3446	3759	59360	64600	73.28	79.75
840	5908	6445	7.03	7.67	3545	3867	61040	66400	72.67	79.05
870	6073	6625	6.98	7.61	3644	3975	62720	68200	72.09	78.39
900	6238	6805	6.93	7.56	3743	4083	64400	70000	71.56	77.78
930	6403	6985	6.88	7.51	3842	4191	66080	71800	71.05	77.20
960	6568	7165	6.84	7.46	3941	4299	67760	73600	70.58	76.67
990	6733	7345	6.80	7.42	4040	4407	69440	75400	70.14	76.16
1020	6898	7525	6.76	7.38	4139	4515	71120	77200	69.73	75.69
1050	7063	7705	6.73	7.34	4238	4623	72800	79000	69.33	75.24
1080	7228	7885	6.69	7.30	4337	4731	74480	80800	68.96	74.81
1110	7393	8065	6.66	7.27	4436	4839	76160	82600	68.61	74.41
1140	7558	8245	6.63	7.23	4535	4947	77840	84400	68.28	74.04
1170	7723	8425	6.60	7.20	4634	5055	79520	86200	67.97	73.68
1200	7888	8605	6.57	7.17	4733	5163	81200	88000	67.67	73.33
1230	8053	8785	6.55	7.14	4832	5271	82880	89800	67.38	73.01
1260	8218	8965	6.52	7.12	4931	5379	84560	91600	67.11	72.70
1290	8383	9145	6.50	7.09	5030	5487	86240	93400	66.85	72.40
1320	8548	9325	6.48	7.06	5129	5595	87920	95200	66.61	72.12
1350	8713	9505	6.45	7.04	5228	5703	89600	97000	66.37	71.85
1380	8878	9685	6.43	7.02	5327	5811	91280	98800	66.14	71.59
1410	9043	9865	6.41	7.00	5426	5919	92960	100600	65.93	71.35
1440	9208	10045	6.39	6.98	5525	6027	94640	102400	65.72	71.11
1470	9373	10225	6.38	6.96	5624	6135	96320	104200	65.52	70.88
1500	9538	10405	6.36	6.94	5723	6243	98000	106000	65.33	70.67
1530	9703	10585	6.34	6.92	5822	6351	99680	107800	65.15	70.46
1560	9868	10765	6.33	6.90	5921	6459	101360	109600	64.97	70.26
1590	10033	10945	6.31	6.88	6020	6567	103040	111400	64.81	70.06
1620	10198	11125	6.29	6.87	6119	6675	104720	113200	64.64	69.88
1650	10363	11305	6.28	6.85	6218	6783	106400	115000	64.48	69.70
1680	10528	11485	6.27	6.84	6317	6891	108080	116800	64.33	69.52
1740	10858	11845	6.24	6.81	6515	7107	111440	120400	64.05	69.20
1800	11188	12205	6.22	6.78	6713	7323	114800	124000	63.78	68.89

SECONDARY age 12 - 18 Key Stages 3:2, 4:2, 16+:2

	School Building Areas (m2)						Site Areas (m2)			
School	Total		Per Pupil		Teaching Area		Total		Per Pupil	
Roll	Min	Max	Min	Max	Min	Max	Min	Max	Min	Max
600	4786	5203	7.98	8.67	2871	3122	47600	52000	79.33	86.67
630	4953	5385	7.86	8.55	2972	3231	49280	53800	78.22	85.40
660	5120	5568	7.76	8.44	3072	3341	50960	55600	77.21	84.24
690	5288	5750	7.66	8.33	3173	3450	52640	57400	76.29	83.19
720	5455	5933	7.58	8.24	3273	3560	54320	59200	75.44	82.22
750	5623	6115	7.50	8.15	3374	3669	56000	61000	74.67	81.33
780	5790	6297	7.42	8.07	3474	3778	57680	62800	73.95	80.51
810	5957	6480	7.35	8.00	3574	3888	59360	64600	73.28	79.75
840	6125	6662	7.29	7.93	3675	3997	61040	66400	72.67	79.05
870	6292	6845	7.23	7.87	3775	4107	62720	68200	72.09	78.39
900	6460	7027	7.18	7.81	3876	4216	64400	70000	71.56	77.78
930	6627	7209	7.13	7.75	3976	4326	66080	71800	71.05	77.20
960	6794	7392	7.08	7.70	4077	4435	67760	73600	70.58	76.67
990	6962	7574	7.03	7.65	4177	4545	69440	75400	70.14	76.16
1020	7129	7757	6.99	7.60	4277	4654	71120	77200	69.73	75.69
1050	7297	7939	6.95	7.56	4378	4763	72800	79000	69.33	75.24
1080	7464	8121	6.91	7.52	4478	4873	74480	80800	68.96	74.81
1110	7631	8304	6.88	7.48	4579	4982	76160	82600	68.61	74.41
1140	7799	8486	6.84	7.44	4679	5092	77840	84400	68.28	74.04
1170	7966	8669	6.81	7.41	4780	5201	79520	86200	67.97	73.68
1200	8134	8851	6.78	7.38	4880	5311	81200	88000	67.67	73.33
1230	8301	9033	6.75	7.34	4981	5420	82880	89800	67.38	73.01
1260	8468	9216	6.72	7.31	5081	5529	84560	91600	67.11	72.70
1290	8636	9398	6.69	7.29	5181	5639	86240	93400	66.85	72.40
1320	8803	9581	6.67	7.26	5282	5748	87920	95200	66.61	72.12
1350	8971	9763	6.64	7.23	5382	5858	89600	97000	66.37	71.85
1380	9138	9945	6.62	7.21	5483	5967	91280	98800	66.14	71.59
1410	9305	10128	6.60	7.18	5583	6077	92960	100600	65.93	71.35
1440	9473	10310	6.58	7.16	5684	6186	94640	102400	65.72	71.11
1470	9640	10493	6.56	7.14	5784	6296	96320	104200	65.52	70.88
1500	9808	10675	6.54	7.12	5885	6405	98000	106000	65.33	70.67
1530	9975	10857	6.52	7.10	5985	6514	99680	107800	65.15	70.46
1560	10142	11040	6.50	7.08	6085	6624	101360	109600	64.97	70.26
1590	10310	11222	6.48	7.06	6186	6733	103040	111400	64.81	70.06
1620	10477	11405	6.47	7.04	6286	6843	104720	113200	64.64	69.88
1650	10645	11587	6.45	7.02	6387	6952	106400	115000	64.48	69.70
1680	10812	11769	6.44	7.01	6487	7062	108080	116800	64.33	69.52
1740	11147	12134	6.41	6.97	6688	7281	111440	120400	64.05	69.20
1800	11482	12499	6.38	6.94	6889	7499	114800	124000	63.78	68.89

SECONDARY age 13 - 18 Key Stages 3:1, 4:2, 16+:2

School Roll	School Building Areas (m2)						Site Areas (m2)			
	Total		Per Pupil		Teaching Area		Total		Per Pupil	
	Min	Max	Min	Max	Min	Max	Min	Max	Min	Max
600	5108	5525	8.51	9.21	3065	3315	47600	52000	79.33	86.67
630	5279	5711	8.38	9.07	3167	3427	49280	53800	78.22	85.40
660	5450	5897	8.26	8.93	3270	3538	50960	55600	77.21	84.24
690	5621	6083	8.15	8.82	3372	3650	52640	57400	76.29	83.19
720	5792	6269	8.04	8.71	3475	3761	54320	59200	75.44	82.22
750	5963	6455	7.95	8.61	3578	3873	56000	61000	74.67	81.33
780	6134	6641	7.86	8.51	3680	3985	57680	62800	73.95	80.51
810	6305	6827	7.78	8.43	3783	4096	59360	64600	73.28	79.75
840	6476	7013	7.71	8.35	3885	4208	61040	66400	72.67	79.05
870	6647	7199	7.64	8.27	3988	4319	62720	68200	72.09	78.39
900	6818	7385	7.58	8.21	4091	4431	64400	70000	71.56	77.78
930	6989	7571	7.51	8.14	4193	4543	66080	71800	71.05	77.20
960	7160	7757	7.46	8.08	4296	4654	67760	73600	70.58	76.67
990	7331	7943	7.40	8.02	4398	4766	69440	75400	70.14	76.16
1020	7502	8129	7.35	7.97	4501	4877	71120	77200	69.73	75.69
1050	7673	8315	7.31	7.92	4604	4989	72800	79000	69.33	75.24
1080	7844	8501	7.26	7.87	4706	5101	74480	80800	68.96	74.81
1110	8015	8687	7.22	7.83	4809	5212	76160	82600	68.61	74.41
1140	8186	8873	7.18	7.78	4911	5324	77840	84400	68.28	74.04
1170	8357	9059	7.14	7.74	5014	5435	79520	86200	67.97	73.68
1200	8528	9245	7.11	7.70	5117	5547	81200	88000	67.67	73.33
1230	8699	9431	7.07	7.67	5219	5659	82880	89800	67.38	73.01
1260	8870	9617	7.04	7.63	5322	5770	84560	91600	67.11	72.70
1290	9041	9803	7.01	7.60	5424	5882	86240	93400	66.85	72.40
1320	9212	9989	6.98	7.57	5527	5993	87920	95200	66.61	72.12
1350	9383	10175	6.95	7.54	5630	6105	89600	97000	66.37	71.85
1380	9554	10361	6.92	7.51	5732	6217	91280	98800	66.14	71.59
1410	9725	10547	6.90	7.48	5835	6328	92960	100600	65.93	71.35
1440	9896	10733	6.87	7.45	5937	6440	94640	102400	65.72	71.11
1470	10067	10919	6.85	7.43	6040	6551	96320	104200	65.52	70.88
1500	10238	11105	6.83	7.40	6143	6663	98000	106000	65.33	70.67
1530	10409	11291	6.80	7.38	6245	6775	99680	107800	65.15	70.46
1560	10580	11477	6.78	7.36	6348	6886	101360	109600	64.97	70.26
1590	10751	11663	6.76	7.34	6450	6998	103040	111400	64.81	70.06
1620	10922	11849	6.74	7.31	6553	7109	104720	113200	64.64	69.88
1650	11093	12035	6.72	7.29	6656	7221	106400	115000	64.48	69.70
1680	11264	12221	6.70	7.27	6758	7333	108080	116800	64.33	69.52
1740	11606	12593	6.67	7.24	6963	7556	111440	120400	64.05	69.20
1800	11948	12965	6.64	7.20	7169	7779	114800	124000	63.78	68.89

SECONDARY (upper) age 14 - 18 Key Stages 4:2, 16+:2

School Roll	School Building Areas (m2)						Site Areas (m2)			
	Total		Per Pupil		Teaching Area		Total		Per Pupil	
	Min	Max	Min	Max	Min	Max	Min	Max	Min	Max
600	5758	6175	9.60	10.29	3455	3705	47600	52000	79.33	86.67
630	5936	6369	9.42	10.11	3562	3821	49280	53800	78.22	85.40
660	6115	6562	9.26	9.94	3669	3937	50960	55600	77.21	84.24
690	6293	6756	9.12	9.79	3776	4053	52640	57400	76.29	83.19
720	6472	6949	8.99	9.65	3883	4169	54320	59200	75.44	82.22
750	6650	7143	8.87	9.52	3990	4286	56000	61000	74.67	81.33
780	6829	7336	8.75	9.41	4097	4402	57680	62800	73.95	80.51
810	7007	7530	8.65	9.30	4204	4518	59360	64600	73.28	79.75
840	7186	7723	8.55	9.19	4311	4634	61040	66400	72.67	79.05
870	7364	7917	8.46	9.10	4418	4750	62720	68200	72.09	78.39
900	7543	8110	8.38	9.01	4526	4866	64400	70000	71.56	77.78
930	7721	8304	8.30	8.93	4633	4982	66080	71800	71.05	77.20
960	7900	8497	8.23	8.85	4740	5098	67760	73600	70.58	76.67
990	8078	8691	8.16	8.78	4847	5214	69440	75400	70.14	76.16
1020	8257	8884	8.09	8.71	4954	5330	71120	77200	69.73	75.69
1050	8435	9078	8.03	8.65	5061	5447	72800	79000	69.33	75.24
1080	8614	9271	7.98	8.58	5168	5563	74480	80800	68.96	74.81
1110	8792	9465	7.92	8.53	5275	5679	76160	82600	68.61	74.41
1140	8971	9658	7.87	8.47	5382	5795	77840	84400	68.28	74.04
1170	9149	9852	7.82	8.42	5489	5911	79520	86200	67.97	73.68
1200	9328	10045	7.77	8.37	5597	6027	81200	88000	67.67	73.33
1230	9506	10239	7.73	8.32	5704	6143	82880	89800	67.38	73.01
1260	9685	10432	7.69	8.28	5811	6259	84560	91600	67.11	72.70
1290	9863	10626	7.65	8.24	5918	6375	86240	93400	66.85	72.40
1320	10042	10819	7.61	8.20	6025	6491	87920	95200	66.61	72.12
1350	10220	11013	7.57	8.16	6132	6608	89600	97000	66.37	71.85
1380	10399	11206	7.54	8.12	6239	6724	91280	98800	66.14	71.59
1410	10577	11400	7.50	8.08	6346	6840	92960	100600	65.93	71.35
1440	10756	11593	7.47	8.05	6453	6956	94640	102400	65.72	71.11
1470	10934	11787	7.44	8.02	6560	7072	96320	104200	65.52	70.88
1500	11113	11980	7.41	7.99	6668	7188	98000	106000	65.33	70.67
1530	11291	12174	7.38	7.96	6775	7304	99680	107800	65.15	70.46
1560	11470	12367	7.35	7.93	6882	7420	101360	109600	64.97	70.26
1590	11648	12561	7.33	7.90	6989	7536	103040	111400	64.81	70.06
1620	11827	12754	7.30	7.87	7096	7652	104720	113200	64.64	69.88
1650	12005	12948	7.28	7.85	7203	7769	106400	115000	64.48	69.70
1680	12184	13141	7.25	7.82	7310	7885	108080	116800	64.33	69.52
1740	12541	13528	7.21	7.77	7524	8117	111440	120400	64.05	69.20
1800	12898	13915	7.17	7.73	7739	8349	114800	124000	63.78	68.89

PRIMARY CLASSROOMS

Group size	AREA OF PRIMARY CLASSROOMS: Classroom Area (m2)			Classroom Area (m2)		
	- 5%	1.82 m2	+ 5%	- 5%	2.10 m2	+ 5%
5	8.65	9.10	9.56	9.98	10.50	11.03
6	10.37	10.92	11.47	11.97	12.60	13.23
7	12.10	12.74	13.38	13.97	14.70	15.44
8	13.83	14.56	15.29	15.96	16.80	17.64
9	15.56	16.38	17.20	17.96	18.90	19.85
10	17.29	18.20	19.11	19.95	21.00	22.05
11	19.02	20.02	21.02	21.95	23.10	24.26
12	20.75	21.84	22.93	23.94	25.20	26.46
13	22.48	23.66	24.84	25.94	27.30	28.67
14	24.21	25.48	26.75	27.93	29.40	30.87
15	25.94	27.30	28.67	29.93	31.50	33.08
16	27.66	29.12	30.58	31.92	33.60	35.28
17	29.39	30.94	32.49	33.92	35.70	37.49
18	31.12	32.76	34.40	35.91	37.80	39.69
19	32.85	34.58	36.31	37.91	39.90	41.90
20	34.58	36.40	38.22	39.90	42.00	44.10
21	36.31	38.22	40.13	41.90	44.10	46.31
22	38.04	40.04	42.04	43.89	46.20	48.51
23	39.77	41.86	43.95	45.89	48.30	50.72
24	41.50	43.68	45.86	47.88	50.40	52.92
25	43.23	45.50	47.78	49.88	52.50	55.13
26	44.95	47.32	49.69	51.87	54.60	57.33
27	46.68	49.14	51.60	53.87	56.70	59.54
28	48.41	50.96	53.51	55.86	58.80	61.74
29	50.14	52.78	55.42	57.86	60.90	63.95
30	51.87	54.60	57.33	59.85	63.00	66.15

SECONDARY CLASSROOMS

Group size	AREA OF SECONDARY CLASSROOMS: Small (m2)		Large (m2)	
	Min	Max	Min	Max
5	11.00	14.00	14.00	25.00
6	12.40	15.60	15.60	26.80
7	13.80	17.20	17.20	28.60
8	15.20	18.80	18.80	30.40
9	16.60	20.40	20.40	32.20
10	18.00	22.00	22.00	34.00
11	19.40	23.60	23.60	35.80
12	20.80	25.20	25.20	37.60
13	22.20	26.80	26.80	39.40
14	23.60	28.40	28.40	41.20
15	25.00	30.00	30.00	43.00
16	26.40	31.60	31.60	44.80
17	27.80	33.20	33.20	46.60
18	29.20	34.80	34.80	48.40
19	30.60	36.40	36.40	50.20
20	32.00	38.00	38.00	52.00
21	33.40	39.60	39.60	53.80
22	34.80	41.20	41.20	55.60
23	36.20	42.80	42.80	57.40
24	37.60	44.40	44.40	59.20
25	39.00	46.00	46.00	61.00
26	40.40	47.60	47.60	62.80
27	41.80	49.20	49.20	64.60
28	43.20	50.80	50.80	66.40
29	44.60	52.40	52.40	68.20
30	46.00	54.00	54.00	70.00

16 PLUS CLASSROOMS

Group size	AREA SECONDARY CLASSROOMS (16+): Small (m2)		Large (m2)	
	Min	Max	Min	Max
3	10.45	13.20	13.20	24.10
4	12.60	15.60	15.60	26.80
5	14.75	18.00	18.00	29.50
6	16.90	20.40	20.40	32.20
7	19.05	22.80	22.80	34.90
8	21.20	25.20	25.20	37.60
9	23.35	27.60	27.60	40.30
10	25.50	30.00	30.00	43.00
11	27.65	32.40	32.40	45.70
12	29.80	34.80	34.80	48.40
13	31.95	37.20	37.20	51.10
14	34.10	39.60	39.60	53.80
15	36.25	42.00	42.00	56.50
16	38.40	44.40	44.40	59.20
17	40.55	46.80	46.80	61.90
18	42.70	49.20	49.20	64.60
19	44.85	51.60	51.60	67.30
20	47.00	54.00	54.00	70.00

SECONDARY - SIZES OF PRACTICAL ROOMS (m2) - age 11 to 16 years

Group size	Zone B		Zone C		Zone D		Zone E		Zone F		Zone G		Zone H		Special	
	Min	Max	Min	Max	Min	Max	Min	Max	Min	Max	Min	Max	Min	Max	Min	Max
15	30.00	43.00	43.00	56.50	56.50	64.00	64.00	71.50	71.50	79.00	79.00	86.50	86.50	94.00	94.00	101.50
16	31.60	44.80	44.80	58.00	58.00	65.80	65.80	73.60	73.60	81.40	81.40	89.20	89.20	97.00	97.00	104.80
17	33.20	46.60	46.60	59.50	59.50	67.60	67.60	75.70	75.70	83.80	83.80	91.90	91.90	100.00	100.00	108.10
18	34.80	48.40	48.40	61.00	61.00	69.40	69.40	77.80	77.80	86.20	86.20	94.60	94.60	103.00	103.00	111.40
19	36.40	50.20	50.20	62.50	62.50	71.20	71.20	79.90	79.90	88.60	88.60	97.30	97.30	106.00	106.00	114.70
20	38.00	52.00	52.00	64.00	64.00	73.00	73.00	82.00	82.00	91.00	91.00	100.00	100.00	109.00	109.00	118.00
21	39.60	53.80	53.80	65.50	65.50	74.80	74.80	84.10	84.10	93.40	93.40	102.70	102.70	112.00	112.00	121.30
22	41.20	55.60	55.60	67.00	67.00	76.60	76.60	86.20	86.20	95.80	95.80	105.40	105.40	115.00	115.00	124.60
23	42.80	57.40	57.40	68.50	68.50	78.40	78.40	88.30	88.30	98.20	98.20	108.10	108.10	118.00	118.00	127.90
24	44.40	59.20	59.20	70.00	70.00	80.20	80.20	90.40	90.40	100.60	100.60	110.80	110.80	121.00	121.00	131.20
25	46.00	61.00	61.00	71.50	71.50	82.00	82.00	92.50	92.50	103.00	103.00	113.50	113.50	124.00	124.00	134.50
26	47.60	62.80	62.80	73.00	73.00	83.80	83.80	94.60	94.60	105.40	105.40	116.20	116.20	127.00	127.00	137.80
27	49.20	64.60	64.60	74.50	74.50	85.60	85.60	96.70	96.70	107.80	107.80	118.90	118.90	130.00	130.00	141.10
28	50.80	66.40	66.40	76.00	76.00	87.40	87.40	98.80	98.80	110.20	110.20	121.60	121.60	133.00	133.00	144.40
29	52.40	68.20	68.20	77.50	77.50	89.20	89.20	100.90	100.90	112.60	112.60	124.30	124.30	136.00	136.00	147.70
30	54.00	70.00	70.00	79.00	79.00	91.00	91.00	103.00	103.00	115.00	115.00	127.00	127.00	139.00	139.00	151.00

SECONDARY - SIZES OF PRACTICAL ROOMS (m2) - 16 Plus

Group size	Zone B		Zone C		Zone D		Zone E		Zone F		Zone G		Zone H		Special	
	Min	Max	Min	Max	Min	Max	Min	Max	Min	Max	Min	Max	Min	Max	Min	Max
10	29.50	43.00	43.00	56.00	56.00	64.00	64.00	72.00	72.00	78.00	78.00	87.00	87.00	93.40	93.40	100.00
11	31.85	45.70	45.70	58.20	58.20	66.70	66.70	75.20	75.20	81.50	81.50	91.10	91.10	97.84	97.84	104.80
12	34.20	48.40	48.40	60.40	60.40	69.40	69.40	78.40	78.40	85.00	85.00	95.20	95.20	102.28	102.28	109.60
13	36.55	51.10	51.10	62.60	62.60	72.10	72.10	81.60	81.60	88.50	88.50	99.30	99.30	106.72	106.72	114.40
14	38.90	53.80	53.80	64.80	64.80	74.80	74.80	84.80	84.80	92.00	92.00	103.40	103.40	111.16	111.16	119.20
15	41.25	56.50	56.50	67.00	67.00	77.50	77.50	88.00	88.00	95.50	95.50	107.50	107.50	115.60	115.60	124.00
16	43.60	59.20	59.20	69.20	69.20	80.20	80.20	91.20	91.20	99.00	99.00	111.60	111.60	120.04	120.04	128.80
17	45.95	61.90	61.90	71.40	71.40	82.90	82.90	94.40	94.40	102.50	102.50	115.70	115.70	124.48	124.48	133.60
18	48.30	64.60	64.60	73.60	73.60	85.60	85.60	97.60	97.60	106.00	106.00	119.80	119.80	128.92	128.92	138.40
19	50.65	67.30	67.30	75.80	75.80	88.30	88.30	100.80	100.80	109.50	109.50	123.90	123.90	133.36	133.36	143.20
20	53.00	70.00	70.00	78.00	78.00	91.00	91.00	104.00	104.00	113.00	113.00	128.00	128.00	137.80	137.80	148.00

PRIMARY - SIZES OF PRACTICAL ROOMS (m2)

Group size	GENERAL		STUDIOS	
	Min	Max	Min	Max
5	14.50	20.50	7.50	12.50
6	16.20	22.60	8.40	13.80
7	17.90	24.70	9.30	15.10
8	19.60	26.80	10.20	16.40
9	21.30	28.90	11.10	17.70
10	23.00	31.00	12.00	19.00
11	24.70	33.10	12.90	20.30
12	26.40	35.20	13.80	21.60
13	28.10	37.30	14.70	22.90
14	29.80	39.40	15.60	24.20
15	31.50	41.50	16.50	25.50
16			17.40	26.80
17			18.30	28.10
18			19.20	29.40
19			20.10	30.70
20			21.00	32.00
21			21.90	33.30
22			22.80	34.60
23			23.70	35.90
24			24.60	37.20
25			25.50	38.50
26			26.40	39.80
27			27.30	41.10
28			28.20	42.40
29			29.10	43.70
30			30.00	45.00

Roll No.	HALLS	
	Min	Max
90	80.00	150.00
120	80.00	150.00
150	80.00	150.00
180	80.00	150.00
210	86.00	150.00
240	92.00	150.00
270	98.00	150.00
300	104.00	150.00
330	110.00	150.00
360	116.00	150.00
390	122.00	157.50
420	128.00	165.00
450	134.00	172.50
480	140.00	180.00
510	146.00	187.50
540	152.00	195.00
570	158.00	202.50
600	164.00	210.00
630	170.00	217.50

SIZES OF MUSIC ROOMS - age 11 to 16 yrs

Group size	Zone B		Zone C		Zone D	
	Min	Max	Min	Max	Min	Max
15	30.00	43.00	43.00	56.50	56.50	64.00
16	31.60	44.80	44.80	58.00	58.00	65.80
17	33.20	46.60	46.60	59.50	59.50	67.60
18	34.80	48.40	48.40	61.00	61.00	69.40
19	36.40	50.20	50.20	62.50	62.50	71.20
20	38.00	52.00	52.00	64.00	64.00	73.00
21	39.60	53.80	53.80	65.50	65.50	74.80
22	41.20	55.60	55.60	67.00	67.00	76.60
23	42.80	57.40	57.40	68.50	68.50	78.40
24	44.40	59.20	59.20	70.00	70.00	80.20
25	46.00	61.00	61.00	71.50	71.50	82.00
26	47.60	62.80	62.80	73.00	73.00	83.80
27	49.20	64.60	64.60	74.50	74.50	85.60
28	50.80	66.40	66.40	76.00	76.00	87.40
29	52.40	68.20	68.20	77.50	77.50	89.20
30	54.00	70.00	70.00	79.00	79.00	91.00

SIZES OF MUSIC ROOMS - 16 Plus

Group size	Zone B		Zone C		Zone D	
	Min	Max	Min	Max	Min	Max
10	29.50	43.00	43.00	56.00	56.00	64.00
11	31.85	45.70	45.70	58.20	58.20	66.70
12	34.20	48.40	48.40	60.40	60.40	69.40
13	36.55	51.10	51.10	62.60	62.60	72.10
14	38.90	53.80	53.80	64.80	64.80	74.80
15	41.25	56.50	56.50	67.00	67.00	77.50
16	43.60	59.20	59.20	69.20	69.20	80.20
17	45.95	61.90	61.90	71.40	71.40	82.90
18	48.30	64.60	64.60	73.60	73.60	85.60
19	50.65	67.30	67.30	75.80	75.80	88.30
20	53.00	70.00	70.00	78.00	78.00	91.00

WEEKLY TIME TABLE	**FREQUENCY IN USE OF TEACHING SPACES (% per Week)**																				
40 periods	100.00	97.50	95.00	92.50	90.00	87.50	85.00	82.50	80.00	77.50	75.00	72.50	70.00	67.50	65.00	62.50	60.00	57.50	55.00	52.50	50.00
35 periods						100.00	97.14	94.29	91.43	88.57	85.70	82.86	80.00	77.14	74.29	71.43	68.57	65.71	62.86	60.00	57.14
30 periods											100.00	96.66	93.33	90.00	86.66	83.33	80.00	76.66	73.33	70.00	66.66
	TEACHING PERIODS (Numbers per Week)																				
	40	39	38	37	36	35	34	33	32	31	30	29	28	27	26	25	24	23	22	21	20
GROUP SIZE	**PUPIL PERIODS (Numbers per Week)**																				
40 nr	1600	1560	1520	1480	1440	1400	1360	1320	1280	1240	1200	1160	1120	1080	1040	1000	960	920	880	840	800
39 nr	1560	1521	1482	1443	1404	1365	1326	1287	1248	1209	1170	1131	1092	1053	1014	975	936	897	858	819	780
38 nr	1520	1482	1444	1406	1368	1330	1292	1254	1216	1178	1140	1102	1064	1026	988	950	912	874	836	798	760
37 nr	1480	1443	1406	1369	1332	1295	1258	1221	1184	1147	1110	1073	1036	999	962	925	888	851	814	777	740
36 nr	1440	1404	1368	1332	1296	1260	1224	1188	1152	1116	1080	1044	1008	972	936	900	864	828	792	756	720
35 nr	1400	1365	1330	1295	1260	1225	1190	1155	1120	1085	1050	1015	980	945	910	875	840	805	770	735	700
34 nr	1360	1326	1292	1258	1224	1190	1156	1122	1088	1054	1020	986	952	918	884	850	816	782	748	714	680
33 nr	1320	1287	1254	1221	1188	1155	1122	1089	1056	1023	990	957	924	891	858	825	792	759	726	693	660
32 nr	1280	1248	1216	1184	1152	1120	1088	1056	1024	992	960	928	896	864	832	800	768	736	704	672	640
31 nr	1240	1209	1178	1147	1116	1085	1054	1023	992	961	930	899	868	837	806	775	744	713	682	651	620
30 nr	1200	1170	1140	1110	1080	1050	1020	990	960	930	900	870	840	810	780	750	720	690	660	630	600
29 nr	1160	1131	1102	1073	1044	1015	986	957	928	899	870	841	812	783	754	725	696	667	638	609	580
28 nr	1120	1092	1064	1036	1008	980	952	924	896	868	840	812	784	756	728	700	672	644	616	588	560
27 nr	1080	1053	1026	999	972	945	918	891	864	837	810	783	756	729	702	675	648	621	594	567	540
26 nr	1040	1014	988	962	936	910	884	858	832	806	780	754	728	702	676	650	624	598	572	546	520
25 nr	1000	975	950	925	900	875	850	825	800	775	750	725	700	675	650	625	600	575	550	525	500
24 nr	960	936	912	888	864	840	816	792	768	744	720	696	672	648	624	600	576	552	528	504	480
23 nr	920	897	874	851	828	805	782	759	736	713	690	667	644	621	598	575	552	529	506	483	460
22 nr	880	858	836	814	792	770	748	726	704	682	660	638	616	594	572	550	528	506	484	462	440
21 nr	840	819	798	777	756	735	714	693	672	651	630	609	588	567	546	525	504	483	462	441	420
20 nr	800	780	760	740	720	700	680	660	640	620	600	580	560	540	520	500	480	460	440	420	400
19 nr	760	741	722	703	684	665	646	627	608	589	570	551	532	513	494	475	456	437	418	399	380
18 nr	720	702	684	666	648	630	612	594	576	558	540	522	504	486	468	450	432	414	396	378	360
17 nr	680	663	646	629	612	595	578	561	544	527	510	493	476	459	442	425	408	391	374	357	340
16 nr	640	624	608	592	576	560	544	528	512	496	480	464	448	432	416	400	384	368	352	336	320
15 nr	600	585	570	555	540	525	510	495	480	465	450	435	420	405	390	375	360	345	330	315	300
14 nr	560	546	532	518	504	490	476	462	448	434	420	406	392	378	364	350	336	322	308	294	280
13 nr	520	507	494	481	468	455	442	429	416	403	390	377	364	351	338	325	312	299	286	273	260
12 nr	480	468	456	444	432	420	408	396	384	372	360	348	336	324	312	300	288	276	264	252	240
11 nr	440	429	418	407	396	385	374	363	352	341	330	319	308	297	286	275	264	253	242	231	220
10 nr	400	390	380	370	360	350	340	330	320	310	300	290	280	270	260	250	240	230	220	210	200

Ref	Activity Area	Pupils in Year	Period Taught	Maximum Group Size	Number of Groups	Pupil Periods	Teacher Periods	Average Group Size	AGE GROUP ANALYSIS												Average Group Size	Number of Spaces	Number of Spaces Adjusted	Frequency in use	Maximum Group Size	Area per Place	TOTAL AREAS
									Year 7		Year 8		Year 9		Year 10		Year 11		TOTALS								
									Pupil Periods	Teacher Periods	Pupil Periods	Teacher Periods	Pupil Periods	Teacher Periods	Pupil Periods	Teacher Periods	Pupil Periods	Teacher Periods	Pupil Periods	Teacher Periods							
		K	L	M	N	O	P	Q	R	S	T	U	V	W	X	Y	Z	AA	AB	AC	AD	AE	AF	AG	AH	AI	AJ
	Formulae	A/I or select	Planned	Planned	K/M rounded up	K	LxN rounded up	O/P	(As Columns K and P for display purposes only on this spread sheet only) (In practice the data in these columns should relate to the curriculum breakdown for year group)										sum of age groups	sum of age groups	AB/AC	AC/B	AE Rounded up	AE/AF	Curriculum Breakdown	Section 4 Guidelines	AF x AI
M	MANAGEMENT																										
1	Offices																						See Page 22 for area guide lines				45
2	Reception																						See Page 26 for area guide lines				128
3	Staff Room																						See Page 30 for area guide lines				80
4	Medical Centre																						See Page 34 for area guide lines				60
5	Boiler House																						See Page 36 for area guide lines				21
6	Laundry Room																						See Page 36 for area guide lines				25
H	HOSPITALITY																										
7	Assembly Hall																						See Page 40 for area guide lines				260
8	Library																						See Page 44 for area guide lines				142
9	Conferencing																						See Page 48 for area guide lines				
10	Dining Room																						See Page 52 for area guide lines				135
11	Kitchen																						See Page 56 for area guide lines				143
A	TEACHING AREAS - GENERAL																										
12	English	180	5	27	7	900	35	25.71	900	35	900	35	900	35	900	**40**	900	**40**	4,500	185	24.32	4.625					
13	Mathematics	180	5	27	7	900	35	25.71	900	35	900	35	900	35	900	**40**	900	**40**	4,500	185	24.32	4.625					
14	Modern Languages	180	5	27	7	900	35	25.71	900	35	900	35	900	35	**720**	**32**	**720**	**32**	4,140	169	24.50	4.225					
15	Humanities	180	**4**	27	7	720	28	25.71	720	28	720	28	720	28	**720**	**32**	**720**	**32**	3,600	148	24.32	3.700					
16	Religion	180	**2**	27	7	360	14	25.71	360	14	360	14	360	14	**360**	**16**	**360**	**16**	1,800	74	24.32	1.850					
17	Personal and Social	180	**1**	27	7	180	7	25.71	180	7	180	7	180	7					540	21	25.71	0.525					
18	General Studies	180	**2**	27	7	360	14	25.71							**360**	**16**	**360**	**16**	720	32	22.50	0.800					
A	**Total - General**	**1260**				**4320**			**3960**		**3960**		**3960**		**3960**		**3960**		19,800	814	24.3	20.350					
	Analysis of Classroom sizes																										
	Standard size	**77.50%**																	15,345	631	24.3	15.771	18	87.62%	30	**50**	900
	Large size	**22.50%**																	4,455	183	24.3	4.579	5	91.58%	30	**62**	310
T	TEACHING AREAS - PRACTICAL																										
19	Information Technology	180	1	20	9	180	9	20.00	180	9	180	9	180	9	180	9	180	9	900	45	20.00	1.125	2	56.25%	25	**72**	144
20	Science	180	5	27	7	900	35	25.71	900	35	900	35	900	35	**1440**	**64**	**1440**	**64**	5,580	233	23.95	5.825	7	83.21%	30	**85**	595
21	Design & Technology																										
	Food	180	1	20	9	180	9	20.00	180	9	180	9	180	9	180	9	180	9	900	45	20.00	1.125	1	112.50%	20	**103**	103
	m-m/ graphics	180	1	20	9	180	9	20.00	180	9	180	9	180	9	180	9	180	9	900	45	20.00	1.125	2	56.25%	20	**103**	206
	PECT	90	1	20	5	90	5	20.00	90	5	90	5	90	5	90	5	90	5	450	25	18.00	0.625	1	62.50%	25	**88**	88
	Textiles	90	1	20	4	90	4	30.00	90	4	90	4	90	4	90	4	90	4	450	20	22.50	0.500	1	50.00%	25	**84**	84
22	Art																										
	2D Art	180	1	30	6	180	6	30.00	180	6	180	6	180	6	**112**	**5**	**112**	**5**	764	28	27.29	0.700	1	70.00%	20	**91**	91
	3S Art/ Textiles	180	1	30	6	180	6	30.00	180	6	180	6	180	6	**68**	**3**	**68**	**3**	676	24	28.17	0.600	1	60.00%	30	**109**	109
23	Music	180	2	30	6	360	12	30.00	360	12	360	12	360	12	**180**	**8**	**180**	**8**	1,440	52	27.69	1.300	2	65.00%	15	**70**	140
24	Drama	180	1	30	5	180	6	30.00	180	6	180	6	180	6	**180**	**8**	**180**	**8**	900	34	26.47	0.850	1	85.00%	20	**91**	91
25	PE Indoor	180	2	30	6	360	12	30.00	360	12	360	12	360	12	270	12	270	12	1,620	60	27.00	1.500	1	150.00%	30	**260**	260
26	PE Outdoor	180	2	30	6	360	12	30.00	360	12	360	12	360	12	270	12	270	12	1,620	60	27.00	1.500	External		35		
	Total - Practical	**1980**				**3240**			**3240**		**3240**		**3240**		**3240**		**3240**		**16200**								3,121
	Total - All Teaching Areas								**7200**		**7200**		**7200**		**7200**		**7200**		**36000**	**1,485**							
H	CLOAKROOMS																										
27	Toilets																						See Page 58 for area guide lines				105
28	Cloakroom																						See Page 60 for area guide lines				120
29	Changing Rooms																						See Page 61 for area guide lines				83
S	SPORTING FACILITIES																										
30	Gymnasium																						See Page 92 for area guide lines				600
31	Squash Court																						See Page 93 for area guide lines				
32	Swimming Pool																						See Page 94 for area guide lines				
33	Sports Centre																										
M	ANCILLARY SPACES																										
34	Careers																						Choice by School				13
35	Local Resources																						Choice by School				50
36	Special Activities																						6.5% of Teaching Area				79
37	Storage																						6 % of Nett Area				500
38	Circulation																						See Page 37 for area guide lines				1,766
	TOTALS								**To enlarge this spreadsheet ; zoom to 150% in landscape format on your photocopying machine.**																		6,437

Ref	Activity Area	TOTAL AREAS	AREA ANALYSIS (M2)				ANALYSIS OF CAPITAL COSTS						
							New Buildings		Extension		Re-Modelled		TOTAL BUILDING COST
			Existing Premises	New School	Extension	Re-Modelled Premises	Rate perm2	Capital Cost	Rate perm2	Capital Cost	Rate perm2	Capital Cost	
		as AJ	*AK*	*AL*	*AM*	*AN*	*AO*	*AP*	*AQ*	*AR*	*AS*	*AT*	*AU*
	Formulae	*AF x AI*	*Allocation of area requirements*				*See Page 6*	*ALxAO*	*See Page 6*	*AMxAQ*	*See Page 6*	*ANxAS*	*Total*
M	**MANAGEMENT**												
1	Offices	45	30.00	15			670	10,050	605		490		10,050
2	Reception	128	85.00			43.00	560		495		375	16,125	16,125
3	Staff Room	80				80.00	630		570		445	35,600	35,600
4	Medical Centre	60		60			660	39,600	600		485		39,600
5	Boiler House	21	21.00				485		420		275		
6	Laundry Room	25		25			630	15,750	565		445		15,750
H	**HOSPITALITY**												
7	Assembly Hall	260		260			595	154,700	520		350		154,700
8	Library	142				142.00	635		570		430	61,060	61,060
9	Conferencing						790		740		515		
10	Dining Room	135				135.00	610		540		390	52,650	52,650
11	Kitchen	143				143.00	990		920		815	116,545	116,545
A	**TEACHING AREAS - GENERAL**												
12	English												
13	Mathematics												
14	Modern Languages												
15	Humanities												
16	Religion												
17	Personal and Social												
18	General Studies												
	Total - General												
	Analysis of Classroom sizes												
	Standard size	900	380.00	300		220.00	626	187,800	553		417	91,740	279,540
	Large size	310		205		105.00	626	128,330	553		417	43,785	172,115
T	**TEACHING AREAS - PRACTICAL**												
19	Information Technology	144		144			650	93,600					93,600
20	Science	595		395		200.00	650	256,750					256,750
21	Design & Technology												
	Food	103				206.00	655		585		465	95,790	95,790
	m-m/ graphics	206				206.00	630		565		455	93,730	93,730
	PECT	88	88.00				630		600		455		
	Textiles	84	84.00				630				455		
22	Art												
	2D Art	91	91.00				645		580		440		
	3S Art/ Textiles	109	109.00				645		580		440		
23	Music	140		140			675	94,500	605		470		94,500
24	Drama	91				260.00	626		553		417	108,420	108,420
25	PE Indoor	260											
26	PE Outdoor												
	Total - Practical												
	Total - All Teaching Areas												
H	**CLOAKROOMS**												
27	Toilets	105				105	900		870		830	87,150	87,150
28	Cloakroom	120				120	575		510		350	42,000	42,000
29	Changing Rooms	83				83	850		805		740	61,420	61,420
S	**SPORTING FACILITIES**												
30	Gymnasium	600		600			610	366,000	530		345		366,000
31	Squash Court												
32	Swimming Pool												
33	Sports Centre												
M	**ANCILLARY SPACES**												
34	Careers	13				13	640		570		435	5,655	5,655
35	Local Resources	50				50	640						
36	Special Activities	79		79			640	50,560					50,560
37	Storage	500	216	35		62					335	20,770	20,770
38	Circulation	1,766	875	116		116					335	38,860	38,860
	TOTALS	6,437	1,979	2,374		2,289		1,397,640				971,300	2,368,940

Ref	Activity Area	MAINTENANCE WORKS - *per Annum*					ADDITIONAL REVENUE FROM ACCOMMODATION - PER ANNUM					
		Existing Premises		New Build		TOTAL MAINT'CE COST	Total Operational Cost				Total Revenue	
		Rate per m2	Total Cost	Rate per m2	Total Cost		Revenue Unit	Uses per annum	Cost per unit	Total Cost per annum	Revenue Unit per unit	Total Revenue per annum
		AV	*AW*	*AX*	*AY*	*AZ*	*BA*	*BB*	*BC*	*BD*	*BE*	*BF*
	Formulae	*See Page 11*	*AKxAV*	AV less *15%*	*ALXAT*	*Total*	*Area Models Table A*		*Area Models Table A*	*(BA X BB x BC)*	*See Page 5*	*(BA X BB x BE)*
M	**MANAGEMENT**											
1	Offices	16.67	500	14.17	213	713						
2	Reception	16.67	1,417	14.17		1,417	43	80	3.50	12,040	5.00	17,200
3	Staff Room	16.67		14.17								
4	Medical Centre	16.67		14.17	850	850						
5	Boiler House	16.67	350	14.17		350						
6	Laundry Room	16.67		14.17	354	354						
H	**HOSPITALITY**											
7	Assembly Hall	16.67		14.17	3,684	3,684	72	100	3.50	25,358	5.00	36,225
8	Library	16.67		14.17			48	120	1.50	8,640	2.00	11,520
9	Conferencing	16.67		14.17								
10	Dining Room	16.67		14.17			72	60	2.50	10,800	3.00	12,960
11	Kitchen	16.67		14.17			72	80	3.50	20,160	5.00	28,800
A	**TEACHING AREAS - GENERAL**											
12	English											
13	Mathematics											
14	Modern Languages											
15	Humanities											
16	Religion											
17	Personal and Social											
18	General Studies											
	Total - General											
	Analysis of Classroom sizes											
	Standard size	16.67	500	14.17	4,251	4,751	180	52	3.50	33,062	5.00	47,232
	Large size	16.67	500	14.17	2,905	3,405						
T	**TEACHING AREAS - PRACTICAL**											
19	Information Technology	16.67	500	14.17	2,040	2,541	48	52	3.50	8,817	5.00	12,595
20	Science	16.67	500	14.17	5,597	6,097	165	52	3.50	30,307	5.00	43,296
21	Design & Technology											
	Food	16.67	500	14.17		500	68	52	3.50	12,490	5.00	17,843
	m-m/ graphics	16.67	500	14.17		500	56	52	3.50	10,286	5.00	14,694
	PECT	16.67	500	14.17		500	30	52	3.50	5,510	5.00	7,872
	Textiles	16.67	500	14.17		500	28	52	3.50	5,143	5.00	7,347
22	Art											
	2D Art	16.67	500	14.17		500	30	52	3.50	5,510	5.00	7,872
	3S Art/ Textiles	16.67	500	14.17		500	36	52	3.50	6,612	5.00	9,446
23	Music	16.67	500	14.17	1,984	2,484	62	52	3.50	11,388	5.00	16,269
24	Drama	16.67	500	14.17		500	72	52	3.50	13,225	5.00	18,893
25	PE Indoor	16.67	500	14.17		500						
26	PE Outdoor											
	Total - Practical											
	Total - All Teaching Areas											
H	**CLOAKROOMS**											
27	Toilets	16.67	500	14.17		500						
28	Cloakroom	16.67	500	14.17		500						
29	Changing Rooms	16.67	500	14.17		500	30	52	1.50	2,362	2.00	3,149
S	**SPORTING FACILITIES**											
30	Gymnasium	16.67	500	14.17	8,502	9,002	264	300	2.00	158,400	3.00	237,600
31	Squash Court	16.67	500	14.17		500						
32	Swimming Pool	16.67	500	14.17		500						
33	Sports Centre	16.67	500	14.17		500						
M	**ANCILLARY SPACES**											
34	Careers	16.67	500	14.17		500						
35	Local Resources	16.67	500	14.17		500						
36	Special Activities	16.67	500	14.17	1,119	1,619	35	300	2.00	21,000	3.00	31,500
37	Storage	16.67	500	14.17	496	996						
38	Circulation	16.67	500	14.17	1,644	2,144						
	TOTALS		14,770		33,638	48,408				401,111		582,314

THE CONDITION SURVEY

This survey is the key to determining both the cost of the necessary repairs and upkeep when considering the future use and convenience of existing school premises and deciding upon their refurbishment and/or new building programme.

A suggested layout for the whole reporting process is suggested in three parts set out as follows:

Part (i) The Accommodation Survey
Provides a record of the size, floor area, height and potential usage of the premises and a general statement describing its condition and any problems with regard to Health and Safety, insulation, condensation, accoustics, heating and lighting.

Part (ii) The Condition Survey with Guide to Costs
Provides a report on the condition of each element of the building and an appraisal of the probable costs of any remedial or remodelling works to be undertaken.

The example shows how the estimates of cost can be phased to suit the anticipated building programme and available finance and allows the cost of repairs and remodelling to be compared with the capital cost of building new accommodation.

Part (iii) Specification of the Repair Work
Provides a detailed specification of the repairs necessary for each element of the work for tendering purposes.

The survey should be completed by an experienced professional consultant because it will form the basis of the decisions you make in the future about the existing premises. The instruction of the consultant can be phased if required, but the accuracy of the costings will also depend on the content of the detailed condition report in Part (ii).

The content of each survey will vary according to the premises and the school's particular requirements and the costs in the table in Part (ii) are shown only to describe how they can be phased and/or compared with the capital cost of a new building.

Surrey Secondary School

CONDITION SURVEY

Part (i) Accommodation and Usage

Block Ref	B	Activity Area	16 Female Toilets	Date of Inspection	17 Aug 97

Accommodation

16.01 Floor area 11.25 x 5.75 metres = 64.69 m2

16.02 Ceiling height 2.85 metres

Facilities

White glazed sanitary appliances:

16.02	WC suites with high level water waste preventers	16 nr
16.03	Wash basins with brass taps	12 nr
16.04	Drinking fountains	1 nr
16.05	Cleaner's sink with hot and cold taps	1 nr

Cubicles; Metal framed with melamine faced panels

16.06 Cubicles 16 nr

General Condition

16.07 The toilet accommodation was built over 30 years ago and the sanitary appliances and heating and lighting have now reached the end of their useful life and do not meet present day standards for health and hygiene.

16.08 Re-decoration on its own would be a superficial and inadequate element of the repairs required to the finishings. Both the floor and wall tiling are badly worn and require renewal.

16.09 Heavy condensation has caused severe damage to the windows and the doors and cubicle partitions are also badly worn and require renewal.

Surrey Secondary School

CONDITION SURVEY

Part (ii) Condition and Guide to Costs

Block Ref	Activity Area	Date of Inspection
A	12 Staff Room	17 Aug 97

Schedule of Conditions — **Approximate Cost of Repair**

12.00	**Floors**	
12.01	**The vinyl floor tiling is badly worn and should be taken up and replaced with fine ribbed bonded carpet tile with a 3mm latex levelling screed.**	**£ 600.00**
12.02	**Redecorate walls with wall paper.**	**£ 275.00**
12.03	**Redecorate ceiling with emulsion paint.**	**£ 120.00**

Approximate Cost Forecast

Element	97/98	98/99	99/2000	2000/01	Total
Finishings					
Floors		£ 600	0	0	£ 600
Walls	£ 275	0	0	0	£ 275
Ceilings	£ 120	0	0	0	£ 120
Redecorations	£ 210	0	0	0	£ 210
Fixtures					
Doors	£ 45	0	0	0	£ 45
Windows	£ 1,250	0	0	0	£ 1,250
Fittings					
Cupboards	£ 35	0	0	0	£ 35
Furniture	£ 0	£ 1,350	£ 850	£ 650	£ 2,850
Services					
Accoustics	£ 450	0	0	0	£ 450
Heating		600	0	0	£ 600
Lighting	£ 100	450	0	0	£ 550
Total Cost Repairs	**£ 2,485**	**£ 3,000**	**£ 850**	**£ 650**	**£ 6,985**
Total if Rebuilt					**£ 14,500**

Surrey Secondary School

CONDITION SURVEY

Part (iii) Schedule of Repairs

Block Ref	Activity Area	Date of Inspection
C	48 Roof Coverigs	17 Aug 97

SCHEDULE OF REPAIRS

Block C **Two storey classrom block size 64 metres x 17.5 metres and 6.85 metres high to eaves above ground level with annex one side size 2.85 metres x 6 metres and 6.20 metres high to parapet to flat roof above ground level.**

	Roof Coverings.	Rate	Total
	Temporary Roof		
A	Provide and erect temporary roof over existing two storey classroom block. (An area of 1350 m2)		
B	Provide and erect scaffolding to perimeter of Block C as necessary for roof repairs and replacement of windows and re-pointing brickwork.		£ 18.560
	Demolition Work		
C	Strip off and clear away existing defective slating to pitched roof complete with wood battens, sarking felt and any ridge and hip tiles. (An area of 1645 m2)		£ 3,750
	New Welsh Slate Roof Coverings.		
D	Allow the Provisional Sum of £ 80,000.00 for new Slate roof coverings complete with battens, sarking felt and insulation to be carried out by a Specialist Subcontractor.		£80,000
E	Allow for builder's profit		£ 2.000

Letter to the Architect;

Tele. 0181 191 4567: Facsimile. 0181 190 45672: EMail Secondary. school@BT internet.com

SURREY SECONDARY SCHOOL
Sutton Road, Purley, Surrey

John Thompson RIBA
Architectural Partnership
Purley
Surrey

Our Ref; JL/45
17 July 1998

Dear Mr Thompson,

Thank you for visiting the school yesterday evening and we confirm that the Board of Governors are now considering plans for the future development of our school premises to meet the continuing extension of the school's curriculum and social activities and in order to attract the investment of private finance.

As discussed, the Governors wish to appoint an 'Architect' to lead the design team for this work and the attached Appendix sets out an outline of the services we believe it would be appropriate for you to provide.

Your outline proposals should be submitted to this office not later than noon on Tuesday 8 September with your fee proposals together with a brief method statement on how you would tackle the assignment and if possible, an indication of your design intentions.

You should also include a portfolio showing examples of your work, preferably of a similar nature in educational premises. You must also provide three references and a copy of your professional indemnity insurance certificate.

The Governors anticipate that they will be able to select a short list from the applicants and if you are selected we should be pleased to ask you to attend an interview during the course of October in this school.

Yours sincerely

J Smith
Head Teacher

Appendix which should be attached to the letter of appointment.

Proposed Five Year Development Brief
for
Surrey Secondary School.

Outline Development Requirements

It is anticipated that the school roll will increase from 850 to 1100 during this five year period and which will require an increase of 30% in teaching spaces and 20% in sports and hospitality facilities.

The development of the curriculum will also make it necessary to provide four extra science laboratories together with a music centre for 45 pupils and our research indicates that we shall also have to refurbish our hospitality facilities and provide for a larger gymnasium/sport facility and indoor swimming pool.

Professional Services to be provided by the Architect.

a) Prepare a schedule of accommodation and survey the existing premises and report upon its condition and the cost of any necessary repairs and renovations to meet current Health and Safety standards;

b) To assist the Governors in the preparation of an ' accommodation schedule' to take account of the proposed phased development plans and, in conjunction with the Quantity Surveyors, to prepare a bench mark analysis for the future plans;

c) To prepare, when instructed, outline and detailed designs in accordance with our phased development programme;

d) To obtain competitive lump sum tenders for each phase of the works;

e) Assist in the preparation of an outline turnkey agreement to form the basis of competitive partnering bids from private investors or developers for the construction and/or operation and management of each phase of the project.

Invitation to Tender - Small Works

Tele. 0181 190 4567: Facsimile. 0181 190 45672: EMail Secondary. school@BT internet.com

SURREY SECONDARY SCHOOL
Whyteleaf Road, Dunmail Drive, Purley, Surrey

A Painter Ltd
Brighton Road
Sutton
Surrey

Our Ref; JL/45
17 July 1998

Dear Mr Painter,

Ref;- Re-decoration of Dining Hall, Assembly Hall and Kitchen Areas

You are invited to submit your fixed price lump sum quotation for the redecoration of the above hospitality areas strictly in accordance with the attached Specification.

The works shall be commenced on 10 August and completed in their entirety on or before the 28 August 98 and the rooms shall be thoroughly cleaned and left ready for immediate re-occupation and use by the school. The school furniture will be removed before you take possession of these areas.

If, in the opinion of the Head Teacher, the works are satisfactorily completed, you will be paid 95% of the value of your quotation plus any value added tax which may be due and the balance will be released after a further period of 3 months after any snagging works have been put right.

Your quotation should be submitted not later than noon on Friday 31 August stating clearly the total amount of your fixed lump sum and together with a copy of your Tax Exemption and Insurance Certificates. Should you wish to inspect the premises, please contact the Caretaker, Mr Ford, regarding all matters concerning access, safety and security.

Yours sincerely

J Smith
Head Teacher

Appendix which may be enclosed with the Invitation to Tender.

Brief Specification of Re-decoration Work
to
Dining Room, Kitchen and Assembly Hall
at
Surrey Secondary School.

Preliminaries

1 The Contractor shall provide all necessary protection to the works and adjoining premises including dust sheets and shall allow for removing all paint spillage marks.

2 The Contractor shall remove all ironmongery before painting work is commenced and he shall allow for easing doors and windows, cleaning, oiling and refixing ironmongery and leaving in good working order at completion.

3 The paint shall be Dulux and the colours shall be agreed with the Head Master prior to commencement of the work.

Materials and Workmanship

4 Rub down all wood and metal surfaces, remove loose or flaking paint and prime exposed patches prior to re-decoration.

5 Wash down all walls and ceilings, remove loose or flaking paint, seal where required to receive new paintwork.

6 Fill all cracks and make good defective areas of plasterwork where necessary.

7 Apply one mist coat and two full coats of vinyl silk emulsion paint to surfaces of existing previously decorated plastered walls and ceilings.

8 Apply one undercoat and one finishing coat of gloss paint to previously decorated wood and metal surfaces.

Invitation to Tender - Major Works

Tele. 0181 763 8899: Facsimile. 0181 763 8456: EMail Barnsley. qs. dial pipex.com

Barnsley and Partners
850, Brighton Road, Purley, Surrey

A Builder Ltd
London Road
Sutton
Surrey

Our Ref; DB/41268
17 July 1998

Dear Sirs,

PROJECT:- Tender for New Classroom Block at Surrey Secondary School, Purley, Surrey.

Further to your interview with the Governors on Wednesday last we are pleased to invite you to submit a fixed price tender in limited competition for the above.

You will find enclosed one complete set of tender documents including;

a) Drawings as follows;

Architects drawing Nos; B/02I, B/03E, B/04I, B/05G,B/06D, B/08C, B/09F,
Structural Engineers drawing nos; G2, G3 and G4;

b) Bills of Quantities consisting of;

part i; Conditions of Contract and Description of Materials and Workmanship;

part ii; Schedule of Works and Main Summary;

c) Health and Safety Document

d) Form of Tender with two envelopes.

Invitation to Tender - Major Works (Cont'd)

Your Fixed Price Tender must be submitted on the form provided.

One completed copy of the Form of Tender must be sent in the envelope provided to the Chairman of the Board of Governors, Surrey Secondary School, Purley, Surrey **no later than noon on Wednesday 25 August 1998**.

The second copy of the Form of Tender must be sent together with the fully priced Bills of Quantities in the envelope provided **together with all supporting documents no later than noon on Wednesday 25 August 1998 to these offices at 850, Brighton Road, Purley, Surrey, CR8 2BH.**

If you are appointed as the Contractor for this project you will be required to provide a programme for the works within a period of two weeks and to commence work on site immediately thereafter and you will be required to complete the whole of the works on or before 28 August 1999 in accordance with the agreed Contract Programme.

Your attention is drawn to the requirements of the Construction Act 1998 and you should take note of the special clauses in this contract related to payments and disputes with which you will be expected to comply.

If you have any queries, please telephone the Architect or ourselves immediately.

Yours faithfully,

D. Barnsley
Director

Encl; Architects Drawings as listed above;
Engineers Drawings listed above;
Bills of Quantities, Parts 1 and 2;
Pre-Tender Health and Safety Document;
Form of Tender and envelopes

FORM OF TENDER

To; Chairman of the Board of Governors
Surrey Secondary School
Purley
Surrey

Date..................

Dear Sirs,

PROJECT:- New Classroom Block at Surrey Secondary School Purley, Surrey.

Having carefully examined the tender documents as prepared by your Quantity Surveyors on behalf of the Board of Governors for the construction of a new classroom block and having familiarised ourselves with the site conditions and the conditions upon which the tender is based, the undersigned does hereby agree to carry out and complete the whole of the works in accordance with the said terms and conditions.

Our Fixed Lump Sum Price for carrying out and completing the whole of the

works is in the amount of...

....

(£..................................) excluding value added tax.

If our tender offer is accepted and we are selected as the main Contractor we undertake to start work immediately upon the acceptance of our tender price and to complete the whole of the works not later than 19 September 1999

It is understood that the Contract Administrator, without invalidating the contract, may order extra work or make changes by adding to, or deducting from the work, the contract sum being altered accordingly. The value of such extra work shall be determined in accordance with Terms and Conditions of the Contract.

Yours faithfully,

to be signed by the Contractor.
Director

Formal Acceptance - Major Works

- Signature of Contract by the Employer
- Issue of confirmation by the Architect:

Architect's Instruction

Job Ref: 45
Instruction No: 1
Issue Date: 26 August 98

Architect;	**John Thompson................** **Address**
Employer;	**Surrey Secondary School** **Address**
Contractor;	**Contracting Builders Ltd** **Address**
Works;	**New Classroom Block at Surrey Secondary School**

Under the terms of the Contract dated......... we issue the following instructions.
Where applicable the Contract Sum will be adjusted in accordance with the terms of the relevant Condition;

Instructions;

Confirmation of the acceptance of the Fixed Price Contract amount in the Sum of ....(£.........................) excluding value added tax for the works shown on the tender drawings and described in the Bills of Quantities .

It is also confirmed that works shall commence on site on Monday 31 August 1998 and complete the whole of the works on or before 19 September 1999.

Signed by the Architect/ Supervising Officer.

Copies to; Contractor
Quantity Surveyor
Employer

OFFICE

Element	Qty	Cost	Total
Chair- Standard	2	55	110
Chair - Operators	4	165	660
Chair - Managerial	1	300	300
Chair - Visitor	2	110	220
Table Standard	2	80	160
Desk	4	250	1,000
Workstation	1	220	220
Bookcase	2	255	510
Filing Cabinet	4	750	3,000
Cupboards	4	195	780
Pinboard	4	58	232
Safe	1	700	700
			£ 7,892

STAFF ROOM

Element	Qty	Cost	Total
Bookcase	2	255	510
Chair - Arm	6	200	1,200
Chair - Lounge	6	175	1,050
Chair - Operators	1	165	165
Chair - Sofa; Two Seater	2	350	700
Chair - Standard	8	55	440
Cupboards	4	195	780
Pinboards	6	58	348
Plants	2	80	160
Storage - Filing Cabinet	2	750	1,500
Storage - Lockers	12	80	960
Table - Standard	4	80	320
Table - Coffee	2	200	400
			£ 8,533

ASSEMBLY HALL

Element	Qty	Cost	Total
Standard Chairs	200	15	3,000
Stage	1	1,860	1,860
			£ 4,860

CONFERENCE HALL

Element	Qty	Cost	Total
Chairs - Lecture	200	85	17,000
Conference Boards	6	197	1,182
Mobile Flip Boards	4	350	1,400
Projection screen - Tripod	1	95	95
Projection screen - Wall Mount	1	450	450
Tables - General	18	96	1,728
			£ 21,855

RECEPTION AREA

Element	Qty	Cost	Total
Chair - Operators	1	165	165
Chair - Visitor	2	110	220
Cupboards	2	195	390
Display Case - Upright	1	910	910
Display Case - Wall	2	200	400
Display Screen	2	195	390
Noticeboard - Glazed	2	280	560
Pinboard	3	58	174
Plants	3	80	240
Reception Desk - 5 piece	1	2,500	2,500
Reception Seating - 3 piece	2	1,000	2,000
Table - Coffee	2	200	400
			£ 8,349

MEDICAL ROOM

Element	Qty	Cost	Total
Bed	1	189	189
Bookcase	1	255	255
Chairs - Arm	2	200	400
Chairs - Lounge	2	175	350
Chairs - Standard	2	55	110
Cupboards	2	195	390
Desk	1	250	250
Pinboards	3	58	174
Storage - Filing Cabinets	2	750	1,500
Storage - Lockers	4	80	320
Storage - Medical Cabinet	1	150	150
Table- Bed	1	189	189
Table - Standard	1	80	80
Trolleys	2	175	350
Wardscreen	1	137	137
			£ 4,844

LIBRARY

Element	Qty	Cost	Total
Boockcases - Double Sided	12	400	4,800
Boockcases - High Level	10	260	2,600
Boockcases - Low Level	10	255	2,550
Book Trolleys	4	160	640
Chairs	35	20	700
Coat and Bag Storage	2	250	500
Computer Workstations	4	250	1,000
Display Cases - Wall	2	200	400
Display Screens	2	195	390
Pinboards	6	58	348
Tables - General	10	80	800
Reception Counter	1	845	845
			£ 15,573

KITCHEN

Element	Qty	Cost	Total
Aluminium Ventilation Canopy	1	3,710	3,710
Boiling Pan	1	3,800	3,800
Chef's Table	1	583	583
Cold Room	1	5,800	5,800
Double Pan Fryer tops	2	1,690	3,380
Food Mixer (with stand)	1	2,710	2,710
Freezers - Upright	2	2,380	4,760
Insect Control Unit	1	128	128
Lockers	5	78	390
Refrigerator	1	2,080	2,080
Service Counters	3	1,200	3,600
Service Spine	1	1,375	1,375
Slicer	1	2,060	2,060
Solid Top Six Burner Range	1	1,633	1,633
Staff Changing Lockers	2	88	176
Stainles Steel Condense Canopy	1	858	858
Stainless Steel Sink (Single)	1	553	553
Stainless Steel Work Benches	3	280	840
Sterilizing Sinks (Double Bowl)	3	555	1,665
Ten Tray Combination Ovens	2	6,930	13,860
Wash hand Basin	1	175	175
Water Softener	1	518	518
Worktop/base Units (350mm)	2	400	800
Worktop/base Units (1400mm)	2	950	1,900
Zinc Chromate Four Tier Racks	12	220	2,640
			£ 59,994

DINING ROOM

Element	Qty	Cost	Total
Chairs - Pupil	200	15	3,000
Chairs - Teacher	50	20	1,000
Tables - Dining	65	90	5,850
			£ 9,850

CLASSROOMS

Element	Qty	Cost	Total
Blackboards	1	80	80
Book Trolley	1	160	160
Bookcases	2	255	510
Chairs - Standard	35	20	700
Coat & Bag Storage	2	250	500
Computer Workstations	4	250	1,000
Cupboards - High Level	1	350	350
Cupboards - Low Level	2	195	390
Cupboards - Paper Storage	1	180	180
Dry Wipe Board	1	120	120
Mobile Swivel Board	1	240	240
Pinboards	8	58	464
Projection Screen - Wall Mounted	1	120	120
Roller Board - Floor standing	1	400	400
Tables - Standard (two person)	20	80	1,600
			£ 6,814

STUDY AREAS

Element	Qty	Cost	Total
Bookcases - High Level	4	260	1,040
Bookcases - Low Level	4	255	1,020
Book Trolleys	2	160	320
Carrels - Double Sided	5	500	2,500
Carrels - Single Sided	20	300	6,000
Chairs	30	20	600
Chairs - Operators	10	75	750
Coat & Bag Storage	2	250	500
Display Screens	2	195	390
Pinboards	8	58	464
Tables - General	5	80	400
			£ 13,984

ART AND CRAFT ROOM

Element	Qty	Cost	Total
Air Filter Trolley	1	380	380
Art Desks	35	95	3,325
Blackboards	1	80	80
Book Trolleys	4	160	640
Bookcases	2	255	510
Chairs - Standard	35	20	700
Coat & Bag Storage	2	250	500
Computer Workstations	2	250	500
Cupboards - High Level	2	350	700
Cupboards - Low Level	2	195	390
Cupboards - Paper Storage	1	180	180
Drawingboards	4	240	960
Dry Wipe Boards	1	120	120
Pinboards	8	58	464
Projection Screen - Wall Mount	1	120	120
Roller Board - Floor Standing	1	400	400
Sink Units	2	380	760
Tables - Crafts	2	170	340
Tables - Preparation	2	245	490
Table - Silkscreen	1	480	480
Workstations - Art	2	460	920
Workstation - Lightbox	1	340	340
			£ 13,299

POTTERY ROOM

Element	Qty	Cost	Total
Banding Wheel (Floor standing)	4	69	276
Banding Wheel (Table mounted)	10	29	290
Blackboards	1	80	80
Chairs - Standard	8	20	160
Coat & Bag Storage	2	250	500
Cupboards - High level	2	350	700
Cupboards - Low level	2	195	390
Drawing Boards	4	240	960
Dry Wipe Boards	1	120	120
Kilns (Front loading)	2	2,000	4,000
Kilns (Top loading)	2	1,000	2,000
Pinboards	4	58	232
Potters Wheel (Electric)	2	795	1,590
Potters Wheel (Manual)	2	443	886
Sink Unit	1	380	380
Stools	8	32	256
Tables - Craft	2	170	340
Tables - Preparation	2	245	490
			£ 13,650

MUSIC CENTRE

ELement	Qty	Cost	Total
Blackboards	1	80	80
Book Trolleys	2	160	320
Bookcases	2	255	510
Chairs - Cellist	10	155	1,550
Chairs - Musician Upholstered	10	116	1,160
Chairs - Student Musicians	35	66	2,310
Chairs - Symphony	10	230	2,300
Coat & Bag Storage	2	250	500
Coductor's Podium - with Rail	1	520	520
Coductor's Podium - Folding	1	170	170
Computer Workstation	4	250	1,000
Cupboards - High level	2	350	700
Cupboards - Low level	2	195	390
Cupboards - Paper Storage	1	180	180
Dry Wipe Board	1	120	120
Instrument Storage	4	900	3,600
Mobile Swivel Board	1	240	240
Music Stands	20	50	1,000
Pinboards	8	58	464
Projection Screen - Wall Mounted	1	120	120
Roller Board - Floor Standing	1	400	400
Tables - Standard (two person)	20	80	1,600
			£ 19,234

LABORATORY

Element	Qty	Cost	Total
Benching (Modular Octagonal)	5	750	3,750
Benching (Wall)	6	150	900
Blackboards	2	80	160
Bookcases	2	255	510
Coat & Bag Storage	2	250	500
Cupboards - Hazardous Chemicals	2	245	490
Cupboard - Paper Storage	1	180	180
Cupboards - Under Bench	4	210	840
Dry Wipe Boards	2	120	240
Fume Cupboards	2	1,500	3,000
Gas taps (Four-way)	5	45	225
Gas taps (Two-way)	4	32	128
Mobile Swivel Board	1	240	240
Pinboards	8	58	464
Projection Screen - Wall Mounted	1	120	120
Roller Board - Floor Standing	1	400	400
Service Pedestal with Sinks	2	450	900
Sink Unit (Eye wash)	1	130	130
Sink Unit (Hand wash)	1	130	130
Sink Unit (Standard)	1	200	200
Sink Unit (Wash-up)	1	450	450
Workstations - Computer	4	250	1,000
Stools	35	32	1,120
			£ 16,077

INFORMATION TECHNOLOGY

Element	Qty	Cost	Total
Benching (Wall)	10	96	960
Benching (Workstation)	6	250	1,500
Blackboard	1	80	80
Bookcases	2	255	510
Chairs - Operators	35	75	2,625
Coat & Bag Storage	2	250	500
Cupboards - High level	2	350	700
Cupboards - Low level	2	195	390
Cupboard - Paper Storage	1	180	180
Cupboards - Under Bench	4	250	1,000
Dry Wipe Board	1	120	120
Mobile Swivel Board	1	240	240
Pinboards	8	58	464
Printer Trolleys	3	200	600
Projection Screen - Wall Mounted	1	120	120
Roller Board - Floor Standing	1	400	400
Stools	10	32	320
Tables - Standard (two person)	2	80	160
Workstations - Mobile	4	250	1,000
			£ 11,869

FOOD TECHNOLOGY

Element	Qty	Cost	Total
Benching (With under cupboard)	10	360	3,600
Benching - Wall	8	150	1,200
Blackboards	2	80	160
Bookcases	2	255	510
Coat & Bag Storage	2	250	500
Cookers - Domestic	12	350	4,200
Cupboards - Under Bench	6	210	1,260
Cupboards - Wall	8	120	960
Dishwashers - Domestic	2	300	600
Dry Wipe Boards	2	120	240
Fridges - Domestic	4	250	1,000
Microwaves - Domestic	6	150	900
Pinboards	6	58	348
Projection Screen - Wall Mounted	1	120	120
Roller Board - Floor Standing	1	400	400
Sink Units - Hand wash	2	130	260
Sink Units - Standard	6	200	1,200
Sink Units - Wash-up	2	450	900
Stools	35	32	1,120
Tables - Standard	6	160	960
			£ 20,438

MULTI MATERIALS

Element	Qty	Cost	Total
Benching	10	750	7,500
Blackboards	1	80	80
Bookcases	2	255	510
Coat & Bag Storage	2	250	500
Computer Workstations	4	250	1,000
Cupboards - Tool Storage	8	550	4,400
Cupboards - Wall	6	180	1,080
Dry Wipe Boards	1	120	120
Pinboards	6	58	348
Projection Screen -Wall Mounted	1	120	120
Roller Board - Floor Standing	1	400	400
Stools	20	32	640
			£ 16,698

CLOAKROOM

Element	Qty	Cost	Total
Benching	17	195	3,315
Coat Hooks (1.5 meter length)	17	80	1,360
Lockers	35	80	2,800
			£ 7,475

CHANGING ROOM

Element	Qty	Cost	Total
Benching	12	195	2,340
Coat Hooks (1.5 meter length)	12	80	960
Lockers	12	80	960
			£ 4,260

THE IMPERIAL SYSTEM

Length		
12 inches	equal	1 foot
3 feet	equal	1 yard
1 chain	equal	22 yards
1 furlong	equal	220 yards
1 furlong	equal	10 chains
1 mile	equal	1760 yards
1 mile	equal	8 furlongs
Nautical		
1 fathom	equal	6.08 feet
1 nautical mile	equal	1000 fathoms
Area		
1 square foot	equal	144 square inches
1 square yard	equal	9 square feet
1 acre	equal	4840 square yards
1 square mile	equal	640 acres
Volume		
1 cubic foot	equal	1728 cubic inches
1 cubic yard	equal	27 cubic feet
Weights		
1 pound	equal	16 ounces
1 stone	equal	14 pounds
1 hundredweight	equal	8 stone
1 ton	equal	2240 pounds
1 ton	equal	20 hundredweights
1 hundredweight (US)		100 pounds (Imp)
1 ton (US)		2000 pounds (Imp)
Capacity		
1 gill	equal	5 fluid ounces
1 pint	equal	20 fluid ounces
1 quart	equal	2 pints
1 gallon	equal	8 pints
1 bushel	equal	8 gallons
1 gallon (US)	equal	4/5ths gallon (Imp)

THE METRIC SYSTEM

Length		
10 millimetres	equal	1 centimetre
10 centimetres	equal	1 decimetre
1000 millimetres	equal	1 metre
100 centimetres	equal	1 metre
10 decimetres	equal	1 metre
10 metres	equal	1 dekametre
10 dekametres	equal	1 hectometre
10 hectometres	equal	1 kilometre
1000 kilometres	equal	1 megametre
Area		
100 sq millimetres	equal	1 sq centimetre
100 sq centimetres	equal	1 sq decimetre
100 sq decimetres	equal	1 sq metre
100 sq metres	equal	1 are
100 ares	equal	1 hectare
100 hectares	equal	1 sq kilometre
Volume		
1000 cu millimetres	equal	1 cu centimetre
1000 cu centimetres	equal	1 cu decimetre
1000 cu decimetres	equal	1 cu metre
1000 cu metres	equal	1 cu dekametre
Weights		
1000 milligrams	equal	1 gram
10 grams	equal	1 dekagram
10 dekagrams	equal	1 hectogram
10 hectograms	equal	1 kilogram
100 kilograms	equal	1 quintal
1000 kilograms	equal	1 tonne
Capacity		
10 millilitres	equal	1 centilitres
10 centilitres	equal	1 decilitres
10 decilitres	equal	1 litre
1 litre	equal	1 cu decilitre
10 litres	equal	1 dekalitre
10 dekalitres	equal	1 hectolitres
10 hectolitres	equal	1 kilolitre
1 kilolitre	equal	1 cu metre

WEIGHTS

Imperial to Metric

from -	to -	multiply by;
Grains	Grams	0.06480
Ounces	Grams	28.34950
Pounds	Grams	453.59200
Pounds	Kilograms	0.45360
Tons	Kilograms	1,016.05000

Metric to Imperial

from -	to -	multiply by;
Grams	Grains	15.43000
Grams	Ounces	0.03527
Grams	Pounds	0.00220
Kilograms	Pounds	2.20462
Kilograms	Tons	0.00098

LIQUIDS

Imperial to Metric

from -	to -	multiply by;
Litres	Cubic Inches	61.02400
Gallons	Litres	4.54600

Metric to Imperial

from -	to -	multiply by;
Cubic Inches	Litres	0.01639
Litres	Gallons	0.22000

CUBIC MEASUREMENTS

Imperial to Metric

from -	to -	multiply by;
Cubic Inches	Cubic Centimetres	16.3871
Cubic Inches	Cubic Metres	0.000016
Cubic Feet	Cubic Centimetres	27,129.8075
Cubic Feet	Cubic Metres	0.0283
Cubic Yards	Cubic Metres	0.7646

Metric to Imperial

from -	to -	multiply by;
Cubic Centimetres	Cubic Inches	0.0610
Cubic Centimetres	Cubic Feet	105.4216
Cubic Centimetres	Cubic Yards	2,847.1067
Cubic Metres	Cubic Inches	61,021.5180
Cubic Metres	Cubic Feet	35.3134
Cubic Metres	Cubic Yards	1.3079

CUBIC MEASUREMENTS

Imperial to Metric

from -	to -	multiply by;
Square Inches	Square Millimetres	645.1600
Square Inches	Square Centimetres	6.4516
Square Inches	Square Metres	0.0006
Square Feet	Square Millimetres	90,288.2304
Square Feet	Square Centimetres	902.8823
Square Feet	Square Metres	0.0929
Square Yards	Square Millimetres	836,127.36
Square Yards	Square Centimetres	8,361.2736
Square Yards	Square Metres	0.8361
Square Miles	Square Metres	2,589,975.24
Square Miles	Square Kilometres	2.5900
Acres	Hectares	0.4047

Metric to Imperial

from -	to -	multiply by;
Square Millimetres	Square Inches	0.001550005
Square Millimetres	Square Feet	0.000010764
Square Millimetres	Square Yards	0.000001196
Square Centimetres	Square Inches	0.1550
Square Centimetres	Square Feet	22.3162
Square Centimetres	Square Yards	200.8796
Square Metres	Square Inches	1,549.9654
Square Metres	Square Feet	10.7636
Square Metres	Square Yards	1.1960
Square Kilometres	Square Feet	10,763,903.76
Square Kilometres	Square Yards	1,195,989.31
Square Kilometres	Square Miles	0.3861
Hectares	Acres	2.4710
Hectares	Square Yards	1,581.4464

LINEAR MEASUREMENTS

Imperial to Metric

from -	to -	multiply by;
Inches	Millimetres	25.4000
Inches	Centimetres	2.5400
Inches	Metres	0.0254
Feet	Millimetres	300.4800
Feet	Centimetres	30.0480
Feet	Metres	0.3048
Yards	Millimetres	914.4000
Yards	Centimetres	91.4400
Yards	Metres	0.9144
Miles	Metres	1,609.3400
Miles	Kilometres	1.6093

Metric to Imperial

from -	to -	multiply by;
Millimetres	Inches	0.0394
Millimetres	Feet	0.0033
Millimetres	Yards	0.0011
Centimetres	Inches	0.3937
Centimetres	Feet	4.7240
Centimetres	Yards	14.1732
Metres	Inches	39.3696
Metres	Feet	3.2808
Metres	Yards	1.0936
Kilometres	Feet	3,280.84
Kilometres	Yards	1,093.61
Kilometres	Miles	0.6214